T0340192

Chemical Engineering Process Simulation

Chemical Engineering Process Simulation

Dominic Chwan Yee Foo, Nishanth Chemmangattuvalappil,
Denny K.S. Ng, Rafil Elyas, Cheng-Liang Chen, René D. Elms,
Hao-Yeh Lee, I-Lung Chien, Siewhui Chong, Chien Hwa Chong

Elsevier
Radarweg 29, PO Box 211, 1000 AE Amsterdam, Netherlands
The Boulevard, Langford Lane, Kidlington, Oxford OX5 1GB, United Kingdom
50 Hampshire Street, 5th Floor, Cambridge, MA 02139, United States

Library of Congress Cataloging-in-Publication Data
A catalog record for this book is available from the Library of Congress

British Library Cataloguing-in-Publication Data
A catalogue record for this book is available from the British Library

ISBN: 978-0-12-803782-9

For information on all Elsevier publications visit our website at https://www.elsevier.com/books-and-journals

Publisher: Joe Hayton
Acquisition Editor: Fiona Geraghty
Editorial Project Manager: Gabriela Capille
Production Project Manager: Mohana Natarajan
Designer: Matthew Limbert

Typeset by TNQ Books and Journals

Contents

Part 2
UniSim Design

5. Basics of Process Simulation With UniSim Design

Dominic C.Y. Foo

6. Modeling of a Dew Point Control Unit With UniSim Design

Rafil Elyas, Zhi Hong Li

Part 3
PRO/II

7. Basics of Process Simulation With SimSci PRO/II

Chien Hwa Chong

List of Contributors

Zi Jie Ai, National Taiwan University, Taipei, Taiwan

Joanne W.R. Chan, Taylor's University, Subang Jaya, Malaysia

Nishanth Chemmangattuvalappil, University of Nottingham Malaysia Campus, Semenyih, Malaysia

Cheng-Liang Chen, National Taiwan University, Taipei, Taiwan, ROC

I-Lung Chien, National Taiwan University, Taipei, Taiwan

Chien Hwa Chong, Taylor's University, Subang Jaya, Malaysia

Siewhui Chong, University of Nottingham Malaysia Campus, Semenyih, Malaysia

René D. Elms, Texas A&M University, College Station, TX, United States

Rafil Elyas, East One-Zero-One Sdn Bhd, Shah Alam, Malaysia

Dominic C.Y. Foo, University of Nottingham Malaysia Campus, Semenyih, Malaysia

Tyng-Lih Hsiao, National Taiwan University of Science and Technology, Taipei, Taiwan

Denny K.S. Ng, University of Nottingham Malaysia Campus, Semenyih, Malaysia

Hao-Yeh Lee, National Taiwan University of Science and Technology, Taipei, Taiwan

Zhi Hong Li, East One-Zero-One Sdn Bhd, Shah Alam, Malaysia

Wei-Jyun Wang, National Taiwan University, Taipei, Taiwan, ROC

Bor-Yih Yu, National Taiwan University, Taipei, Taiwan

How to Use This Book

This book presents some of the most popular commercial steady-state process simulation software packages in the market. We believe that this will be a good self-learning guide for students and working professionals in learning process simulation software, as well as for university instructors who are conducting lectures on process simulation. Even though the chapters are interconnected, they are mostly written independently; this allows readers to read each chapter without having to read the preceding chapters. Note, however, that the four chapters in Part 1 of the book serve as an important guide on the various basic principles behind all commercial simulation software packages, e.g., physical properties estimation (thermodynamic models), new component registration, recycle stream simulation. Readers are encouraged to read these chapters before proceeding to other parts of the book (see Fig. 1). This is particularly important for a novice in process simulation.

Parts 2–5 of the book cover four different families of commercial software in the market, i.e., UniSim Design, PRO/II, ProMax, and aspenONE Engineering. Each of these software packages has an introductory chapter (with step-by-step guide) to allow new users in mastering the usage of the software, before going into advanced topics. Table 1 shows the level of difficulties for all chapters in the book, in which the readers may refer to in selecting the topic for reading.

Apart from basic simulation knowledge and the use of simulation tools in designing various processes (gas sweetening, acrylic acid production, etc.), several advanced topics on process synthesis and design are also included in this book. These advanced topics on reactive and azeotropic distillation systems, heat exchanger network, and utility planning are found in Part 5 of the book.

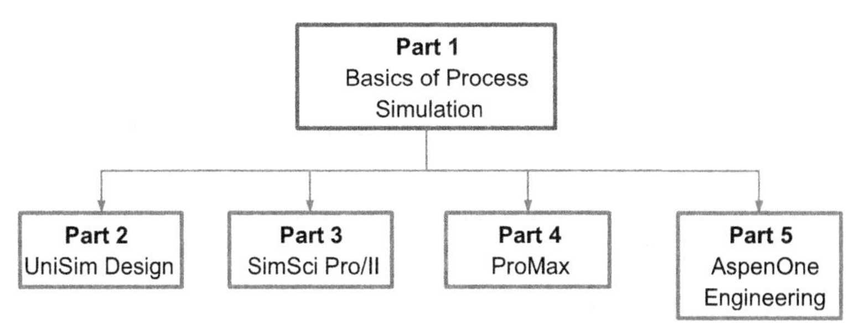

FIGURE 1 Suggested flow in reading this book.

To assist readers in better understanding the chapters, process simulation files are made available on author support website, which is found in the following URL: https://www.elsevier.com/books-and-journals/book-companion/9780128037829. For university instructors, the solution for various exercises in each chapter is also made available in the password-protected author support website, in which permission to access will be granted to university instructors who adopt the book for their lecture.

It is hope that this book will serve as a useful guide for a good learning experience in process simulation knowledge. Have fun in your simulation exercises!

TABLE 1 Level of Difficulty for Each Chapters

	Chapters	Level
Part 1—Basics of Process Simulation		
1	Introduction to process simulation	Basic
2	Registration of new components	Basic
3	Physical property estimation for process simulation	Basic
4	Simulation of recycle streams	Basic
Part 2—UniSim Design		
5	Basics of process simulation with UniSim Design	Basic
6	Modeling of a dew point control unit with UniSim Design	Advanced
Part 3—SimSci PRO/II		
7	Basics of process simulation with SimSci PRO/II	Basic
8	Modeling for biomaterial drying, extraction, and purification technologies	Advanced
Part 4—ProMax		
9	Basics of process simulation with ProMax	Basic
10	Modeling of sour gas sweetening with MDEA	Advanced
Part 5—aspenONE Engineering		
11	Basics of process simulation with Aspen HYSYS	Basic
12	Process simulation for VCM production	Advanced

TABLE 1 Level of Difficulty for Each Chapters—cont'd

	Chapters	Level
13	Process simulation and design of acrylic acid production	Advanced
14	Design and simulation of reactive distillation processes	Advanced
15	Design of azeotropic distillation systems	Advanced
16	Simulation and analysis of heat exchanger networks with Aspen energy analyzer	Advanced
17	Simulation and analysis of steam power plants with Aspen utility planner	Advanced

Dominic C.Y. Foo
April 2017

Part 1

Basics of Process Simulation

Chapter 1

Introduction to Process Simulation

Dominic C.Y. Foo[1], Rafil Elyas[2]
[1]*University of Nottingham Malaysia Campus, Semenyih, Malaysia;* [2]*East One-Zero-One Sdn Bhd, Shah Alam, Malaysia*

Process simulation is the representation of a chemical process by a mathematical model, which is then solved to obtain information about the performance of the chemical process (Motard et al., 1975). It is also known as *process flowsheeting*. Westerberg et al. (1979) also defined flowsheeting as *the use of computer aids to perform steady-state heat and mass balancing, sizing, and costing calculations for a chemical process*. In this chapter, some basic information about simulation will be presented. This includes the historical developments, basic architectures, and solving algorithms. Besides, 10 good habits of process simulation are also provided at the end of the chapter, so to guide readers in nurturing some good practices in using process simulation software.

1.1 PROCESS DESIGN AND SIMULATION

Many regard process simulation being equivalent to process design, which is indeed a misleading understanding. In fact, process simulation and *process synthesis* are two important and interrelated elements in chemical process design, which may be used to achieve optimum process design. The aim for process simulation is to predict how a defined process would actually behave under a given set of operating conditions. In other words, we aim to predict the outputs of the process when the process flowsheet and its inputs are given (Fig. 1.1). In the modern days, commercial process simulation software packages are often used for such exercises.

On the other hand, when an unknown process flowsheet is to be created for given process input and output streams, this entails the exercise on process synthesis (Fig. 1.2). Process synthesis has been an active area of research in the past five decades, with some significant achievements in specific applications, e.g., heat recovery system, material recovery system, and reaction network (El-Halwagi and Foo, 2014). Process synthesis and process simulation

Chemical Engineering Process Simulation. http://dx.doi.org/10.1016/B978-0-12-803782-9.00001-7

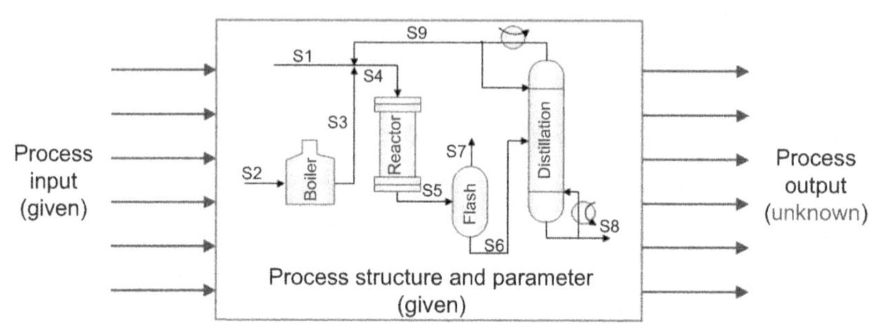

FIGURE 1.1 A process analysis problem (El-Halwagi, 2006; Foo, 2012).

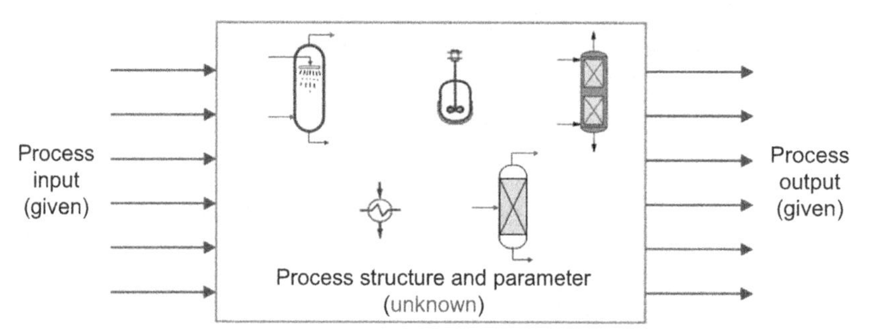

FIGURE 1.2 A process synthesis problem (El-Halwagi, 2006; Foo, 2012).

supplement each other well. In most cases, once a process flowsheet is synthesized, its detailed characteristics (e.g., temperature, pressure, and flowrates) may be predicted using various process simulation tools, to develop into an optimum flowsheet.

Within process synthesis, one of the important models to guide flowsheet synthesis is the *onion model* first reported in Linnhoff et al. (1982). As shown in Fig. 1.3, process design exercise begins from the core of the process and moves outward. In the center of the onion, the reactor system is first designed. The reactor design influences the separation and recycle structures at the second layer of the onion. Next, the reactor and separator structures dictate the overall heating and cooling requirement of the process. Hence, the heat recovery system is designed next, in the third layer. A utility system at the fourth layer is next designed, to provide additional heating and cooling requirements, which cannot be satisfied through heat recovery system. At the final layer, the waste treatment system is designed to handle various emissions/effluents from the process, prior to final environmental discharge.

In a later section of this chapter, the onion model will be used to guide the simulation of the integrated flowsheet of chemical processes.

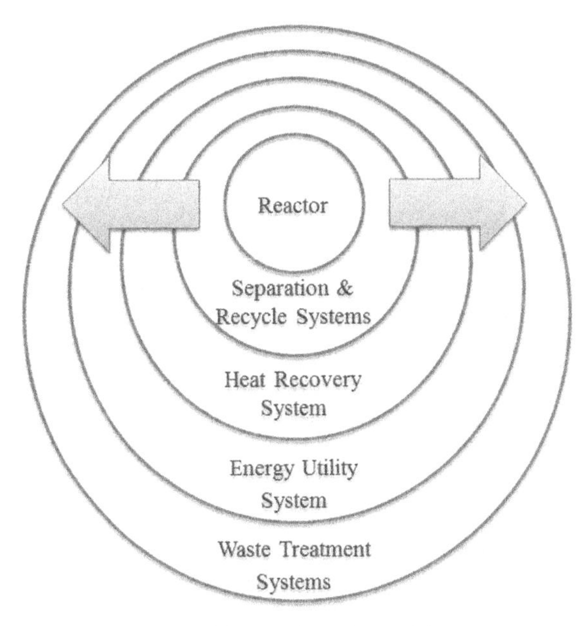

FIGURE 1.3 The onion model.

1.2 HISTORICAL PERSPECTIVE FOR PROCESS SIMULATION

The history of commercial process simulation software was dated in the 1960s. The first generic process simulation software known as PROCESS was launched by Simulation Science based at Los Angeles (US) in 1966, for the simulation of distillation columns (Dimian et al., 2014). This software evolved into PRO/II and is marketed by Schneider Electric in recent years (Schneider Electric, 2017). Another commercial software for gas and oil applications, known as DESIGN, was launched in 1969 by ChemShare Corporation based at Houston (US) (Dimian et al., 2014). This software is marketed as DESIGN II for Windows by WinSim Inc. since 1995 (WinSim, 2017).

Stepping into the 1970s, which was generally known as the "golden age" of scientific computing, several important historical milestones mark the active developments of process simulation tools. First, FORTRAN programming language became the de facto standard among scientists and engineers (Evans et al., 1977). Two important books on process simulation (Crowe et al., 1971; Westerberg et al., 1979) described some important developments in the 1970s. Of particularly important is the formal introduction of *sequential modular* (SM) approach by Westerberg et al. (1979), which is commonly utilized in most software. Next, the first oil crisis in 1973 simulated the development of simulation tools that can be used for solid handling, i.e., power generation with coal. Following this was the important ASPEN (Advanced System for Process ENgineering) project at Massachusetts Institute of Technology (MIT) between years 1976 and 1979, sponsored by the US Department of Energy (Evans et al., 1979, Gallier et al., 1980). Aspen Technology Inc. (AspenTech) was then formed in 1981 to commercialize the technology, with Aspen Plus software being released in 1982 (AspenTech, 2017). Another important achievement in 1970s is the development of software based on *equation-oriented* (EO) approach. Important EO-based software includes SPEEDUP developed at Imperial College, London (UK) (Hernandez and Sargent, 1979; Perkins et al., 1982), which was later succeeded by gPROMS (PSE, 2017). Note that in the 1970s, simulation was mainly executed on fast but expensive mainframe systems, where user was connected via a remote terminal.

With the arrival of personal computer (PC) in the 1980s, several other important software packages such as ChemCAD (developed by ChemStations) and HYSYS (originally by Hyprotech) were launched. These software packages no longer operate on the mainframe systems but are PC based. Late 1980s also saw the needs of developing simulation software for biochemical processes (Petrides et al., 1989). This leads to the introduction of BioPro Designer (Petrides, 1994), which later evolved into SuperPro Designer marketed by Intelligen Inc. in the 1990s (Intelligen, 2017). It is worth mentioning that up to the 1980s, most basic developments of process simulation architectural are quite established, with review papers outlining their state-of-the-art techniques (Rosen, 1980; Evans, 1981).

In the 1990s, with the dominant of Microsoft Windows, most software packages were migrated from their previous Macintosh version into the more attractive graphical user interface (GUI). Another important milestone happened in the early 21st century. In 2002, Hyprotech was acquired by AspenTech, which resulted with the ownership of HYSYS software. However, the US Federal Trade Commission judged that acquisition of Hyprotech was anticompetitive and ruled AspenTech to divest its software to the approved buyer—Honeywell (Federal Trade Commission, 2003). This later leads to the introduction of UniSim Design (Honeywell, 2017), which shares the same GUI as HYSYS (of Hyprotech), while AspenTech continues to market its Aspen HYSYS software (with different GUI in year 2016).

Some of the commonly used process simulation software packages are listed in Table 1.1.

1.3 BASIC ARCHITECTURES FOR COMMERCIAL SOFTWARE

Fig. 1.4 shows the basic structure of a process simulation software and sequential steps in performing the simulation task (Turton et al., 2013). As shown at the upper side of the figure, a typical commercial simulation software includes the following components, i.e., component database, thermodynamic model database, flowsheet builder, unit operation model database, and flowsheet solver. Note that some other elements, e.g., subflowsheet, financial analysis tools, and engineering units option, are software dependent and hence are excluded in this figure.

TABLE 1.1 Commercial Process Simulation Software

Corporations	Software	Websites
AspenTech	AspenONE Engineering (consists of Aspen Plus, Aspen HYSYS, Aspen Economic Evaluation, Aspen Exchanger Design & Rating, Aspen Energy Analyzer, Aspen Utilities Planner)	www.aspentech.com
Honeywell	UniSim Design	www.honeywellprocess.com
Schneider Electric	SimSci PRO/II	http://software.schneider-electric.com
Chemstations	ChemCAD	www.chemstations.com
WinSim	DESIGN II for Windows	www.winsim.com
Intelligen	SuperPro Designer, SchedulePro	www.intelligen.com
Bryan Research & Engineering	ProMax	www.bre.com
Process Systems Enterprise	gPROMS	www.psenterprise.com

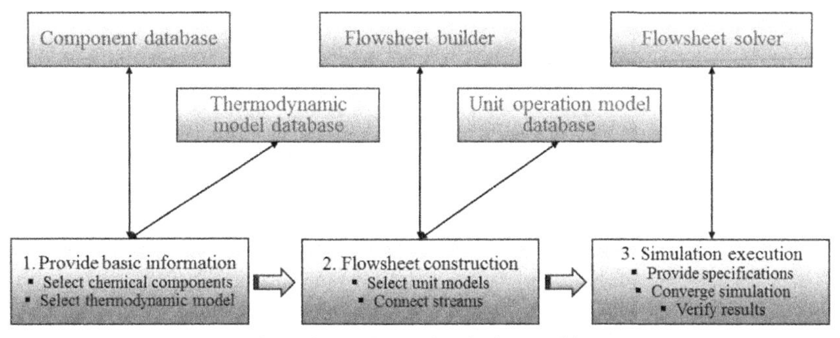

FIGURE 1.4 Basic structure of a commercial simulation software and sequence in solving a simulation model (adapted from Turton et al., 2013).

The bottom side of Fig. 1.4 presents a list of sequential steps in solving a simulation problem. In step 1, the basic information for a simulation problem is first provided. This includes chemical components and thermodynamic model selection, which can be done easily through their associated databases of the software. Note that it is advisable to select all components needed for the flowsheet, even though some components will only be used at the later part of the flowsheet. The selection of thermodynamic model is a crucial step, as different thermodynamic models will lead to very different mass and energy balances for some processes. Next in step 2, the process flowsheet is constructed using the flowsheet builder. This involves the selection of appropriate unit operation models (from the unit model database) and the connections among them with the process streams (some software may need to have energy streams connected too). In step 3, specifications are to be provided for the unit models as well as important inlet streams (e.g., flowrate, temperature, pressure) to execute the simulation. Note that in all modern simulation software, users may choose to display the simulation results in various forms. Finally, it is important to cross-check the simulation results, either through some empirical model, mass, and energy balances or through reported plant/experimental data. Doing so will increase the confident level of the simulation model.

1.4 BASIC ALGORITHMS FOR PROCESS SIMULATION

Two main classical techniques used in solving process simulation models are SM and EO approaches. Most commercial simulation software packages in the market, e.g., Aspen Plus, ChemCAD, and PRO/II, are using SM approach and hence will be discussed more in-depth in the following sections.

1.4.1 Sequential Modular Approach

The term "sequential modular" was formally introduced in late 1970s (although commercial software packages were found in the market prior to that) by Westerberg et al. (1979). The concept of SM may be explained using Fig. 1.5. Each of the unit modules contains some algorithms that are utilized to solve a set of process models, provided that the inlet stream information and unit specifications are given. Once a module is solved to convergence, it will generate the results for the outlet stream(s). The latter is then connected as a feed stream for the following unit module, which is then solved for convergence (Turton et al., 2013). The same process is repeated until all process units in the flowsheet are solved and converged. Note that certain unit modules may require iterative solution algorithm to achieve convergence; the overall process is, however, sequential in nature, i.e., no iteration is required (Turton et al., 2013).

For process flowsheet that contains recycle stream(s), *tear stream* strategy is commonly used with SM approach to converge the recycle stream. As

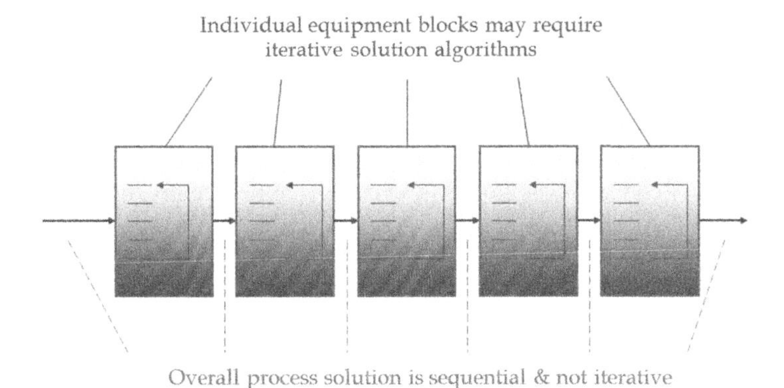

FIGURE 1.5 Concept of sequential modular approach (Turton et al., 2013).

shown in Fig. 1.6A, the flowsheet consists of six operations, i.e., units A–F and a recycle stream that connects units C and F. If SM approach in Fig. 1.5 is adopted for the simulation exercise, one will start to simulate and converge unit A. This is followed by unit B and then unit C. However, because of the existence of the recycle stream, unit C can only be simulated once the recycle stream contains the necessary properties (e.g., pressure, temperature, and flowrates) after unit F is converged. However, unit F cannot be simulated without first converging unit C. In other words, the convergence of units C and F involves iterative steps.

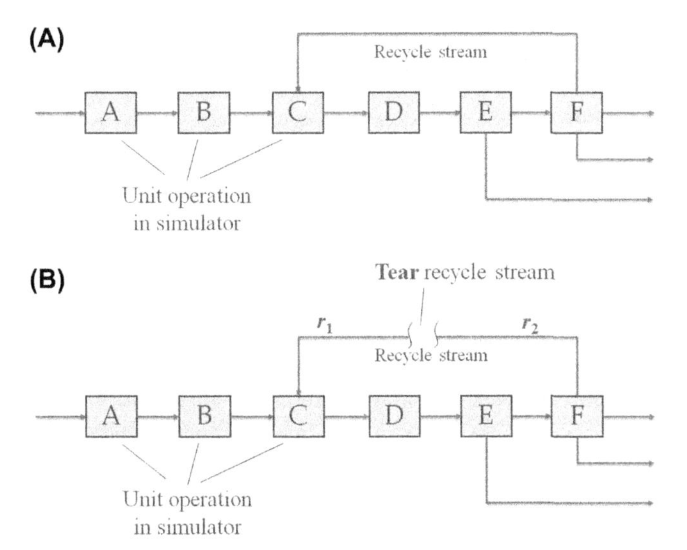

FIGURE 1.6 (A) A process consists of a recycle stream; (B) concept of tear stream (Turton et al., 2013).

To cater the iterative procedure, a tear-stream strategy is used. As shown in Fig. 1.6A, the recycle stream is virtually "torn" into two parts—r_1 (inlet for unit C) and r_2 (outlet from unit F). Some estimated data (e.g., temperature, pressure, and flowrate) are provided for r_1 to simulate unit C. Once unit C is converged, simulation then proceeds to unit D, E, and finally unit F. Once unit F is converged, the simulated results from outlet stream r_2 are then compared to those estimated data given to r_1 earlier on. If their values agree to the specified convergence tolerance (typically given in terms of percentage difference), the simulation is converged, or else the simulated results from r_2 will substitute the estimated data in r_1 and the iterative procedure is carried on[1].

The main advantage of SM is that it is intuitive and easy to understand. It allows the interactions of users as the model develops. However, large problems (with many recycle streams) may be difficult to converge.

1.4.2 Equation-Oriented Approach

In EO approach, a set of equations are solved simultaneously for a simulation problem. For instance, for a problem with n design variables, p equality constraints, and q inequality constraints, the problem is formulated as follows (Smith, 2016):

$$\text{Solve } h_i\,(x_1, x_2, x_3, \ldots, x_n) = 0\ (i = 1, \ldots, p) \tag{1.1}$$

$$\text{subject to } g_j\,(x_1, x_2, x_3, \ldots, x_n) \leq 0\ (i = 1, \ldots, q) \tag{1.2}$$

The main advantage of EO approach is its ability to be formulated as an optimization problem. However, complex EO problems are difficult to solve and diagnose. It is also not as robust as SM approach (Smith, 2016). Hence, it has not been favored among commercial simulation software in the past few decades, until recent years where it is embedded in solving complex models in SM-based software.

1.5 INCORPORATION OF PROCESS SYNTHESIS MODEL AND SEQUENTIAL MODULAR APPROACH

In simulating an integrated flowsheet, when many units are involved, it is always good to break down the complex flowsheet into small systems that are manageable. A useful way of doing so is to make use of the onion model to guide the simulation tasks (Foo et al., 2005). As discussed in Section 1.1, a chemical process is designed from the core of the onion and moves outward. Typically, at each of the layer, decision is made by the designer after detailed analysis. Hence, SM approach can be incorporated with the onion model in

1. See Chapter 4 for a detailed discussion on handling recycle systems with sequential modular approach.

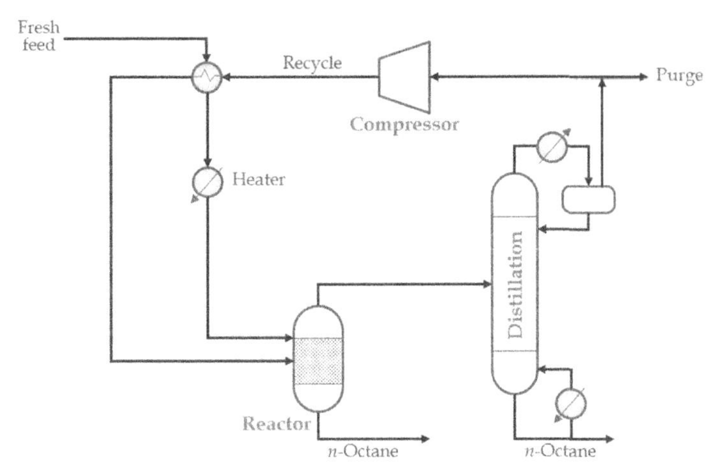

FIGURE 1.7 Process flow diagram for *n*-octane production (Foo et al., 2005).

guiding the design process. This is particularly useful in evaluating new process pathways or generating alternatives for new process development (Foo et al., 2005).

An example on the production of *n*-octane is next demonstrated to illustrate this concept.

Example 1.1: *n*-Octane Production Example

Fig. 1.7 shows the process flow diagram for the production of *n*-octane (C_8H_{18}) (Foo et al., 2005). The fresh feed stream, containing ethylene (C_2H_6), *i*-butane (*i*-C_4H_{10}), and some trace amounts of impurities [i.e., nitrogen and *n*-butane (*n*-C_4H_{10})], is preheated before being fed to the reactor, along with the recycle stream. The reactor operates isothermally with a high conversion rate (98%) following the reaction stoichiometry in Eq. (1.3).

$$2C_2H_4 + C_4H_{10} \rightarrow C_8H_{18} \tag{1.3}$$

A big portion of *n*-octane product exits from the reactor bottom stream. The vapor effluent of reactor is then fed into a distillation column, where more *n*-octane product is recovered as bottom stream; while the unconverted reactant from the distillation top stream is being recycled to the reactor. The recycle stream passes through a compressor where its pressure is adjusted to match with that of the reactor and then exchanges its heat to preheat the fresh feed stream.

Following the concept of the onion model, one shall first simulate the reactor, which is the core of the process (see Fig. 1.3). Once the reactor model is converged, simulation is moved on to the distillation column and next the recycle systems; both of these are located at the second layer of the onion model. This is illustrated in Fig. 1.8.

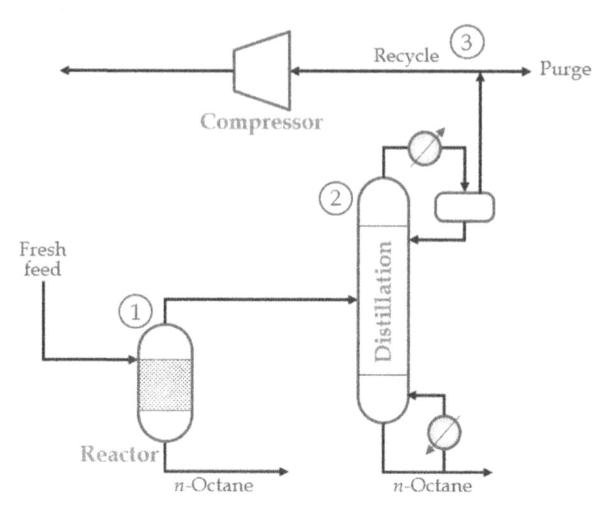

FIGURE 1.8 Sequence in simulating the *n*-octane case guided by onion model (numbers indicate the sequence of simulation).

Note that from the simulation perspective, recycle may be further classified as *material* and *heat recycle systems*. For the case of *n*-octane production, these subsystems for the complete flowsheet are shown in Fig. 1.9. The main challenge for this case is that these subsystems are interrelated, as they are connected by the process-to-process heat exchanger. The latter has two inlet streams, i.e., fresh feed and the outlet stream from the compressor. However,

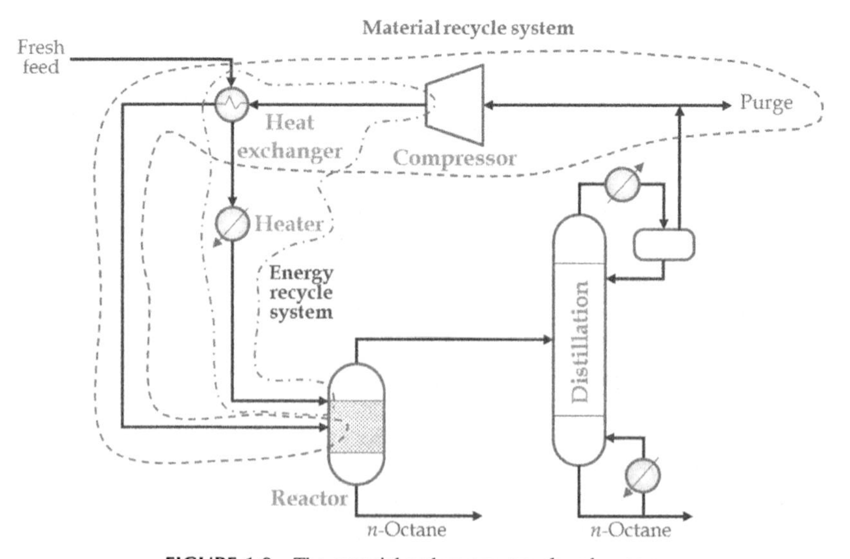

FIGURE 1.9 The material and energy recycle subsystems.

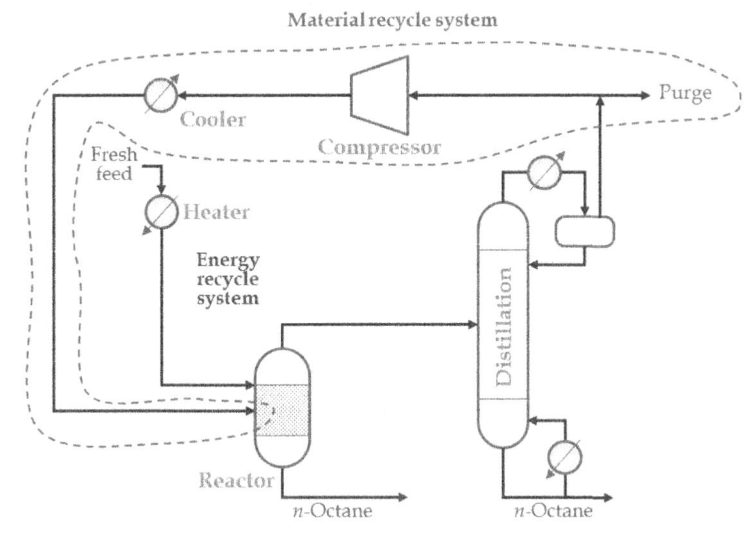

FIGURE 1.10 A decoupled material and energy recycle subsystems.

only the fresh feed stream contains the data (e.g., pressure, temperature, component, and flowrates) necessary for simulation to take place. The outlet stream from the compressor will have no result until the compressor is converged. Without complete data for its inlet streams, the heat exchanger simulation cannot be carried out, which means that its two outlet streams will have no results. This also leads to the unconvergence of the subsequent units, i.e., reactor, distillation, etc.

A more convenient way to simulate material and heat recycle systems is to decouple them. This can be done by replacing the process-to-process heat exchanger with other equivalent heating and cooling units[2]. As shown in the revised flowsheet in Fig. 1.10, the material recycle stream is cooled using a utility cooler before being fed into the reactor, while the fresh feed stream is heated by the utility heater prior being sent to the reactor. In other words, energy recovery (and hence the energy recycle system) is not being considered at this stage. One then can make use of the tear stream concept to converge the material recycle stream.

Upon the convergence of the material recycle system, we then move on to the third layer of the onion model, where heat recovery system (i.e., the heat recycle system) is considered. In the converged flowsheet in Fig. 1.10, the heating duty from the utility heater, as well as the cooling duty of utility cooler may be extracted. One can make use of the well-established *heat pinch analysis* technique to design the heat recovery system. For instance, for the

2. See detailed discussion for various strategies in converging recycle systems in Chapter 4.

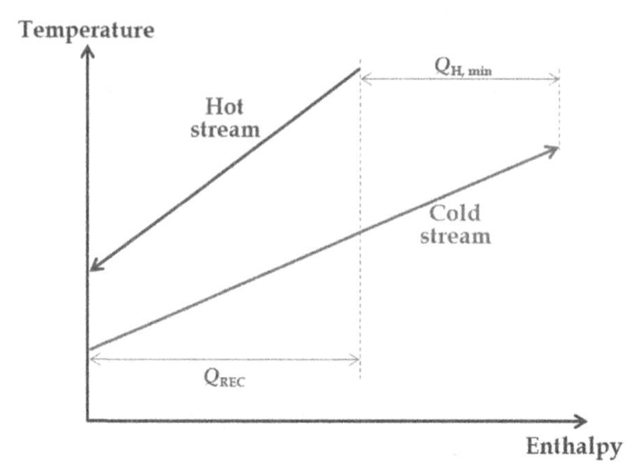

FIGURE 1.11 Temperature versus enthalpy diagram for *n*-octane case.

flowsheet in Fig. 1.10, cooling and heating duties of the cooler and heater are extracted to plot a temperature versus enthalpy diagram (similar to the *heat transfer composite curve*) in Fig. 1.11. With this diagram, the minimum amount of hot ($Q_{H,\ min}$) and cold ($Q_{C,\ min}$) utility targets may be determined. Doing this also determines how much energy are recovered (Q_{REC}) between the *hot* (compressor outlet) and *cold* (fresh feed) streams[3]. For the case in Fig. 1.11, the duty of the hot stream is completely recovered to the cold stream, resulting in a *threshold* case where no cold utility (i.e., $Q_{C,\ min} = 0$) is needed.

With the heat duties ($Q_{H,\ min}$ and Q_{REC}) identified, we then replace the cooler in the material recycle system with the process-to-process heat exchanger, while the heater is kept to provide additional heating duty ($Q_{H,\ min}$) required by the fresh feed stream. This concludes the simulation of the *n*-octane case study. Note that the last two layers of the onion model are not applicable in this case. The *n*-octane case is utilized as work examples throughout this book (see Chapters 5, 7, 9, and 11).

1.6 TEN GOOD HABITS FOR PROCESS SIMULATION

Habit 1: You build a simulation model to meet an objective

As a process designer, the objective of building a simulation model should be understood well. In other words, the purpose of running a simulation should be clearly defined upfront. For instance, if one were to perform mass and energy balances for a preliminary flowsheet, it may be acceptable to use simplified unit operation models available in the simulation software. A good example is the use of a shortcut distillation model that uses the Fenske–Underwood–Gilliland

3. See Chapter 16 for details on heat exchanger network simulation and analysis.

method to provide a first-pass distillation model before constructing a rigorous tray-by-tray distillation model. The latter would require some detailed information, such as number of trays, top and bottom temperature. On the other hand, a shortcut model would normally require the definition of light and heavy keys, top and bottom column pressures, along with reflux ratio to converge the column model. The shortcut distillation computation will provide useful information needed for building a rigorous tray-by-tray distillation model[4].

Please also note that no single mathematical model can represent all fluids and processes. Hence, the simulation model must be purpose-built.

Habit 2: Identify the system or process and draw an envelope around it

It is important to identify the system that we wish to simulate. In some cases, not the entire flowsheet will be simulated. This could be due to several reasons. One typical example is the limitation of the simulation software. An example is shown in Fig. 1.12, where biomass is utilized as feedstock for a biorefinery. In the pretreatment section, biomass will go through some physical treatments for size reduction and moisture removal, before the biomass is fed into the gasification reactor (and other downstream separation system). If one were to utilize commercial software dedicated for the hydrocarbon-based industry (e.g., Aspen HYSYS and UniSim Design) for the simulation task, it would be best to leave out the treatment section, which is a very much solid-based operation. In other words, the simulation task will focus on the gasification reactor and other downstream separation systems, but exclude the pretreatment section.

Habit 3: Imagine what is going on physically

The engineers who perform a simulation tasks should have good imagination. For instance, the engineer should imagine the state and flow pattern/regime for an inlet stream heading to a reactor/flash unit or an effluent stream emitted from a reactor. For the latter case, if the reactor effluent stream contains compressed liquid with light gases, an adiabatic separator (i.e., flash unit) may be added once the reactor is converged, to separate the light gases from the liquid components.

Habit 4: Translate the physical model to a mathematical model

Engineers should use their basic knowledge in reaction engineering, thermodynamics, separation processes, etc., to translate a physical process in the plant into an equivalent mathematical model. He/she then need to pick the right process simulation software to perform the simulation tasks. One should note that most commercial simulation software packages are dedicated for continuous processes. Hence, having to simulate batch processes (e.g., biofermentation, and polymerization) using a commercial software that is meant

4. See Chapter 12 for an example on the use of shortcut distillation model for generating data for rigorous distillation model on VCM production.

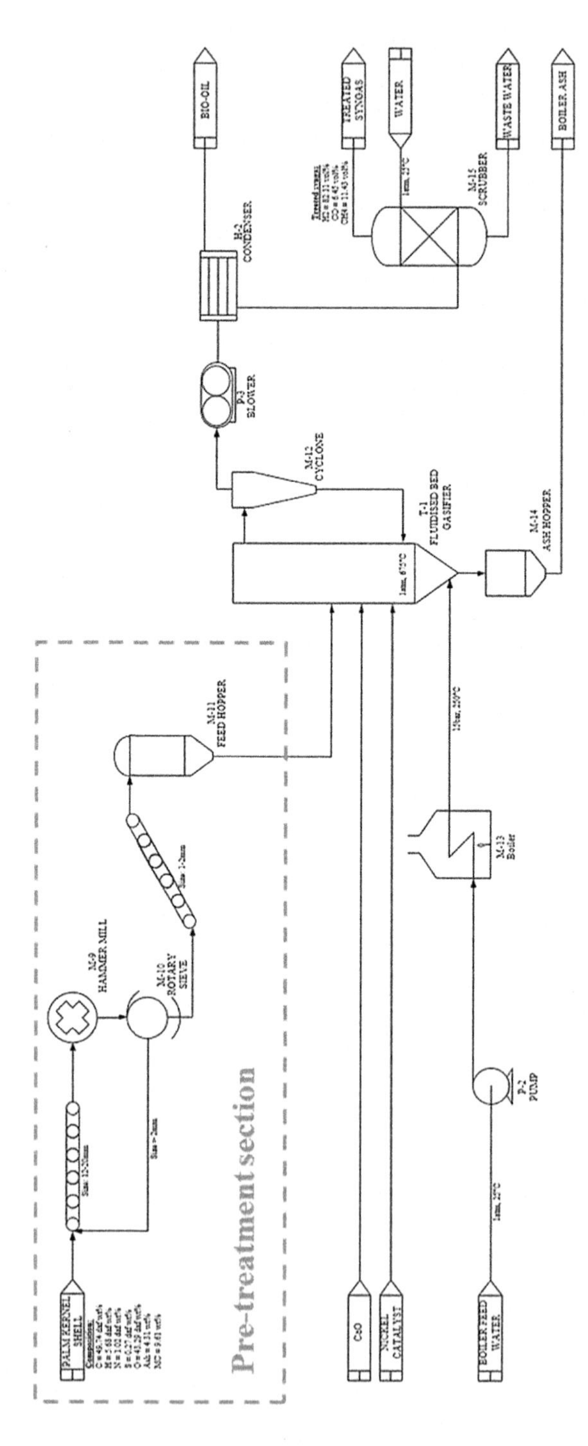

FIGURE 1.12 A biorefinery problem where pretreatment section is excluded from process simulation.

for continuous processes is indeed a challenging task. Even if one may be able to approximate the model to represent the process behavior, there is no way to model the time-related elements (e.g., production scheduling). Besides, many commercial software packages (e.g., Aspen HYSYS and UniSim Design) are more applicable to hydrocarbon-based chemical process industry, e.g., oil and gas and petrochemical, with component and thermodynamic databases that are suited for those industrial sectors. If one were to use these software packages for biochemical process modeling (e.g., production of yeast, vitamin, antibody, food, and beverage), one will have to customize the component as well as its associated thermodynamic models, which is time-consuming and yet inaccurate. For the modeling of biochemical processes, it would be wiser to consider the use of dedicated software packages such as SuperPro Designer (www.intelligen.com) or Aspen Batch Process Developer (www.aspentech.com).

A similar situation also applies for environmental applications, where a high degree of accuracy and very small numbers need to be represented in stream composition (e.g., ppm level). If one were to use the hydrocarbon-based software for wastewater treatment plant design, the stream compositions will hardly be traced. Hence, the advice is to use the right software for the right applications.

Habit 5: Know your components

It is important to know the chemicals that are present in the system you are working on. Furthermore, it is important to understand the type of intermolecular interactions that may exist in that system. Understanding the components and their interactions will enable the engineer to choose the right property estimation methods[5]. In some cases, there may be some components that are not represented in the simulator's component database. In that case, it may be necessary to either find an equivalent component or create a user-defined or "hypothetical" component[6].

One will also have to be aware of the presence of water in a predominantly hydrocarbon process stream. Some thermodynamic models may have to lump the organic and aqueous phases into a single liquid phase, even though water is immiscible with the hydrocarbon. Hence, the selection of thermodynamic model is crucial for this kind of system.

In some cases, azeotropic mixtures introduce challenges in distillation process. The presence of binary or tertiary azeotrope points leads to the existence of a distillation boundary that limits the degree of separation. An example is given in Fig. 1.13, which shows the temperature–composition (T-xy) plot for a binary mixture containing isopentane (iC5) and methanol. The T-xy plot

5. Refer to Chapter 3 for details on physical properties estimation.
6. Refer to Chapter 2 for new component registration.

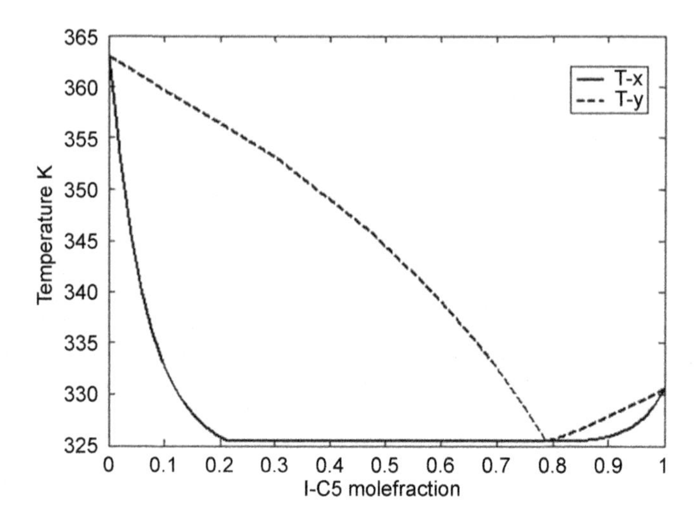

FIGURE 1.13 A binary mixture with minimum boiling azeotrope.

shows a minimum boiling azeotrope occurs for this binary mixture, which prevents the separation of high-purity products with normal distillation. One will have to utilize the enhanced distillation techniques[7] to perform such separation tasks.

Habit 6: Know the context of your feed streams

It is important to know the characteristics of the feed stream, e.g., its origin, composition, sampling conditions, and the presence of impurities.

In some cases, a composition for a gas stream that comes from a three-phase (gas, oil, and water) separator may not contain any water. This is because the stream composition has been determined using gas chromatography (GC). In a GC analysis, a carrier gas, typically helium, transports the gas sample into a length of tubing (called a "column"), which is packed with a polymer or solid support. Water will damage this polymer and is removed before the gas enters the column.

In this case, it is necessary to saturate the gas with water prior to using it for any computations; otherwise, the contribution of water in the gas, particularly the heat of vaporization, will not be accounted for and will result in an inaccurate heat and material balance.

Habit 7: Know your components boiling points

In performing simulation for a separation system (e.g., flash and distillation), it is important to know the boiling points, and hence the relative volatility of the chemicals involved. It is good practice to have the chemicals arranged in

7. Refer to Chapter 15 for a detailed discussion on azeotropic distillation systems.

TABLE 1.2 Chemicals Arranged in the Ascending Order of Their Boiling Points

Components	Boiling Points (°C)
Ethane (C_2H_6)	−88.6
Propane (C_3H_8)	−42.1
n-Butane (C_4H_{10})	−0.5
n-Pentane (C_5H_{12})	36.2
n-Hexane (C_6H_{14})	68.8

the ascending order of their boiling points (i.e., volatilities). An example is shown in Table 1.2. As shown, a process stream consists of five hydrocarbon components, arranged in the ascending order of their boiling points. If these components were to be used for a distillation computation, it would be easy to draw a line between the heavy and light keys and perform a quick material balance to determine distillate and bottom product flowrates.

Besides, we should also watch out for polar molecules and those chemicals that have hydrogen bonds. This may give rise to azeotropic systems where high boiling point components boil and vaporize before low boiling point components.

Habit 8: Keep track of the units of measure in all calculations

Errors on units of measurement are very serious and the easiest errors to avoid with some simple discipline. The best practice is always to keep track of the units of measurement in all computation or simulation exercises. All modern commercial simulation software packages are equipped with unit conversion functions for all unit operations and stream parameters. These simulation packages allow for global units of measure profiles to be defined and saved. Local definitions of units are also possible to allow for flexibility, but when possible it is a good idea to preprocess any data to ensure consistency.

Prior to conducting any engineering analysis or project, it is necessary to first define what units of measure are to be used for the analysis and reporting purposes and ensure that this is adopted by all personnel working on the analysis or project.

Habit 9: Always do a simple material and energy balance first

An important good habit is to perform some manual calculations (or "hand calculations"). For example, performing a quick material and energy balances prior to executing a process simulation exercise enables the engineer to have a better "feel" for the orders of magnitude in the numbers that may be encountered. In addition to that, these hand calculations can also be used as initial guesses for more complex types of computations. For instance, in designing a depropanizer column, we can perform a quick material balance by assuming most (if not all) propane is recovered at the top stream of the column and heavier components are recovered at the bottom stream. This will provide an estimate for overhead and bottom flows and can be used as an initial guess for a rigorous tray-by-tray distillation model.

Performing mass and energy balances for some unit operations may be tedious, especially for those that involve nonlinear equations. Hence, some linearized models are always useful for preliminary flowsheet, especially for conceptual design stage. Linear models for some commonly used unit operations (distillation, absorption, etc.) may be found in Biegler et al. (1997).

Habit 10: Plot the phase envelope for important streams

It is extremely important to know in which state the fluid is in, e.g., is it a subcooled or saturated liquid? For a process stream with a subcooled liquid, temperature rise is expected when sensible heat is added; the latter is the product of mass flowrate, heat capacity, and temperature rise. On the other hand, no temperature rise will be reported when a saturated liquid is heated, as latent heat is involved. Also, it is necessary to check if the fluid is near the critical point or is it at the supercritical stage?

It is also important to know how close the process stream is to the dew point (either hydrocarbon or water). When adsorption beds or fuel gas systems were designed, liquids formed at the dew point may damage the adsorption bed or combustion chamber. Another important practice is to identify the retrograde region of the fluid. In the retrograde region, compressing a fluid may result in its vaporization instead of liquefaction[8].

REFERENCES

AspenTech, 2017. A History of Innovation. www.aspentech.com.

Biegler, L., Grossmann, I.E., Westerberg, A.W., 1997. Systematic Methods of Chemical Process Design. Prentice-Hall.

Crowe, C.M., Hamielec, A.E., Hoffman, T.W., Johnson, A.I., Woods, D.R., Shannon, P.T., 1971. Chemical Plant Simulation. Prentice Hall, Englewood Cliffs.

Dimian, A., Bildea, C., Kiss, A., 2014. Integrated Design and Simulation of Chemical Processes, second ed. Elsevier Science, Amsterdam.

El-Halwagi, M.M., 2006. Process Integration. Elsevier, San Diego.

El-Halwagi, M.M., Foo, D.C.Y., 2014. Process synthesis and integration. In: Seidel, A., Bickford, M. (Eds.), Kirk-Othmer Encyclopedia of Chemical Technology. John Wiley & Sons.

Evans, L.B., 1981. Advances in process flowsheeting systems. In: Mah, R.S., Seider, W.D. (Eds.), Foundations of Computer-Aided Chemical Process Design. Engineering Foundation, New York.

Evans, L.B., Joseph, B., Seider, W.D., 1977. System structures for process simulation. AIChE Journal 23 (5), 658–666.

8. See detailed discussion in Chapter 3.

Evans, L.B., Boston, J.F., Britt, H.I., Gallier, P.W., Gupta, P.K., Joseph, B., Mahalec, V., Ng, E., Seider, W.D., Yagi, H., 1979. Computers & Chemical Engineering 3 (1–4), 319–327.

Federal Trade Commission, 2003. FTC Charges Aspen Technologys Acquisition of Hyprotech, LTD. Was Anticompetitive. www.ftc.gov.

Foo, D.C.Y., 2012. Process Integration for Resource Conservation. CRC Press, Boca Raton, Florida, US.

Foo, D.C.Y., Manan, Z.A., Selvan, M., McGuire, M.L., October 2005. Integrate process simulation and process synthesis. Chemical Engineering Progress 101 (10), 25–29.

Gallier, P.W., Evans, L.B., Boston, J.F., Britt, H.I., Boston, J.F., Guupta, P.K., 1980. ASPEN: advanced capabilities for modeling and simulation of industrial processes. In: Squires, R.G., Reklaitis, G.V. (Eds.), Computer Applications to Chemical Engineering. ACS Symposium Series, vol. 124. American Chemical Society, Washington, DC, pp. 293–308.

Hernandez, R., Sargent, R.W.H., 1979. A new algorithm for process flowsheeting. Computers & Chemical Engineering 3 (1–4), 363–371.

Honeywell, 2017. www.honeywellprocess.com.

Linnhoff, B., Townsend, D.W., Boland, D., Hewitt, G.F., Thomas, B.E.A., Guy, A.R., Marshall, R.H., 1982. A User Guide on Process Integration for the Efficient Use of Energy. IChemE, Rugby.

Motard, R.L., Shacham, M., Rosen, E.M., 1975. Steady state chemical process simulation. AIChE Journal 21 (3), 417–436.

Perkins, J.D., Sargent, R.W.H., Thomas, S., 1982. SPEEDUP: a computer program for steady-state and dynamic simulation of chemical processes. IChemE Symposium Series 73, H78–H86.

Petrides, D., 1994. BioPro designer: an advanced computing environment for modeling and design of integrated biochemical processes. Computers & Chemical Engineering S18, 621–625.

Petrides, D., Cooney, C.L., Evans, L.B., 1989. Bioprocess simulation: an integrated approach to process development. Computers & Chemical Engineering 13 (4–5), 553–561.

Process System Enterprise (PSE) 2017. www.psenterprise.com.

Rosen, E.M., 1980. Steady state chemical process simulation: a state-of-the-art review. Computer applications to chemical engineering. ACS Symposium Series 124, 3–36.

Schneider Electric, 2017. www.schneider-electric.com.

Smith, R., 2016. Chemical Process Design and Integration, second ed. John Wiley & Sons.

Turton, R., Bailie, R.C., Whiting, W.B., Shaeiwitz, J.A., 2013. Analysis, Synthesis and Design of Chemical Processes, fourth ed. Prentice Hall, New Jersey.

Westerberg, A.W., Hutchison, H.P., Motard, R.L., Winter, P., 1979. Process Flowsheeting. Cambridge University Press, Cambridge.

WinSim, I., 2017. www.winsim.com.

Chapter 2

Registration of New Components

Denny K.S. Ng[1], Chien Hwa Chong[2], Nishanth Chemmangattuvalappil[1]
[1]*University of Nottingham Malaysia Campus, Semenyih, Malaysia;* [2]*Taylor's University, Subang Jaya, Malaysia*

As mentioned in Chapter 1, the first step of setting up a process simulation is to define chemical components that involve in the entire process. Note, however, that not all components are available in the database of the simulation software. Hence, one will have to define/register the chemical component(s) before one can make use of them in the simulation flowsheet. This may involve various types of hypothetical and oil components. In this chapter, the registration of components in several important process simulation software packages, i.e., Aspen HYSYS and PRO/II, are illustrated.

2.1 REGISTRATION OF HYPOTHETICAL COMPONENTS

Components that are not available in simulator library can be defined in the simulation software as "hypothetical" components based on the properties of the chemicals. Different simulation software packages treat this step differently. For example, in Aspen HYSYS and UniSim Design, it is possible to define such components in the library based on their properties. On the other hand, components are defined based on the molecular structure in Aspen Plus. In this section, registration of hypothetical components is demonstrated with Aspen HYSYS and PRO/II.

2.1.1 Hypothetical Component Registration With Aspen HYSYS

To define the "hypothetical" components in Aspen HYSYS library, important properties of the chemicals such as molecular weight and boiling point are first defined in the simulator. Aspen HYSYS will then estimate the other missing properties based on UNIFAC group contribution models (AspenTech, 2015). However, it is necessary to provide the boiling point and molecular weight to estimate the rest of the properties. If more properties are made available, the

Chemical Engineering Process Simulation. http://dx.doi.org/10.1016/B978-0-12-803782-9.00002-9

accuracy of the prediction of the other properties would be much higher. Steps involved in building hypothetical components are illustrated in Example 2.1.

Example 2.1

This example involves the registration of a hydrocarbon component as a "hypothetical" component in Aspen HYSYS. For this case, the hydrocarbon has a molecular weight of 86 and boiling point of 64°C. Fig. 2.1 shows the detailed steps for registering this component in the software. Based on the boiling point and molecular weight, the other physical properties are then estimated, as shown in Fig. 2.2.

2.1.2 Hypothetical Component Registration With PRO/II

Two methods can be used to estimate components in PRO/II (Schneider Electric, 2015). A user can create a new component using the user-defined

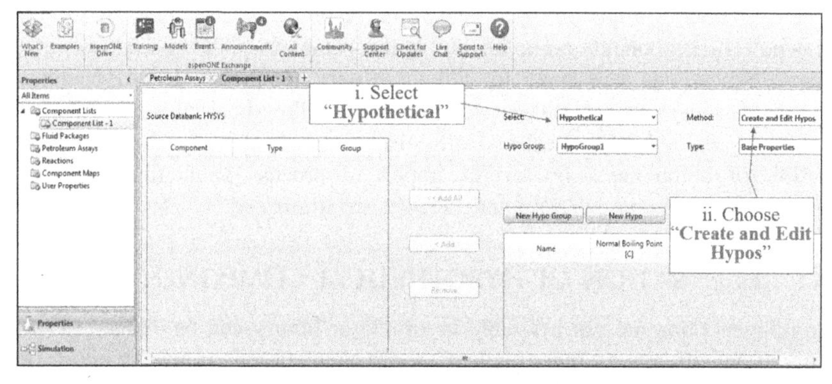

FIGURE 2.1 Hypo selection in Aspen HYSYS.

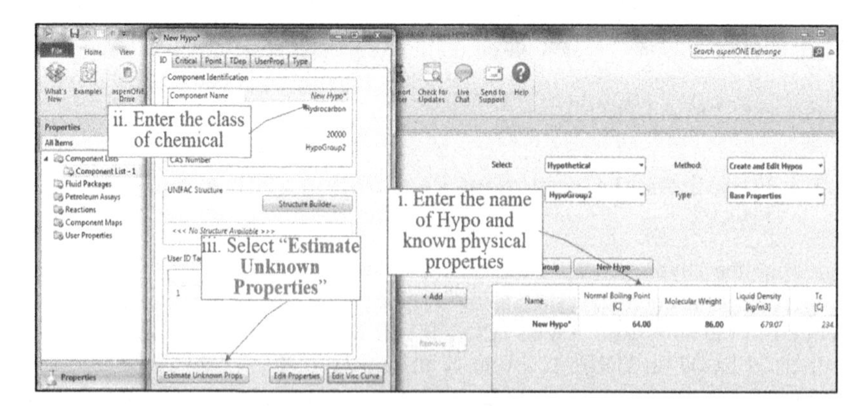

FIGURE 2.2 Property estimation of Hypo components in Aspen HYSYS.

library and generate chemical structure using the UNIFAC structure and finally fill options to generate all the component properties of user-defined components. The second option is defining the component using the PROPRED property prediction methodology and then estimating with and without measured normal boiling point (NBP). Example 2.2 shows the first method used in defining a component.

Example 2.2

Estimating a component in PRO/II. In this case, gallic acid is selected to be developed as a new user-defined component using PRO/II. To create a new component, the user needs to create a user-defined component name from the component selection section. The steps are shown in Figs. 2.3 and 2.4.

Next, the user is required to define UNIFAC structure. In this case, acids, aromatics, and alcohol category structures are selected for the gallic acid component. Detailed steps for creating the structures are shown in Fig. 2.5.

Molecular weight, critical temperature, critical pressure, NBP, and miscellaneous properties data can be modified in "fixed" thermophysical properties section. In this case, the user can modify the NBP value based on steps in Fig. 2.6.

To conduct a simulation, a heat exchanger can be added to the stream following detailed steps in Fig. 2.7.

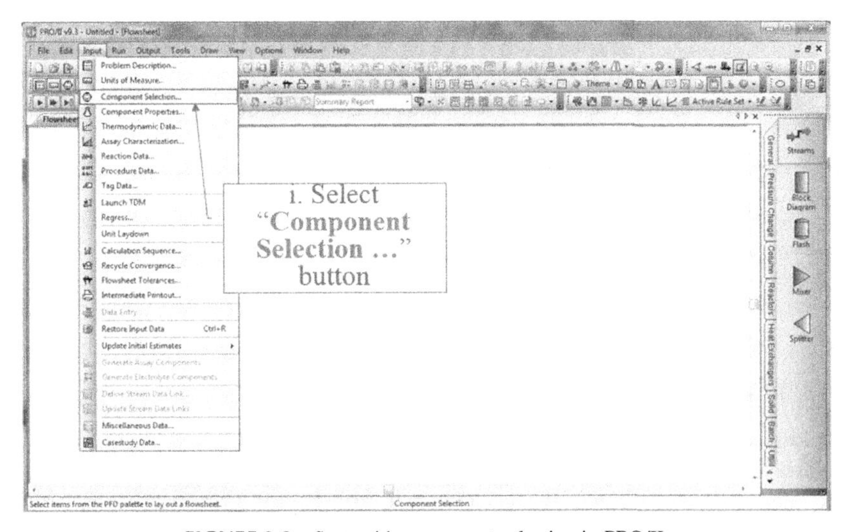

FIGURE 2.3 Start with component selection in PRO/II.

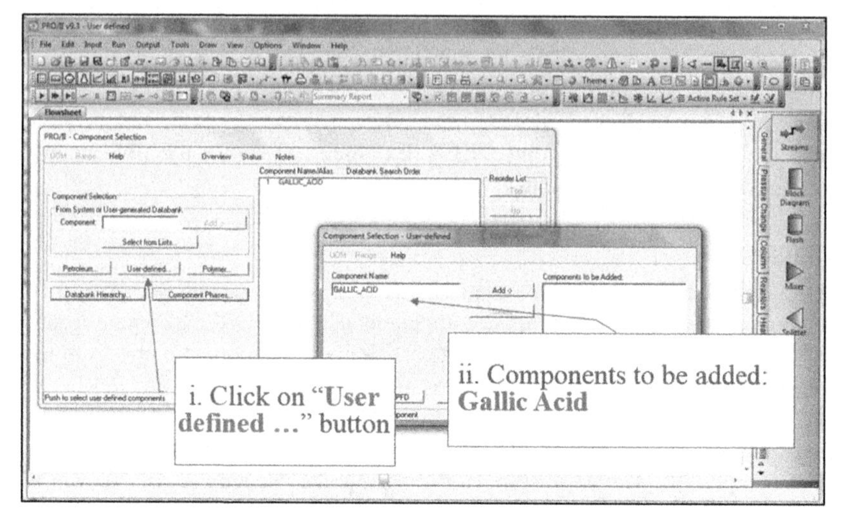

FIGURE 2.4 Define the name of a new component.

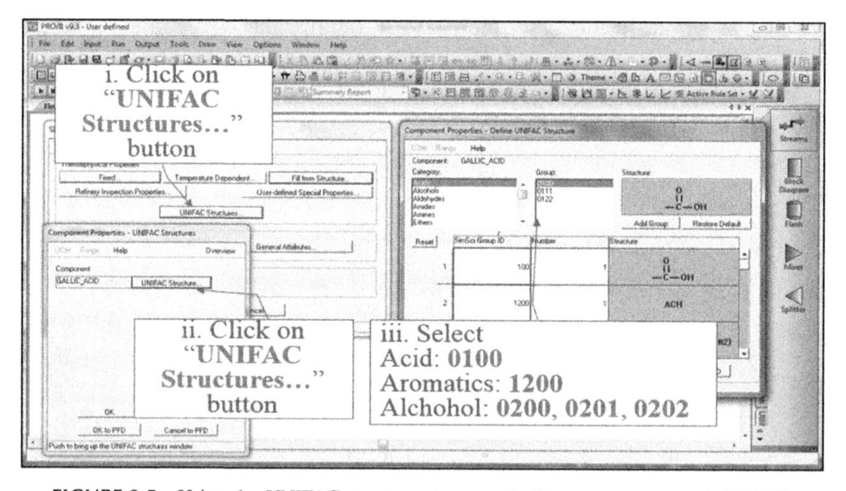

FIGURE 2.5 Using the UNIFAC structures to generate the new component in PRO/II.

2.2 REGISTRATION OF CRUDE OIL

Crude oil is a naturally occurring, unrefined petroleum product composed of hydrocarbon deposits and other organic materials, which can be refined to produce other usable products such as gasoline, diesel, and various forms of petrochemicals. As crude oil is a mixture of multiple components, it is difficult to identify individual components present in the oil. In industry practice, characteristics of a crude oil sample is normally determined via laboratory

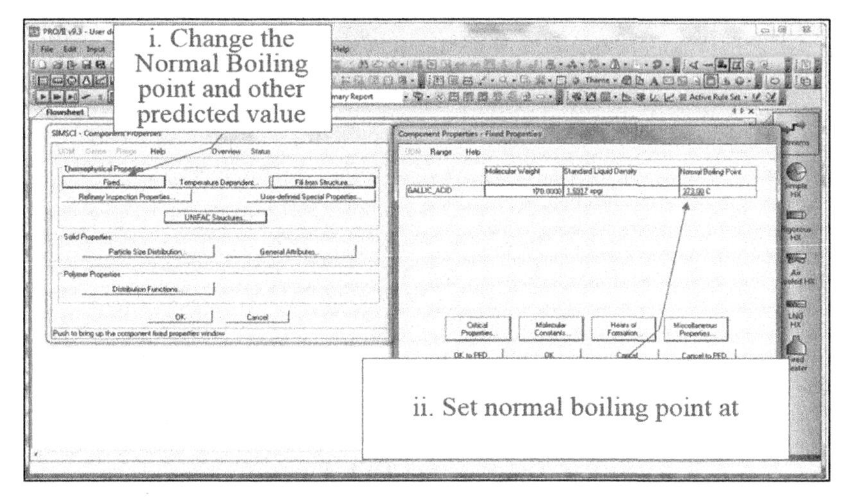

FIGURE 2.6 Specifying normal boiling point and other properties.

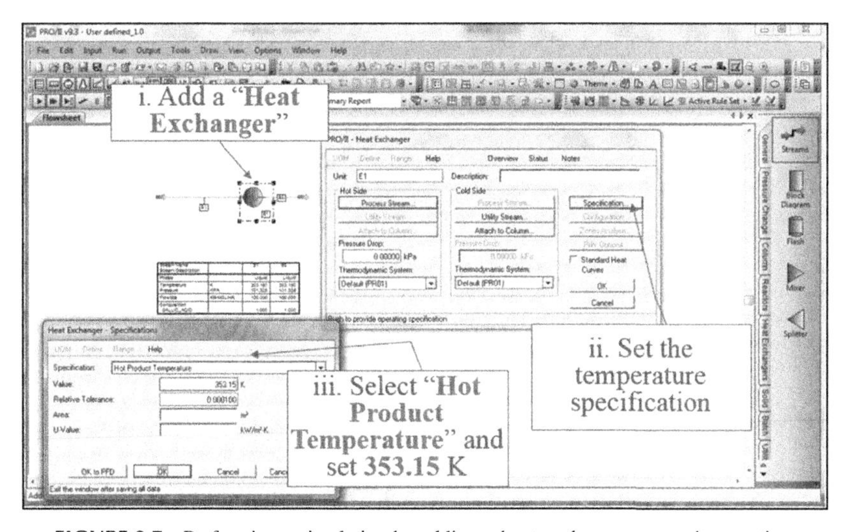

FIGURE 2.7 Performing a simulation by adding a heat exchanger as a unit operation.

distillation tests (see list in Table 2.1). Similarly, rather than defining individual oil components in process simulation, it is commonly defined based on the results of common laboratory distillation tests.

Fig. 2.8 shows the sample of true boiling point (TBP) of a crude oil. The TBP distillation (ASTM D2892) uses a 15-theoretical plate column, and 5:1 flux ratio is a classical method to obtain the distillation curve for the crude oil sample. The distillation fractionates the crude oil into a number of narrow fractions up to 400°C atmospheric equivalent temperature.

TABLE 2.1 Common Laboratory Distillation Tests

Test Name	Reference	Main Applicability
ASTM (atmospheric)	ASTM D86	Petroleum fractions (products do not decompose when vaporized at 1 atm)
ASTM (vaccum, 1.3 kPa)	ASTM D1160	Heavy petroleum fractions or products that tend to decompose in ASTM D86 test but can be partially or completely vaporized at a maximum liquid temperature of 400°C at 0.13 kPa
TBP (atmospheric or 1.3 kPa)	Nelson, ASTM 2892	Crude oil petroleum fractions
Simulated TBP (gas chromatography)	ASTM D2887	Crude oil petroleum fractions

ASTM, American Society for Testing and Materials; *TBP*, true boiling point.

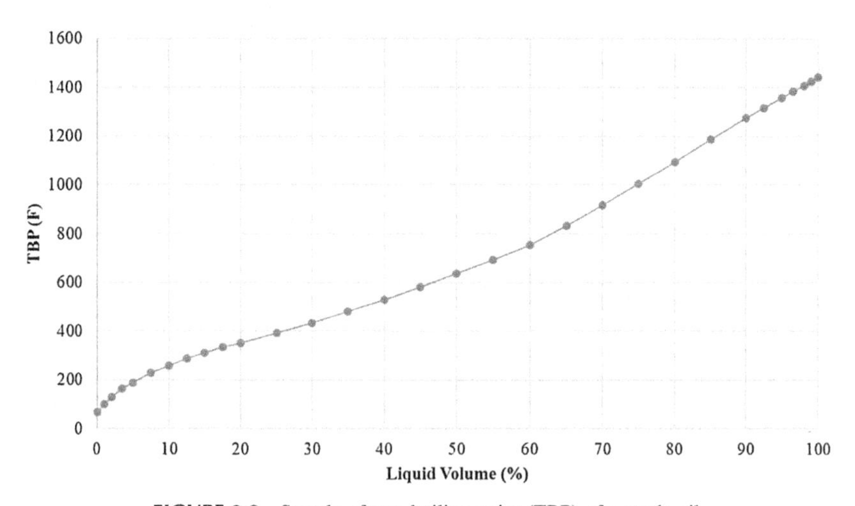

FIGURE 2.8 Sample of true boiling point (TBP) of a crude oil.

Example 2.3 Crude Oil Registration With Aspen HYSYS

To model crude oil in Aspen HYSYS, the oil needs to be defined based on its characteristics via Oil Manager. Based on the laboratory results, the characteristics of the crude oil are included in Aspen HYSYS library and hypothetical components. This involves the following three main steps:

- Characterization of crude assay
- Generate pseudocomponents—create cut and blend
- Install the oil in the flowsheet

In this example, 477,000 kg/h of crude oil is required to heat up to 482°F (523.15K) under ambient conditions, 86.18°F and 16.7 psia (303.15K and 101.325 kPa). To illustrate the proposed procedure in detail, the following oil properties will be installed in Aspen HYSYS. Thermodynamic package to be used is Peng–Robinson, with "Auto Cut" as the option.

Step 1: Characterization of Crude Assay Before defining the oil, the components that exist in simulator databank "HYSYS" are first defined. Next, the thermodynamic package (also known as fluid package in Aspen HYSYS) is also defined. Based on the information given, i-Butane, n-Butane, i-Pentane, and n-Pentane are installed via HYSYS databank and Peng–Robinson is installed as the fluid package in the simulator, as shown in Figs. 2.9 and 2.10.

Next, the oil properties are defined via "Oil Manager" in the simulator. As discussed, the characteristics of oil are determined based on the laboratory test. Therefore, the result of the test will be used as input information for the simulator. Detailed steps for characterizing the crude assay are shown in Figs. 2.11–2.15.

As mentioned previously, there are few types of laboratory distillation tests; therefore, the user needs to select the respective test in Oil Manager, as shown in Fig. 2.13. In Aspen HYSYS, few options of Assay Data Type are available, i.e., TBP, ASTM D86, ASTEM D1160, ASTM D86–D1160, ASTM D2887, Chromatograph, and EFV.

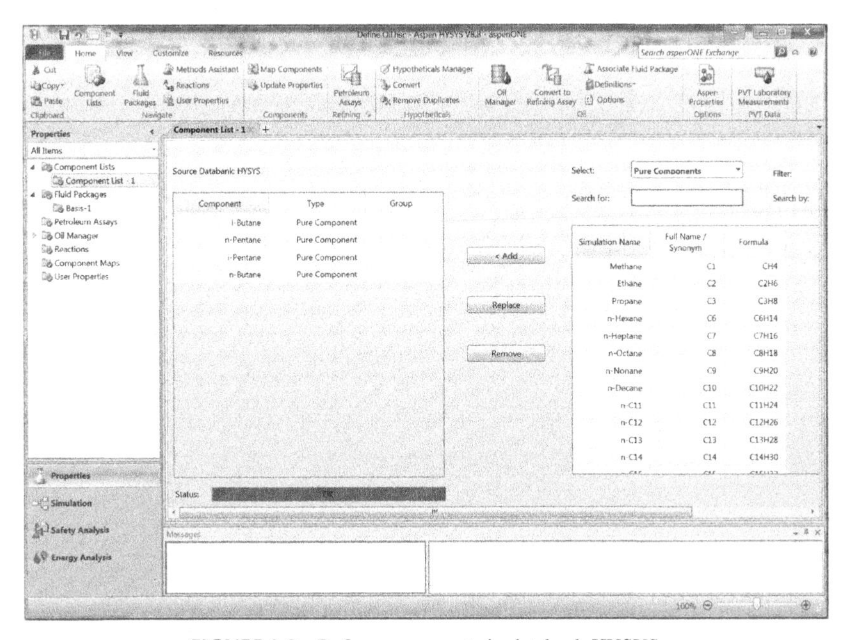

FIGURE 2.9 Define components in databank HYSYS.

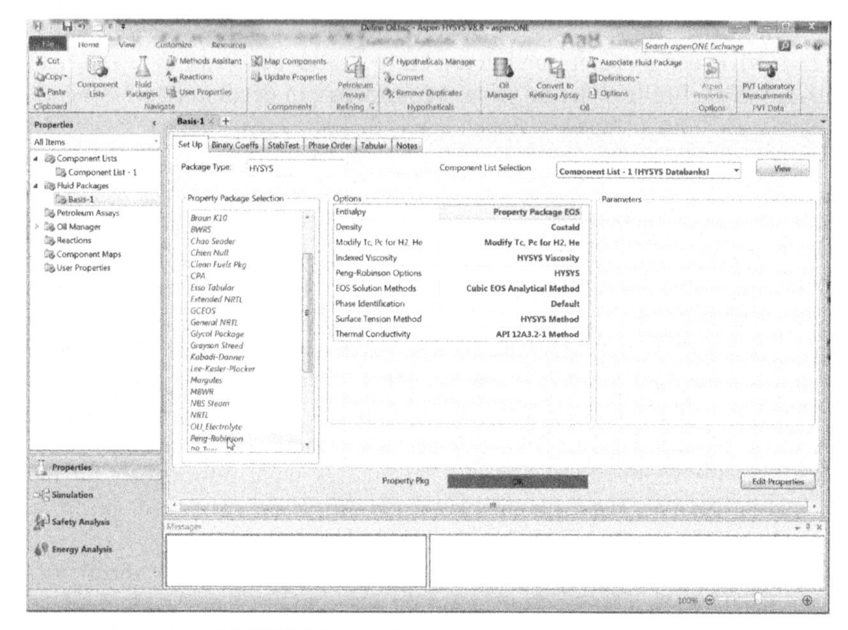

FIGURE 2.10 Define thermodynamic package.

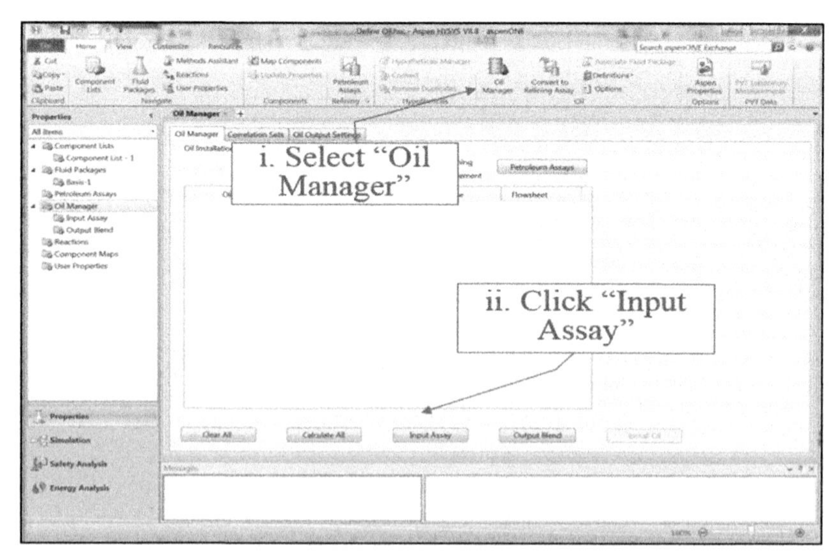

FIGURE 2.11 Defining crude oil in Oil Manager of Aspen HYSYS.

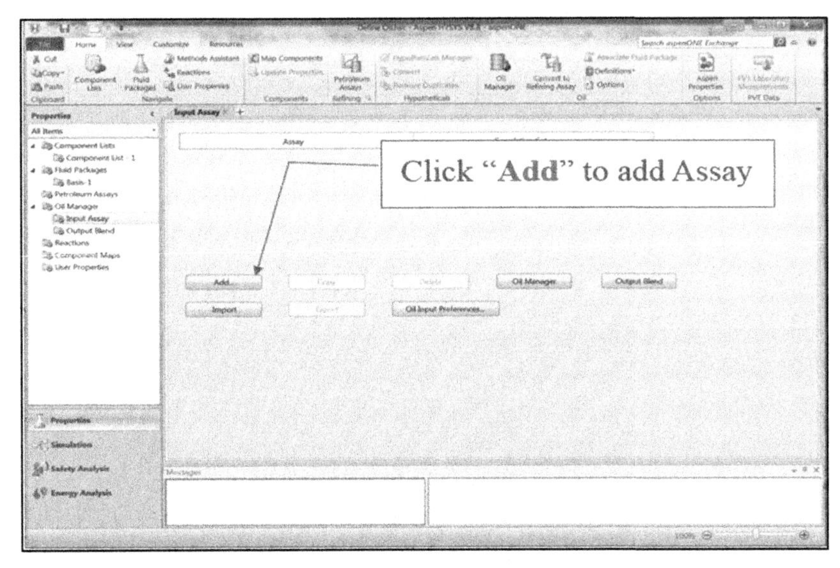

FIGURE 2.12 Add assay in Oil Manager.

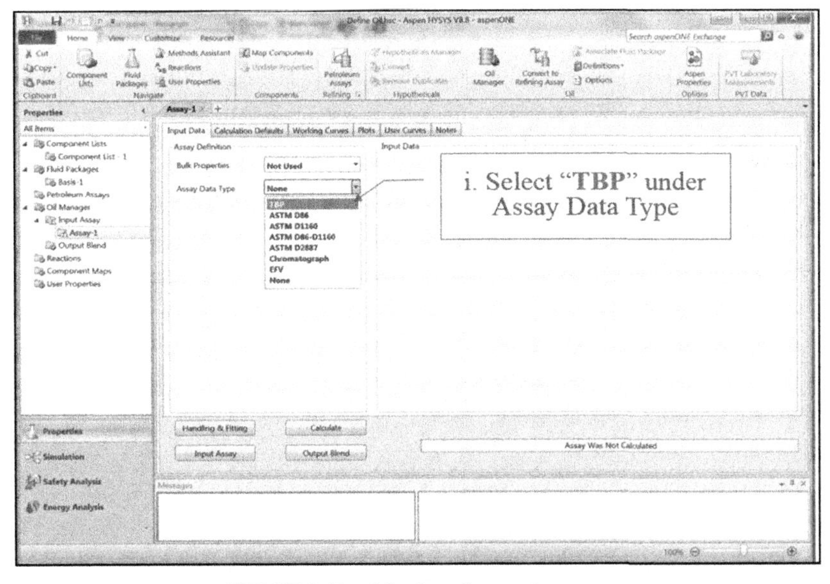

FIGURE 2.13 Selection of assay data type.

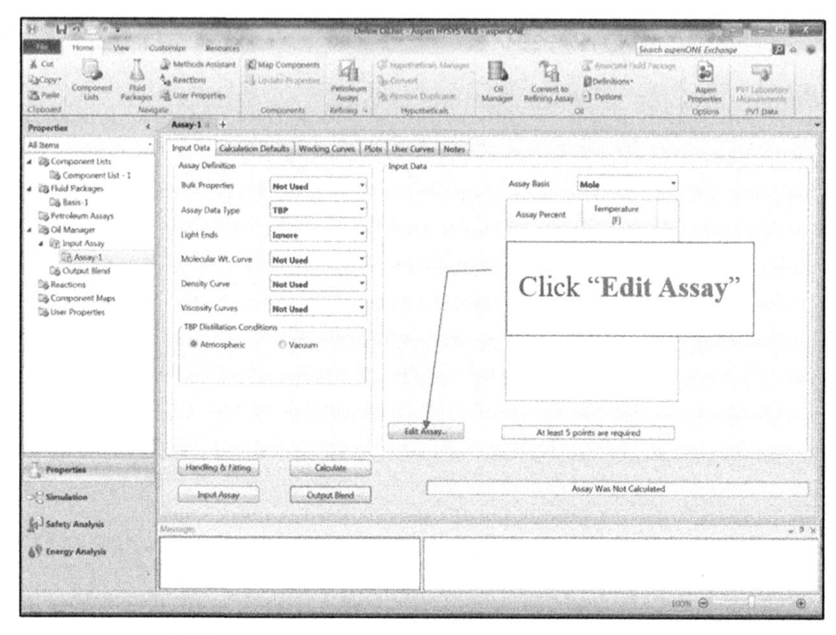

FIGURE 2.14 Editing assay and input of assay information.

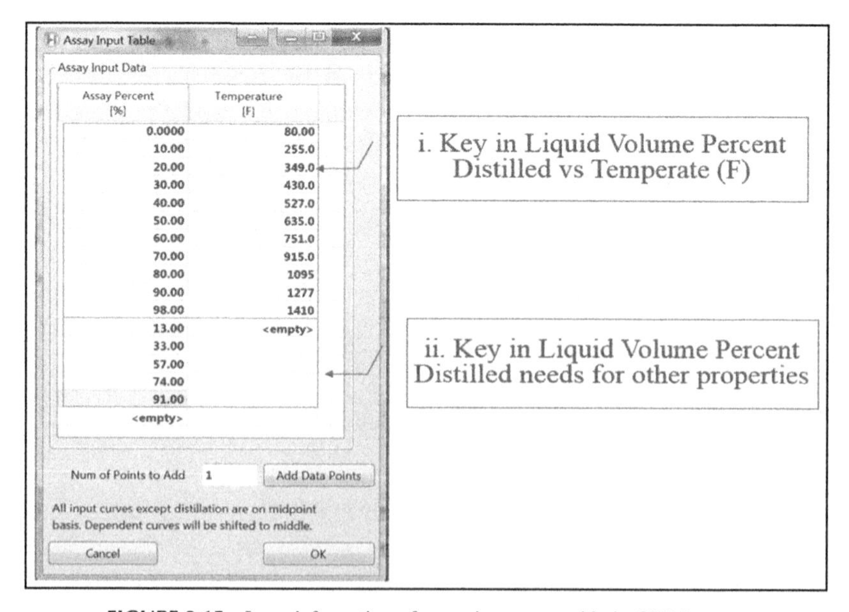

FIGURE 2.15 Input information of assay into assay table in Oil Manager.

Next, the user should select "Edit Assay" to key in the assay information that is obtained from the laboratory results (see Figs. 2.14 and 2.15). Note that liquid volume percent distilled versus temperature is first entered in the table. Next, liquid volume percent distilled, which requires other properties, such as density and molecular curve, is also included in the table (as shown in Fig. 2.15).

Once the assay table is completed, information about other properties such as bulk properties, molecular weight curve, light ends, density curve, and viscosity curve can also be included. Fig. 2.16 shows the input of bulk properties into the simulator. Besides, based on the given information, light ends, molecular weight, API gravity, viscosity (cP) at 100°F, and viscosity (cP) at 210°F are available; therefore, information about these properties is included to estimate the crude oil properties accurately in the simulator. However, in the event where the information is absent, the user can select the option of "Not Used" in the Assay window. The simulator will then estimate the properties based on the available information accordingly.

Note that the following information is required to estimate the crude oil properties:

- Mass density
- Bulk viscosity at 100°F and at 210°F

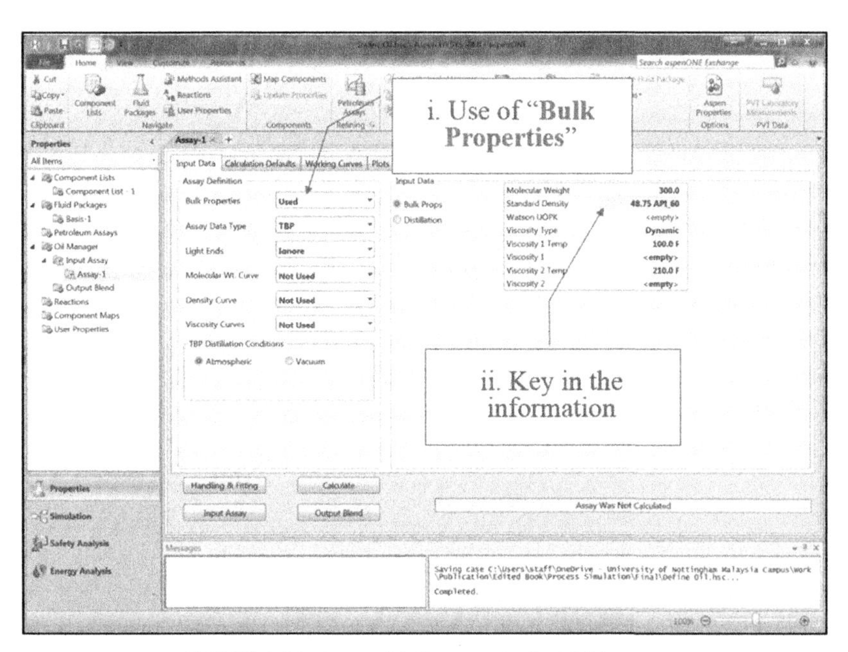

FIGURE 2.16 Input of bulk properties in Oil Manager.

Meanwhile, the other optional properties are recommended to be included. These include the following:

- Molecular weight curve
- Light ends
- Density curve
- Viscosity curve

Figs. 2.16–2.20 show the input of bulk properties, light ends, molecular weight, API gravity, viscosity (cP) at 100°F, and viscosity (cP) at 210°F.

Once all the information is included into the simulator, the assay can be determined by a click on "Calculate" (see Fig. 2.21). This has shown that the assay is ready to generate pseudocomponents in the following step.

Step 2: Generate Pseudocomponents—Create Cut and Blend When more information is provided to the simulation software, more accurate properties of the crude oil assay can be generated. In Aspen HYSYS, blend and cut functions of the assay are required to generate the pseudocomponents. Fig. 2.21 shows the blend of assay to generate the general presentation of the whole curve. Next, the calculated blend is required to add into cut function (Fig. 2.21). After the blend is added (Figs. 2.22 and 2.23), the array is ready to perform cut function.

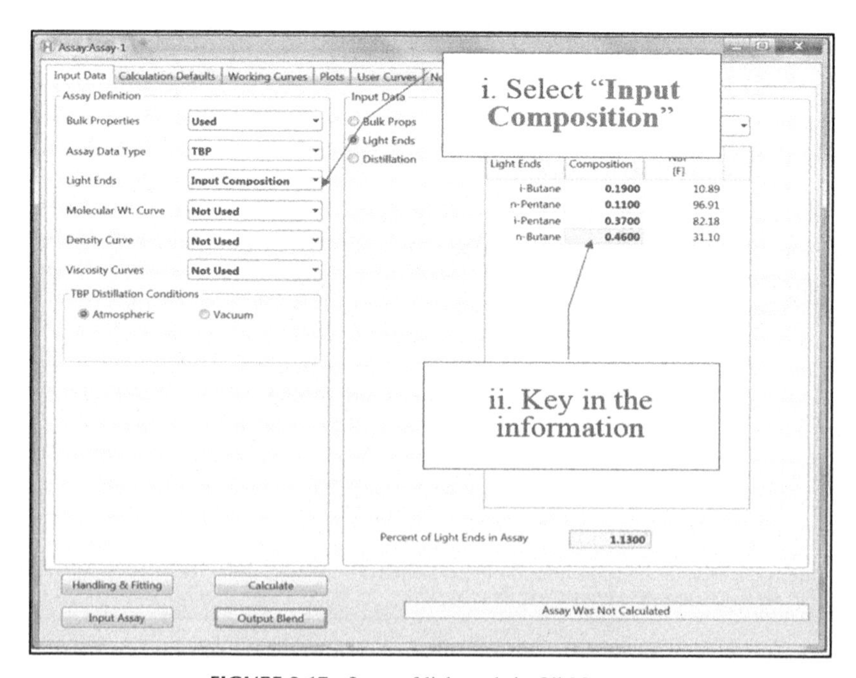

FIGURE 2.17 Input of light ends in Oil Manager.

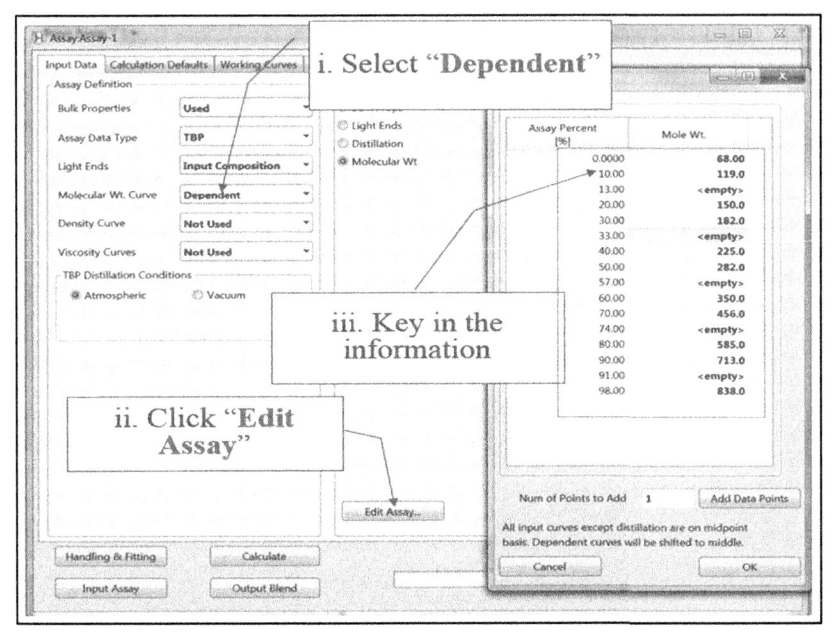

FIGURE 2.18 Input of molecular weight curve in Oil Manager.

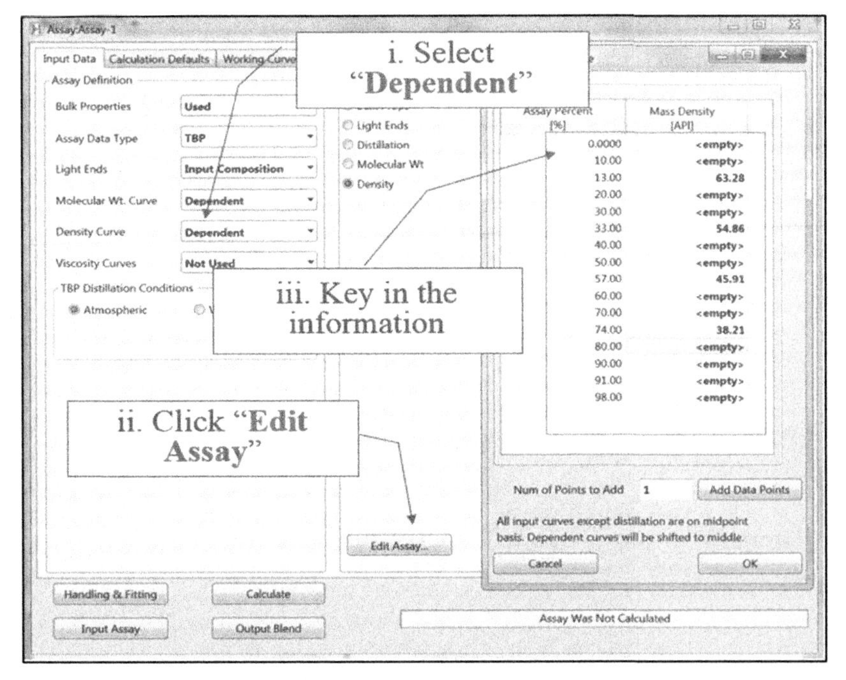

FIGURE 2.19 Input of density curve in Oil Manager.

(A)

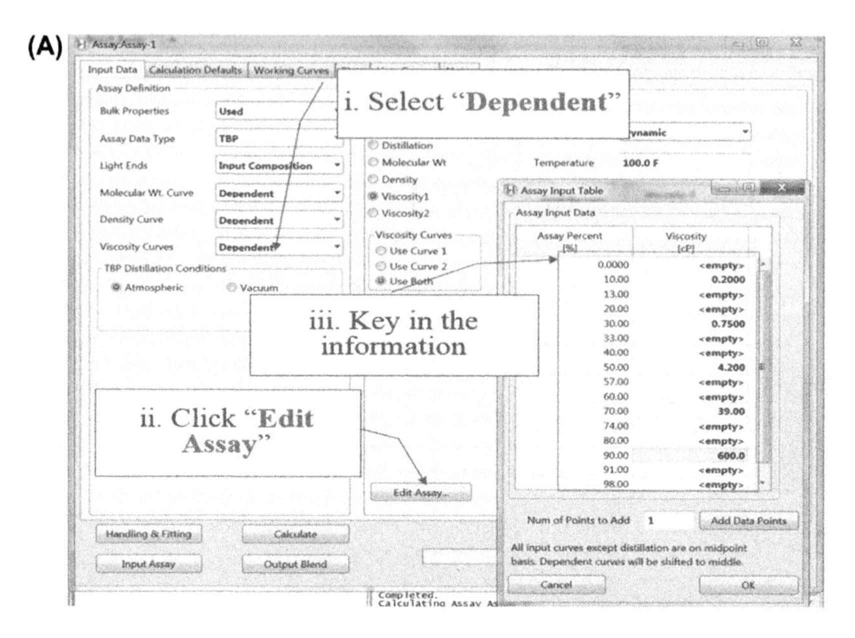

(B)

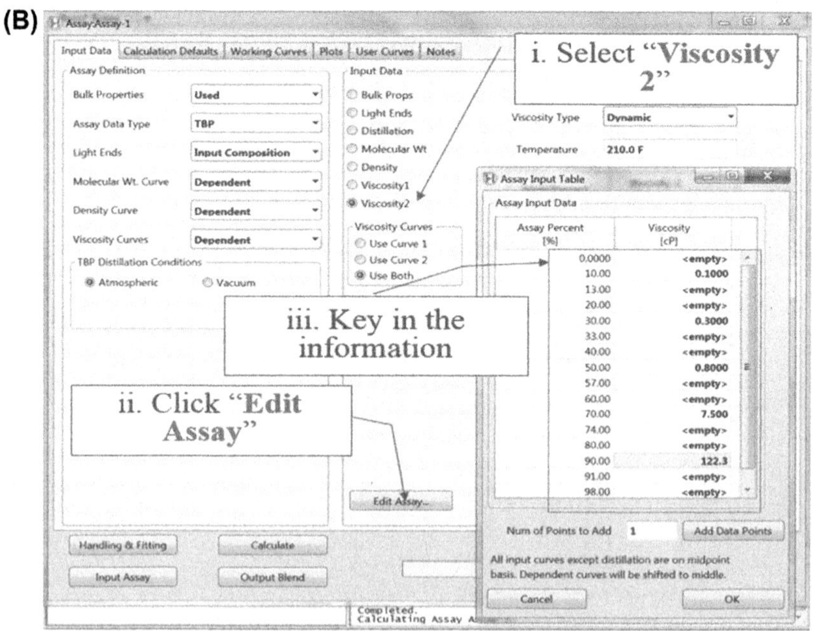

FIGURE 2.20 Input of viscosity curves (A) at 100°F (B) at 210°F in Oil Manager.

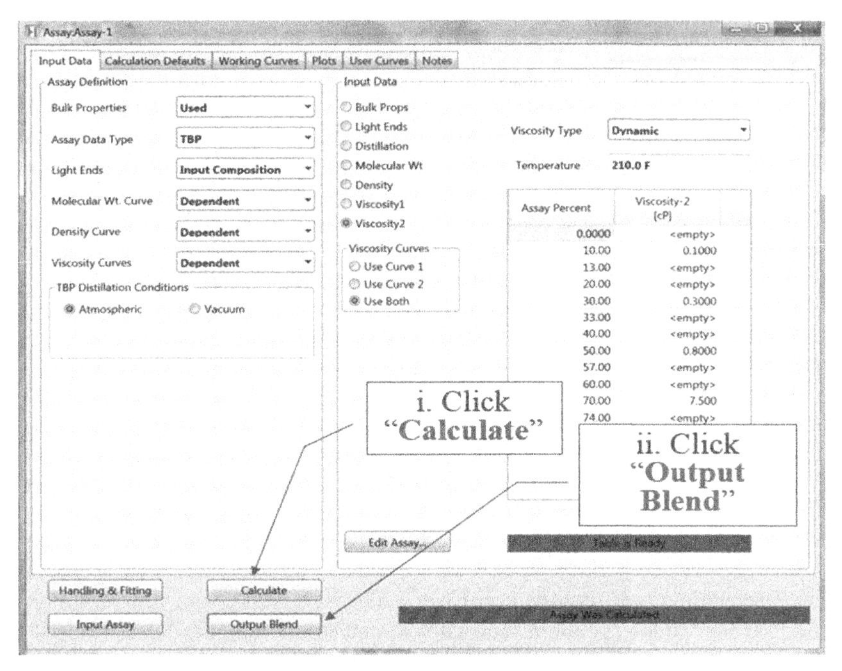

FIGURE 2.21 Generating the blend based on the input assay.

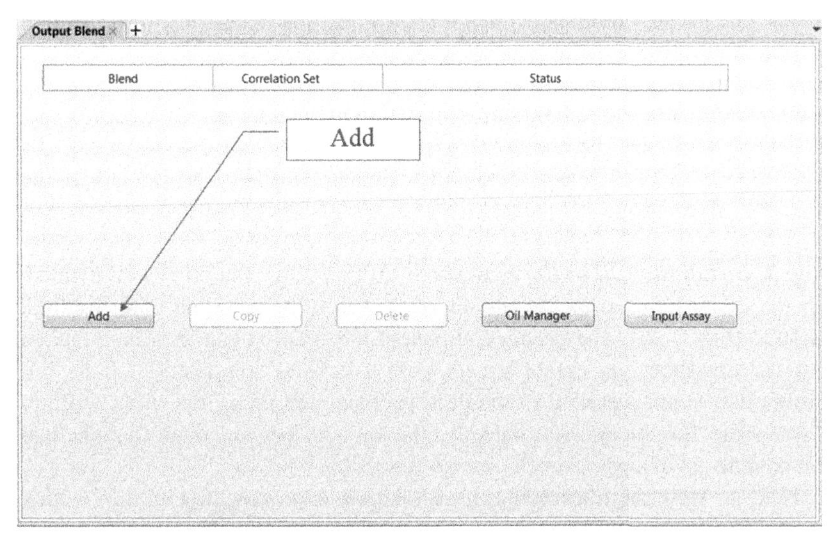

FIGURE 2.22 Add calculated blend into the simulator.

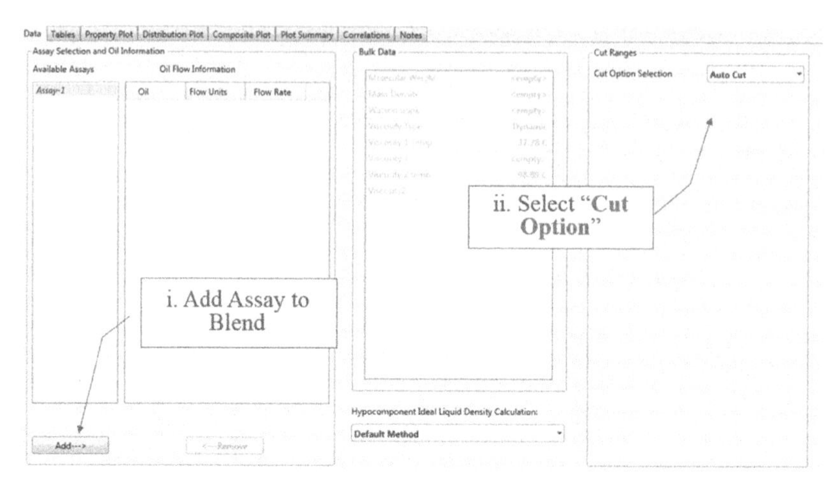

FIGURE 2.23 Selection of option for cuts.

In Aspen HYSYS, three types of cut, which are "Auto Cut," "User Ranges," and "User Points" are provided (see Fig. 2.23). Note that "Auto Cut" option is based on the values specified internally to determine the cut; "User Ranges" specified boiling point ranges and then number of cuts per range; and "User Points" specified cut points that are proportioned based on internal weight scheme (Fig. 2.23).

Once the blend and cut are ready, the pseudocomponents can be determined. The pseudocomponents that represent the characteristics of the oil will be generated. To view the generated oil information, click at tabs "Tables," "Composite Plot," and others as well. Figs. 2.24 and 2.25 show the information of oil. Note that in Fig. 2.25, both information of calculated oil (red) and input of user (green) can be viewed clearly. Note that both curves should be superimposed with each other when the calculated oil is very similar with the input oil information.

Step 3: Install the Oil in the Flowsheet At this stage, the oil is well defined. For the oil to be used in the simulation, it has to be installed with the fluid package (Fig. 2.26). The pseudocomponent information can then be included into the flowsheet via create stream with a defined composition. If the oil information is not installed in the fluid package, the oil composition will not exist in the flowsheet. Alternatively, the oil can be added as hypothetical components as discussed in the previous section.

Next, to view the information in simulation environment (Fig. 2.27), click on the "Crude Oil," which is defined previously. Note that the composition of the hypothetical components will be included in the stream. In the event where the oil composition is not included in the "Crude Oil," the stream should be redefined based on the previous steps.

Blend-1 +

Data | Tables | Property Plot | Distribution Plot | Composite Plot | Plot Summary | Correlations | Notes

Table Type: Component Properties

Table Control: Main Properties / Other Properties

Oil: Blend-1

Click at "Tables" tab

Component Physical Properties

Comp Name	NBP [F]	Mole Wt.	Density [API]	Viscosity1 [cP]	Viscosity2 [cP]
NBP_173	172.7	89.11	76.57	0.19152	8.0875e-002
NBP_194	193.8	93.20	75.83	0.21350	8.8518e-002
	218.9	96.89	74.87	0.24640	9.9103e-002
	244.7	101.3	73.78	0.29074	0.11541
	267.0	109.4	72.39	0.36100	0.14346
	292.3	114.3	71.13	0.44170	0.17683
NBP_319	319.0	117.7	69.89	0.53994	0.21772
NBP_342	342.4	124.9	68.34	0.69939	0.28109
NBP_367	367.1	132.5	66.78	0.90767	0.34226
NBP_392	391.9	140.8	65.30	1.1733	0.39834
NBP_416	416.4	149.9	63.91	1.5141	0.45776
NBP_441	441.2	159.3	62.63	1.9296	0.51593
NBP_466	466.1	168.7	61.49	2.4254	0.57445
NBP_491	490.9	178.5	60.41	3.0365	0.64719
NBP_516	515.7	188.8	59.38	3.8035	0.74210
NBP_540	540.5	199.3	58.40	4.7905	0.92568
NBP_565	565.2	210.2	57.45	6.0684	1.1851
NBP_590	590.0	220.9	56.51	7.7049	1.5070
NBP_615	614.9	231.4	55.59	9.8174	1.9787
NBP_640	639.7	242.8	54.66	12.615	2.5866
NBP_664	664.3	255.2	53.74	16.357	3.3184

Install Oil | Output Blend | Input Assay

Blend Was Calculated

FIGURE 2.24 Component physical properties of calculated oil.

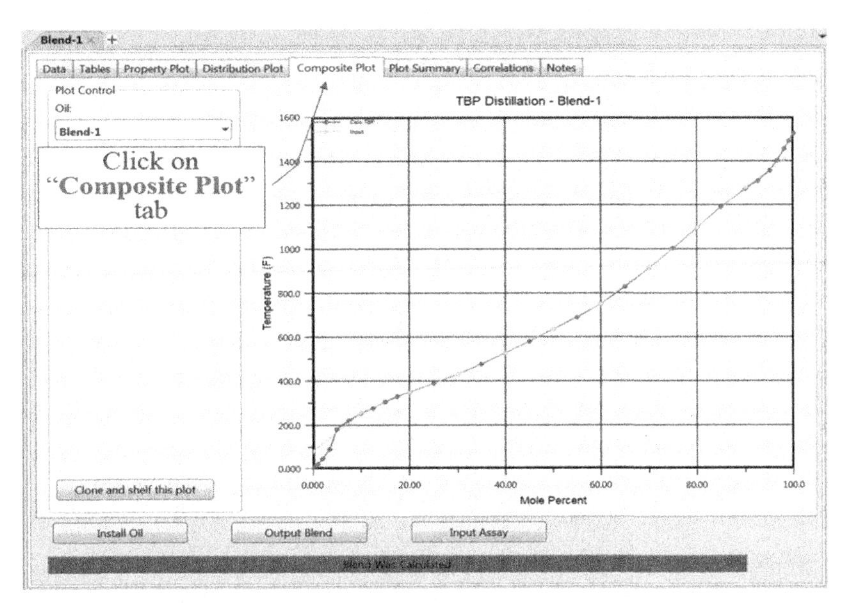

FIGURE 2.25 Composite plot of calculated oil versus the input oil information [calculated oil (red) and input of user (green)].

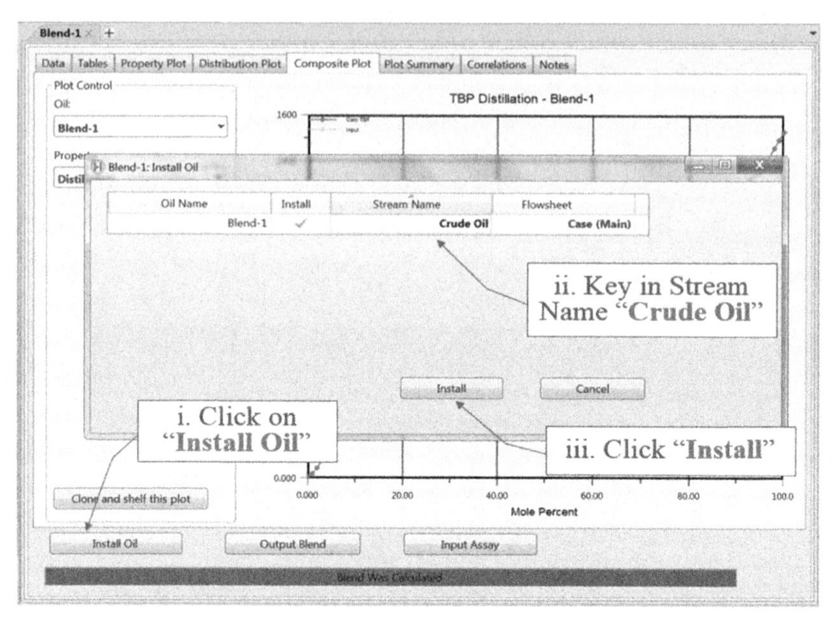

FIGURE 2.26 Install oil in fluid package and stream name.

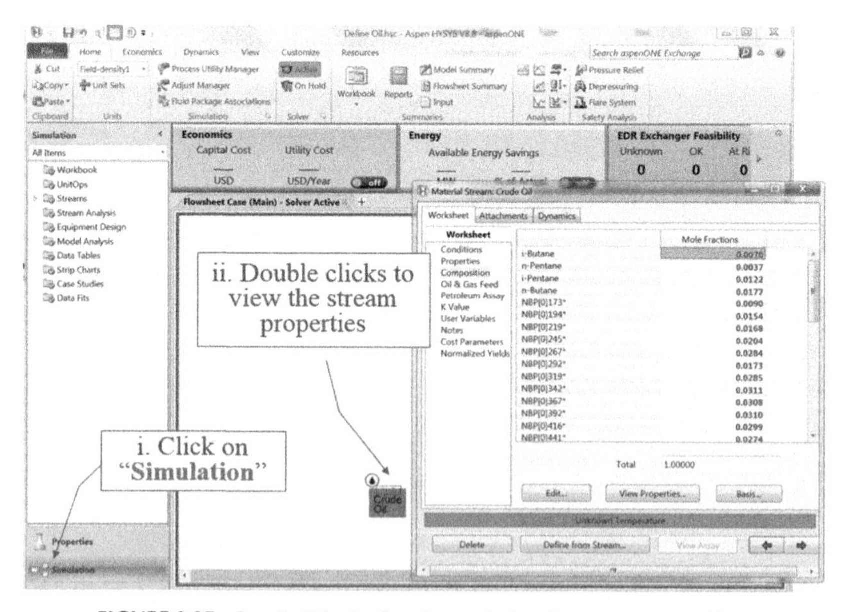

FIGURE 2.27 Install oil in the flowsheet and view the stream composition.

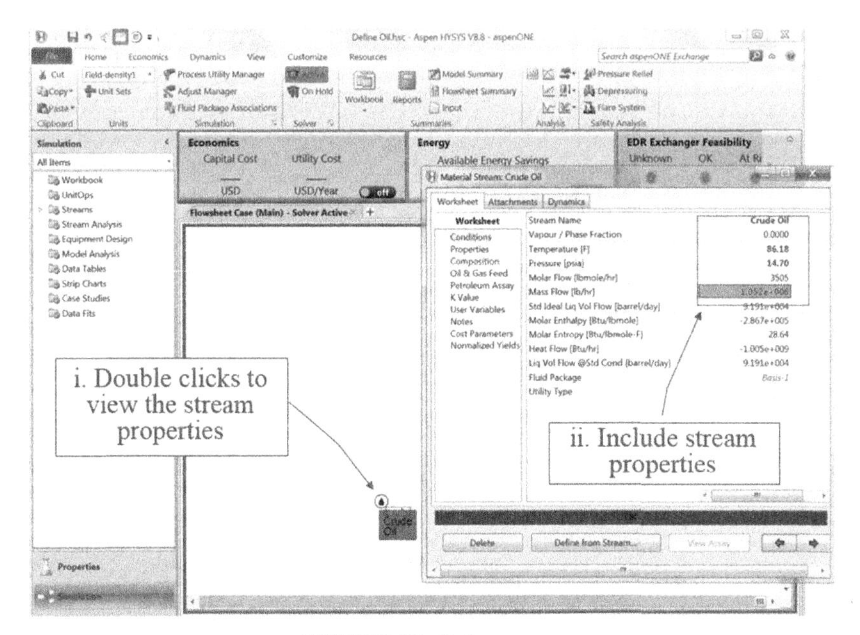

FIGURE 2.28 Define stream.

When "Crude Oil" stream is ready, stream properties (e.g., temperature, pressure, and flowrate) are included as shown in Fig. 2.28. Next, the crude oil is heated up via heater in simulator. Setting of heater and heated crude oil are shown in Fig. 2.29.

Example 2.4 Crude Oil Registration in PRO/II

To model crude oil in PRO/II, the oil needs to be defined based on its characteristics via assay characterization. Based on the laboratory results, the characteristics of the crude oil are included in PRO/II library and hypothetical components. This involves the similar steps shown in Example 2.3.

An example of registration of crude oil is created based on information provided in Table 2.2, but the thermodynamic method selected is Soave–Redlich–Kwong. In PRO/II, the user is required to define light ends components first prior to modify the assay characterization data. Fig. 2.30 shows components required for this example. In terms of thermodynamic method, Soave–Redlich–Kwong is selected (Fig. 2.31).

Next, modify the assay Primary TBP Cutpoints Definition based on the feed data (Fig. 2.32). Make sure the minimum temperature for the first interval is below the temperature of the initial point of distillation boundaries value. Referring to the feed data, initial and end point boundaries value are set at 5% and 95%, respectively (Fig. 2.33). Besides, the user is required to define assay and light ends data including distillation test, percentage distilled, temperature,

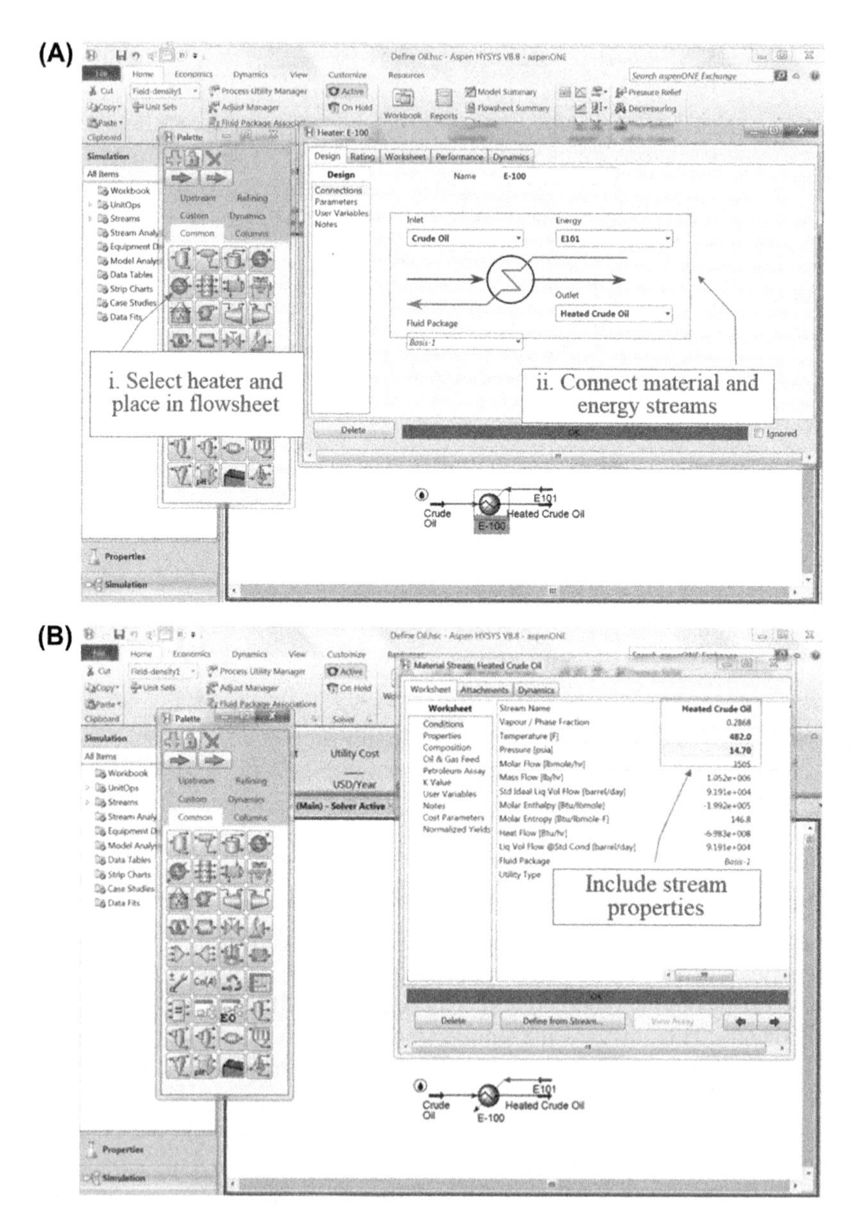

FIGURE 2.29 Heat up crude oil via heater: (A) setting of heater and (B) setting of stream properties of heated crude oil.

TABLE 2.2 Information of Crude Oil

(a) Bulk Properties		
Bulk Crude Properties	**Reference**	
Molecular weight	300	
API gravity	48.75	
(b) Light Ends Liquid		
Light Ends Liquid Volume Percent Distilled		
i-Butane	0.19	
n-Butane	0.11	
i-Pentane	0.37	
n-Pentane	0.46	
(c) True Boiling Point Distillation Assay		
Liquid Volume Percent Distilled	**Temperature (°F)**	**Molecular Weight**
0.0	80.0	68.0
10.0	255.0	119.0
20.0	349.0	150.0
30.0	430.0	182.0
40.0	527.0	225.0
50.0	635.0	282.0
60.0	751.0	350.0
70.0	915.0	456.0
80.0	1095.0	585.0
90.0	1277.0	713.0
98.0	1410.0	838.0
(d) API Gravity Assay		
Liquid Volume Percent Distilled	**API Gravity**	
13.0	63.28	
33.0	54.86	
57.0	45.91	

Continued

TABLE 2.2 Information of Crude Oil—cont'd

(d) API Gravity Assay		
Liquid Volume Percent Distilled	**API Gravity**	
74.0	38.21	
91.0	26.01	
(e) Viscosity Assay		
Liquid Volume Percent Distilled	**Viscosity (cP) 100°F**	**Viscosity (cP) 210°F**
10.0	0.20	0.10
30.0	0.75	0.30
50.0	4.20	0.80
70.0	39.00	7.50
90.0	600.00	122.30

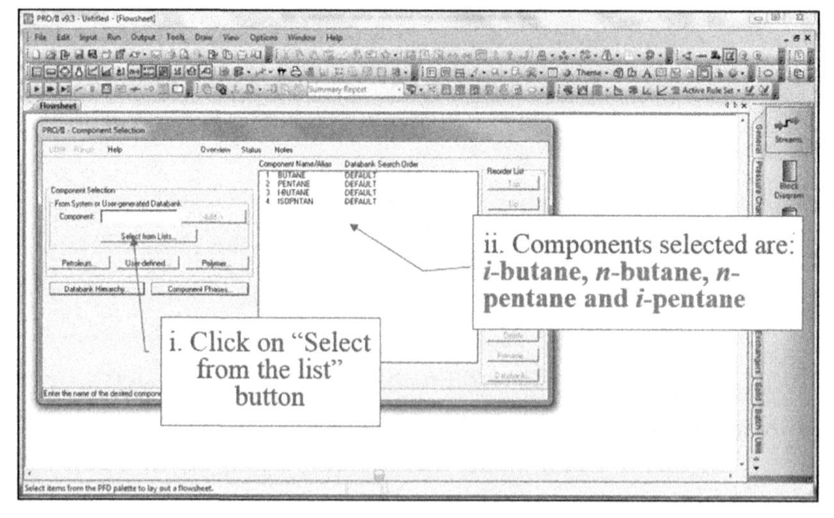

FIGURE 2.30 Define components for the light ends.

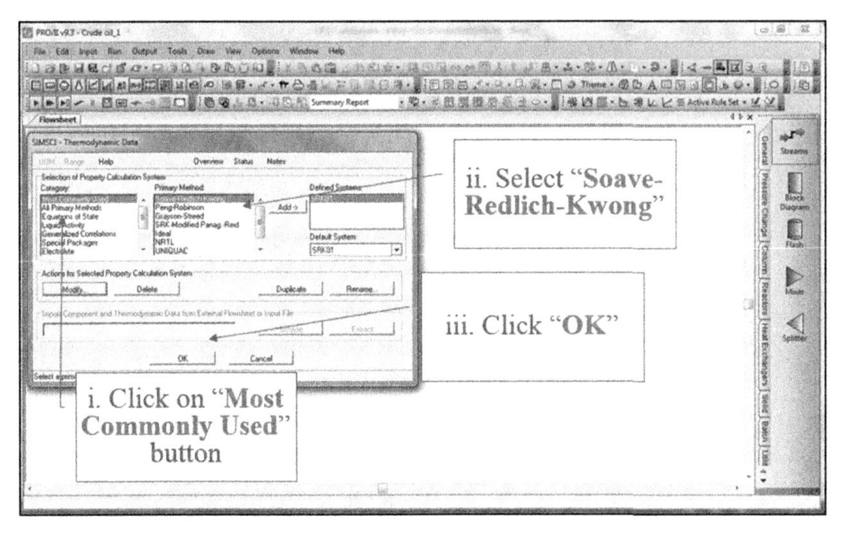

FIGURE 2.31 Thermodynamic properties for crude assay.

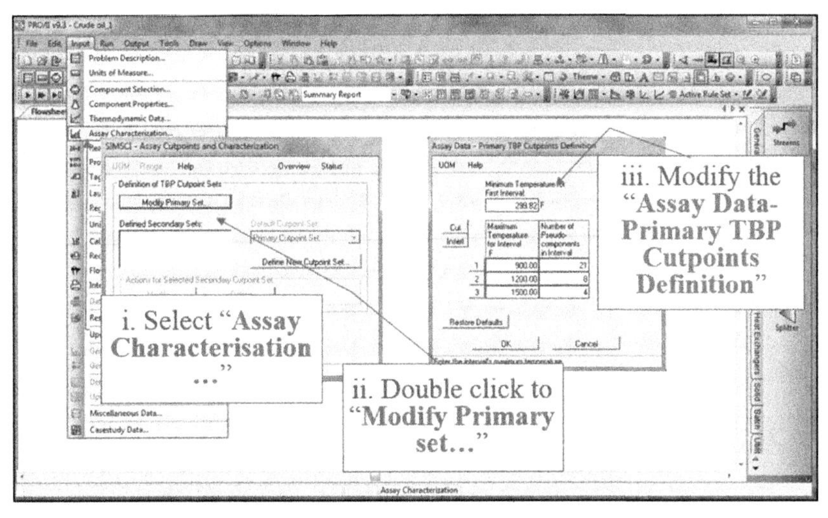

FIGURE 2.32 Modify the assay characterization data.

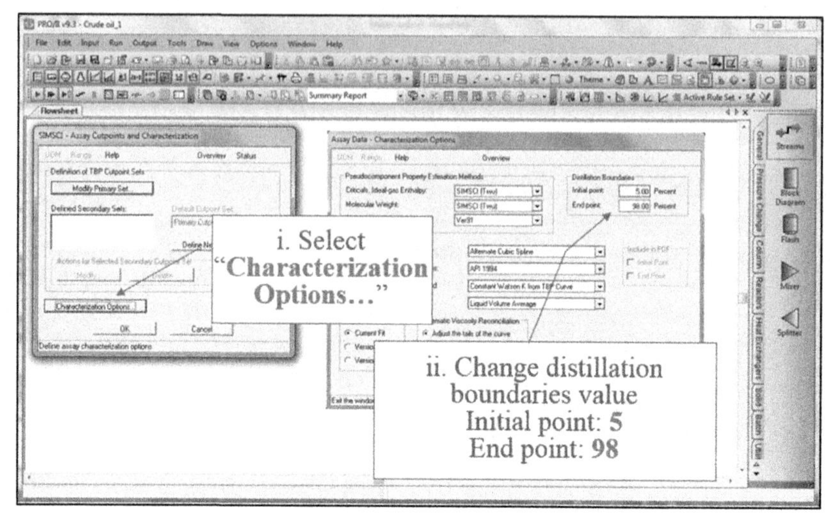

FIGURE 2.33 Specifying distillation initial and end boundaries value.

composition, and gravity data. Detailed steps for defining assay, flowrate, and assay light ends data for crude oil are shown in Figs. 2.34 and 2.35.

Finally, generate the assay component output report and distillation curve to view the profile of crude oil assay. Detailed steps are shown in Figs. 2.36 and 2.37.

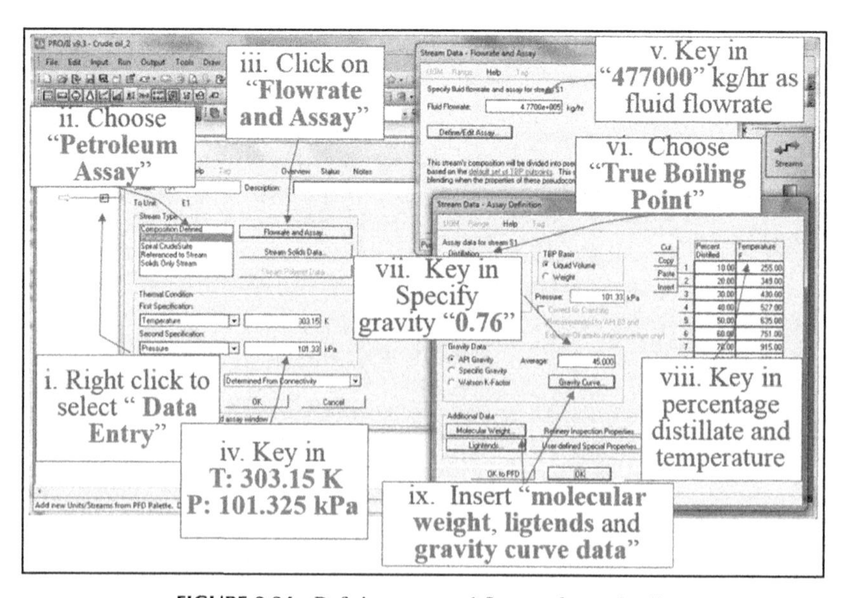

FIGURE 2.34 Defining assay and flowrate for crude oil.

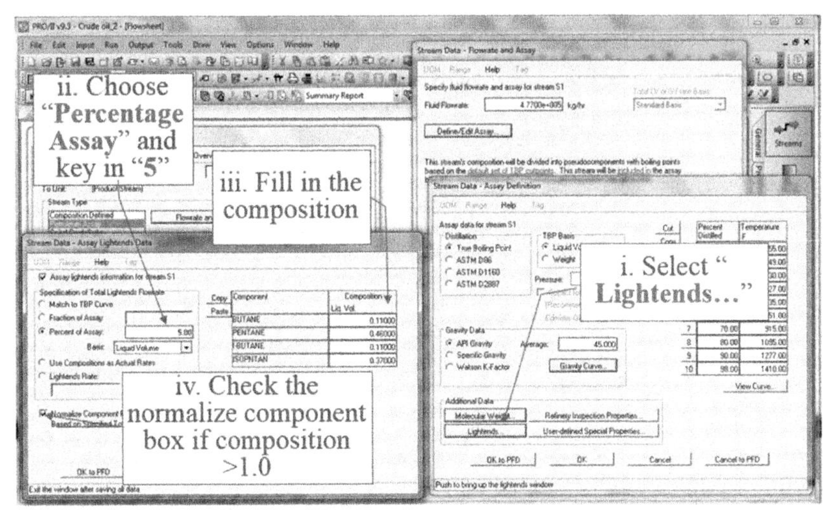

FIGURE 2.35 Specification of total light ends flowrate and compositions of components.

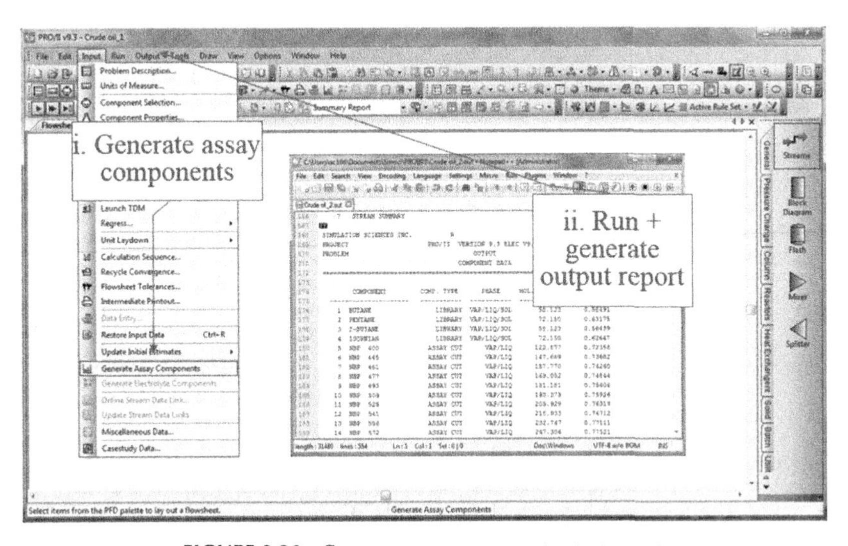

FIGURE 2.36 Generate assay component output report.

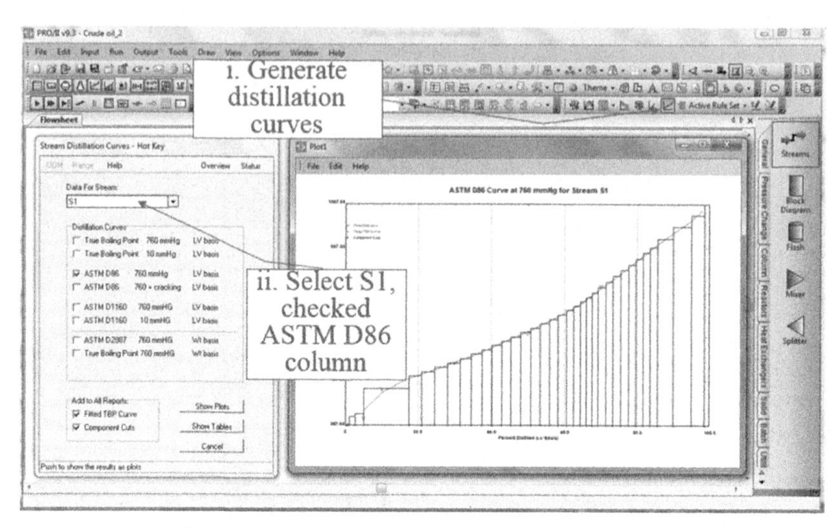

FIGURE 2.37 Distillation curve for defined crude oil assay.

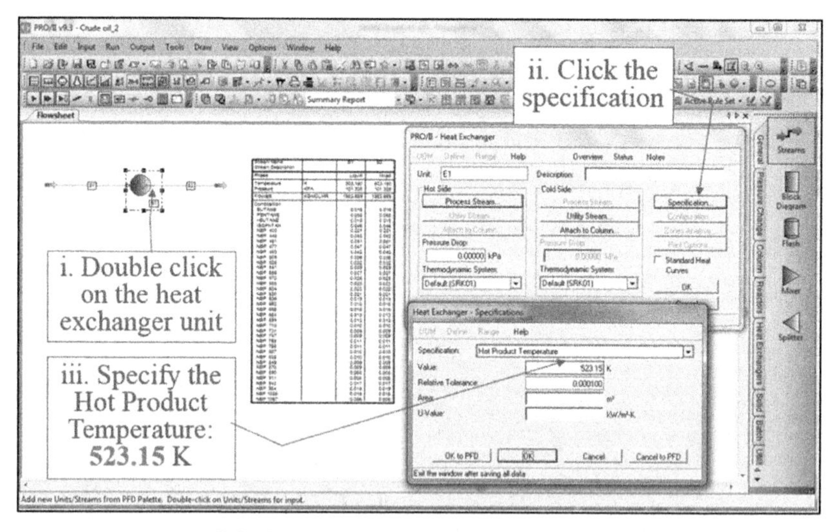

FIGURE 2.38 Heating of registered crude oil.

A heat exchanger unit is added to the feed stream to preheat the registered crude oil to 523.15K. Detailed steps of adding a heat exchanger is shown in Fig. 2.38.

Exercise

Define the oil stream based on the properties given in Table E1. The oil is to be heated to 100°C. Perform the simulation task using:

1. Aspen HYSYS
2. PRO/II

TABLE E1 ASTM D86 and properties of distillate oil

Properties	Unit	Distillate Oil
Mass flowrate	kg/h	3.925×10^5
Temperature	°C	15
Pressure	kPa	500
Density at 15°C	kg/m^3	828
Assay basis		Liquid volume
ASTM D86 0%	°C	−5
ASTM D86 5%	°C	166
ASTM D86 10%	°C	226
ASTM D86 30%	°C	265
ASTM D86 50%	°C	283
ASTM D86 70%	°C	301
ASTM D86 90%	°C	330
ASTM D86 95%	°C	341
ASTM D86 100%	°C	359
Light ends		Auto calculate
		i-Butane, *i*-pentane, *n*-butane, *n*-pentane, thiophene, *m*-mercaptan
Cut ranges		Auto Cut
Thermodynamic properties		SRK

SRK, Soave–Redlich–Kwong.

REFERENCES

AspenTech, 2015. Aspen HYSYS User Guide.
Schneider Electric, 2015. SimSci PRO/II v9.3.2 Reference Manual.

Chapter 3

Physical Property Estimation for Process Simulation

Rafil Elyas
East One-Zero-One Sdn Bhd, Shah Alam, Malaysia

Like the foundation of a building, the methods used for physical property estimation determine the integrity of a chemical engineering computation. These days, most engineers rely on commercial simulators to perform their computations, and all commercial simulators these days come with a myriad of property packages, where various *property estimation methods* have been combined into *property packages* such as Peng–Robinson, Soave–Redlich–Kwong (SRK), BWRS, Grayson–Streed, Braun-K10, NRTL, UNIQUAC, and the list goes on. It is critical to know which property package would be applicable for one's computation. The objective of this chapter is to provide some insight into the workings of those property packages and enable the reader to make the correct selection.

3.1 CHEMICAL ENGINEERING PROCESSES

One is often "advised" by "seniors" to use the Peng–Robinson property package in commercial simulators such as UniSIM Design or Aspen HYSYS, or the SRK property package in PRO/II. It is a common misconception that the property package is a single equation used to calculate "everything." Each property package consists of *several* computational methods that are used to estimate *thermophysical* and *transport properties of interest.* What sort of thermophysical and transport properties are of interest?

If we take a look at most processes a chemical engineer works with, we find that a majority of these processes involve vaporizing, condensing, separating, compressing, and moving fluids. Let us take a look at a typical process, the accompanying computational requirements, and the corresponding properties required. To simplify things, we shall look at a typical upstream oil and gas production facility in Fig. 3.1. In this facility, oil, gas, and water are separated in the three-phase separator. Gas stream is then compressed, while

Chemical Engineering Process Simulation. http://dx.doi.org/10.1016/B978-0-12-803782-9.00003-0

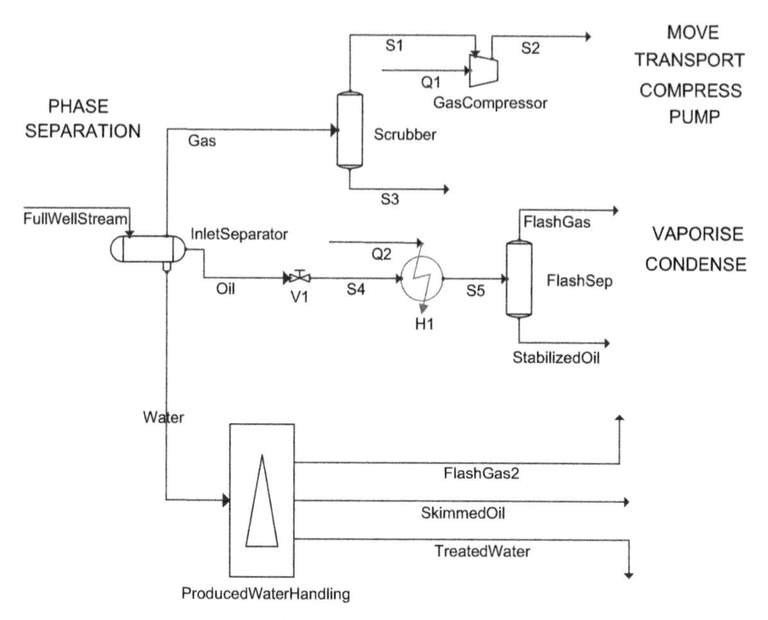

FIGURE 3.1 A typical upstream oil and gas production facility.

the oil is pumped and stabilized. Let us analyze some equipment in this facility and identify what properties are required to execute some typical computations.

3.1.1 Separator

Starting with the gas wells, the first process that is encountered is a separator. In the design and sizing of this separator, it is necessary to understand *how much vapor and liquid are produced* at their operating pressure and temperature and *how the components partition themselves between vapor and liquid phases* and to establish the volume–mass relationships of the phases. Hence, the required thermophysical properties include *vapor and liquid densities, enthalpies, and vapor pressures.*

3.1.2 Heat Exchanger

Heat exchangers allow fluids to either be heated or cooled. In this example, it is necessary to stabilize the oil by removing the lighter or more volatile components, to allow the oil to be stored in atmospheric tanks or transported in single-phase liquid pipelines. To represent the heating or cooling process, the following thermophysical properties are required: *vapor and liquid heat capacities and densities, liquid vapor pressure, and heat of vaporization.*

3.1.3 Compressor

A compressor moves a fluid by adding work to it. There are positive displacement compressors that simply displace the volume using a piston, or centrifugal compressors that transfer momentum to a gas, which increases its velocity and converts it to pressure by reducing its velocity through a diffuser. Compression computations require the properties of *gas density, compressibility, and heat capacities (Cp/Cv).*

In summary, the basic thermophysical properties required for engineering computations of these oil and gas production facilities are *density, vapor pressure, and energy characteristics*. In addition, it is also necessary to estimate the transport properties of the fluids because the fluids will obviously have to flow across these processes. Some of the typical transport properties would be *viscosity and surface tension.*

3.2 THERMODYNAMIC PROCESSES

One of the most basic requirements when processing fluids is to understand how a state of a fluid changes with respect to pressure and temperature. This is largely addressed by thermodynamics. Most chemical engineering processes such as those in Fig. 3.1 are expressed as thermodynamic processes in commercial process simulators. Processes are represented by changes in state of a fluid. Gas compression is used to illustrate this. As discussed earlier, the following *properties* were identified for gas compression: density, compressibility, and *Cp*/*Cv*; these properties are used to determine the *thermodynamic state variables* of the fluid.

The thermodynamic path is described in the phase diagram in Fig. 3.2.

An actual compression is an irreversible process. Most simulators employ a two-step procedure for this computation. First, a reversible compression, A-B, is used along isentrope $S1$. $H_2 - H_1$ is the reversible work for this process. This work is then multiplied (or divided) by an efficiency to give the actual work (point C). Total work imparted to the fluid is given by the difference of H_1 and H_3. Hence, the difference between H_2 and H_3 may be regarded as the

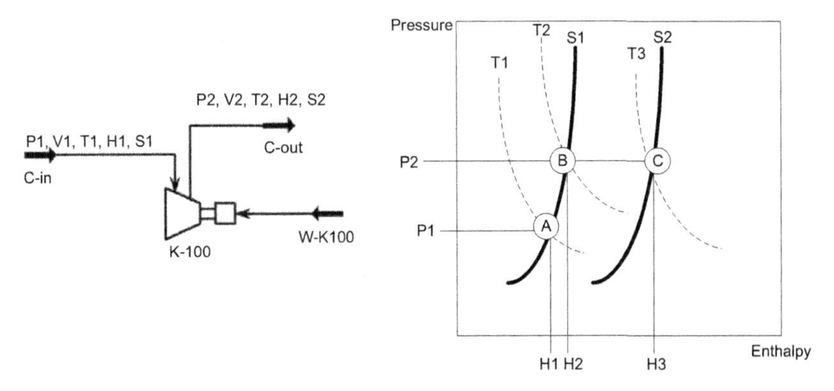

FIGURE 3.2 Gas compression.

"lost work." The adiabatic efficiency for this process is hence defined as the ratio of these two differences, i.e.,

$$\eta_{ad} = \frac{H_2 - H_1}{H_3 - H_1} \tag{3.1}$$

where η_{ad} is the adiabatic efficiency, $H_2 - H_1$ is the reversible work for the process, and $H_3 - H_1$ is the total work.

For this process, the thermodynamic state variables required in this computation are *temperature* (T), *pressure* (P), *volume* (V), *enthalpy* (H), *and entropy* (S). In addition to the above, other state variables are also required to represent other processes; for example, Gibbs (G) and Helmholtz (A) free energies are used to describe thermodynamic potential, which in turn is used for phase equilibrium computations. These state variables are conveniently related to each other from characteristic thermodynamic definitions and the Maxwell relationships. These are discussed in the following subsections.

3.2.1 Characteristic Thermodynamic Relationships (Smith et al.)

3.2.1.1 Internal Energy (U)

The internal energy of a substance can be seen as the energy possessed by the molecules or atoms making up the substance. Most physical chemistry models describe atoms and molecules in ceaseless vibrational, rotational, or translational modes. In addition to that, there are energies that hold the atoms and molecules together in the form of bonds.

It is impossible to know what the absolute value of internal energy of a substance is, but this is not a problem as most thermodynamic analysis deals with changes in energy (ΔU). For example, adding heat (Q) to the system increases this molecular activity; work (W) is extracted when this activity is made to interact with external forces. Hence, this can be described by Eq. (3.1).

$$\Delta U = Q - W \tag{3.2}$$

Note that Eq. (3.1) is also the mathematical statement of the first law of thermodynamics, i.e., the conservation of energy.

3.2.1.2 Enthalpy (H)

Enthalpy is defined as the summation of the internal energy system and a work term, which is given by the product of pressure (p) and volume (V).

$$H = U + pV \tag{3.3}$$

3.2.1.3 Entropy (S)

In simple terms, entropy is used to quantify "lost work." For example, one first puts in 100 W of work to compress a gas from V_1 to V_2. Expanding the gas from V_2 to V_1, however, does not produce 100 W of work. Some of the work is lost because of intermolecular "friction" or other phenomena. The change in entropy provides an indication of the amount of "lost work." In Fig. 3.2 the amount of lost work (difference between H_2 and H_3) corresponds to an increase in entropy from $S1$ to $S2$.

The mathematical definition of entropy is given by

$$dS = \frac{dQ}{T} \tag{3.4}$$

or

$$\Delta S = \int \frac{dQ}{T} \tag{3.5}$$

Eq. (3.4) is also a statement of the second law of thermodynamics, which states that total entropy of an isolated system increases with time. It only remains constant in a system that remains in steady state or that undergoes a reversible process.

3.2.1.4 Gibbs Free Energy (G)

The Gibbs free energy (G) is defined as

$$G = H - TS \tag{3.6}$$

From this definition, it can be seen as the "available" energy a fluid may have. It can be used to describe the maximum or net amount of extractable work from a fluid (TS can be seen as the work that will be lost). However, Gibbs free energy is commonly used in chemistry and chemical engineering to determine the chemical potential of a system.

3.2.1.5 Helmholtz Free Energy (A)

The Helmholtz free energy is defined as:

$$A = U - TS \tag{3.7}$$

Similar to the Gibbs free energy, the Helmholtz free energy, too, can be used to describe the maximum or net amount of extractable work from a fluid (TS can be seen as the work that will be lost), and it can also be used to determine the chemical potential of a system.

3.2.2 Maxwell Relationships

Now that the characteristic thermodynamic relationships have been developed, it is necessary to relate them to each other. This is made possible using the mathematical framework of the Maxwell relationships.

$$dU = TdS - PdV \quad \rightarrow \quad \left(\frac{\partial T}{\partial V}\right)_S = -\left(\frac{\partial P}{\partial S}\right)_V \tag{3.8}$$

$$dA = -SdT - PdV \quad \rightarrow \quad \left(\frac{\partial S}{\partial V}\right)_T = -\left(\frac{\partial P}{\partial T}\right)_V \tag{3.9}$$

$$dH = TdS + VdP \quad \rightarrow \quad \left(\frac{\partial T}{\partial P}\right)_S = \left(\frac{\partial V}{\partial S}\right)_P \tag{3.10}$$

$$dG = -SdT + VdP \quad \rightarrow \quad -\left(\frac{\partial S}{\partial P}\right)_T = \left(\frac{\partial V}{\partial T}\right)_P \tag{3.11}$$

The equations in this section were not intended to induce thermodynamic class nightmares but to highlight a very important point; that is, the characteristic energy functions that are essential for the definition of thermodynamic processes [i.e., internal energy (U), Helmholtz free energy (A), enthalpy (H), Gibbs free energy (G), and entropy (S)] may be related to temperature, pressure, and volume. Temperature, pressure, and volume in turn are related by the equations of state.

3.3 EQUATIONS OF STATE

3.3.1 The Ideal Gas Law (c.1834)

One of the most basic relationships to describe the state of a gas with respect to pressure and temperature, or an equation of state, is the ideal gas law.

$$PV = nRT \tag{3.12}$$

where P is the absolute pressure of the gas, V is the volume of the gas, n is the amount of substance of the gas, usually measured in moles, R is the gas constant, and T is the absolute temperature.

This is sometimes rewritten in terms of molar volume as

$$Pv = RT \tag{3.13}$$

where v is the molar volume, i.e., V/n.

The ideal gas law provides reasonable estimates for gasses at low pressure and high temperature, for at these conditions the distance between the gas molecules is large enough and the kinetic energy is sufficiently high to eliminate any effects of the size of the molecules or any possible molecular interactions.

Pushing the ideal gas law to low-temperature and high-pressure limits yields the following impossible results:

$$\lim_{T \to 0} v = \frac{RT}{P} = 0 \tag{3.14}$$

$$\lim_{P \to \infty} v = \frac{RT}{P} = 0 \tag{3.15}$$

Observing Eqs. (3.13) and (3.14) leads to a potential mistake. Because matter occupies space and it is highly unlikely that materials shrink to nothingness at absolute zero or under high pressure! This leads to the corrections needed for the ideal gas law, which is discussed next.

3.3.2 Corrections to the Ideal Gas Law (Cubic Equations of State)

The objective of this section is to provide a basic overview of the ideal gas law corrections to provide the reader a basis for understanding how the equation of state is implemented in commercial process simulators. This section is by no means a detailed analysis of the various equations of states; there are many thermodynamic texts available that provide descriptions of much greater rigor.

The first set of corrections came in 1873 from Johannes Diderik van der Waals.

The first correction addresses the fact that molecules have volume. The effect of molecular volume would be realized at lower temperatures and higher pressures when molecules were "closer" together and "less energetic."

To compensate for this, a volume offset term, b, was added to the ideal gas equation.

$$v = \frac{RT}{P} + b \tag{3.16}$$

Eq. (3.15) may be rewritten as

$$P = \frac{RT}{v - b} \tag{3.17}$$

Now, the problem of material vanishing at low temperatures and high pressures has been addressed.

The next correction takes into account cases in which molecules are attracted to one another at certain distances. This effect becomes observable when a liquid is vaporized. To vaporize a liquid, it is necessary to add energy to it. Therefore, there must be some sort of attraction force among the molecules.

Attractive forces are believed to be proportional to concentration or inverse molar volume. The attraction term, a, is appended to Eq. (3.17) and yields the familiar van der Waal's equation:

$$P = \frac{RT}{v - b} - \frac{a}{v^2} \tag{3.18}$$

Most work following van der Waals continued to address the attraction term component, and the general form of a two-parameter cubic equation of state can be expressed as follows:

$$P = \frac{RT}{(v - b)} - \frac{a}{v^2 + ubv + wb^2} \tag{3.19}$$

From this structure, the following common equations of state can be defined:

van der Waals: $u = 0$, $w = 0$ (c.1873)
Redlich–Kwong: $u = 1$, $w = 0$ (c.1949)
Peng–Robinson: $u = 2$, $w = -1$ (c.1976)

The attraction (a) and molecular volume (b) terms are expressed as functions of pure component critical pressure and temperature. In the case of the Peng–Robinson equation, the attraction term (a) is multiplied by a temperature-dependant $\alpha(T)$.

3.3.2.1 Van der Waals

$$a = \frac{27R^2T_c^{\ 2}}{64P_c} \ ; \ b = \frac{RT_c}{8P_c} \tag{3.20}$$

3.3.2.2 Redlich–Kwong

$$a = \frac{0.42748R^2T_c^{\ 2.5}}{P_c} \ ; \ b = \frac{0.08664RT_c}{P_c} \tag{3.21}$$

3.3.2.3 Peng–Robinson

$$a = \frac{0.45724R^2T_c^{\ 2}}{P_c}\alpha(T) \ ; \ b = \frac{0.07780RT_c}{P_c}$$

$$\kappa = 0.37464 + 1.54226\omega = 0.26992\omega^2; T_r = \frac{T}{T_c} \tag{3.22}$$

where $\alpha(T)$ is dependent on temperature.

However, fluids consist of mixtures of components in most applications. It is then necessary to use mixing rules for the a and b terms. These rules are essentially mole weighted averages a and b terms of the pure component, which can be expressed as follows.

The mixture b term is a simple mole weighted average of the pure component b terms.

$$b = \sum_i y_i b_i \tag{3.23}$$

where y_i and b_i are the mole fraction and b term of the ith component, respectively.

The mixture a term is addressed differently as it represents the attraction between molecules. It is an average of the attraction terms between all possible component pairs in the mixture.

$$a = \sum_i \sum_j y_i y_j a_{ij} \tag{3.24}$$

where a_{ij} is referred to as the cross-parameter between components i and j in the mixture. The cross-parameter is simply the root mean square of the pure components a_i and a_j with a tunable "gain" term k_{ij}.

$$a_{ij} = \sqrt{a_i a_j}\,(1 - k_{ij}) \tag{3.25}$$

The k_{ij} term in Eq. (3.24) is referred to as the *binary interaction parameter* (BIP) in most commercial simulators and is allowed to change by users.

Here is a simple way to understand how the value of BIP affects the molar volume. We know that the a term accounts for intermolecular attraction. When the attraction force is large, the molecules get "closer" together and the overall molar volume decreases. On the other hand, when the attraction force is small, the molecules get "farther" from each other and the overall molar volume increases. This is next demonstrated with a simple example.

3.3.2.4 Reducing the "Attractive Force"

Starting with a k_{ij} of zero for a fixed pair of components (with fixed a_i and a_j) and increasing the number to approach 1, one can see that a_{ij} decreases as k_{ij} starts to approach 1. This means that the attraction term is weakened as k_{ij} is increased, and the molar volume of the mixture increases.

3.3.2.5 Increasing the "Attractive Force"

One can also go the other way around and reduce the attractive force by starting with a k_{ij} of zero and reducing the k_{ij} to -1. This has the effect of "strengthening" the attraction between components i and j, resulting in a smaller molar volume of the mixture.

Example 3.1

A gas mixture consisting of equimolar methane and ethane at 25°C and 10 barg. Its molar volume is to be determined using the Peng–Robinson equation of state, with values of the BIP ranging from -0.5 to 0.5. The default BIP provided in the simulator for this mixture was 0.00224. The results for this computation are presented in Table 3.1.

It should be noted that the values of the BIP were *arbitrarily spanned* from -0.5 to 0.5 to demonstrate the effect this tuning parameter has on the molar volume of the gas mixture. In reality, the tuning of the cubic equation BIPs requires experimental data for binary systems (and in some cases ternary systems if available).

TABLE 3.1 Comparison of Molar Volumes for a Range of Binary Interaction Parameters

K_{ij}	V	Δ	% Δ
−0.5	2.0913	−0.0352	−1.66%
−0.1	2.1226	−0.0039	−0.18%
0.00224	**2.1265**	**0.0000**	**0.00%**
0.1	2.1380	0.0115	0.54%
0.5	2.1680	0.0415	1.95%

Most commercial simulators have done this work for some common component pairs. However, it is impossible to cover every single component pair that may ever be encountered in industry. If no experimental data were used to estimate the BIP for a component pair, the BIP value is either set to zero (for the case of SimSci PRO/II) or estimated using extrapolation, or group contribution method (for the case of Aspen HYSYS and UniSim Design). Note that some process simulators provide tuning functions or packages that allow the users to calculate BIPs from their own data sets of P-XY[1], T-XY[2], bubble, and dew points. Note also that in Peng–Robinson and SRK equations, the k_{ij} is symmetric, $k_{ij} = k_{ji}$. However, there are "advanced" implementations of these equations of state (e.g., Peng–Robinson Stryjek Vera); the BIPs can be asymmetric, $k_{ij} \neq k_{ji}$ essentially doubling the number of tuning parameters.

3.4 LIQUID VOLUMES (WALAS, 1985)

The cubic equations of state described in Section 3.3.2 are not applicable for determining liquid densities and must be avoided as they will result in grossly incorrect estimates of the liquid density. Remember that all these equations of state come from one common "ancestor," i.e., the ideal *gas* law.

The liquid phase is far more complicated than the gas phase because molecules are in "closer quarters" and intermolecular interactions begin to dominate. This makes it difficult to predict how these molecules may "arrange" themselves in a liquid continuum, particularly if there are different functional groups present, which may give rise to complexes or exhibit hydrogen bonding.

It should be noted that many users select the Peng–Robinson property package when using software such as Aspen HYSYS or UniSim Design,

1. Pressure–liquid composition and gas composition at constant temperature.
2. Temperature–liquid composition and gas composition at constant pressure.

thinking that the Peng–Robinson equation is used to determine gas and liquid densities. This is (fortunately) incorrect. As mentioned earlier in this chapter, a property package is not a single equation but a collection of methods used to estimate various properties.

Some of the typical methods used to estimate liquid densities in commercial simulators are as follows:

1. Volume translated equation of state methods, where the calculated volume by the equation of state is "corrected." An example of this is the PRF method by Peneloux et al. This is used in PVTSim[3].
2. Correlations or stand-alone equations. Examples would be the Rackett equation and corresponding state liquid density (COSTALD[4]) method by Hankinson–Brobst–Thomson.

In the case of Aspen HYSYS and UniSim Design, Peng–Robinson COSTALD is the default method for liquid density estimation. There are other methods for estimating liquid volumes. The COSTALD equation is an empirical equation that has the following form for a pure component:

$$\frac{v_s}{v^*} = v_r^o\left(1 - \omega_{SRK} v_r^\delta\right) \tag{3.26}$$

$$v_r^o = 1 + a(1 - T_R)^{\frac{1}{3}} + b(1 - T_R)^{\frac{2}{3}} + c(1 - T_R) + d(1 - T_R)^{\frac{4}{3}} \tag{3.27}$$

$$v_r^\delta = \frac{e + fT_R + gT_R^2 + hT_R^3}{T_R - 1.00001} \tag{3.28}$$

where v_s is the molar volume of saturated liquid at some temperature; v_r^o is a characteristic of spherical molecules and v_r^δ is a correction factor, both correlated in terms of the reduced temperature; ω_{SRK} is the acentric factor determined by the reduced pressure and temperature of the system; T_R is the reduced temperature; $a, b, c, d, e, f, g,$ and h are fixed equation parameters, and v^* is an empirical "characteristic volume"; this has been determined for various substances. It has also been correlated in terms of the SRK acentric factor, ω_{SRK}, for nine substance groups.

$$v^* = \frac{RT_c}{P_c}\left(k_1 + k_2\omega_{SRK} + k_3\omega_{SRK}^2\right) \tag{3.29}$$

where T_c is the critical temperature and P_c is the critical pressure.

Empirical values for k_1, k_2, and k_3 are provided for nine substance groups:

1. Paraffins
2. Olefins and diolefins

---

3. PVTSim Help Calsep A/S 2014.
4. An Improved Correlation for Densities of Compressed Liquids and Liquid Mixtures, G.H. Thomson, K.R. Brobst, and R.W. Hankinson. AIChE Journal (Vol. 28, No. 4) July 1982.

3. Cycloparaffins
4. Aromatics
5. Other hydrocarbons
6. Sulfur compounds
7. Fluorocarbons
8. Cryogenic liquids
9. Condensable gasses

The mixing rules for mixtures are highly complicated and are not shown for brevity. The purpose of this section was to highlight that liquid densities and corresponding volumes are generally determined using a completely different description than gas densities and volumes.

3.5 VISCOSITY AND OTHER PROPERTIES

Equations of state cannot be used to calculate transport properties such as viscosity. The latter is highly dependent on the chemical nature of the substance, fluid mechanics, and surface chemistry. For example, paraffinic change entanglement and wax formation in oils or colloidal/emulsion formation in water and oil mixtures. The methodologies to estimate viscosities are generally empirical in nature. Some examples of these methods are summarized in Table 3.2.

3.6 PHASE EQUILIBRIA

When phases are said to be in thermodynamic equilibrium, then their potential or driving forces to transfer material from one phase to the other are equal. To illustrate this concept, consider a two-component (A and B), two-phase (gas and liquid) system as per Fig. 3.3.

The gas-side driving force to move component A to the liquid phase is referred to as the gas phase fugacity of component A and for now can be expressed simply as its partial pressure Py_A. The liquid driving force exerted on component A to move it to the liquid phase is the liquid phase fugacity and for now can be considered its prorated vapor pressure $P_A^{sat}x_A$.

TABLE 3.2 Examples for Viscosity Estimation

Phase/Fluid	Method
Gas	Lucas, Jossi–Stiel–Thodos, Ely–Hanley
Liquid	Lucas, Letsou–Stiel, Twu
Oil/water emulsion	Woelflin

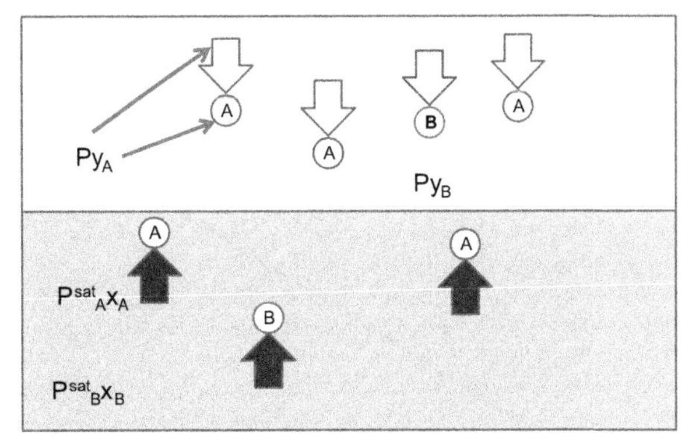

FIGURE 3.3 Two-phase system with two components, A and B.

At equilibrium, gas and liquid fugacities are equal:

$$f_A^V = f_A^L \tag{3.30}$$

or in this example:

$$Py_A = P_A^{sat} x_A \tag{3.31}$$

Eq. (3.31) is the familiar Raoult's law, in which: A is the more volatile component, x_A and y_A are the mole fractions of component A in the liquid and gas phases, respectively, P is the total system pressure, P_A^{sat} is the vapor pressure of component A and is a function of temperature.

The vapor pressure of a component is a function of its temperature. The higher the temperature, the more energy a molecule has and the higher its tendency to "escape" to the vapor phase. This can be seen in the equilibrium plot of methane and ethane mixture at 20 bara in Fig. 3.4. As the temperature increases, the composition of methane in the gas phase increases monotonically.

3.6.1 Vapor Phase Correction

The gas phase driving force in Eq. (3.31) is expressed as the partial pressure. The partial pressure of a component in a mixture is simply a mole fraction prorated pressure. This is derived from Dalton's law:

$$P = \sum_{i}^{n} Py_i \tag{3.32}$$

where Py_i is the partial pressure of the $i$thcomponent, n is the total number of components in the mixture.

Assuming the gas phase driving force or fugacity of a component to be its mole fraction prorated pressure is a rather ideal representation. Consequently, it is often necessary to correct the gas phase fugacity by implementing the gas phase fugacity coefficient.

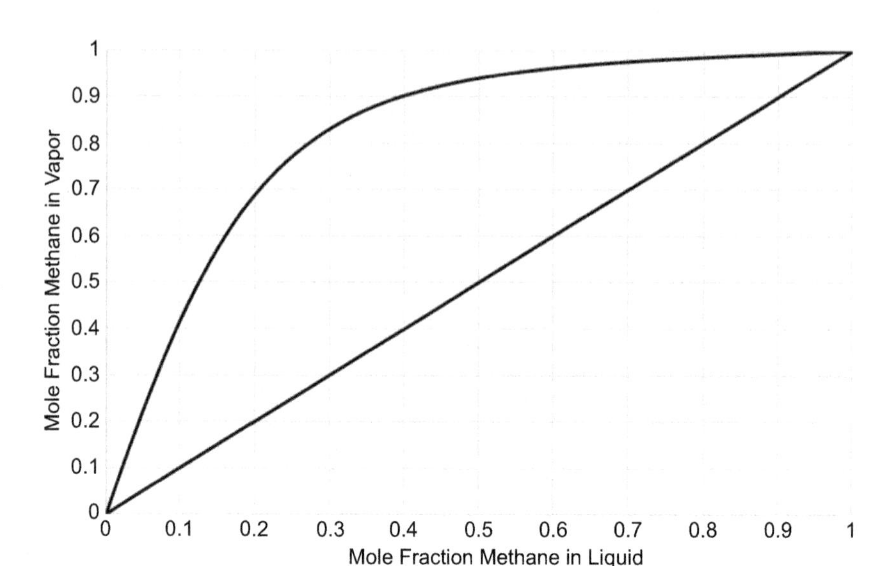

FIGURE 3.4 Methane and ethane equilibrium plot at 20 bara.

$$\varnothing_i P y_i = P_i^{sat} x_i \tag{3.33}$$

where $\varnothing_i$ is the gas phase fugacity coefficient of component i.

The gas phase fugacity coefficient's purpose is to correct for nonidealities in the gas phase. To illustrate this, the definition of the fugacity coefficient of a pure substance will be analyzed.

The fugacity coefficient for a pure substance is defined as:

$$\ln \varnothing = -\frac{1}{RT}\int_0^P \left(\frac{RT}{P} - V_d\right) dP \tag{3.34}$$

where $\varnothing$ is the pure component fugacity, V_d is either the volume determined by an equation of state or an experimental value.

From Eq. (3.34), it can be seen that if:

1. The pressure approaches 0 (vacuum), $\ln \varnothing$ approaches 0, and $\varnothing$ approaches 1.
2. If the ideal gas law is used as the equation of state, the fugacity coefficient also reduces to 1.
3. At higher pressures using a nonideal equation of state will yield a fugacity coefficient less than or greater than 1.

A fugacity coefficient greater than 1 results when $\frac{RT}{P} - V_d$ is negative (and the entire term on the right side of Eq. (3.33) is positive). This happens when the volume of the ideal gas volume is less than that of the real or nonideal gas. This means that the molecules in the real gas may be "repelling" each other. Applying this in Dalton's law will result in a higher system pressure for a given volume.

Analogously, a fugacity coefficient less than 1 results when $\frac{RT}{P} - V_d$ is positive (and the entire term on the right side of Eq. (3.33) is negative). This happens when the volume of the ideal gas is greater than that of the real or nonideal gas. This means that the molecules in the real gas may have a strong attractive force. Applying this in Dalton's law will result in a lower system pressure for a given volume.

For mixtures, the fugacity coefficient is expressed in the following manner:

$$\ln \varnothing_i = \int_V^{\infty} \left(\left(\frac{\partial P}{\partial n_i} \right)_{T,\, V, n_j} - \frac{RT}{V} \right) dV - \ln \left(\frac{PV}{RT} \right) \tag{3.35}$$

where: $\left(\frac{\partial P}{\partial n} \right)_{T,\, V, n_j}$ is the partial derivative of pressure with respect to the ith component and is computed using the equation of state.

From here, we see that in addition to being used to determine gas densities and thermodynamic energy state functions, the equation of state is also used to calculate the gas phase correction, or gas phase fugacity coefficient.

3.6.2 Liquid Phase Corrections

In an ideal liquid, with components A and B, the interactions between A–A, B–B, and A–B are considered the same. From Raoult's law, the vapor pressure contribution of component A was simply defined as its mole fraction prorated vapor pressure, $P_A^{sat} x_A$. The total vapor pressure exerted by components in an ideal multicomponent mixture is:

$$P = \sum_i^n x_i P_i^{sat} \tag{3.36}$$

If the real solution vapor pressure is greater than the ideal pressure, then the fluid is said to exhibit a positive deviation from Raoult's law, the different molecules can be seen to be "repelling" each other. And vice versa, if the real solution vapor pressure is less than the ideal pressure, then the fluid is said to exhibit a negative deviation from Raoult's law, the different molecules can be seen to be "attracting" each other. In a situation like this, the equilibrium line would no longer be monotonic.

Fig. 3.5 shows a mixture of ethyl acetate and water. It can be seen that initially the more volatile ethyl acetate (normal boiling point 77.1°C) boils off. However, at a concentration of around 0.6 mol fraction of ethyl acetate in the vapor, water (normal boiling point 100°C) begins boiling off and diluting ethyl acetate in the vapor phase. This is an example of an azeotropic mixture that results from hydrogen bonding in the liquid phase.

In cases like this, it is necessary to add a correction factor to the vapor pressure term to account for this behavior, the activity coefficient, γ_i.

$$P = \sum_i^n \gamma_i x_i P_i^{sat} \tag{3.37}$$

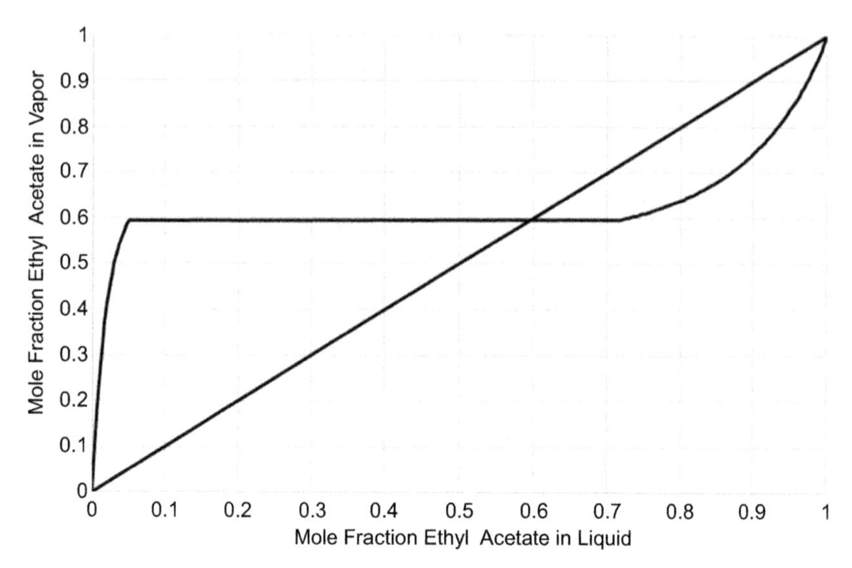

FIGURE 3.5 Water and ethyl acetate at 5 bara.

When implemented in the equilibrium equation:

$$Py_i = \gamma_i P_i^{sat} x_i \tag{3.38}$$

The activity coefficient can be related to the excess Gibbs energies in the following manner:

$$\ln \gamma_i = \frac{G^{ex}}{RT} - \sum_{k \neq i} x_k \left(\frac{\partial \left(\frac{G^{ex}}{RT} \right)}{\partial x_k} \right)_{T,P,x_j \neq i,\, k} \tag{3.39}$$

Commercial simulators typically have a large selection of activity coefficient methods. The most commonly used activity coefficient would be the nonrandom two liquid (NRTL), which has the following form:

$$\ln(\gamma_j) = \frac{\sum_j \tau_{ji} G_{ji} x_j}{\sum_k G_{ki} x_k} + \sum_j \frac{xjGij}{\sum_k G_{kj} x_k} \left[\tau_{ij} - \frac{\sum_k x_k \tau_{kj} G_{kj}}{\sum_k G_{kj} x_k} \right]$$

$$\tau_{ij} = a_{ij} + \frac{b_{ij}}{T} + \frac{c_{ij}}{T^2}$$

$$Gij = \exp(\alpha_{ij} \tau_{ij})$$

$$\alpha_{ji} = \alpha_{ji}^1 + \beta_i^1 T$$

The tuning parameters are $a_{ij/ji}$, $b_{ij/ji}$, and $c_{ij=ji}$. This equation can be used for strongly nonideal and polar mixtures, as well as partially miscible systems. The concept of NRTL is based on the concept of a "local" composition, where the local concentration around a molecule can be different from the bulk concentration.

In addition to NRTL, other activity coefficient models are also made available in commercial simulators. These include Wilson, van Laar, and the UNIQUAC method.

It should be noted that in commercial simulators, unlike the tuning parameters (BIPs) for the equations of state, which are available for almost all of the library component combinations, activity coefficient tuning parameters can be somewhat scarce. Hence it is necessary to either

1. acquire binary T-XY, P-XY, and vapor pressure data to tune the activity coefficient models or
2. estimate the activity coefficient tuning parameters computationally. Most commercial simulators have a UNIFAC group contribution method tuning parameter estimation algorithm. This algorithm generally estimates the tuning parameters for a user-specified temperature at atmospheric pressure.

Because most engineers or companies do not have access to binary data or expertise for tuning the activity coefficient models, it may be necessary to outsource this to specialist organizations. In a majority of cases, option 2 using the built-in tuning parameter generator is typically used. Caution should be exercised and results should be checked for vapor pressure accuracy.

3.6.3 Bringing It All Together

In Section 3.6.1, vapor phase corrections were addressed, and in Section 3.6.2, liquid phase corrections were discussed. Implementing both vapor and liquid phase corrections to the equilibrium equation yields:

$$\varnothing_i P y_i = \gamma_i P_i^{sat} x_i \tag{3.41}$$

This can also be written as:

$$y_i = K_i x_i \tag{3.42}$$

K_i is referred to as the K-value or the distribution coefficient for component i.

$$K_i = \frac{\gamma_i P_i^{sat}}{\varnothing_i P} \tag{3.43}$$

From Eq. (3.43), one can see that implementing *both* vapor and liquid phase corrections (a dual model) would lead to a very complex endeavor. There would be two groups of tuning parameters that would have to be determined. In most cases, details will be put into the liquid activity coefficient model, while the gas phase fugacity coefficient is assumed ideal, or estimated using an untuned equation of state. In addition to that solving for both the fugacity and liquid activity coefficients would place a greater demand on computational requirements particularly if the simulation were large, or when dynamic simulation were to be performed. Table 3.3 summarizes the methods used in industry for typical vapor-liquid equilibrium and vapor-liquid-liquid equilibrium computations.

3.7 FLASH CALCULATIONS (SMITH ET AL.)

All commercial simulators perform vapor–liquid and vapor–liquid–liquid (not addressed in this chapter) equilibria computations for fluid systems. This is generally referred to as a flash calculation. This is a common computation that is used to predict phase splits, bubble and dew points, and distribution/partitioning of components across the phases. The calculations involved are described in the following subsections.

TABLE 3.3 Common Practices for Vapor-Liquid Equilibrium Method Selection

Application/System	Gas Phase Fugacity Coefficient Method	Liquid Activity Coefficient Method
Oil and gas processing. Estimation of bulk flow rates of oil, gas, and water. Accurate compositional tracking and partitioning of CO_2, H_2S, and water not critical	Peng–Robinson Soave–Redlich–Kwong Kabadi Danner Lee–Kessler–Plocker	Ideal ($\gamma_i = 1$)
Gas dehydration or hydrate suppression with glycol or alcohols	Peng–Robinson Soave–Redlich–Kwong Specialty glycol or alcohol packages	Ideal ($\gamma_i = 1$)
Refining (traditional computations)	Empirical vapor pressure models, Grayson Streed, Braun-K10	
Refining (recent developments)	Peng–Robinson Soave–Redlich–Kwong	Ideal ($\gamma_i = 1$)
Sour Gas. H_2S partitioning across vapor, hydrocarbon, and aqueous predictions critical	Specialty packages. Typically enhanced cubic equations of state	
Gas sweetening using amines/solvents	Specialty packages	
Nonideal liquids, highly polar, self-associating	Ideal Peng–Robinson Soave–Redlich–Kwong	NRTL UNIQUAC
Steam systems (only water)	ASME or NBS steam tables	

3.7.1 "MESH" Equations

The flash calculation involves the solving of four main sets of equations, which are known by the "MESH" equations:

1. **M**aterial Balance
2. **E**quilibrium
3. **S**ummation
4. **H**eat Balance

3.7.1.1 Material Balance

For a simple flash unit in Fig. 3.6, the material balance consists of the overall balance and component balance.

Overall balance:

$$F = V + L \tag{3.44}$$

Component balance:

$$Fz_i = V_i y_i + L_i x_i \tag{3.45}$$

where F is the total molar flow rate of material, V is the vapor molar flow rate, L is the liquid molar flow rate, z_i is the mole fraction of component i in the combined phases, y_i is the mole fraction of component i in the gas phase, and x_i is the mole fraction of component i in the gas phase.

3.7.1.2 Equilibrium

Equilibrium and component distribution across phases are determined by the equilibrium equation introduced in Section 3.6.3:

$$y_i = K_i x_i$$

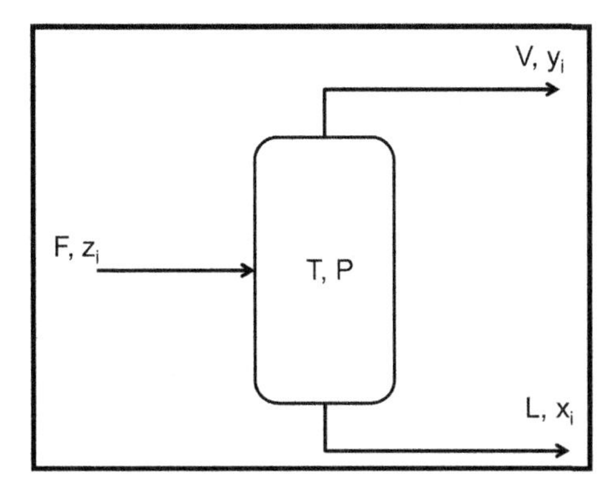

FIGURE 3.6 Flash calculation at a given temperature and pressure.

3.7.1.3 Summation

The sum of all mole fractions must equal 1

$$\sum_i z_i = 1 \ ; \sum_i x_i = 1 \ ; \sum_i y_i = 1 \ ; \tag{3.46}$$

3.7.1.4 Heat Balance

$$Fh_f = Vh_{vap} + Lh_{liq} \tag{3.47}$$

where h_f is the molar enthalpy of the combined phases; h_{vap} and h_{liq} are the molar enthalpies of the vapor and liquid phases, respectively.

3.7.2 Bubble Point Flash

The bubble point or saturated liquid flash is typically performed to determine the bubble point temperature *or* pressure. Here, the composition of the liquid is the composition of the entire fluid.

3.7.2.1 Methodology

$$\sum_i K_i x_i = \sum_i y_i = 1$$

- Assume a temperature for the known pressure or vice versa.
- Find K_i at the pressure and temperature known and assumed.
- Check that the summation approaches your desired tolerance (close to zero).
- If not, repeat until satisfactory tolerance has been achieved.

In commercial simulators such as Aspen HYSYS or UniSim Design, this would be performed by setting the vapor fraction equal to zero, and specifying a temperature (to calculate a bubble point pressure) *or* pressure (to calculate a bubble point temperature).

3.7.3 Dew Point Flash

The dew point or saturated vapor flash is typically performed to determine the dew point temperature *or* pressure. Here, the composition of the vapor is the composition of the entire fluid. Procedure of dew point flash calculation consists of the following steps:

- Set $\sum_i \frac{y_i}{K_i} = \sum_i x_i = 1$.

- Assume a temperature for the known pressure or vice versa.
- Find K_i at the pressure and temperature (known or assumed).
- Check that the summation approaches your desired tolerance (close to zero).
- If not, repeat until satisfactory tolerance has been achieved.

In Aspen HYSYS or UniSim Design, this would be performed by setting the vapor fraction equal to 1, and specifying a temperature (to calculate a dew point pressure) *or* pressure (to calculate a dew point temperature).

3.7.4 Two-Phase Pressure–Temperature Flash

The pressure–temperature (PT) flash is used to determine the amount of liquid and vapor and their compositions at a given pressure and temperature.

Equations

1. $F = V + L$
2. $Fz_i = Vy_i + Lx_i$
3. Take a basis of $F = 1$ mol
4. $z_i = Vy_i + Lx_i$
5. $y_i = K_i x_i$
6. $x_i = z_i/(L + VK_i)$
7. $y_i = z_i/(V + L/K_i)$
8. Summation criteria:
   $\sum x_i = \sum(z_i/(L + VK_i)) = 1$
   $\sum y_i = \sum(z_i/(V + LK_i)) = 1$

Procedure

- Find K at T and P
- Assume V or L
- Solve step 8
- Iterate until step 8:
  $\sum x_i - \sum y_i <= \text{Tolerance}$

3.7.5 Other Flash Routines

In addition to the dew point, bubble point, and PT flashes described in this chapter, most commercial simulators include the following flash routines:

1. Pressure enthalpy (used in the valve and mixer computations)
2. Temperature enthalpy (used in heat exchanger computations)
3. Pressure entropy and temperature entropy (used for rotating equipment, compressors, and pumps)

3.8 PHASE DIAGRAMS

Phase envelopes are one of the most useful tools a chemical engineer has at his or her disposal. The phase envelope provides a graphical overview of the possible phases a fluid may have for a range of pressures and temperatures. All commercial simulators come with phase envelope utilities. The previous section addressed flash routines that are commonly built into commercial simulators. We next discuss the phase envelopes generated by running these flash routines for a given pair of variables.

3.8.1 Pressure–Temperature Diagrams of Pure Components and Mixtures

A PT phase diagram is generated by running the bubble point and dew point flash routines over a range of temperatures and pressures. When the routine is run for a single component, a phase diagram or a saturation line is generated. PT diagrams for pure methane are given in Fig. 3.7, while those for pure ethane are given in Fig. 3.8. As shown, methane and ethane have different boiling and critical points.

The PT diagram for pure component can be used to determine if the fluid is a subcooled liquid, superheated gas, or supercritical. It does not, however, give an indication that the fluid is a saturated liquid, saturated vapor, or two-phase as all these phases fall within the one-dimensional saturation line. To determine this, it is necessary to look at the pressure–enthalpy or temperature–enthalpy diagrams, such as those in Fig. 3.9. The maximum temperature and

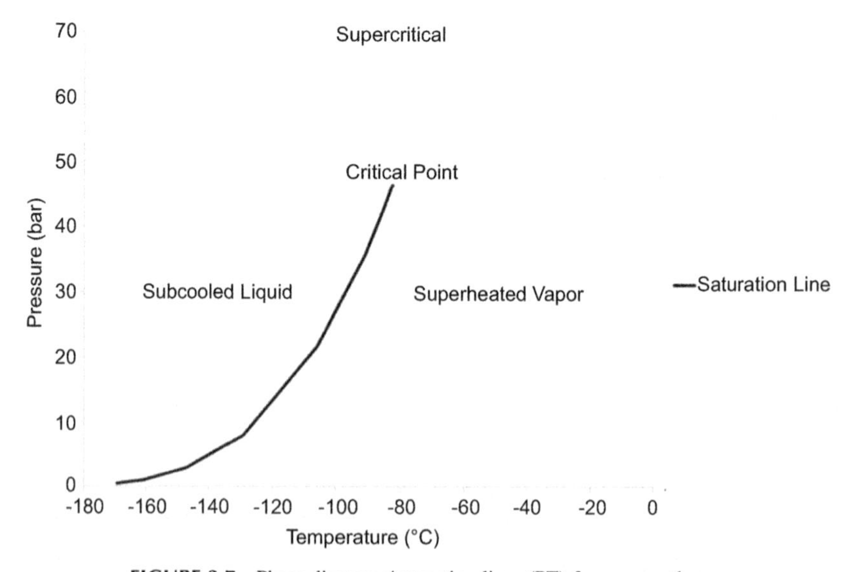

FIGURE 3.7 Phase diagrams/saturation lines (PT) for pure methane.

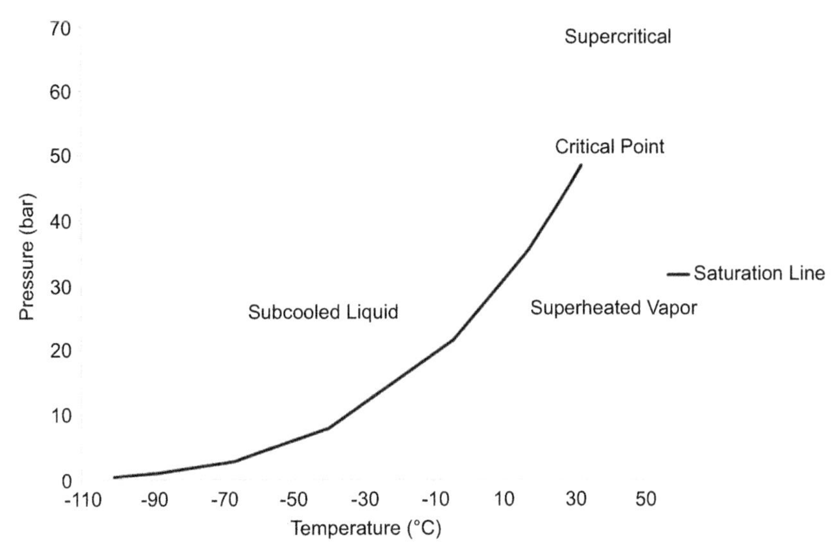

FIGURE 3.8 Phase diagrams/saturation lines (PT) for pure ethane.

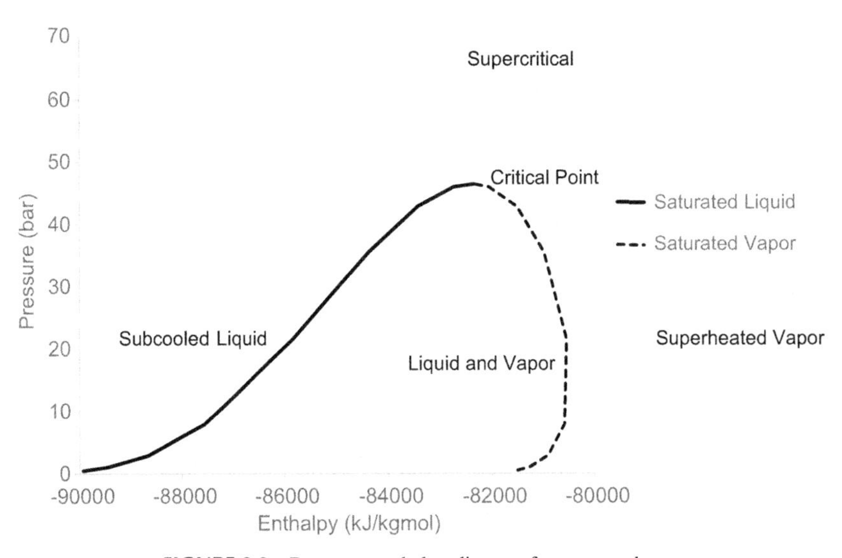

FIGURE 3.9 Pressure—enthalpy diagram for pure methane.

pressure where the fluid can exist as vapor or liquid is the critical point and the terminus of the saturation line.

For an equimolar mixture of methane and ethane, its phase diagram looks quite different than the pure component phase diagrams (e.g., Figs. 3.7 and 3.8). In this case, we see a phase envelope, such as that in Fig. 3.10.

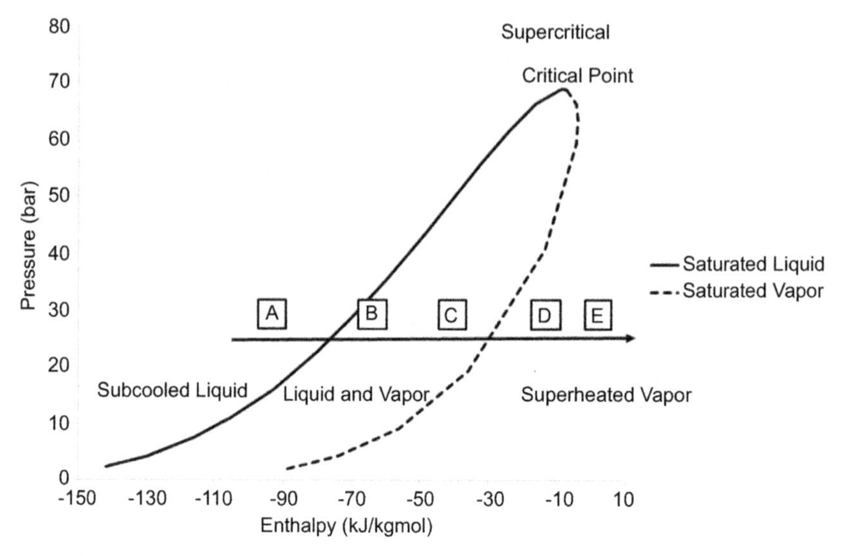

FIGURE 3.10 Phase diagram (PT) for an equimolar mixture of methane and ethane.

Methane is the more volatile component with a normal boiling point of −161.5°C. The normal boiling point of ethane is significantly higher at −89°C. Instead of a single saturation line like those pure component cases in Figs. 7 and 8, two lines are observed for Fig. 3.10. The solid line is the saturated liquid or bubble point line, while the dashed line is the saturated vapor or dew point line. Both these lines meet at the critical point.

Because of the difference in boiling points in this mixture, the saturation line has "spread out." Starting from a subcooled liquid at point *A* and increasing the temperature of the mixture moves the operating point to point B, the bubble point line. Here, a bubble begins to nucleate; the composition of that bubble is predominantly the more volatile component, methane. As the temperature of the mixture is increased, both the methane and ethane vaporize. At point C, the mixture consists of a vapor and a liquid phase. Point *D* is the dew point where the last drop of liquid vaporizes. The composition of this last drop of liquid is mainly the heavier component, ethane. As the temperature is increased, the operating point moves to the superheated gas region.

If all three diagrams (Figs. 3.7, 3.8, and 3.10) are overlaid in one plot, we can gain some insight into the phase envelope of the mixtures, such as that in Fig. 3.11.

The following observations can be made. The bubble point and dew point lines are bound by the saturation lines of the lightest and heaviest components. The mixture critical point is higher than the individual pure component critical points because of the intermolecular interactions between the components.

Phase diagrams for various methane and ethane ratios are overlaid in one plot (Fig 3.12). The dashed line drawn tangent to the critical points of the

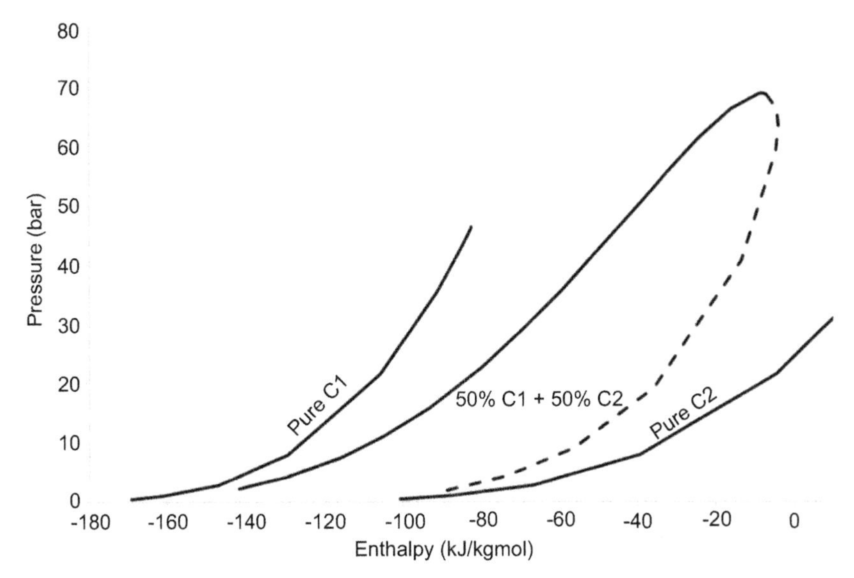

FIGURE 3.11 Phase diagrams for pure methane, pure ethane, and an equimolar mixture of methane and ethane.

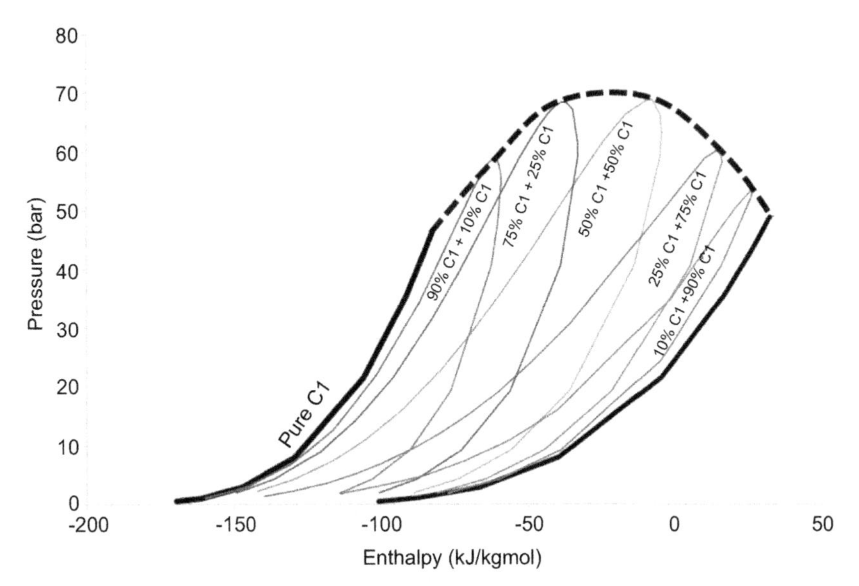

FIGURE 3.12 The critical locus for a binary mixture of methane and ethane.

phase envelopes for compositions at the critical point at each curve is referred to as the critical locus (Campbell, 2004).

3.8.2 Retrograde Behavior

In Fig. 3.9, the phase envelope for a 50:50 mol mixture of methane and ethane exhibits a single critical point, the maximum temperature and pressure at which vapor and liquid may exist.

Fig. 3.13 is a typical phase diagram for multicomponent natural gas mixture.

In this case, there is no single point defining the temperature and pressure maxima; instead there are three points:

1. The cricondenbar, the maximum temperature where liquid and vapor may exist.
2. The cricondentherm, the maximum temperature where liquid and vapor may coexist in equilibrium.
3. The retrograde region, the shaded area in the phase envelope where liquid condensation can occur by increasing temperature or lowering pressure (which is counterintuitive).

This demonstrates the importance of plotting the phase envelope when performing any computation involving a multicomponent mixture to avoid "unexpected" phase behavior.

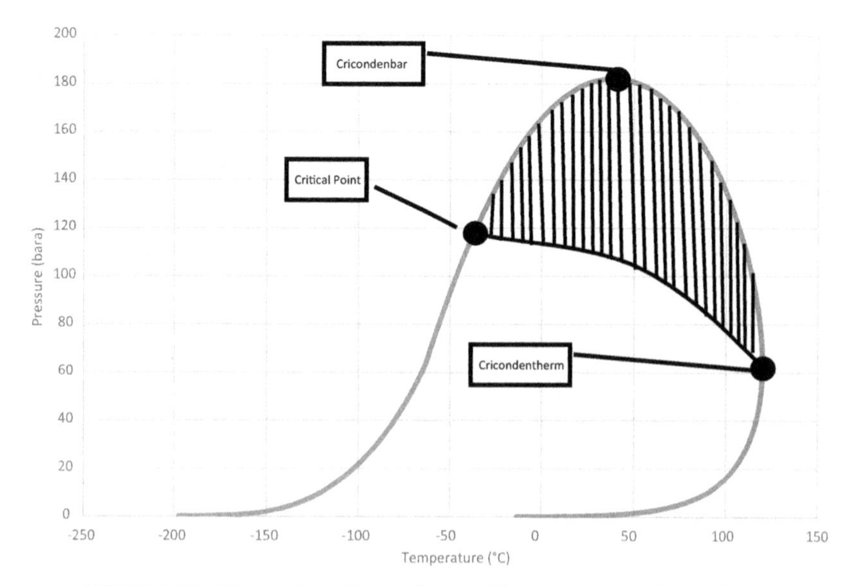

FIGURE 3.13 Typical phase diagram for a multicomponent natural gas mixture.

3.9 CONCLUSIONS

A good chemical engineer must have a strong foundation in thermodynamics!

A significant number of chemical engineering processes are characterized using thermodynamic processes, which are in turn defined by the thermodynamic state functions such as internal energy, enthalpy, entropy, and Gibbs and Helmholtz free energies. These thermodynamic state functions are derived from pressure, volume, and temperature using equations of state.

For nonpolar systems, where liquid phase activity can be neglected, phase equilibrium and component phase (and corresponding distribution or partitioning) may be estimated using equations of state. However, in cases where there are strong molecular interactions in the liquid phase, it would be necessary to use liquid activity coefficient models.

Caution must be exercised when using liquid activity models because in many cases, tuning parameters may not be available in most commercial simulations. They would need to be estimated from group contribution methods or experimentally determined. Always check the vapor pressure prediction and compare that to experimental or field observations.

When working with pure component systems, exercise caution when using PT flashes. It is not possible to determine the phase of a saturated pure component fluid by knowing the pressure and temperature alone. It is necessary to look at the pressure–volume, pressure–enthalpy, and temperature–volume diagrams to determine whether a saturated fluid is a saturated liquid, saturated vapor, or multiphase. It is absolutely critical to plot the phase envelope of the fluid under study. The phase envelope will provide a convenient visual indication of whether a fluid is a subcooled liquid, superheated gas, multiphase, critical, or supercritical. In particular, mixtures with wide boiling point ranges may exhibit retrograde behavior, where the phase behavior is counterintuitive. Having a graphical representation will allow one to flag potential issues like these.

While, all commercial simulation software packages conveniently support most of the equations of state commonly used for engineering computations, it is important to remember that these equations of state have been derived from gas phase relationships. Hence, these process simulation packages will have additional corrections or empirical correlations to describe liquid densities and transport properties, such as viscosities.

EXERCISES

Use your process simulator and favorite equation of state–based property package to perform the following tasks:

- Generate PT, PV, and PH plots for pure $C1$ and pure $C2$. Determine the critical temperatures and pressures for $C1$ and $C2$.

TABLE E1 Stream Condition

Component	Mole%
H_2O	0.00
Nitrogen	0.01
H_2S	4.78
CO_2	0.71
Methane	78.15
Ethane	3.41
Propane	5.92
i-Butane	1.82
n-Butane	0.94
i-Pentane	0.61
n-Pentane	0.77
n-Hexane	1.23
*C7+**	1.64

*C7+** is a hypothetical component (Refer to Chapter 2 for details on registration of new components.) with the following properties:

Molecular Weight (kg/kg mol)	111.00
Normal Boiling Point (°C)	110.00
Ideal Liquid Density (kg/m^3)	745.38

- On one PT diagram, plot (overlay the following curves) (hint: you would have to copy the data from your simulator into a spreadsheet program):
  - pure $C1$
  - pure $C2$
  - 60% $C1$ + 40% $C2$
  - 40% $C1$ + 60% $C2$
  - 20% $C1$ + 80% $C2$

  For these phase envelopes, what can you say about the

1. critical points and
2. bubble and dew point lines?

Create fluids with the condition given in Table E1 using an equation of state property package.

Perform flash calculations and determine the following pressures and temperatures of the stream:

1. Bubble point temperature at 6000 kPa
2. Dew point temperature at 6000 kPa
3. Bubble point temperature at 14,000 kPa
4. Dew point temperature at 14,000 kPa
5. Dew point pressure at 60°C

REFERENCES

Campbell, J.M., 2004. Gas Conditioning and Processing (Vol. 1), eighth ed. BBS.

Smith, J.M., Van Ness, H.C., 1975. Introduction to Chemical Engineering Thermodynamics, third ed. McGraw-Hill.

Walas, S.M., 1985. Phase equilibria in chemical engineering. Butterworth.

FURTHER READING

Poling, B., Prausnitz, J., O'Connell, J. The Properties of Gasses and Liquids, fifth ed. McGraw Hill, 2017.

Chapter 4

Simulation of Recycle Streams

Dominic C.Y. Foo, Siewhui Chong, Nishanth Chemmangattuvalappil
University of Nottingham Malaysia Campus, Semenyih, Malaysia

Recycle system is commonly found in a process flowsheet. However, it is one of the systems that are difficult to achieve convergence, especially for novice in process simulation. In this chapter, strategies to converge recycle systems will be covered. Emphasis is placed on sequential modular (SM) approach, which is commonly used in commercial process simulation software. Chapter 1 discussed how SM approach is used to simulate recycle stream simulation briefly. More tips on recycle stream convergence are given in this chapter.

4.1 TYPES OF RECYCLE STREAMS

From process simulation perspective, recycle system may be generally categorized as material or heat recycle. The former normally involves the recovery of material (e.g., unconverted feedstock) to certain process units (e.g., reactors) for further processing. One such example is shown in Fig. 4.1. When sequential modular (SM) approach is adopted in solving this system, the recycle stream will have no data for the reactor simulation to converge. As discussed in Chapter 1, *tear-stream* strategy can be used to converge this material recycle stream.

On the other hand, heat recycle system involves the recovery of energy between the process streams. Fig. 4.2 shows such a case, where the top product

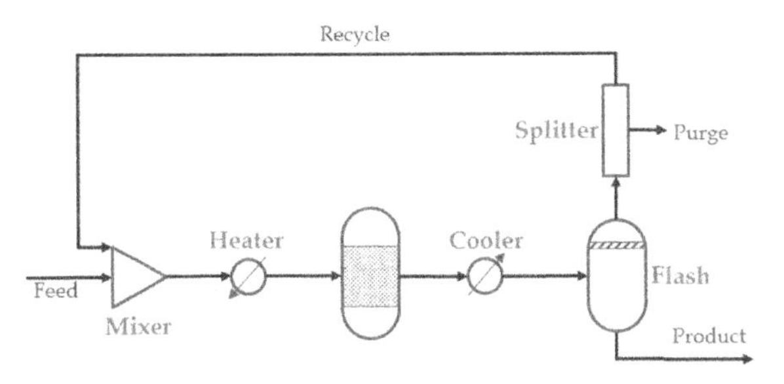

FIGURE 4.1 A material recycle system.

Chemical Engineering Process Simulation. http://dx.doi.org/10.1016/B978-0-12-803782-9.00004-2

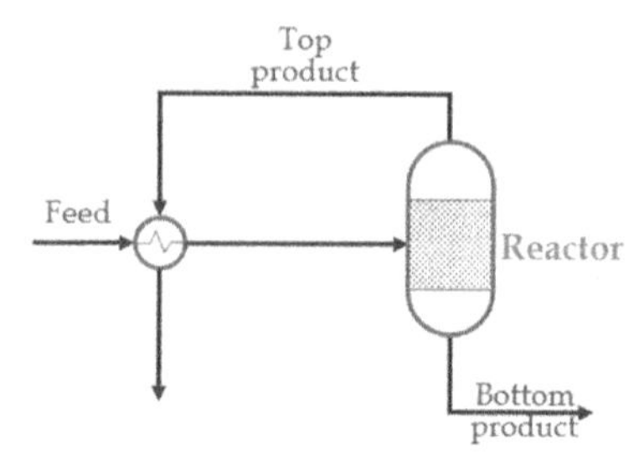

FIGURE 4.2 A heat recycle system.

stream from a reactor is used to preheat the feed stream to the reactor. In this case, the top product stream contains no data to simulate the heat exchanger unit. Similarly, tear-stream concept can be used to handle this heat recycle system.

4.2 TIPS IN HANDLING RECYCLE STREAMS

To converge recycle streams, some of the following strategies (WinSim, 2017) are very useful.

4.2.1 Analyze the Flowsheet

Not all flowsheet contains a recycle system. Fig. 4.3 shows a flowsheet that has no recycle stream, even though it looks like having one. Hence, it is important to analyze the flowsheet carefully before attempting to solve a flowsheet with tear-stream strategy directly.

For the process in Fig. 4.4, if one were to simulate the process with conventional sequence, i.e., from feed stream to product stream, one will have to

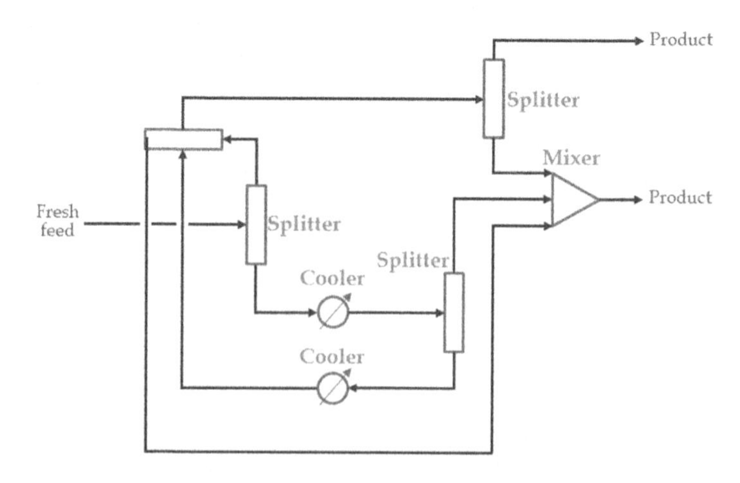

FIGURE 4.3 A flowsheet without recycle stream (WinSim, 2017).

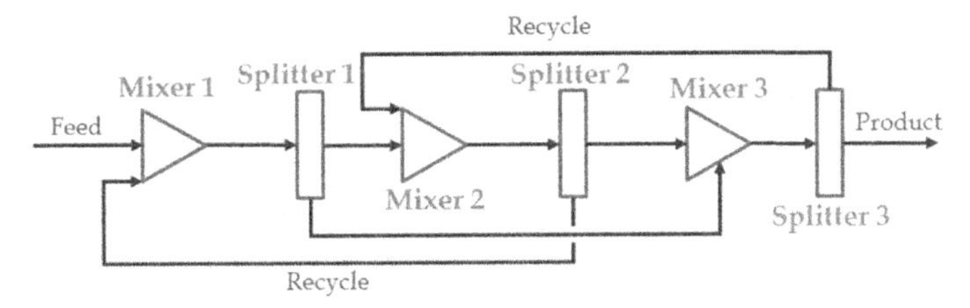

FIGURE 4.4 A flowsheet with two recycle streams (WinSim, 2017).

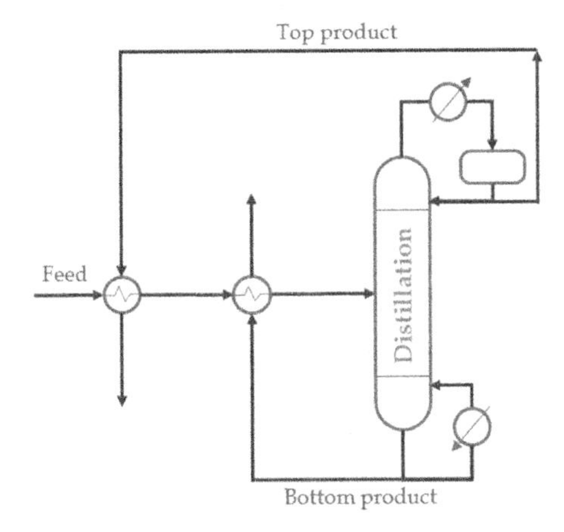

FIGURE 4.5 A distillation column with feed preheat (WinSim, 2017).

employ the tear-stream concept for both the recycle streams. However, if one were to employ the tear-stream concept on a commonly shared stream on both recycle systems, i.e., outlet stream of Mixer 2 (i.e., inlet for Splitter 2), only one tear stream needs to be solved to converge the entire process. This greatly enhances the speed of convergence in this simulation exercise.

The same strategy is applicable for the process in Fig. 4.5, where the top and bottom product streams are used to preheat the feed stream of the distillation. Instead of applying the tear-stream concept on both of these energy recycle streams, one can apply the tear-stream concept on the outlet stream of the second heat exchanger (i.e., inlet to distillation), as it is the commonly shared stream between both the energy recycle streams.

4.2.2 Provide Estimates for Recycle Streams

When applying the tear-stream concept on a stream, it is always good to supply some known data for the stream, to assist its convergence. For instance, for the distillation example in Fig. 4.5, the stream flowrate and compositions of the outlet stream from second heat exchanger (i.e., inlet stream to distillation)

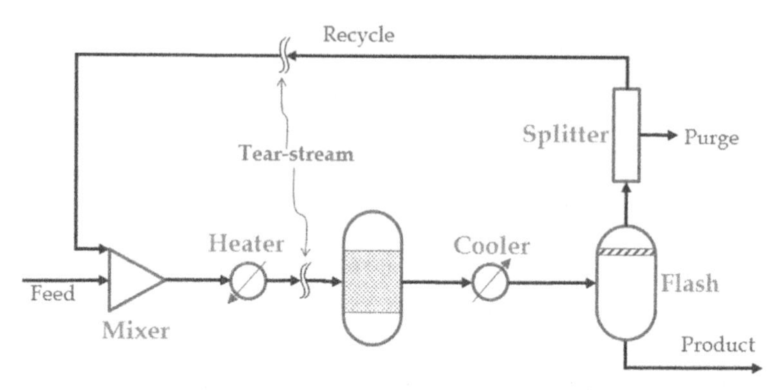

FIGURE 4.6 Specifying the tear stream *Reproduced from Fig. 4.1.*

should stay the same as those of the feed stream. With modern process simulation software, it is often possible to copy the properties of this stream from the fresh feed stream. Besides, one may also set the pressure drop of the heat exchanger, which leaves the outlet stream with an unknown temperature. Specifying these data for the distillation inlet stream will greatly enhance the convergence of simulation.

Another example is the example in Fig. 4.6. Instead of specifying the recycle stream as tear stream, one may also specify the reactor inlet stream if its properties are known. In both examples, one notices that the tear-stream concept needs not be necessarily applied on the recycle stream; it can be applied to any stream within the recycle loop to promote faster convergence.

4.2.3 Simplify the Flowsheet

Before simulating a complex unit, it is always a good practice to perform simulation on a simplified unit. A good example is the modeling of rigorous distillation column. In most simulation software, shortcut distillation model is made available to determine the basic parameters needed for a rigorous column. Hence, it is encouraged to converge a shortcut distillation model to provide some initial guess for the rigorous model[1]. Besides, it is also a good practice to converge the rigorous distillation model as a stand-alone unit, before it is connected to a complete flowsheet (especially for cases where distillation is part of the recycle systems).

For flowsheets that contain both material and energy recycle systems that are interconnected, it is easier to converge the individual recycle system when they are decoupled. The *n*-octane case study in Example 1.1 provides a good illustration for this strategy.

1. Refer to VCM production example in Chapter 12 for better understanding.

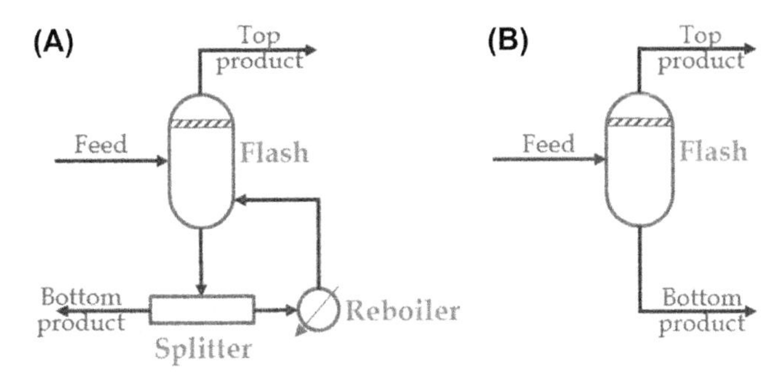

FIGURE 4.7 (A) A flash unit with reboiler and internal recycle system; (B) a simplified system without reboiler (WinSim, 2017).

It is important to remember the objective of a simulation exercise. For instance, for preliminary mass and energy balances at conceptual design stage, it is acceptable to simplify a flowsheet to achieve faster convergence. An example is shown in Fig. 4.7A, where a flash unit contains a reboiler and an internal recycle system. This unit will definitely require some iterative calculation before the unit can converge. This system can be simplified such as that in Fig. 4.7B for easier convergence.

4.2.4 Avoid Overspecifying Mass Balance

When a purge stream is found in a material recycle system (e.g., Fig. 4.1), it is best to set the recycle fraction of the purge stream, rather than the exact material flowrate of the recycle stream. Setting an exact flowrate may prevent the convergence of the recycle stream.

For a distillation train that contains multiple columns (e.g., see Fig. 4.8), specifying the top or bottom product flowrates may be overconstraining the mass balances of the system. For these cases, apart from converging the shortcut distillation model prior to rigorous model (see earlier tip), it will be useful to specify the reflux ratio and the top/bottom product rate (WinSim, 2017).

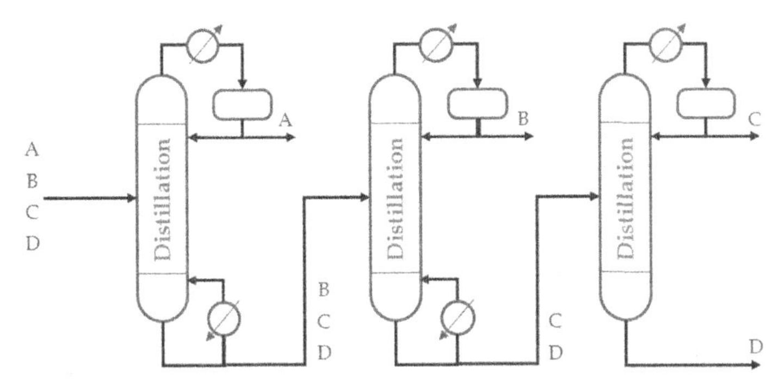

FIGURE 4.8 Distillation train with three columns.

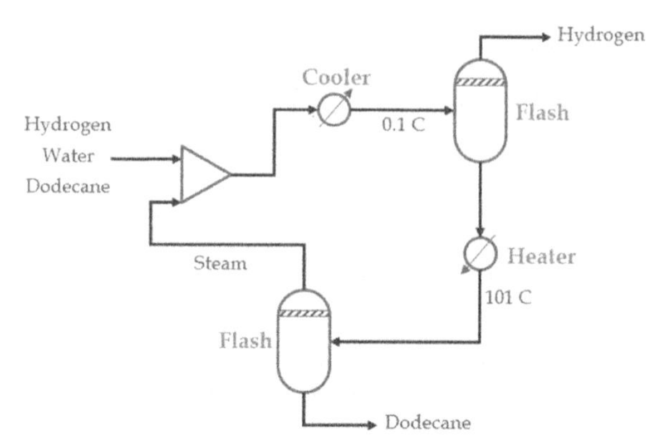

FIGURE 4.9 A recycle system where water is trapped in the recycle loop (WinSim, 2017).

4.2.5 Check for Trapped Material

When a recycle system does not converge, it is useful to check if there exists any unnecessary material buildup in the system. In general, component with medium boiling range is easily trapped. For instance, for the system in Fig. 4.9, water is condensed by the cooler, which is then sent to the heater. The heater vaporizes water into steam, which is then sent back to the cooler. This causes the water component to be trapped in the recycle loop.

In checking an unconverged recycle loop, it is necessary to check the material balance summary to see which components have the largest error. When the component flowrate leaving the recycle loop is less than that in the entering stream, it is likely that the component is being trapped in the recycle loop.

4.2.6 Increase Number of Iterations

Most flowsheets will converge easily within a few iterations. However, when a recycle loop is unconverged after some iterations, it will be good to increase its iterations. When the recycle loop is approaching convergence, properties of the stream (where tear-stream concept is applied) such as pressure, temperature, flowrate, and composition may be updated to provide better guess for the recycle loop to converge (WinSim, 2017).

4.3 RECYCLE CONVERGENCE AND ACCELERATION TECHNIQUES

Chapter 1 describes how SM approach may be used to converge a recycle loop. In most cases, a tear stream is to be chosen at which a convergence solver is to be placed. The convergence solver computes the difference between the calculated and estimated values of the tear stream and then updates the estimated value with the calculated one. The simplest method to converge the recycle calculation is the *direct substitution* method (or *successive substitution*) (Smith, 2016), in which the calculated value of the variable [$G(x)$] simply updates the estimated value of the variable (x). The substitution will stop when the convergence criteria are met, which is given by Eq. (4.1) (known as the scaled residue; Smith, 2016):

$$-Tolerance \leq \frac{G(x) - x}{x} \leq Tolerance \tag{4.1}$$

There are several techniques that can be used to accelerate the convergence of a recycle calculation; the most commonly used one is called the *Wegstein acceleration* method. This method will be illustrated using Example 4.1. For highly nonlinear and interdependent equations, other acceleration methods such as dominant eigenvalue, Newton–Raphson, and Broyden's quasi-Newton methods may also be used (Smith, 2016).

Example 4.1

An isomerization process (Smith, 2016) is used to illustrate the concept of recycle convergence using SM approach. In an isomerization process, component A is converted to component B. As shown in Fig. 4.10A, the mixture from the reactor is then separated into relatively pure A, which is then recycled back to the reactor system, and relatively pure B, which is the product. Fig. 4.10B shows the block structure of the SM approach where the convergence solver is placed for the selected tear stream—the recycle stream, S5. The following assumptions are made (Smith, 2016):

- No by-products are formed.
- The reactor performance can be characterized by its conversion.
- The performance of the separator is characterized by the recovery of A (r_A) and B (r_B) to the recycle stream.

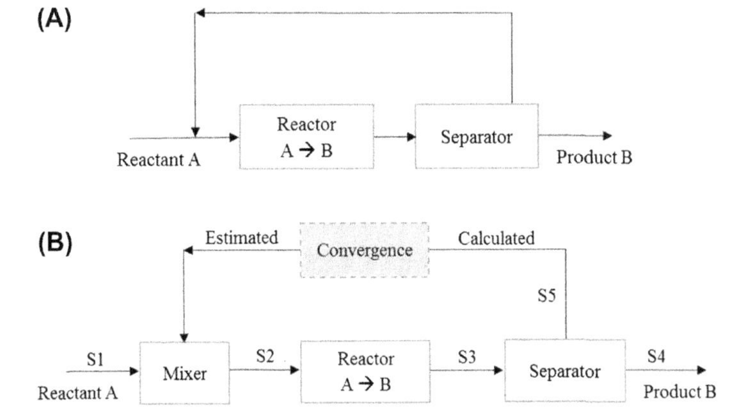

FIGURE 4.10 An isomerization process in: (A) process flowsheet; (B) block structure of sequential modular approach.

Material balances for components A and B are conducted for the isomerization process, which consists of 8 equations and 13 variables.

At the mixer:

$$\dot{m}_{A,S2} = \dot{m}_{A,S1} + \dot{m}_{A,S5} \tag{4.2}$$

$$\dot{m}_{B,S2} = \dot{m}_{B,S1} + \dot{m}_{B,S5} \tag{4.3}$$

At the reactor:

$$\dot{m}_{A,S3} = \dot{m}_{A,S2}(1 - X) \tag{4.4}$$

$$\dot{m}_{B,S3} = \dot{m}_{B,S2} + \dot{m}_{A,S2}X \tag{4.5}$$

At the separator:

$$\dot{m}_{A,S4} = \dot{m}_{A,S3}\ (1 - r_A) \tag{4.6}$$

$$\dot{m}_{A,S5} = \dot{m}_{A,S3}\ (r_A) \tag{4.7}$$

$$\dot{m}_{B,S4} = \dot{m}_{B,S3}\ (1 - r_B) \tag{4.8}$$

$$\dot{m}_{B,S5} = \dot{m}_{B,S3}\ (r_B) \tag{4.9}$$

Where $\dot{m}_{i,j}$ = molar flowrate of component i in stream j; X = reactor conversion of A; and r_i = fractional recovery of component i to the separator top stream.

An example of converging the recycle stream using direct substitution is demonstrated in Table 4.1. A spreadsheet is created for the calculations of material balances for components A and B. The initial conditions $-\dot{m}_{A,S1}$, $\dot{m}_{B,S1}$, X, r_A, and r_B are known and are given as follows:

$$\dot{m}_{A,S1} = 100\ \text{kmol/h}$$

$$\dot{m}_{B,S1} = 0\ \text{kmol/h}$$

$$X = 0.7$$

$$r_A = 0.95;\ r_B = 0.05$$

Using initial estimated values of $\dot{m}_{A,S5} = 50\ \text{kmol/h}$ and $\dot{m}_{B,S5} = 5\ \text{kmol/h}$, the scaled residue is computed for the estimated and the calculated values. Iterations follow the profile as shown in Fig. 4.11A until the specified tolerance is met, for instance, 0.00001 in this case.

The major limitation of this method is the requirement of many iteration steps and that some values may fail to converge to the required tolerance (Smith, 2016). If the direct substitution method is linearized:

$$G(x) = ax + b, \tag{4.10}$$

TABLE 4.1 Construction of Spreadsheet for Calculation of Material Balances Using Direct Substitution Method

Material balance for component A:							
$\dot{m}_{A,S1}$ (Fixed)	$\dot{m}_{A,S2}$ (Eq. 4.2)	$\dot{m}_{A,S3}$ (Eq. 4.4)	$\dot{m}_{A,S4}$ (Eq. 4.6)	$\dot{m}_{A,S5}$			Scaled residue (Eq. 4.1)
				Iteration	Assumed	Calculated with (Eq. 4.7)	
100	150	45	2.25	1	50	42.75	−0.14500
100	142.750	42.825	2.141	2	42.750	40.684	−0.04833
100	140.684	42.205	2.110	3	40.684	40.095	−0.01447
100	140.095	42.028	2.101	4	40.095	39.927	−0.00419
100	139.927	41.978	2.099	5	39.927	39.879	−0.00120
100	139.879	41.964	2.098	6	39.879	39.866	−0.00034
100	139.866	41.960	2.098	7	39.866	39.862	−0.00010
100	139.862	41.959	2.098	8	39.862	39.861	−0.00003
100	**139.861**	**41.958**	**2.098**	**9**	**39.861**	**39.860**	−0.00001 (converged)
Material balance for component B:							
$\dot{m}_{B,S1}$ (Fixed)	$\dot{m}_{B,S2}$ (Eq. 4.3)	$\dot{m}_{B,S3}$ (Eq. 4.5)	$\dot{m}_{B,S4}$ (Eq. 4.8)	$\dot{m}_{B,S5}$			Scaled residue (Eq. 4.1)
				Iteration	Assumed	Calculated with (Eq. 4.9)	
0	5	110.000	104.500	1	5	5.500	0.10000
0	5.500	105.425	100.154	2	5.500	5.271	−0.04159

Continued

TABLE 4.1 Construction of Spreadsheet for Calculation of Material Balances Using Direct Substitution Method—cont'd

0	5.271	103.750	98.562	3	5.271	5.187	−0.01589
0	5.187	103.254	98.091	4	5.187	5.163	−0.00478
0	5.163	103.112	97.956	5	5.163	5.156	−0.00138
0	5.156	103.071	97.917	6	5.156	5.154	−0.00039
0	5.154	103.059	97.906	7	5.154	5.153	−0.00011
0	5.153	103.056	97.903	8	5.153	5.153	−0.00003
0	**5.153**	**103.055**	**97.902**	**9**	**5.153**	**5.153**	−0.00001 (converged)

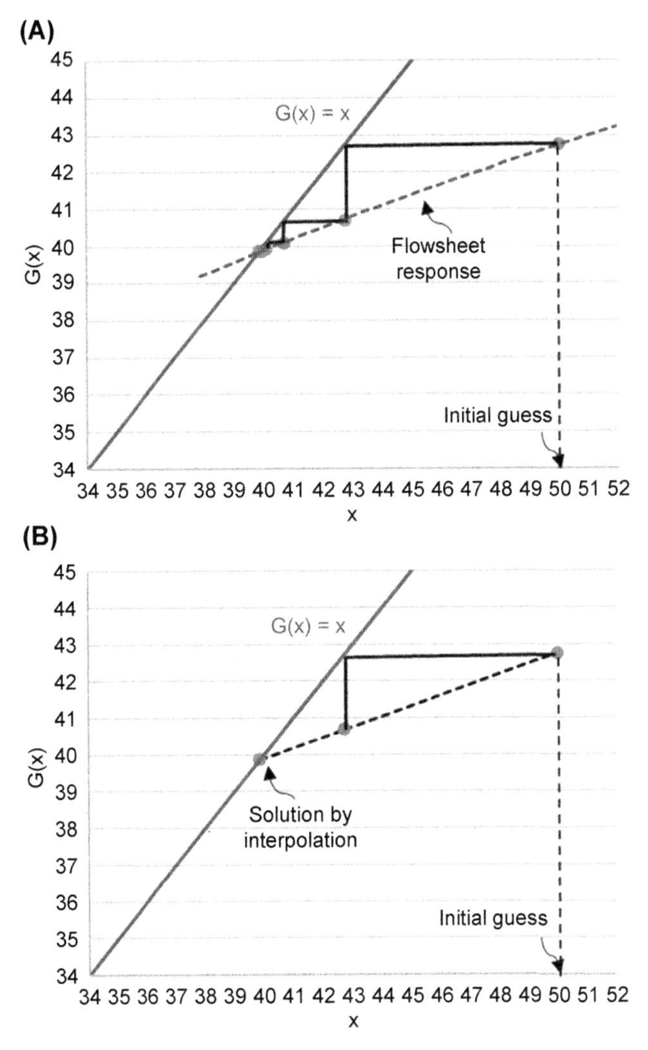

FIGURE 4.11 Convergence of recycle loops using the sequential modular approach via (A) direct substitution or (B) Wegstein method.

with the slope:

$$a = \frac{G(x_k) - G(x_{k-1})}{x_k - x_{k-1}}, \tag{4.11}$$

where x_k and x_{k-1} are the calculated variable values for iterations k and $k-1$, respectively. For iteration k, the intercept can be calculated by:

$$b = G(x_k) - ax_k \tag{4.12}$$

Substituting Eq. (4.12) in Eq. (4.10):

$$G(x_{k+1}) = ax_{k+1} + [G(x_k) - ax_k] \tag{4.13}$$

With the required intersection:

$$G(x_{k+1}) = x_{k+1} \tag{4.14}$$

Eq. (4.13) is then updated to:

$$x_{k+1} = ax_{k+1} + [G(x_k) - ax_k]$$

Or

$$x_{k+1} = \left(\frac{a}{a-1}\right)x_k - \left(\frac{1}{a-1}\right)G(x_k) \tag{4.15}$$

If we define

$$q = \frac{a}{a-1}, \tag{4.16}$$

the final expression becomes:

$$x_{k+1} = x_k + (1-q)[G(x_k) - x_k] \tag{4.17}$$

Eq. (4.17) is known as the Wegstein method (Wegstein, 1958), as interpreted in Fig. 4.11B, which can accelerate the convergence. If $q = 0$, Eq. (4.17) becomes direct substitution; if $q < 0$, acceleration of convergence of the iteration processes occurs, and if $0 < q < 1$, damping occurs. Typically, q is bound between -20 and 0 to ensure stability and reasonable rate of convergence (Seider et al., 2009). If we apply Wegstein method for the second iteration process in Table 4.1:

$$\begin{aligned} a &= \frac{G(x_k) - G(x_{k-1})}{x_k - x_{k-1}} \\ &= \frac{42.75 - 40.684}{50 - 42.750} \\ &= 0.2850 \end{aligned}$$

$$\begin{aligned} q &= \frac{a}{a-1} \\ &= \frac{0.2850}{0.2850 - 1} \\ &= -0.3986 \end{aligned}$$

Substituting in Eq. (4.17), the new estimated value becomes:

$$x_{k+1} = 42.750 + (1 + 0.3986)[40.684 - 42.750] = 39.860 \text{ kmol/h}$$

which is equivalent to the ninth direct substitution iteration process in Table 4.1. Compared to direct substitution, the Wegstein method has accelerated the convergence of solutions.

EXERCISES

Redo the process in Example 4.1, assuming components *A* and *B* are entering at the flowrates of 150 kmol/h and 10 kmol/h, respectively. Further assume that the conversion of *A* is 60% in the reactor and the fractional recovery in the separator is 0.9 for both components. With this information, perform the following tasks.

1. Perform component mass balances for ***A*** and ***B*** for each unit operation in the process.
2. By assuming molar flowrate of the mixture outlet (stream S2) is 400 kmol/h with 80% mole fraction of component A for the first iteration, use MS Excel to find the molar flowrate of components ***A*** and ***B*** in product stream (stream S4) and recycled stream (stream S5). Iteration stops when the scaled residue is smaller than 1×10^{-5}.
3. Apply Wegstein acceleration after the third iteration and calculate flowrates of the streams. Show the complete calculations and values of q, a, and the flowrates. Comment on the acceleration by comparing the results of direct substitution.
4. Determine the purity of the product stream. If the fractional recovery for component ***A*** and component ***B*** are changed to 0.8 and 0.2 respectively, what would be the effect to the purity of the product stream?
5. Assume that a by-product *C* is formed during the conversion of *A* to *B* and the selectivity of *B* over *C* is 4. Component *C* can be separated from *A* and *B* in a unit before the separator. Let the recovery of *C* be 90% and the remaining *C* will be leaving along with product *B* stream. With this information, reestimate the flowrates and purity of the product stream.

REFERENCES

Seider, W.D., Seader, J.D., Lewin, D.R., 2009. Product & Process Design Principles: Synthesis, Analysis and Evaluation (With CD). John Wiley & Sons.

Smith, R., 2016. Chemical Process Design and Integration, second ed. John Wiley, West Sussex, England.

Wegstein, J.H., 1958. Accelerating convergence of iterative processes. Communications of the ACM 1, 9–13.

WinSim, I., 2017. DESIGN II for Windows Training Guide [Online]. Available: www.winsim.com.

Part 2

UniSim Design

Chapter 5

Basics of Process Simulation With UniSim Design

Dominic C.Y. Foo
University of Nottingham Malaysia Campus, Semenyih, Malaysia

This chapter aims to provide a step-by-step guide in simulating an integrated process flowsheet using UniSim Design (Honeywell, 2017). The concept of simulation is based on *sequential modular* approach and follows the Onion model for flowsheet synthesis (see Chapter 1 for details). A simple example involving the production of *n*-octane (C_8H_{18}) is demonstrated, with detailed descriptions given in Example 1.1.

5.1 EXAMPLE ON *N*-OCTANE PRODUCTION

In simulating the integrated flowsheet of *n*-octane production, it will be good to follow the concept of Onion model[1], where simulation starts at the center of the Onion and move outward. In general, the simulation can be carried out in the following stages:

1. Basic simulation setup—This stage involves the setting up of basic information needed for simulation, which includes registration of components, thermodynamic model, and reaction stoichiometry.
2. Reactor system—This involves the setting up of the basic structure of simulation, as well as the selected reactor model.
3. Separation units—This stage involves the simulation of a distillation column.
4. Modeling of recycle system—This involves the simulation of purge stream, compressor, and heat recovery system in the recycle loop.

The basic simulation setup is to be carried out in the *Basis Environments* of UniSim Design, while the other stages are carried out in the *Simulation Environments*. The individual stages are discussed in the following subsections.

1. See detailed discussion in Chapter 1.

Chemical Engineering Process Simulation. http://dx.doi.org/10.1016/B978-0-12-803782-9.00005-4

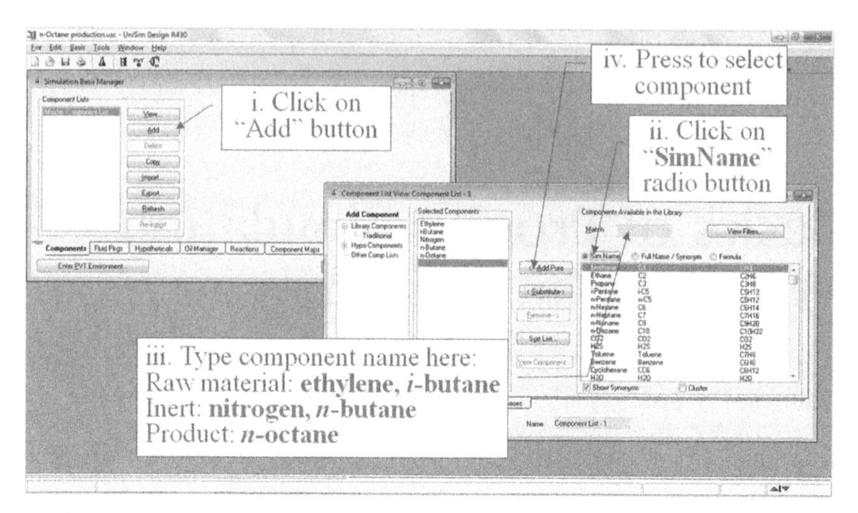

FIGURE 5.1 Component selection using Simulation Basis Manager (Basis Environments).

Tips: Component selection with SimName allows faster identification of necessary components.

5.2 STAGE 1: BASIC SIMULATION SETUP

First, it is necessary to select the components and thermodynamic model for the process. For UniSim Design, these steps were performed using the Simulation Basis Manager in the Basis Environments (Fig. 5.1). All components involved in the simulation, including the reactants, inerts, and the product (see Table 5.1), are to be selected from the component database (see detailed steps for component selection in Fig. 5.1).

Similarly, thermodynamic model (or "Fluid Package" in UniSim Design terminology) for the flowsheet is also selected from the database using the Simulation Basis Manager, however, in the "Fluid Pkgs" tab (see detailed steps in Fig. 5.2). For this process, Peng–Robinson model is to be used.

Next, we move to define the reaction specification for the simulation flowsheet (details given in Table 5.2). This is also carried out using the Simulation Basis Manager, at the "Reaction" tab (see Fig. 5.3). For this example, a simple conversion reaction is modeled. Note that conversion reactor model may be used for preliminary design, when the main purpose of the simulation is to perform basic mass and energy balances. For more detailed modeling, other types of reaction model should be used[2]. Hence, the basic

2. See the use of continuous stirred tank reactor (CSTR) and plug flow reactor (PFR) models in Chapter 12.

TABLE 5.1 Components Needed for Simulation Flowsheet

Components	Role
Ethylene (C_2H_6)	Reactant
i-Butane (*i*-C_4H_{10})	Reactant
n-Octane (*n*-C_8H_{18})	Product
Nitrogen (N_2)	Inert
n-Butane (*n*-C_4H_{10})	Inert

Tips: It is always useful to select all components needed for the entire simulation flowsheet, before a thermodynamic model is selected. Doing this avoids the situation where the earlier selected thermodynamic model does not suit a component that would be selected at a later stage.

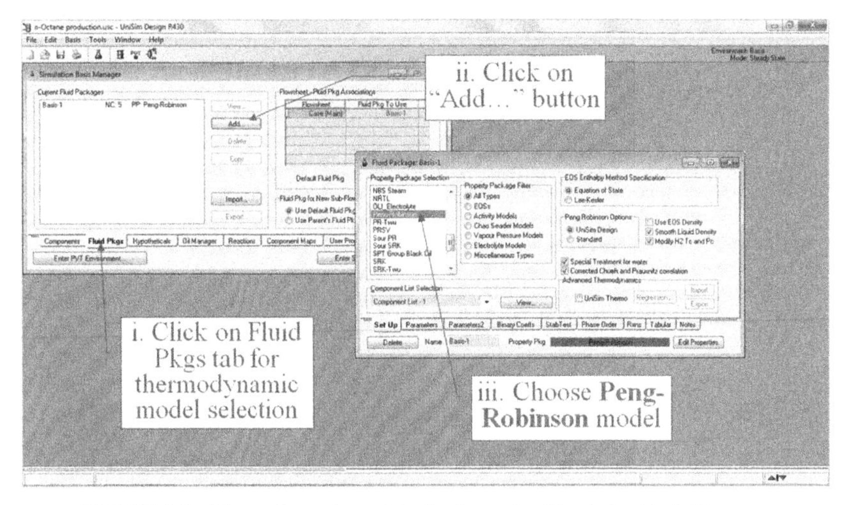

FIGURE 5.2 Thermodynamic model selection using Simulation Basis Manager.

information for reaction stoichiometry and conversion is needed. Note that reaction stoichiometry has to be specified prior to the specification of conversion rate. Fig. 5.3 shows the detailed step in specifying the reaction stoichiometry in the "Stoichiometry" tab. Next, the limiting reactant (i.e., ethylene for this case) and its conversion rate (98%) are defined in the "Basis" tab, following the detailed steps in Fig. 5.4.

TABLE 5.2 Specifications for Conversion Reaction

Specifications	Details
Reaction type	Conversion
Conversion	98%
Limiting reactant	Ethylene

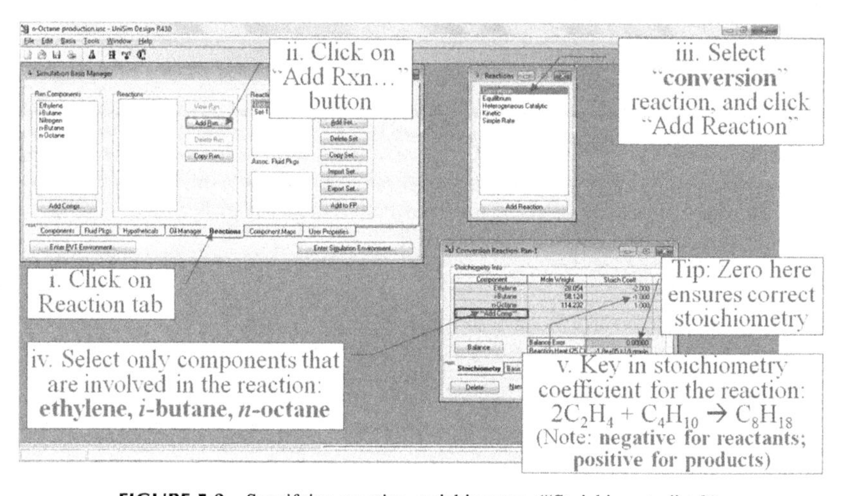

FIGURE 5.3 Specifying reaction stoichiometry ("Stoichiometry" tab).

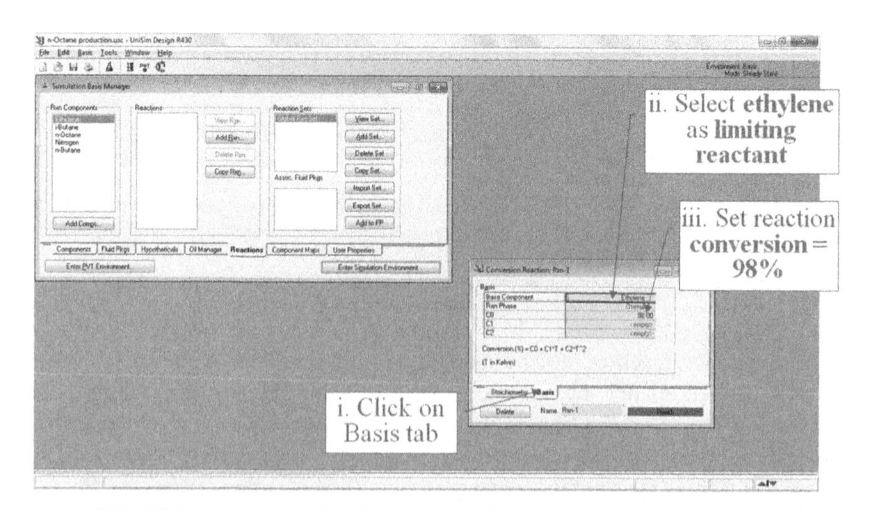

FIGURE 5.4 Specifying limiting reactant and its conversion rate ("Basis" tab).

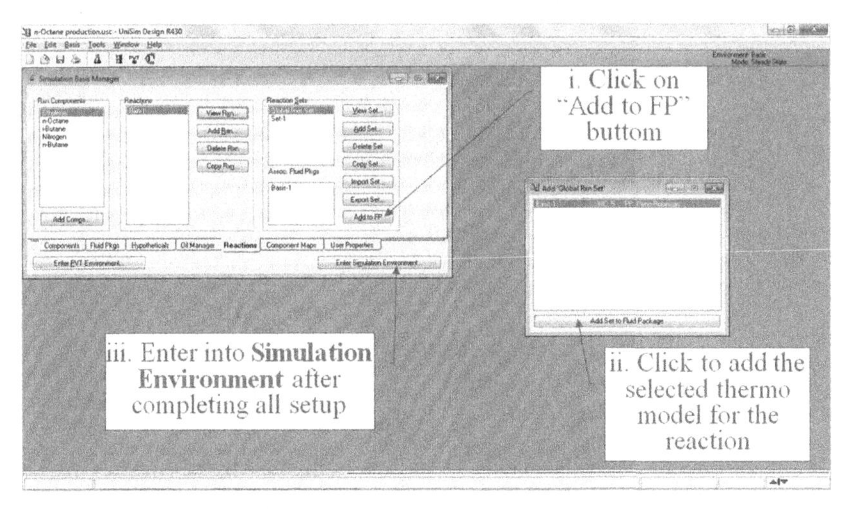

FIGURE 5.5 Associating thermodynamic model for reaction set.

Once the reaction details are specified, we proceed to assign a thermodynamic package (or Fluid Package—FP) for this reaction. For this case, the earlier specified thermodynamic model for the flowsheet, i.e., Peng−Robinson is to be used. Hence, it is associated with this reaction (see detailed steps in Fig. 5.5). In other words, the thermodynamic model will estimate the process condition of all the components during the reaction.

Tips: A good practice in using any computer software is to save the file regularly.

Once the basic information for the simulation flowsheet is specified, we may proceed to the Simulation Environments to perform modeling of the unit operations. To enter the Simulation Environments, we shall press the "Enter Simulation Environment" button on the Simulation Basis Manager (Fig. 5.5).

Tips: To edit basic information of the flowsheet, one can switch back to the Basis Environments by pressing the shake flask icon on the toolbar.

5.3 STAGE 2: MODELING OF REACTOR

The Simulation Environments of UniSim consists of main flowsheet, subflowsheet, and column subflowsheet environments. For the *n*-octane production example, only the main flowsheet is used. We next move on to configure the simulation flowsheet for *n*-octane production. The topology of

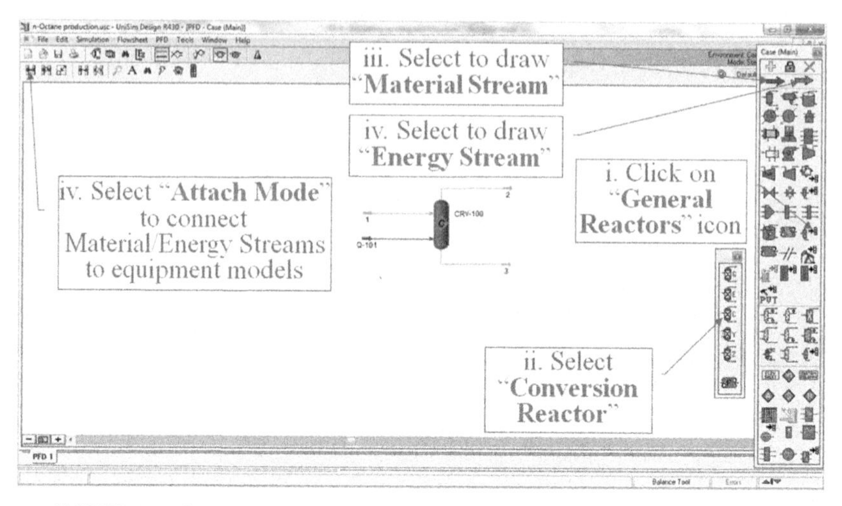

FIGURE 5.6 Construction of flowsheet topology on PFD (Simulation Environments).

the flowsheet is to be built on the simulation interface called the "PFD" in UniSim Design. Different types of reactor models exist within UniSim Design, e.g., continuous stirred tank reactor, plug flow reactor, and other general reactor models (e.g., Gibbs, equilibrium, conversion reactors, etc.). For the *n*-octane production example, we shall utilize the "Conversion Reactor" model. Note that this reaction model should only be used for preliminary flowsheet development, where basic mass and energy balances are to be sorted. Other rigorous reactor models should be utilized for equipment modeling and design. Detailed steps to draw the flowsheet on the PFD are shown in Fig. 5.6, where the conversion reactor consists of two inlet (Streams 1 and Q-101) and two outlet streams (Streams 2 and 3). Note that Streams 1, 2, and 3 are the actual process streams that consist of material (termed as Material Stream in UniSim Design), while Stream Q-101 is actually virtual stream that is used for performing heat balances.

Tips: If you do not see the Object Palette on the PFD (where the unit operation models are found), press the F4 button on the keyboard to launch it.

Tips: You may edit the background color of the PFD, model icons, or text font by visiting Tools/Preferences ("Resources" tab).

We then proceed to define reactor feed stream properties. Component flowrates, temperature (T), and pressure (P) of the stream are given in Table 5.3. Detailed steps for defining stream properties are shown in Fig. 5.7.

TABLE 5.3 Specifications for Process Feed Stream (Foo et al., 2005)

Components	Flowrate (kg mol/h)	Condition
Ethylene (C_2H_6)	20	
i-Butane (*i*-C_4H_{10})	10	T: 93°C
Nitrogen (N_2)	0.1	P: 20 psia
n-Butane (*n*-C_4H_{10})	0.5	

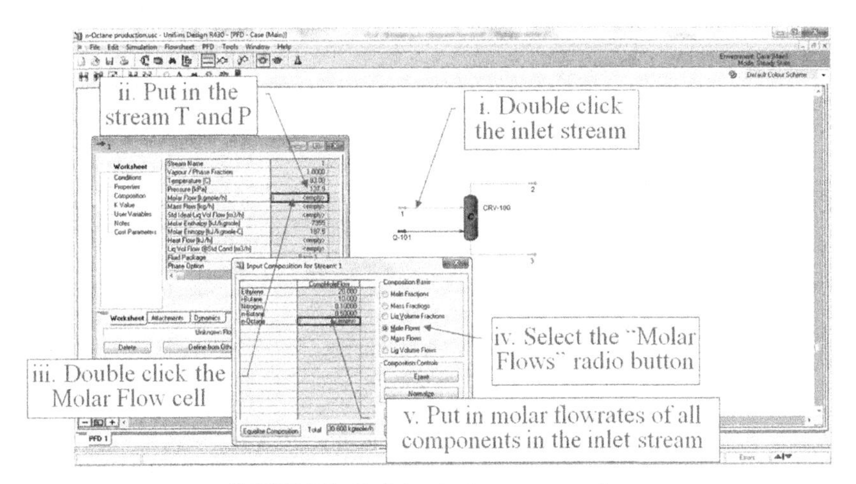

FIGURE 5.7 Defining feed stream properties.

TABLE 5.4 Specifications for Reactor

Equipment	Specifications
Reactor	Delta P: 5 psi
	Operational mode: isothermal

Next, we proceed to provide specifications for the reactor model (given in Table 5.4). To specify pressure drop (known as "Delta P" in UniSim Design terminology) and reaction set for UniSim Design, one would make use of the "Design" and "Reactions" tabs of the reactor model interface (see detailed steps in Fig. 5.8). The isothermal mode of the reactor is specified by setting the temperature for one of the outlet streams of the reactor (Stream 3 for this case), in the "Worksheet" tab (see Fig. 5.9). Once this is specified, the reactor model

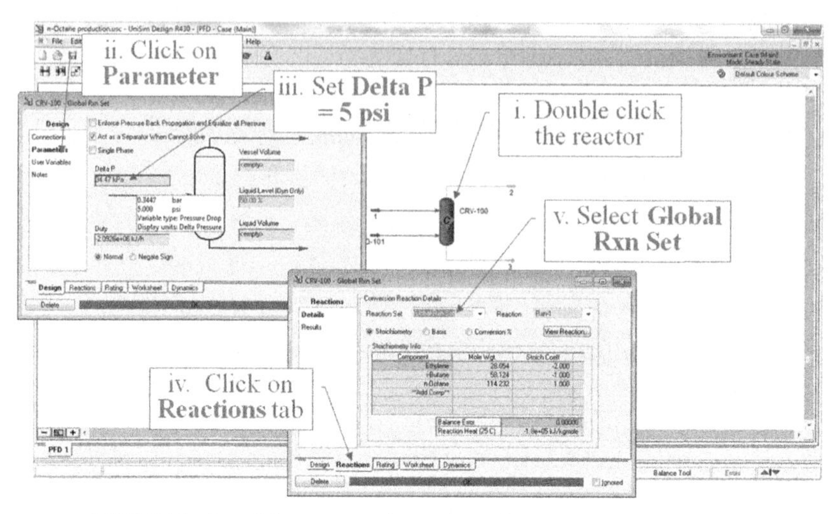

FIGURE 5.8 Specifying pressure drop and reaction set for conversion reactor.

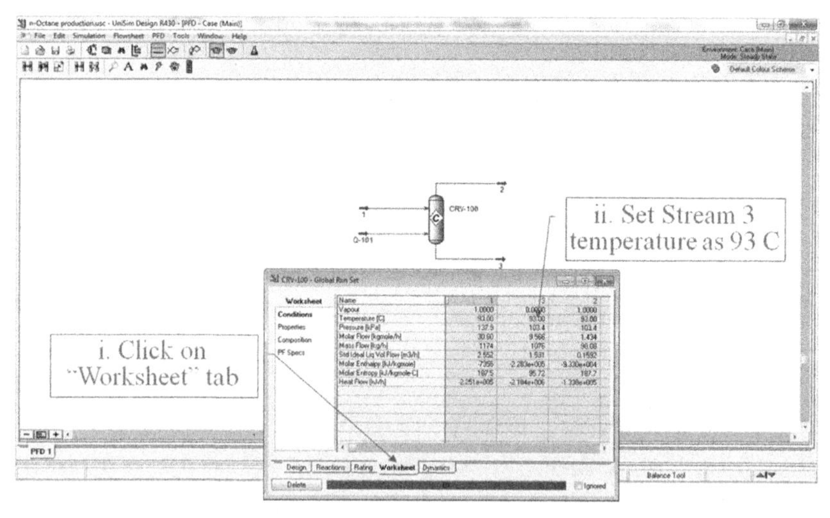

FIGURE 5.9 A converged Conversion Reactor model, after Stream 3 temperature is specified.

changes its color into black, indicating that the simulation model has been converged. Note that UniSim Design software is configured to perform simulation once the necessary data are sufficient. For instance, in Fig. 5.9, the flowsheet is converged once all necessary data for the Conversion Reactor model are complete (i.e., when Stream 3 temperature is specified).

Tips: Red icon and light blue streams indicate that the flowsheet has not been converged, while black icon and royal blue streams indicate a converged flowsheet.

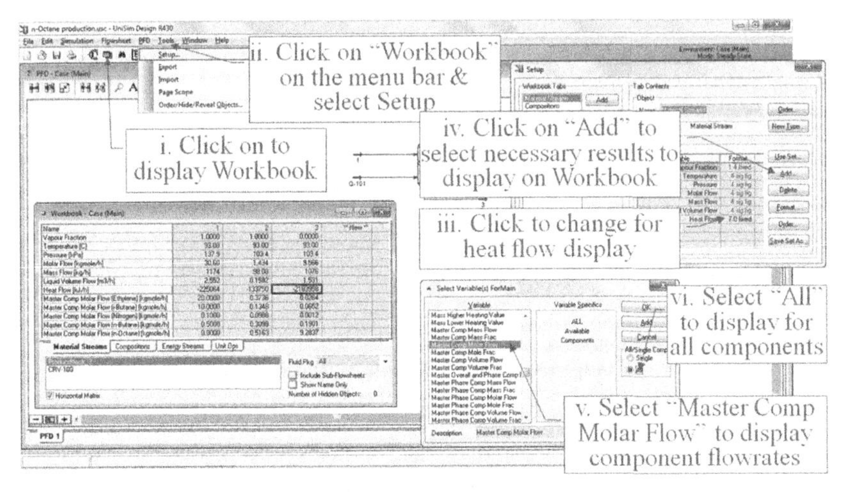

FIGURE 5.10 Displaying results with workbook and steps to display component molar flowrates.

We then move on to display the simulation results. A convenient way of displaying the simulation results in UniSim Design is via the *Workbook*. Fig. 5.10 shows the detailed steps in displaying molar flowrates of all components on the Workbook. One may also insert the *Workbook Table* within the PFD. This is illustrated with Fig. 5.11, where material and energy streams are displayed (one may also choose to display the stream compositions).

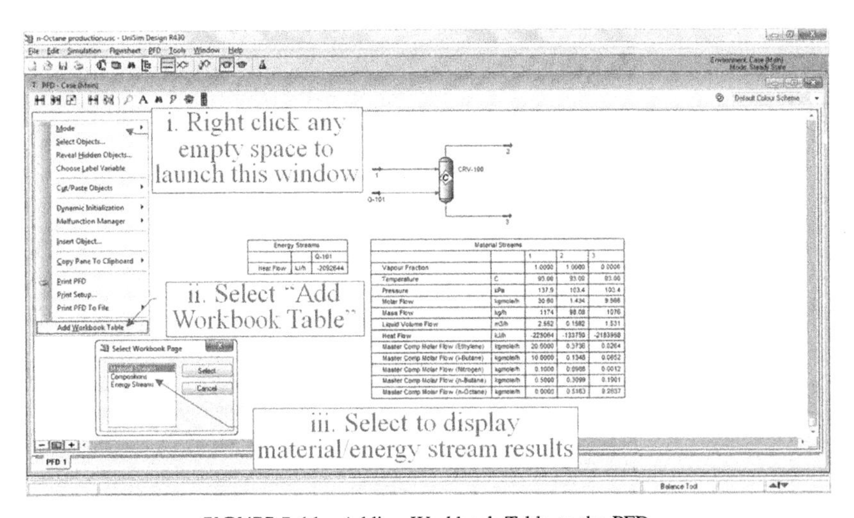

FIGURE 5.11 Adding Workbook Table to the PFD.

Tips: Most commercial simulation software requires users to perform simulation model solving manually.

The energy stream table in Fig. 5.11 indicates that the energy stream of the reactor (Q-101) has an enthalpy of -2.09×10^6 kJ/h, indicating that the conversion of *n*-octane is an exothermic reaction. In other words, cooling utility is needed for heat removal so to operate this reactor in isothermal condition.

5.4 STAGE 3: MODELING OF SEPARATION UNIT

In this stage, we shall model the only separation unit of the flowsheet, i.e., the distillation column. The simulation task is performed using the "shortcut distillation" model. Note that this distillation model is based on the Fenske–Underwood–Gilliland model, which is useful for preliminary flowsheet development. The parameters obtained from shortcut distillation model can be used as initial estimates in the rigorous distillation model, which performs stage-by-stage calculations[3]. To construct the topology of the flowsheet, one may refer to Fig. 5.12 for the alternative steps in connecting the material and energy streams. Fig. 5.12 also shows that the column is set to operate with partial condenser, where the column top stream exists in vapor form.

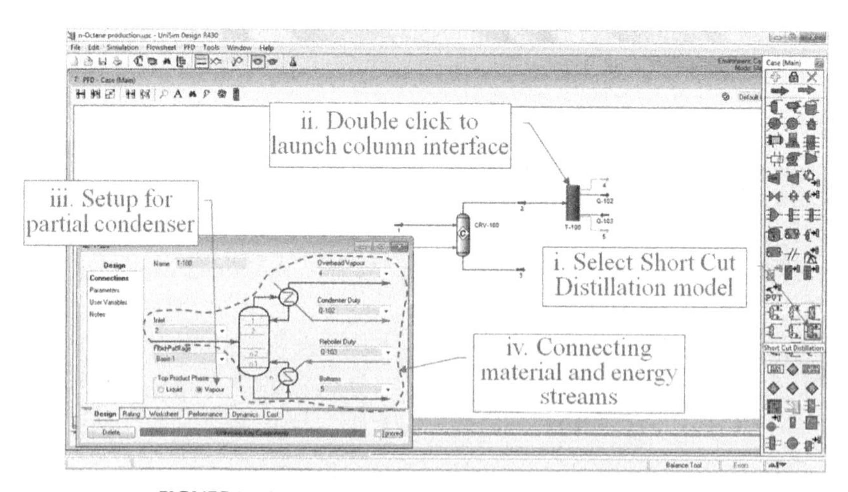

FIGURE 5.12 Adding shortcut distillation model to the PFD.

3. See Chapter 12 for an example on the use of shortcut distillation model for generating data for rigorous distillation model on VCM production.

TABLE 5.5 Specifications for Distillation Model

Equipment	Specifications
Distillation	Condenser type: partial condenser
	Light key in bottom stream: ethylene (0.0001 mole fraction)
	Heavy key in distillate stream: *n*-octane (0.0500 mole fraction)
	Top P: 15 psia
	Bottom P: 25 psia
	External reflux ratio: 1

Tips: You may use the following hot keys to display the important stream properties at the PFD: pressure (SHIFT P), temperature (SHIFT T), mass flowrates (SHIFT M), molar flowrates (SHIFT F). See other hot keys under the Help menu of the main screen.

Once the distillation is added to the flowsheet, with its streams connected, we move on to provide specifications for the distillation model (given in Table 5.5). Detailed steps to provide specifications for the distillation model are shown in Fig. 5.13. The latter also shows the converged distillation model once all specifications are provided. We next move on to display the simulation results using the Workbook Table (Fig. 5.13).

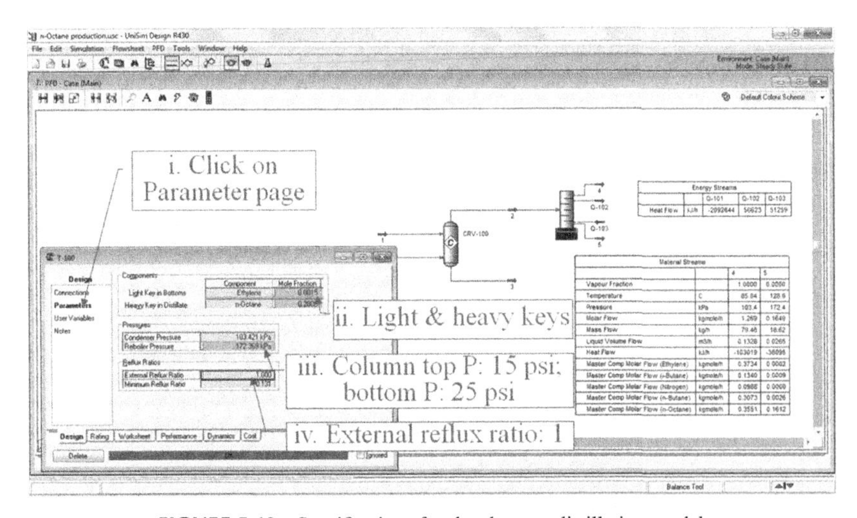

FIGURE 5.13 Specifications for the shortcut distillation model.

5.5 STAGE 4: MODELING OF RECYCLE SYSTEM

As discussed in Chapters 1 and 4, recycle simulation may be further classified as *material* and *energy recycle systems*. For the case of *n*-octane production, material recycle corresponds to the recycling of the unconverted raw material to the reactor (Fig. 5.13 indicates that the distillate stream contains some unconverted ethylene and *i*-butane), while the energy recycle stream corresponds to the heat recovery system. Example 1.1 details out the use of sequential modular approach to converge these recycle loops. Because both of these material and energy recycle systems are connected, it is easier to decouple them for the ease of convergence. This is done by replacing the process-to-process heat exchanger in the heat recovery system with equivalent heater and cooler, so that the cooling requirement of the material recycle stream is taken care by the cooler, while the heating of the fresh feed stream is provided by a heater. Once the material recycle system is converged, the cooler is replaced by the process-to-process heat exchanger to converge the energy recycle system. Doing this allows the flowsheet to be converged easily. This is demonstrated in the following subsections.

5.5.1 Material Recycle System

As described in Example 1.1, the material recycle system involves purge stream unit, compressor, and cooler (see Fig. 1.10). At the purge stream unit, 10% of the distillation top stream is purged, while the remaining is recycled. The compressor and a cooler are then used to adjust the pressure and temperature of the recycle stream to match those of the reactor. Specifications for these units are given in Table 5.6.

To model the purge unit, a stream splitting model (called the "Tee" in UniSim Design) is utilized. Detailed steps to connect the Tee model (to distillation top stream) and to provide its model specifications are shown in Fig. 5.14. As mentioned, 10% of the distillation top stream is purged. Hence, the flow ratio for recycle stream (corresponds to Stream 6 in Fig. 5.14) is set to 0.9.

Because the recycle stream has a pressure of 15 psia (verify this from your flowsheet converged earlier), which is lower than the operating pressure of the

TABLE 5.6 Specifications for Units in the Material Recycle System

Equipment	Specifications
Purge unit	Flow ratio for recycle stream: 0.9
Compressor	Outlet P: 22 psia
Cooler	Outlet T: 93°C
	Delta P: 2 psi

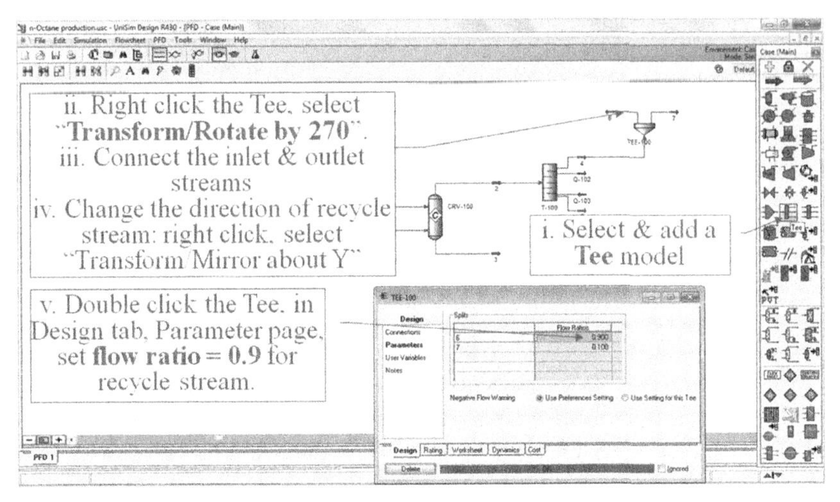

FIGURE 5.14 Steps to connect the purge stream unit and its specifications.

reactor, a compressor is added to raise its pressure to 22 psia. Detailed steps to connect the compressor model and to provide its specifications are shown in Fig. 5.15.

Outlet stream of the compressor has a temperature of 98.9°C (see Fig. 5.15), which is higher than the reactor operating temperature. A cooler unit is next added to reduce the temperature to 93°C. Fig. 5.16 shows the detailed steps to connect the cooler model and to provide its specifications.

With both pressure and temperature of the recycle stream adjusted to match those of the reactor, the recycle stream can now be connected to the reactor. To converge this material recycle stream, we can make use of a useful model in

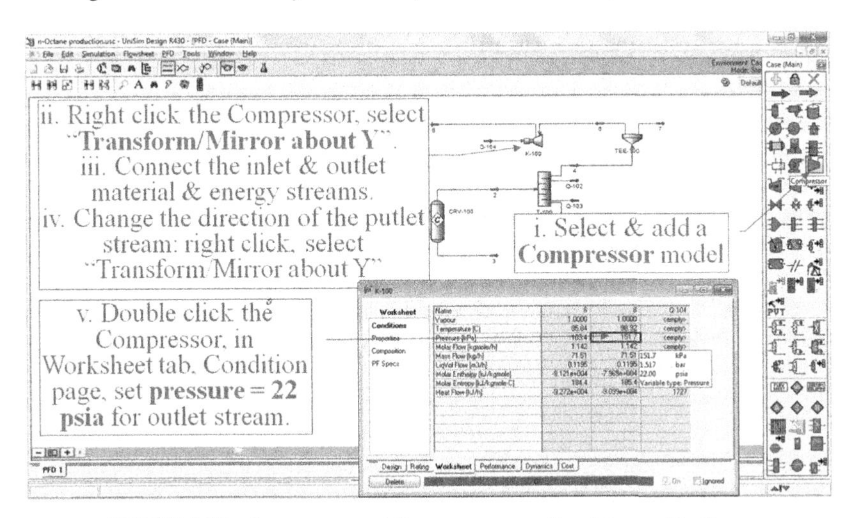

FIGURE 5.15 Steps to connect the compressor unit and its specifications.

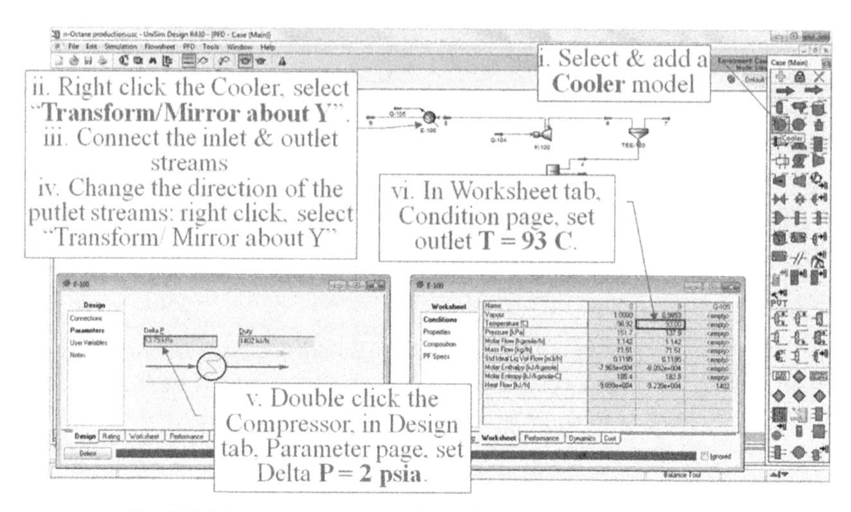

FIGURE 5.16 Steps to connect the cooler unit and its specifications.

UniSim Design—the "Recycle" unit. The latter facilitates the convergence of a recycle stream following the "tear stream" concept[4]. Detailed steps in configuring the "Recycle" unit are shown in Fig. 5.17. Note that the Recycle unit shows a yellow outline when the recycle stream is first connected to the reactor. This means that some parameters are not converged after 20 rounds of iteration (default setting in UniSim Design). Hence more iterations are needed to ensure all parameters are converged completely[5]; this is done by pressing the "Continue" button in its Connections page. The simulation results are also displayed in the Workbook Table in Fig. 5.17.

5.5.2 Energy Recycle System

As discussed in Example 1.1, the process-to-process heat exchanger of the *n*-octane case was replaced by a pair of heater and cooler to decouple the material and energy recycle systems, to facilitate flowsheet converge. After the material recycle system is converged, we next proceed to converge the energy recycle stream. The convergence of the energy recycle stream is to be done using the "tear stream" concept, i.e., without the use of the Recycle unit. Specifications for heat exchanger and heater in the energy recycle system are given in Table 5.7.

In earlier stage, it has been assumed that the fresh feed stream is available at 93°C (see Table 5.3). This assumption is now relaxed. A heater is added to raise the temperature of the fresh feed stream from 30°C. Detailed steps to do so are given in Fig. 5.18. The simulated results indicate that the heater requires

4. See Chapter 4 for detailed discussion on the convergence of recycle stream simulation.
5. See Section 4.2 for tips in handling recycle streams.

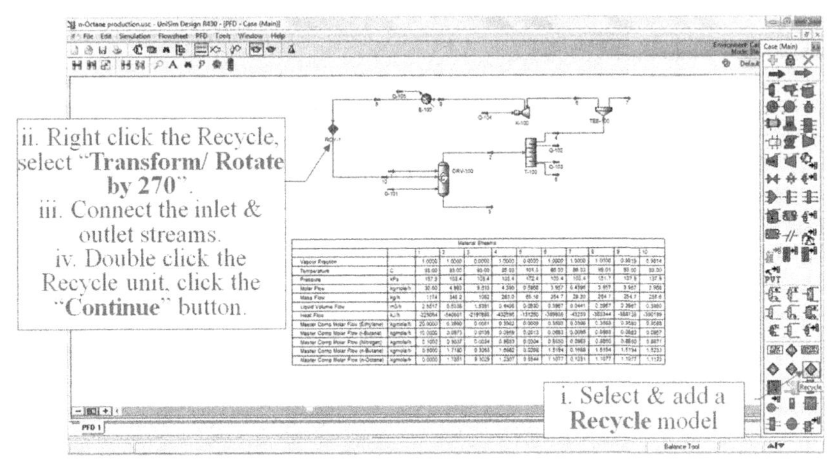

FIGURE 5.17 Steps to converge the material recycle stream with recycle unit.

TABLE 5.7 Specifications for Units in the Energy Recycle System

Equipment	Specifications
Heat exchanger	Delta P: 2 psi (tube side)
	Delta P: 2 psi (shell side)
Heater	Outlet T: 93°C
	Delta P: 2 psi

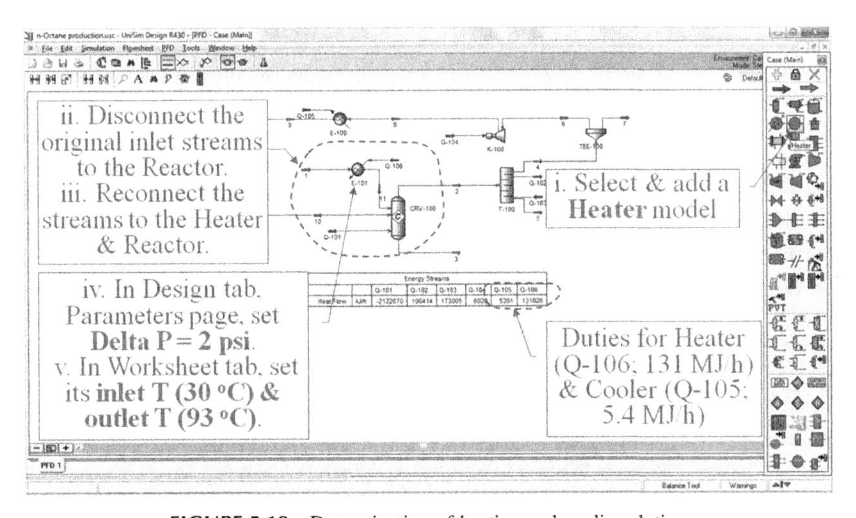

FIGURE 5.18 Determination of heating and cooling duties.

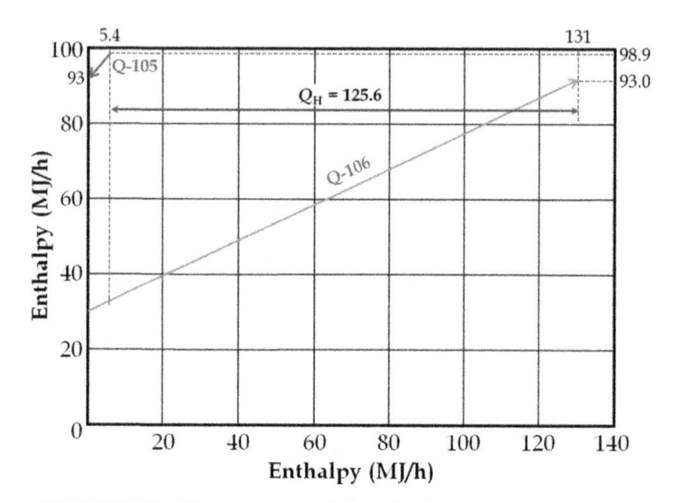

FIGURE 5.19 Temperature—enthalpy plot for heat recovery system.

a total heating duty of 131 MJ/h (indicated by energy stream Q-106), while 5.4 MJ/h of energy needs to be removed by the cooler (indicated by energy stream Q-105).

Fig. 5.19 shows a *temperature—enthalpy plot*[6] for the streams undergoing heating and cooling in the Heater and Cooler. As shown, the temperature profiles of the Cooler (Q-105—the material recycle stream) are higher than those of the Heater (Q-106—fresh feed). Hence, energy released from the heater can be completely recovered to the cold stream. In other words, part of the heating requirement of the Heater (5.4 MJ/h) is to be fulfilled by the cooling duty of the Cooler, through a process-to-process heat exchanger. The remaining heating duty of the cold stream ($Q_H = 125.6$ MJ/h) is to be supplied by the Heater, as shown in Fig. 5.19.

With the heating and cooling requirements identified, the process-to-process heat exchanger will then be included in the PFD. The "Heat Exchanger" model is utilized and added to replace the Cooler model. Because the Recycle model is not utilized in this case, we shall create a *tear stream* for the energy recycle system, for the stream connecting the Heat Exchanger and Heater. Detailed steps for doing so are shown in Fig. 5.20. Note that the pressure of the fresh feed stream has been revised to account for pressure drop across the Heat Exchanger and Heater. Note also that the flowsheet is unconverged at this stage.

We next proceed to provide the missing parameters to converge the flowsheet. These include pressure drop (Delta P) for the Heat Exchanger (see Fig. 5.21), as well as the estimated values for the tear stream. Because this stream is essentially the same fresh feed stream that enters the Heat

6. See Example 1.1 for more details.

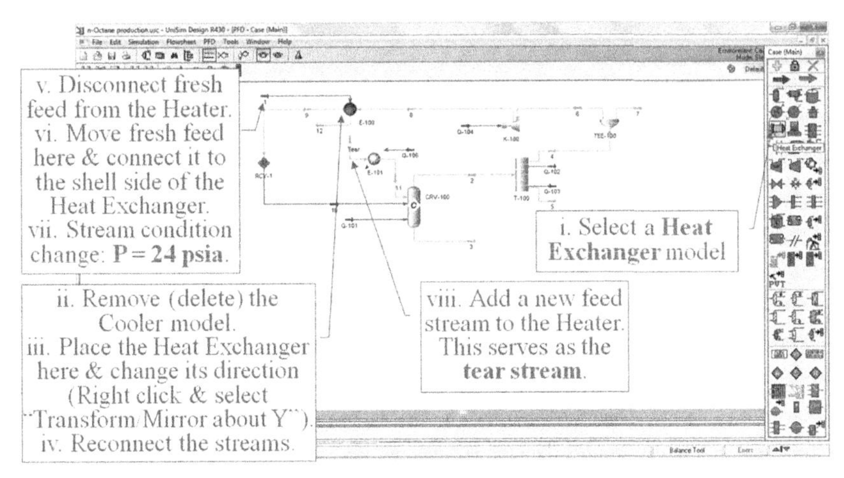

FIGURE 5.20 Introducing a tear stream for energy recycle system.

Exchanger at the shell side, their parameters are very similar (except that with different temperature). A convenient way to define the tear stream condition is to refer that with a source stream[7], with detailed procedure shown in Fig. 5.21. Once the tear stream is specified, the "open loop" flowsheet is converged.

In the final step, the tear stream is removed and the outlet stream from the Heat Exchanger is connected to the Heater. A converged "close loop"

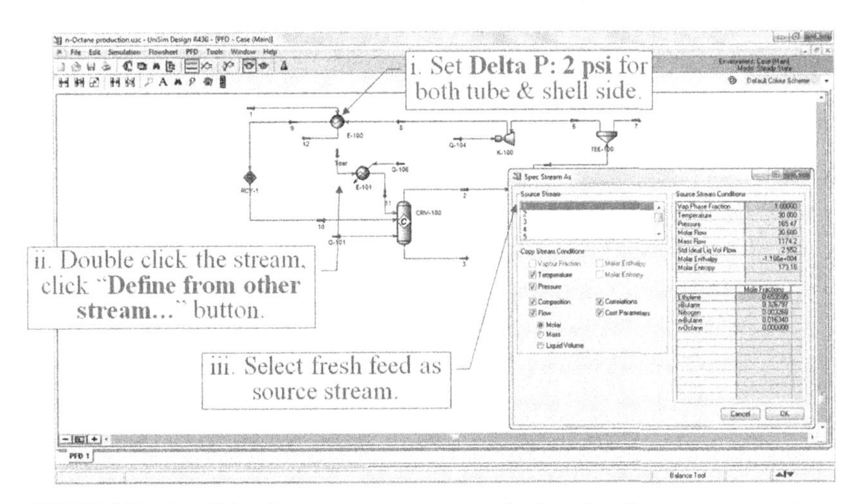

FIGURE 5.21 Specifying the tear stream to converge the "open loop" energy recycle system.

7. See Section 4.2 for tips in handling recycle streams.

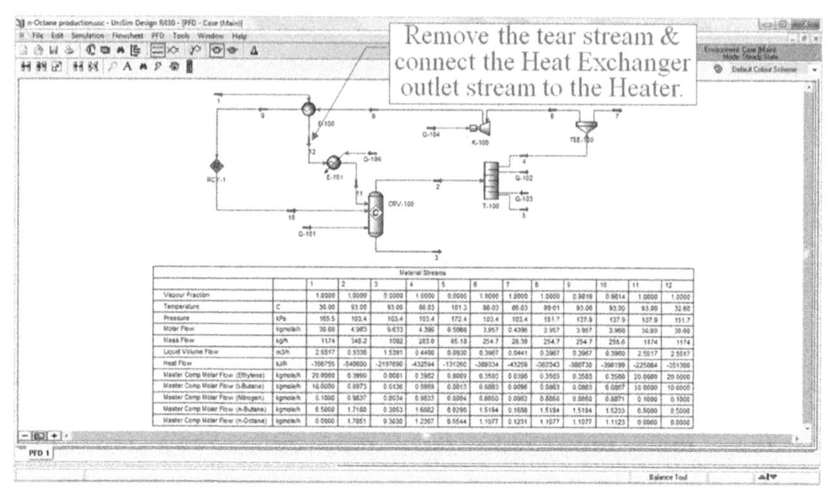

FIGURE 5.22 A converged flowsheet for n-Octane production process.

flowsheet is resulted and is shown in Fig. 5.22. The material stream conditions are shown using the Workbook Table in Fig. 5.22.

5.6 CONCLUSIONS

The n-Octane production in Example 1.1 is simulated using UniSim Design, using the concept of simultaneous modular approach, guided by Onion model. Material and energy recycle systems are handled efficiently when they are decoupled. Note that unit models used in this example are meant for preliminary design. Rigorous models should be used for more detailed engineering design.

EXERCISES

1. Fig. E1 shows the "synthesis loop" for methanol (CH_3OH), in which a mixture of carbon dioxide (CO_2) and hydrogen (H_2) is reacted to form methanol product at high pressure. The reaction stoichiometry is given as follows:

$$CO_2 + 3H_2 \leftrightarrows CH_3OH + H_2O$$

The feed specification is given in Table E1. As shown, the synthesis gas consists of mainly hydrogen and carbon dioxide, but with traces of inert gases.

Additional specifications for the process are given as follows:

- Thermodynamic model: SRK
- Pressure drops in all units are neglected
- The adiabatic converter can be approximated as a Conversion Reactor, operated adiabatically. Reactor conversion is set to 30%, with CO_2 being the limiting reactant.

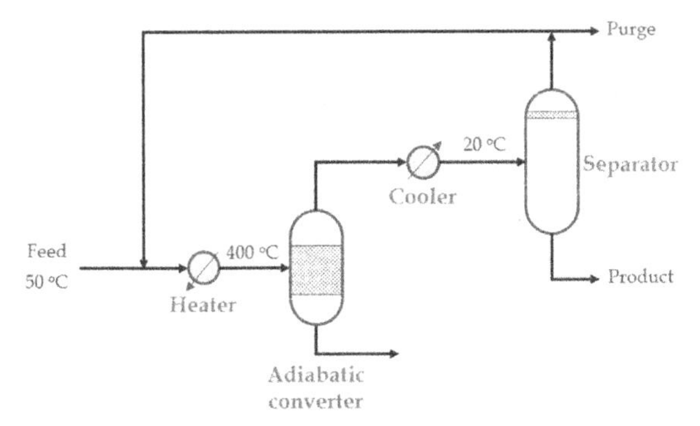

FIGURE E1 Methanol synthesis loop (Seider et al., 2009).

TABLE E1 Feed Stream Specification

Parameter	Value
Temperature (°C)	50
Pressure (MPa)	5, 7.5, 10
Flowrate (kgmol/h)	1000
Composition (mol %)	
Hydrogen	74.85
Carbon dioxide	24.95
Methane (inert)	0.1
Argon (inert)	0.1

- The Separator may be approximated by a flash unit.
- Flow ratio of the purge stream is set to 0.02.

Determine the following:

a. Molar flowrate and mole fraction of the product and purge streams.
b. The effect of purge ratio (range: 0.02–0.08) on the duties of Heater and Cooler.
c. In normal practice, an equilibrium reactor model is used to model a reversible reaction like this case. Justify why in the above case, the reaction can be approximated with an irreversible reaction, modeled by a conversion reactor model?

2. Fig. E2 shows an ammonia production process, where a mixture of nitrogen (N_2) and hydrogen (H_2) is reacted to form the ammonia (NH_3) product at high pressure, with the following stoichiometry:

$$N_2 + 3H_2 \leftrightarrows 2NH_3$$

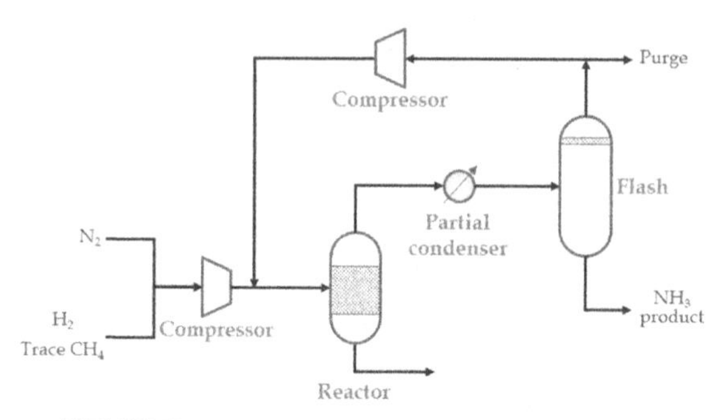

FIGURE E2 Ammonia production (Seider et al., 2009).

The feed gas to the process is largely composed of N_2 and H_2, but with traces of inert CH_4, as shown in Table E2.

Additional specifications for the process are given as follows:

- Thermodynamic model: Chao–Seader
- Both compressors are used to adjust stream pressure according to the reactor operating pressure.
- Conversion Reactor model is used, to be operated isothermally at 500°C and 400 bar. Nitrogen is set as the limiting reactant, with a conversion of 30%. Pressure drop is ignored for the reactor.
- Partial Condenser is approximated by a Cooler, to cool the reactor effluent to −33°C; ignore its pressure drop.
- Model the Flash vessel with a Separator, with pressure drop (Delta P) of 250 bar.
- Flow ratio of the purge stream is set to 0.04.

Determine the following:

a. Molar flowrate and mole fraction of the product and purge streams.

TABLE E2 Feed Stream Specification

Parameter	N_2 Stream	H_2 Stream
Temperature (°C)	25	25
Pressure (bar)	1	1
Flowrate (kgmol/h)		
Nitrogen	24.0	
Hydrogen		74.3
Methane (inert)		1.1

b. The effect of purge ratio (range: 0.02–0.08) on the duties of the compressor.

c. In normal practice, an equilibrium reactor model is used to model a reversible reaction like this case. Justify why in the above case, the reaction can be approximated with an irreversible reaction, modeled by a conversion reactor model?

REFERENCES

Foo, D.C.Y., Manan, Z.A., Selvan, M., McGuire, M.L., October 2005. Integrate process simulation and process synthesis. Chemical Engineering Progress 101 (10), 25–29.

Honeywell, 2017. www.honeywellprocess.com

Seider, W.D., Seader, J.D., Lewin, D.R., 2009. Product & Process Design Principles: Synthesis, Analysis and Evaluation. John Wiley & Sons.

Chapter 6

Modeling of a Dew Point Control Unit With UniSim Design

Rafil Elyas, Zhi Hong Li
East One-Zero-One Sdn Bhd, Shah Alam, Malaysia

6.1 INTRODUCTION

In an upstream hydrocarbon production facility, the produced gas may be exported to a few destinations. These include sales gas pipeline for transportation to gas plants and/or consumers, used as fuel gas to run the facility's power-generating turbines, injected into the reservoir for enhanced oil recovery, injected into well tubing as gas lift to manage the flow regime, and prevent liquid slugging. Before produced gas is exported as sales gas or used as fuel gas, it is generally necessary to first ensure dew point temperature of the gas is below the range of operating temperatures, to prevent hydrocarbon liquids from condensing in the pipeline (which may result in liquid slugs in the pipeline) or fuel injection nozzles in the turbines (liquid formation will result in inefficient combustion). In addition to that, it is also desirable to recover the liquids because they typically consist of propane and heavier components, which have higher economic value.

There are two main methods for reducing the dew point of a gas, i.e., expansion or mechanical refrigeration. In expansion, the gas is expanded by throttling it across a valve (an isenthalpic process) where the Joule–Thompson effect would reduce its temperature, or in some cases by using an expander, in which the temperature reduction of the gas will be achieved by expansion and the production of work. On the other hand, mechanical refrigeration uses refrigeration cycles or "heat pumps" to remove the heat from the gas. The design and configuration of a refrigeration cycle will be covered in this chapter. The decision to select expansion or mechanical refrigeration depends on the availability of pressure and the magnitude of temperature reduction required. In this chapter mechanical refrigeration shall be used as the method for chilling the gas.

Chemical Engineering Process Simulation. http://dx.doi.org/10.1016/B978-0-12-803782-9.00006-6

6.2 PRELIMINARY ANALYSIS

To determine if dew point control is required for a gas stream, it is first necessary to identify the following:

1. The dew point of the gas
2. The range of operating conditions, i.e., temperatures and pressures the gas will encounter

The following example is used to illustrate the preliminary analysis required to select and design a dew point control unit (DPCU). In this example, it was determined that the gas is to be exported as sales gas in a pipeline. The typical inlet and outlet operating pressures for the pipeline are 5500 kPa and 4500 kPa, respectively. The inlet temperature of the pipeline is around 15°C, while the minimum ambient temperature of the pipeline's surroundings is −7°C. Based on this, it is desirable to ensure that the dew point of the gas is well below the ambient temperature of −7°C. A minimum approach temperature of 5°C yields a desired gas dew point of −12°C.

Table 6.1 shows the composition of two hydrocarbon streams that had been dehydrated and mixed. For simplicity, all water content has been excluded in this example. Note, however, that in reality, water content can range between 2 and 5 lb of water per million standard cubic feet (MM scf) of gas.

First, it is desired to determine if the mixed hydrocarbon can be exported as it is. Would it be necessary to process this gas and to reduce its dew point? The

TABLE 6.1 Sample Rich Gas Compositions and Operating Conditions

	Feed 1	Feed 2	Mixed Feed
T (°C)	15.5	15.5	15.5
P (kPa)	4137	4137	4137
Mole flow (kg mol/h)	300	200	500
Component Mole%			
Nitrogen	1.80%	1.00%	1.40%
Carbon dioxide	0.00%	1.00%	0.50%
Methane	62.40%	60.00%	61.20%
Ethane	16.70%	20.00%	18.35%
Propane	11.40%	10.00%	10.70%
Isobutane	4.30%	4.00%	4.15%
n-Butane	3.40%	4.00%	3.70%

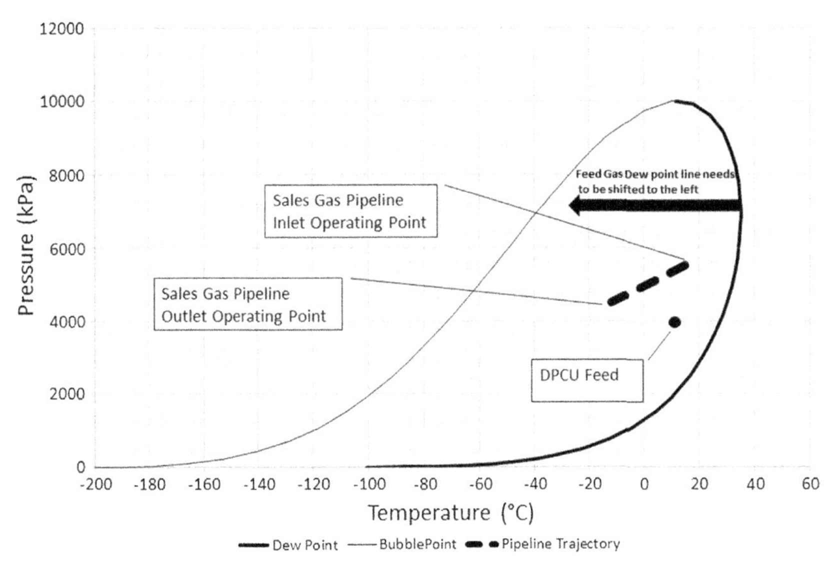

FIGURE 6.1 Phase envelope and pipeline operating trajectory (indicated by the *dashed line*). *DPCU*, dew point control unit.

most efficient way to determine this would be by generating the phase envelope of the hydrocarbon and overlaying the pipeline operating conditions on this diagram. In this example, only the pipeline inlet and outlet pressures were provided, while no information on the length, hydraulic behavior, or heat transfer characteristics of the pipeline was found. For illustrative purposes, the pressure and temperature profile across the pipeline were assumed to be linear. The phase envelope for this case is shown in Fig. 6.1.

From Fig. 6.1, it is clear that the hydrocarbon stream would stay in the two-phase region in the pipeline, as its trajectory lies within the two-phase region of the phase envelope. Based on this, it is necessary to shift the dew point line to the left and to ensure that the operating points stay at the superheated gas region. Hence, a processing strategy needs to be determined to reduce the dew point of this hydrocarbon stream.

6.3 CONCEPTUAL DESIGN FOR DEW POINT CONTROL UNIT

In Chapter 3, it was shown that dew point of a gas is determined by the heaviest component.[1] Hence, to reduce the dew point of a gas, it is necessary to remove the heavier components. It is important to first determine the state of hydrocarbon feed at its operating point. From the compositions presented in

1. See discussion in Fig. 3.11.

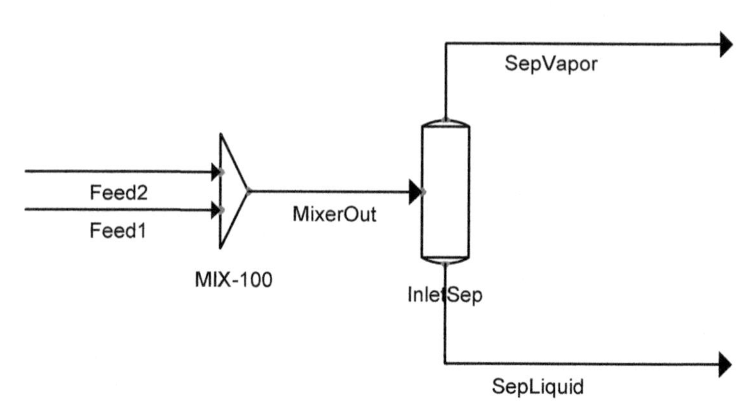

FIGURE 6.2 Inlet separation.

Table 6.1, it can be seen that the feed lies in the two-phase region of the phase envelope. Based on a flash unit simulation with UniSim Design, the feed stream contains 90 mol % of vapor, which is in equilibrium with its liquid (10 mol %).

Feed 1 and Feed 2 are combined using an adiabatic mixer object. The MixerOut stream pressure is set by the mixer, in this example, by taking the lowest inlet pressure. The mixer object then performs a material balance and a pressure–enthalpy flash calculation to determine the temperature of the MixerOut stream. Next, we proceed to separate the vapor and liquid in a knockout drum operated adiabatically (see Fig. 6.2), i.e., with its operating temperature and pressure set to follow those of the feed stream (15.5°C and 4137 kPa).

Once the feed stream (MixerOut) is separated, the heavy hydrocarbons are removed from those vapor components. This resulted with leaner hydrocarbon vapor, with a dew point of 15.5°C. The phase envelopes for all feed and gas outlet streams have been overlaid in Fig. 6.3.

From Fig. 6.3, it is clear that the pipeline operating pressure and temperature trajectory is still within the two-phase region. Thus, it is necessary to further remove heavy components from the hydrocarbon vapor stream, to reduce the dew point of the gas stream. There are two main methods available for doing so, i.e., expansion or mechanical refrigeration. The decision to select expansion or mechanical refrigeration depends on the availability of pressure and the magnitude of temperature reduction required. In this example, a mechanical refrigeration strategy has been selected for this DPCU to minimize any compression requirements. This is achieved through a chiller, where propane is used as coolant for temperature reduction. In this example the pressure drop for the chiller was set to 10 psi.

The conceptual design for this system is illustrated in Fig. 6.4. As shown, the vapor stream is sent to a gas–gas exchanger for heat integration, in which

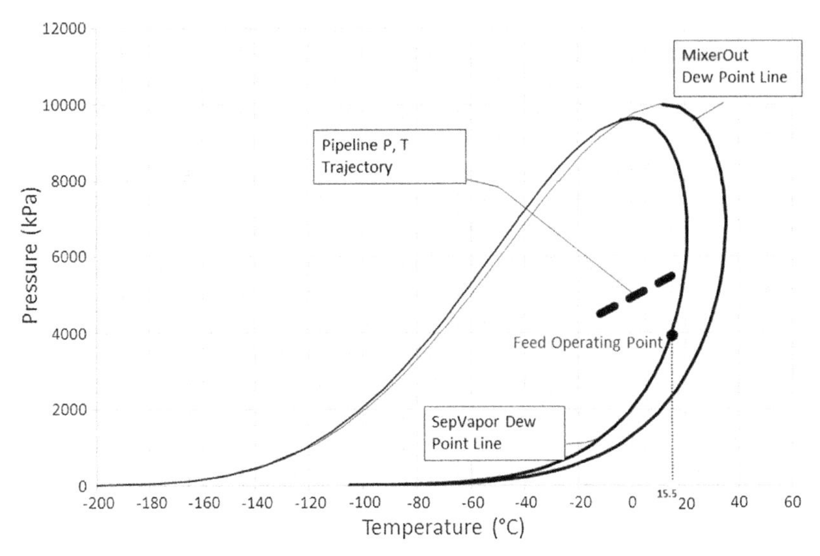

FIGURE 6.3 Feed and gas outlet phase envelopes.

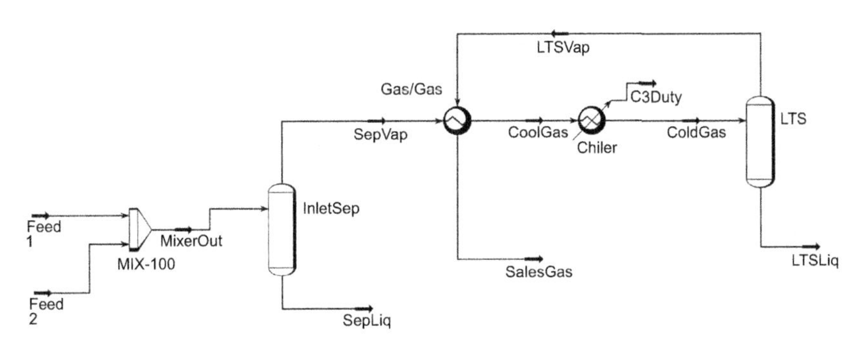

FIGURE 6.4 Conceptual design of the dew point control unit. *LTS,* low-temperature separator.

it precools the gas before entering the chiller; doing this reduces the required chiller duty. In this example, the pressure drop for the hot and cold sides of this exchanger was assumed to be around 10 psi each. The gas–gas exchanger is operated with a minimum temperature approach of 10°C. The heat integration system is considered a *heat recycle* stream, which may be converged following the tear-stream strategy.[2] Once the pressure drops are defined for the heat exchangers and knockout drums (assumed to have zero pressure drop), the pressure of all the streams is calculated. The chiller temperature is set to −15°C, to yield a sales gas dew point of −12°C at 5500 kPa. Simulation

2. See detailed discussion on converging recycle stream simulation in Chapter 4, or the octane production example in Chapter 5.

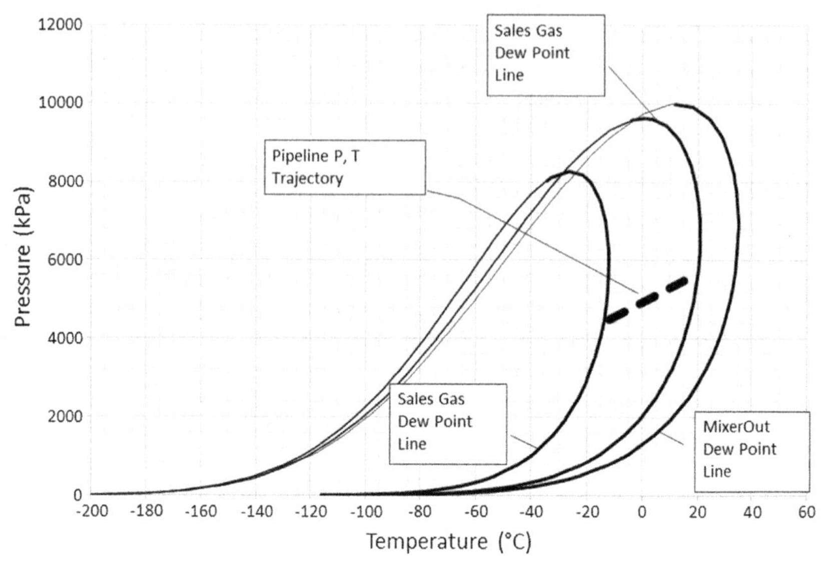

FIGURE 6.5 Original feed (MixerOut), inlet separator vapor (SepVap), and sales gas phase envelopes.

results from UniSim Design reported that the corresponding chiller duty was 340 kW. The condensed hydrocarbon is then removed from the chiller outlet (ColdGas) stream using another knockout drum, i.e., low-temperature separator (LTS).

The phase envelopes for the inlet feed (MixerOut), inlet separator vapor (SepVap), and the final Sales Gas streams are overlaid in Fig. 6.5. It can be seen on this phase diagram that the operating pressures and temperatures of the pipeline now lie in the superheated gas region of the sales gas, thus eliminating the risk of having liquid droplets and potential liquid slugging issues in the pipeline.

6.4 CONCEPTUAL DESIGN OF PROPANE REFRIGERATION SYSTEM

In previous section, mechanical refrigeration is used to reduce the dew point of the gas stream. As discussed, it was determined that the chiller has to fulfill the following duties, that is:

1. Operate at a temperature of −15°C on the process side and
2. Deliver a duty of about 340 kW at the design conditions.

In this section, conceptual design of the refrigeration cycle shall be addressed. For brevity, a simplified refrigeration cycle is demonstrated, as shown in Fig. 6.6.

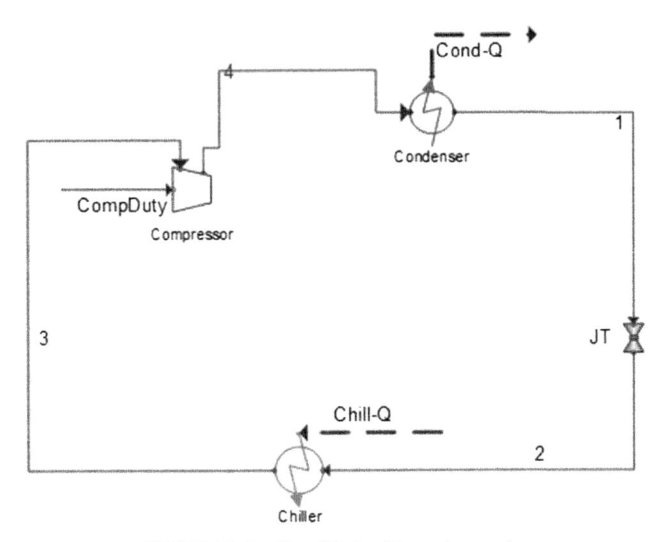

FIGURE 6.6 Simplified refrigeration cycle.

6.4.1 Basic Understanding of a Refrigeration Cycle

The refrigeration cycle functions as a "heat pump," where heat is extracted from the process stream in the chiller at a low temperature by a refrigerant and rejected a higher temperature in the condenser. For most upstream DPCU chillers, propane is utilized as refrigerant because of its availability and low cost. It is hence used in this example for demonstration. Circulation of the refrigerant is maintained by a compressor. The main unit operations in the refrigeration cycle are summarized as follows:

1. Condenser—In this case, either an air- or water-cooled condenser may be used. If the ambient temperature is assumed to be approximately 30–40°C, then it can be expected that the refrigerant to exit the condenser (Stream 1 in Fig. 6.6) is a saturated liquid at around 50°C, based on the minimum approach temperature of 10°C based on the maximum ambient temperature.
2. Joule–Thompson (J-T) valve—The refrigerant (Stream 2) is let down across a J-T valve and cools as it expands. Because of its expansion, the refrigerant is now a two-phase fluid.
3. Chiller—The refrigerant accumulates in the chiller, which may be a kettle-type chiller, and absorbs heat from the process stream. Note that the process stream needs to be cooled to around −15°C. Assuming a minimum approach temperature of 5°C, the refrigerant needs to stay as low as −20°C. This means that the pressure drop across the J-T valve must yield this temperature. Once the liquid refrigerant absorbs sufficient heat, it vaporizes and leaves the chiller as a saturated vapor (Stream 3).

4. Compressor—This vapor refrigerant from the chiller is compressed and condensed, leaving as Stream 4.

The following operating parameters are to be determined for this system:

1. How much propane will be required?
2. What is the duty of the compressor, assuming 100% compression efficiency?
3. How much heat must be removed by the condenser?

6.4.2 Degrees of Freedom Analysis

A degrees of freedom analysis is first performed on this process. Because all streams in the refrigeration cycle consist of pure propane, each stream has two thermal degrees of freedom, or independent intensive variables. Hence, any two of the following parameters may be defined to solve the whole system:

1. Vapor fraction (V_f)
2. Temperature
3. Pressure
4. Molar Enthalpy
5. Molar Entropy

There are four streams in this analysis; hence, there are eight degrees of freedom. These degrees of freedom may be defined by user or determined by a unit operation (which may take the form of a specification or equation). These are summarized in Table 6.2 that follows.

6.4.3 Design of Refrigeration Cycle Using a Mollier Diagram

The thermodynamic processes, variables, and equations are best visualized on a Mollier diagram (Fig. 6.8). The latter provides a graphical relationship between the following thermodynamic state variables:

1. Phase
2. Pressure
3. Enthalpy
4. Entropy
5. Temperature

The refrigeration process is plotted on the Mollier diagram in Fig. 6.7. The information gained by performing the hand calculation using the Mollier diagram is highly recommended as a tool for getting an order of magnitude estimate of the design parameters. Next, we make use of the Mollier diagram to describe the propane refrigeration cycle.

TABLE 6.2 Degrees of Freedom Analysis for the Refrigeration Cycle

Streams	User Specifications	Remarks
1	2	Saturated liquid at 50°C. Hence, vapor fraction (V_f) = 0, temperature (T) = 50°C.
2	0	
3	2	Saturated vapor stream exiting chiller at −20°C. Hence, $V_f = 1$, T = −20°C.
4	0	
Unit Operations	**Equations**	**Remarks**
J-T Valve	1	Isenthalpic device. No heat transfer or work transacted. It introduces a heat balance equation. Molar enthalpy of Stream 2 is equal to that in Stream 1 M.
Chiller	1	The chiller provides an energy balance.
Compressor	2	A polytropic device, where work is performed on the refrigerant. Hence, two equations are needed, i.e., to determine the amount of reversible work, and to determine the amount of "lost" work (based on compressor performance as determined by its efficiency).
Condenser	1	The condenser provides an energy balance.

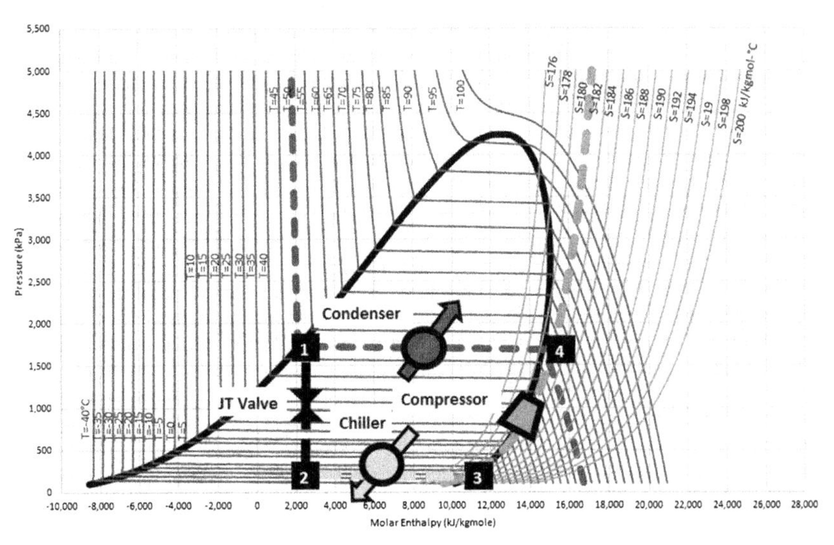

FIGURE 6.7 Refrigeration process illustrated on the Mollier diagram (East101, 2017).

6.4.3.1 Starting Point—Condenser Discharge

Stream 1 represents a saturated propane stream exiting the condenser at 50°C. This is represented on the Mollier diagram by selecting the intersection of the bubble point line and the 50°C isotherm. These points yield the pressure of the stream, around 1700 kPa and its molar enthalpy, around 2000 kJ/kg mol.

6.4.3.2 J-T Valve—Isenthalpic Process

The fluid then passes through the J-T valve. In an ideal valve, no work is done on the fluid and there is no heat transfer into or out of the fluid. Hence, the saturated propane liquid undergoes an isenthalpic (no change in enthalpy) expansion (pressure is reduced) as it passes through the J-T valve, and a line may be drawn going in the direction of decreasing pressure along the corresponding 2000 kJ/kg mol isenthalpic line. The end point of that line will describe the state of the fluid in Stream 2. The temperature in Stream 2 is set at −20°C (assuming a 5°C approach to the desired DPCU process fluid temperature of about −15°C).

The pressure at this state, which is the point of intersection between the line of constant enthalpy and the −20°C isotherm, is around 250 kPa.

In the simulator, the valve object will perform the following computation:

1. Complete the material balance, no accumulation, all material that enters must exit.
2. Complete the energy balance, the outlet and inlet enthalpies to be equal.

6.4.3.3 Chiller—Transfer of Latent Heat

As the propane (Stream 2) passes though the chiller, assuming negligible pressure drop, heat is transferred from the process fluid to the propane at constant pressure, boiling the liquid propane to yield saturated propane vapor at about −20°C in Stream 3. Hence, the amount of propane required by the chiller may be estimated as follows:

$$M = \frac{Q_{chiller}}{H_3 - H_2} \tag{6.1}$$

where M is the mass flowrate of propane required, $Q_{chiller}$ is the chiller duty, H_2 is the molar enthalpy of propane in Stream 2, H_3 is the molar enthalpy of propane in Stream 3, and $H_3 - H_2$ represents the latent heat of vaporization of pure propane.

Recall that the desired chiller duty was about 340 kW (1,224,000 kJ/h).

$H_3 - H_2$ is approximately 9300 kJ/kg mol.

This yields a required propane flowrate of around 132 or 5808 kg/h.

6.4.3.4 Compressor—Isentropic Process

The saturated propane vapor is then compressed to some pressure and temperature at Stream 4. For this example, it will be assumed that there will be negligible pressure drop across the condenser and Stream 4 will be at the same pressure as Stream 1. For the purpose of this example the compression process is assumed to be ideal, reversible, and is 100% efficient. Hence a line may be drawn going in the direction of increasing pressure along the corresponding isentrope (S = 180 kJ/kg K) to represent this process.

The end point of that line will describe the state of the fluid in Stream 4. The duty required to deliver the same pressure as the ideal compressor is greater, as some energy is lost to irreversible processes, such as friction. Hence, the resulting temperature in Stream 4 is dependent on how much energy is lost. The ratio of duties required by an ideal compressor and real compressor to achieve the same pressure is known as the adiabatic efficiency. For the purpose of this exercise, we may assume an adiabatic efficiency of 100%, which yields a temperature of around 60°C.

The compressor duty required may be calculated as follows:

$$Q_{compressor} = M * (H_4 - H_3) \tag{6.2}$$

where $Q_{compressor}$ is the compressor duty, M is the mass flowrate of propane compressed, and H_4 is the specific enthalpy of propane in Stream 4.

This yields a compressor duty of around 160 kW or 576,000 kJ/h.

6.4.3.5 Condenser—Heat Rejection

From the Mollier diagram, the total amount of energy transferred to the refrigerant is the sum of the latent heat transferred in the chiller and the compression work, $H_4 - H_2$.

Hence, the total amount of heat transferred in the condenser is 500 kW (= 160 + 340 kW or 1,800,000 kJ/h).

6.4.4 Design of Refrigeration System With UniSim Design

Even though the Mollier diagram will help the designer in estimating some useful design parameters for the system, it may not be sufficiently accurate for an engineering design computation, as the Mollier diagram is bound with the following assumptions:

1. Negligible pressure drop across chiller
2. Negligible pressure drop across condenser
3. Compressor operating at 100% efficiency

In reality, there are measurable impacts because of the pressure drop across the equipment and it is physically impossible to operate compressors at 100% efficiency. Using the simulation software to solve the problem would reduce any inaccuracies, which resulted from the above assumptions.

The following are the steps to define the simulation model for this computation.

Install the required streams and equipment to set up the process as shown in Fig. 6.6.

1. Define Stream 1 vapor fraction to be 0 (saturated liquid), and its temperature to be 50°C.
2. Define Stream 2 vapor fraction to be 1 (saturated vapor), and its temperature to be −20°C.
3. Some reasonable pressure drops for the chiller and condenser are:
   a. Pressure drop across chiller = 1 psi (7 kPa)
   b. Pressure drop across condenser = 5 psi (35 kPa)
4. Specify the duty for the chiller, 340 kW. Use a heater object instead of a cooler to represent this because the refrigerant will be absorbing heat.
5. Set the compressor efficiency to be 70%.
6. Compare the results of the simulation to the hand calculation using the Mollier diagram.

	Hand Calculation (Mollier Diagram)	Process Simulation
Propane flowrate	5808 kg/h	5832 kg/h
Compressor duty	160 kW	150 kW
Condenser duty	500 kW	490 kW

6.5 CONCLUSIONS

The first part of this chapter illustrated a simple conceptual design procedure for a DPCU. It is clear that one of the most critical aspects of this type of design is the understanding of the operation ranges and the phase behavior of the process fluids. The importance of using the phase envelope is highlighted here as it provides a very clear indication of the process constraints. Once the required chiller temperature has been determined, it will be possible to determine the duties of the chiller and gas−gas exchanger and conduct some preliminary sizing for the separators using the simulator (left as an exercise).

The second part of this chapter highlighted the importance of understanding the relationship of intrinsic thermodynamic state variables in a process by utilizing a Mollier diagram. The benefits of performing this type of hand calculation are to get a first-pass order-of-magnitude estimate for the solution before using the simulator. The Mollier diagram also provides a visual representation of the relationship between the process variables and the degrees of freedom of a system. As described in Section 6.4.2, once the fluid composition of a stream is known, each stream only has two thermal degrees of freedom available for definition. This is easily understood on the Mollier diagram, which is a two-dimensional representation of a fluid's thermal state. Each point on the diagram only requires two variables to be fixed.

In addition to the aforementioned benefits of using a Mollier diagram, doing a computation in this manner is obviously a lot cheaper than purchasing a software license! Software is generally leased based on a given duration; it is desirable to use hand calculations beforehand to identify possible solutions; this will minimize lease durations and reduce your software costs.

EXERCISES

Exercise 1: Dew Point Control Unit: Base Model Setup

These exercises provide a step-by-step tutorial for the conceptual design of the DPCU. The final objective of this analysis is to determine the chiller and gas–gas exchanger duties that yield the desired sales gas dew point temperature.

Steps for constructing the simulation model are given as follows:

1. Create a component list containing the following components:
   - a. Nitrogen
   - b. CO_2
   - c. Methane
   - d. Ethane
   - e. Propane
   - f. Isobutane
   - g. Normal-butane
2. Select an appropriate property package, in this case Peng–Robinson.
3. Install two streams, name them Feed 1 and Feed 2, respectively, and specify their temperatures, pressures, molar flowrates, and compositions as provided in Table E1.
4. Construct the process flow diagram as per Fig. 6.4 based on the following procedure.
   - a. Add the mixer, MIX-100. Pressure specification is "Set Outlet to Lowest Inlet." For the outlet stream (MixerOut), determine the following:
     - i. What is the temperature?
     - ii. What is the pressure?
     - iii. What is the energy flow?
     - iv. What sort of computation did the mixer do?
   - b. Add a separator, i.e., InletSep, with zero pressure drop.
   - c. Add a gas–gas heat exchanger. Provide 10 psi pressure drops across both the shell and tube sides. What information was transmitted to the CoolGas stream? Install a cooler (chiller), with pressure drop of 10 psi. What information was propagated to the ColdGas stream?
   - d. Specify the ColdGas stream temperature as −20°C. Please observe what happened when you entered this temperature, what streams are now known, and what streams remain unknown? What happened to the CoolGas stream? Install the LTS, and connect the remaining streams. What are the resulting flowrates for the LTSVap and LTSLiq streams? Has the flow sheet solved? What information is required?

TABLE E1 Feed Stream Conditions and Compositions

	Feed 1	Feed 2
T (°C)	15.5	15.5
P (kPa)	4137	4137
Mole flow (kg mol/h)	300	200
Component Mole%		
Nitrogen	1.80%	1.00%
Carbon dioxide	0.00%	1.00%
Methane	62.40%	60.00%
Ethane	16.70%	20.00%
Propane	11.40%	10.00%
Isobutane	4.30%	4.00%
n-Butane	3.40%	4.00%

e. Add a minimum approach specification to the gas−gas heat exchanger of 10°C. Has the flow sheet solved?

5. If all the information was defined correct as per steps a−e, you should have a completely solved base model.

Exercise 2: Dew Point Control Unit: Calculate the Dew Point of the Sales Gas

For the case in Exercise 1, the simulation should have calculated a pressure of 3930 kPa at the SalesGas stream. Hence, the SalesGas stream is fully defined. You cannot change any of the variables. In order to allow the dew point temperature to be calculated, the thermal specification of this stream needs to be "removed" using the Balance logical operation. The following instructions may be used to perform such calculation:

1. Install a Balance object. From the Balance object input form, as shown in Fig. E1, connect SalesGas under the Inlet Streams box and create/connect a stream named "Check Dew Point" under the Outlet Streams box.
2. Go to the Parameters tab, as shown in Fig. E2, and select "Mole" as the Balance Type.
3. In the material stream Check Dew Point, specify pressure = 5500 kPa and vapor fraction = 1.0. In UniSim Design, specifying the vapor fraction of a stream to 1 defines it as a saturated vapor, hence the calculated temperature is the temperature at which an infinitesimally small amount of liquid is formed, also known as the dew point.

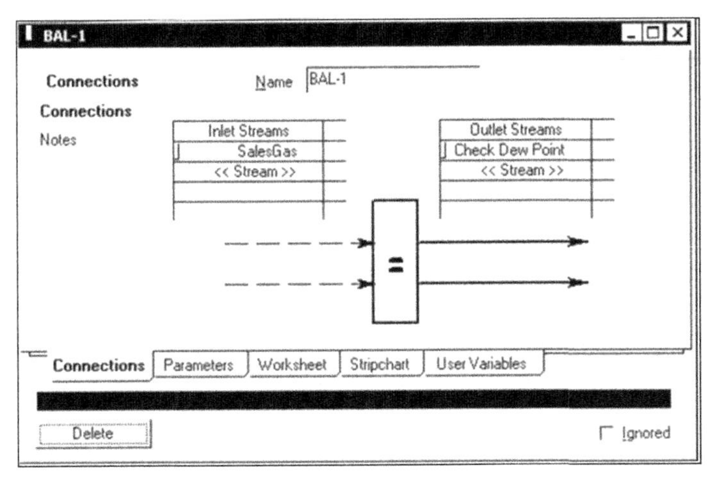

FIGURE E1 Balance object input form.

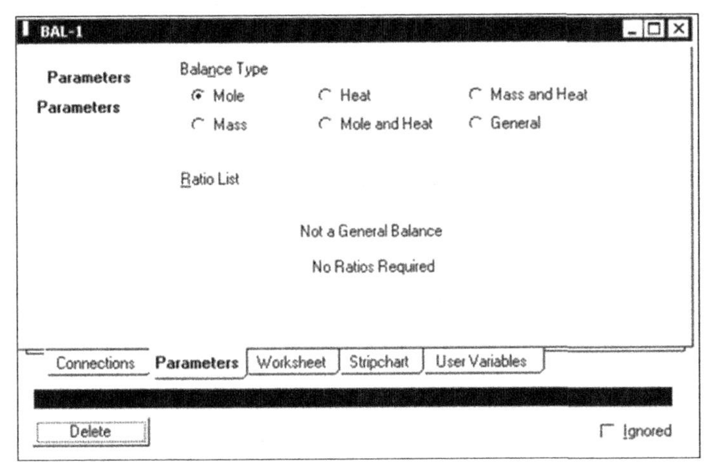

FIGURE E2 Balance object Parameters tab.

Remember that the desired sales gas dew point is −12°C at 5500 kPa. To achieve the desired dew point, the temperature at the chiller outlet needs to be adjusted. Determine the calculated dew point of SalesGas (at 5500 kPa) if the temperature of ColdGas is set to −10, −20 or 0°C

Exercise 3: Dew Point Control Unit: Determining the Chiller Temperature

The Adjust unit in UniSim Design represents an algorithm that adjusts a given variable iteratively, guided by a numerical algorithm, to arrive at a desired value in a given target variable. In this exercise, the Adjust unit is used to automatically determine ColdGas temperature, which yields a Sales Gas dew point of −12°C.

To install the Adjust unit, the following steps are needed:

1. Set the adjusted variable to be the ColdGas temperature and the target variable to be the Check Dew Point stream temperature.
2. Set the target variable to be −12°C
3. In the Parameters page, set the step size anywhere between 1 and 5°C and the tolerance at 0.25°C. Explain why was 0.25°C chosen as a tolerance?
4. Specify the minimum and maximum bounds to be −20°C and 0°C, respectively. Bounding the solution in this way prevents the algorithm from entering a potentially unsolvable solution space, which may occur because of:
   a. The existence of several local minima or
   b. The algorithm "skipping over" the desired solution because of the particular shape of the function.

In general, selecting the appropriate bounds requires the modeler to understand the solution space, typically by performing a hand calculation to identify the approximate order of magnitude of the solution, or generating the phase envelope to identify if the changes in process conditions may lead to infeasible solutions. For example:

1. If the temperature of the chiller is set too low, all the fluid may be condensed into a saturated or subcooled liquid; at this point, no vapor will be generated.
2. If the temperature of the chiller is set too high, all the fluid may be vaporized into a saturated vapor or superheated gas, in which case, changing the temperature in this region will cause the composition to further change.

It is now possible to calculate the following parameters to achieve the desired dew point temperature of the sales gas at −12°C at 5500 kPa:

- Operating temperature of the chiller.
- Chiller duty to achieve the desired temperature.
- The corresponding gas−gas exchanger duty and its required overall heat transfer coefficient (UA) value.

In process design, it is common to build a certain margin into the predicted size and cost of an equipment or process to make predictions more realistic. This design margin helps to account for the fact that the final cost of the equipment may be influenced by one or more of the following, as well as countless possible project-specific constraints:

1. Computational uncertainty
2. Feedstock specification uncertainty
3. Operating range uncertainty

4. Regulatory requirements
5. Equipment design life
6. Cost of off-the-shelf equipment
7. Material availability

The design margin to be used is generally defined by a set of company technical standards or experience. Hence, it is important to have an appreciation of the numerical range or sensitivity a given margin can yield. For the purpose of this tutorial, design margin is added by reducing the dew point temperature from −12 to −16°C. This assumes there may be more light ends in the feed; hence, a lower chiller temperature may be necessary to sufficiently yield the desired sales gas dew point. Determine the corresponding value of the following parameters:

- The required chiller operating temperature?
- The required chiller duty at that temperature?
- The corresponding gas−gas exchanger duty?
- The required gas−gas exchanger UA?

Compare the chiller duty and gas−gas exchanger UA to the earlier result. Determine which property has the largest change and its percentage of change.

Exercise 4: Propane Refrigeration

The propane refrigeration system discussed in Section 6.4 has the assumption that the refrigerant was 100% propane. Note, however, that propane stream normally contains some other hydrocarbons. Try computing the required refrigerant flowrate, compressor duty, and condenser duty for the following refrigerant compositions:

1. 95% propane, 5% ethane
2. 95% propane, 5% isobutane

REFERENCE

East One-Zero-One (East101), 2017. Introduction to Steady State Process Simulation Course Notes.

Part 3

PRO/II

Chapter 7

Basics of Process Simulation With SimSci PRO/II

Chien Hwa Chong
Taylor's University, Subang Jaya, Malaysia

This chapter aims to provide a step-by-step guide in simulating an integrated process flowsheet using SimSci PRO/II (Schneider Electric, 2015). The concept of simulation is based on *sequential modular* approach and follows the onion model for flowsheet synthesis (see Chapter 1 for details). The case study on *n*-octane production (Example 1.1) is used for illustration throughout the chapter.

7.1 EXAMPLE ON *N*-OCTANE PRODUCTION

A simple example that involves the production of *n*-octane (C_8H_{18}) is demonstrated (Foo et al., 2005), with detailed descriptions given in Chapter 1. The basic simulation setup involving the registration of components, thermodynamic model, and reaction stoichiometry is to be carried out in the *Component Selection window*, *Thermodynamic window*, and *Reaction Component window* of PRO/II, respectively, while the other steps are carried out in the *flowsheet*. The individual steps are discussed in the following subsections.

7.2 STAGE 1: BASIC SIMULATION SETUP

7.2.1 Units

A user can select different "Units of Measure" such as English-Set1, Metric-Set1, and SI-Set1 prior to start of a new process flowsheet (refer detailed steps for change of units in Fig. 7.1).

7.2.2 Component Selection

User is required to define the components, viz., nitrogen, ethylene, *i*-butane, *n*-butane, and *n*-octane using the component selection from the "Input" windows features (refer detailed steps for component selection in Fig. 7.2).

Chemical Engineering Process Simulation. http://dx.doi.org/10.1016/B978-0-12-803782-9.00007-8

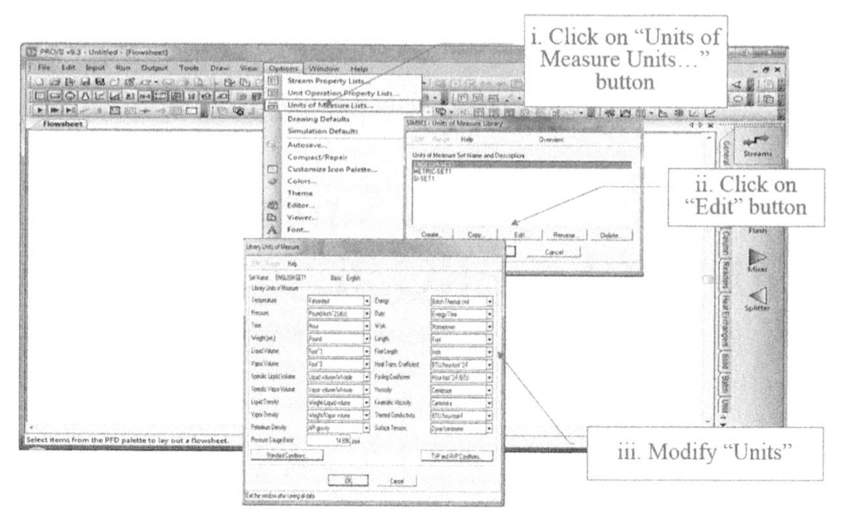

FIGURE 7.1 Units of measure of the PRO/II.

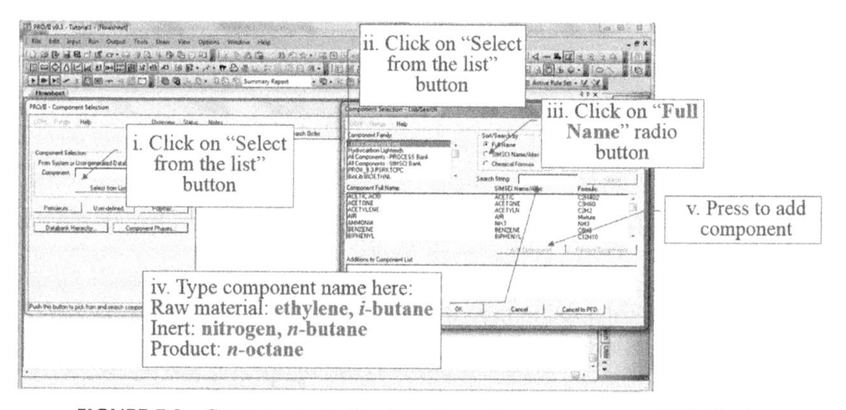

FIGURE 7.2 Component selection from the system or user-generated databank.

7.2.3 Thermodynamics Method

Thermodynamic methods are used to simulate the physical behavior of component system by calculating several physical properties (refer detailed steps for thermodynamic data in Fig. 7.3). For this process, Peng–Robinson model[1] is to be used.

1. See Chapter 3 for detailed discussion on thermodynamic selection.

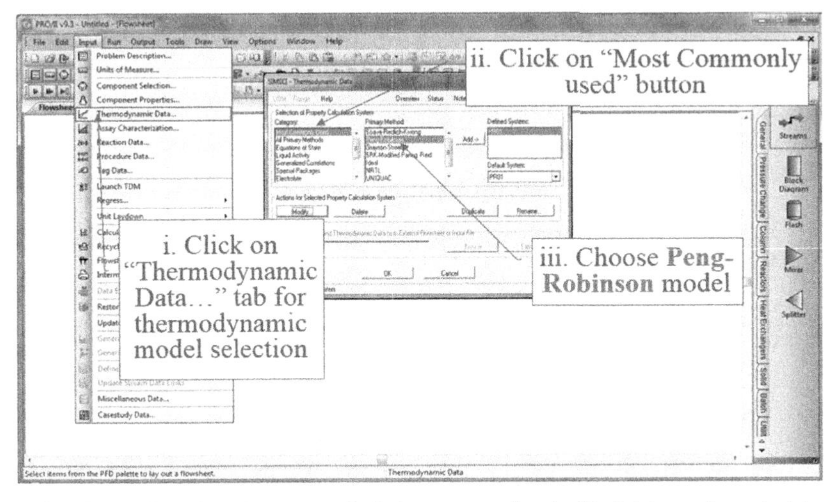

FIGURE 7.3 Selection of property calculation system using the SimSci-thermodynamic data.

7.3 STAGE 2: MODELING OF REACTOR

To select the equipment, a user can choose unit operations from the right side column of the flowsheet. A list of unit operation models are classified as general, pressure change, column, reactors, heat exchanger, solid, batch, utilities, user-aided, classic, and miscellaneous categories.

For this case, a conversion reactor is used for demonstration. The user is required to create an input stream, S1, and an output stream, S2, after selecting the conversion reactor, R1, from the right panel (refer detailed steps Fig. 7.4).

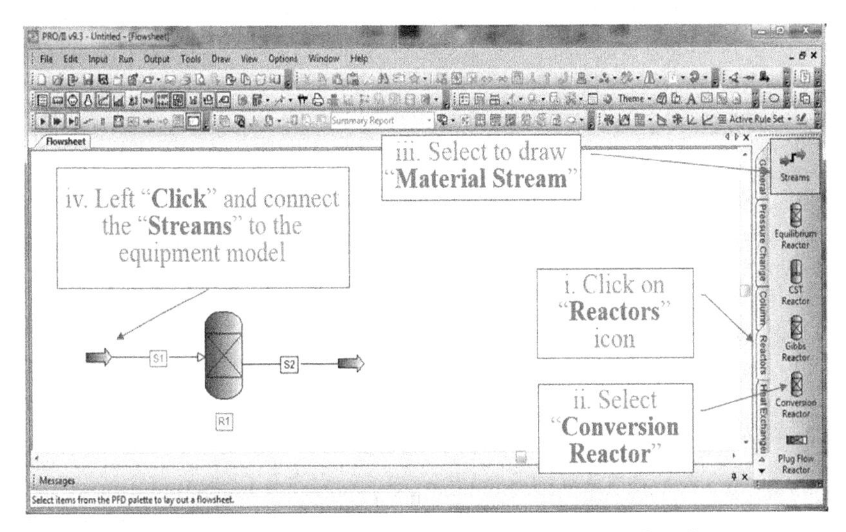

FIGURE 7.4 Construction of flowsheet topology on a process flow diagram.

TABLE 7.1 Specifications for Process Feed Stream (Foo et al., 2005)

Components	Flowrate (kg-mol/h)	Condition
Ethylene (C_2H_6)	20	
i-Butane (*i*-C_4H_{10})	10	T: 93°C
Nitrogen (N_2)	0.1	P: 20 psia
n-Butane (*n*-C_4H_{10})	0.5	

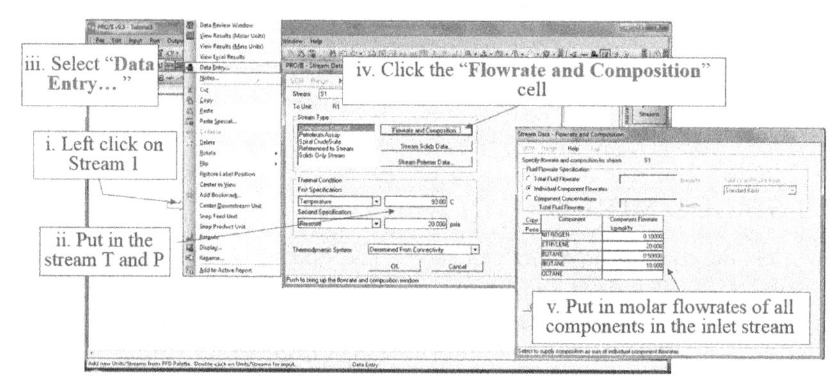

FIGURE 7.5 Defining feed stream properties.

We then proceed to define reactor feed stream properties. Component flowrates, temperature (T), and pressure (P) of the stream are given in Table 7.1. Detailed steps for this step are shown in Fig. 7.5.

For the conversion reactor, the user is required to key in the reaction data. Double-click on the reactor column to key in thermal specification, extent of reaction, and reactor data (see Fig. 7.6 for detailed steps). Reaction stoichiometry and fractional conversion must be supplied for a conversion reactor simulation. It can be defined in the Reaction Data Sets window.

We proceed to define pressure drop and extent of reaction for the reactor (see detailed steps in Fig. 7.7). On the completion of specifying the reactor, the user can proceed to execute the simulation (Fig. 7.8) and examine the simulation results (Fig. 7.9).

7.4 STAGE 3: MODELING OF SEPARATION UNITS

A flash drum is used to separate the reactor effluent stream (S2) into vapor top (S3) and liquid bottom (S4) streams. Pressure drop and temperature of the

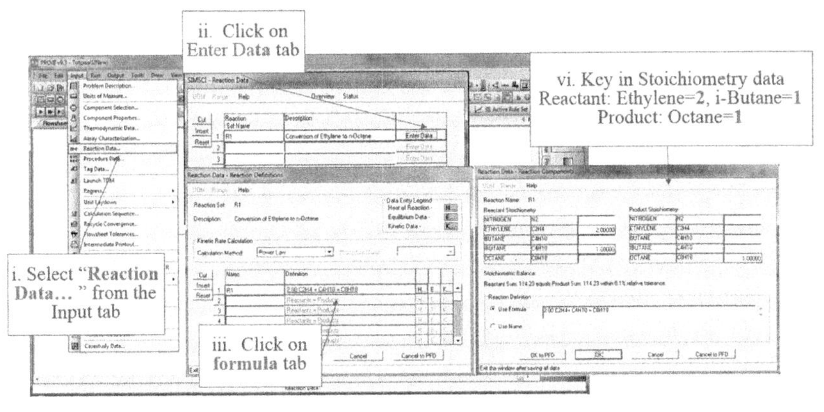

FIGURE 7.6 Specifying reaction data for the process.

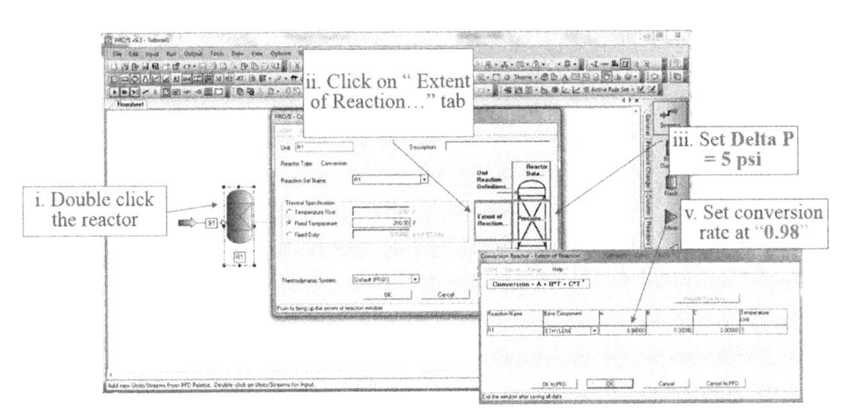

FIGURE 7.7 Specifying pressure drop and extend of reaction for the reactor.

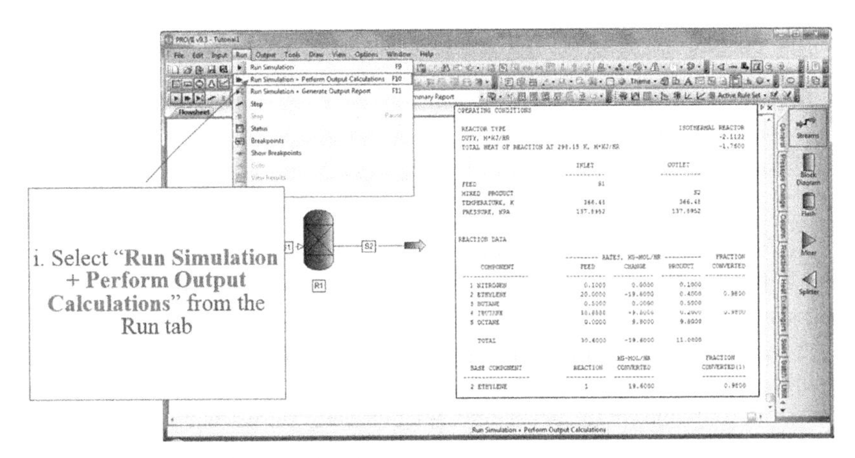

FIGURE 7.8 Simulation report of the conversion reactor.

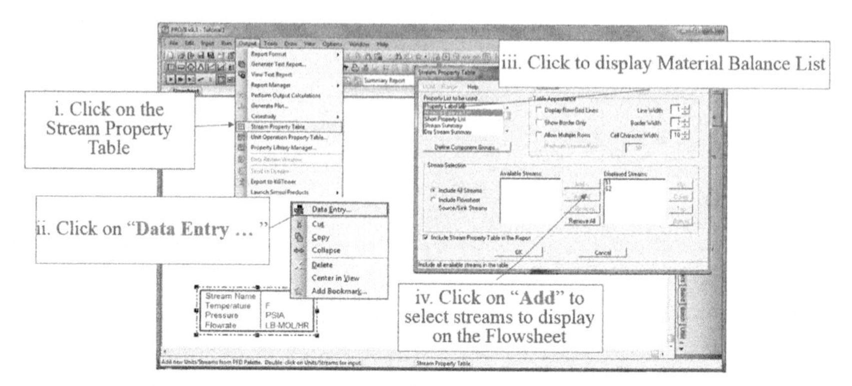

FIGURE 7.9 Displaying component molar flowrates on the process flow diagram.

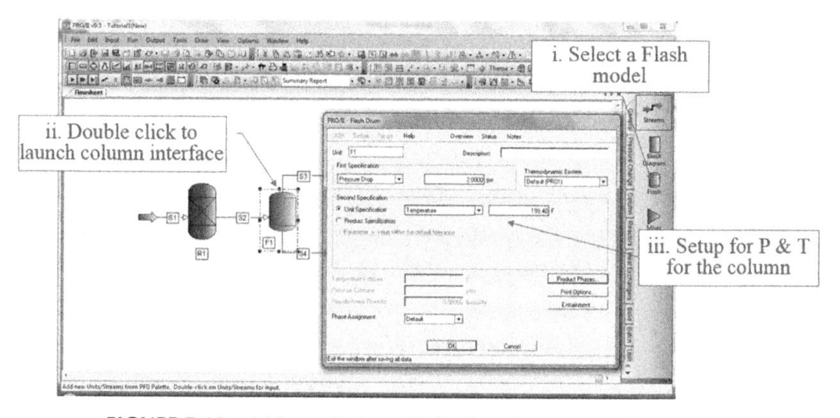

FIGURE 7.10 Adding a flash model for the primary separation process.

flash column are set at 2.0 psi and 93°C°C, respectively. Detailed steps for adding a flash model are shown in Fig. 7.10.

The next step is to simulate a rigorous distillation model. The column is used to simulate any vapor/liquid separation process. When a column condenser checkbox is selected, it is always designated as tray number 1, meanwhile the column reboiler is always designated as the highest numbered tray in the model (refer detailed steps for launching a rigorous distillation model in Fig. 7.11).

Data needed to simulate a rigorous distillation model are shown in Table 7.2. In PRO/II simulator, either a feed or a heat duty on the top and bottom trays must be specified. In addition, the pressure must be defined.

The user is required to select suitable algorithm for a rigorous distillation simulation. Next, the feeds and products phase, tray, and rate are set. There are some commonly used algorithms for modeling of distillation columns in PRO/II, viz., the inside-Out, Sure, Chemdist, Enhanced IO, Electrolytic, and

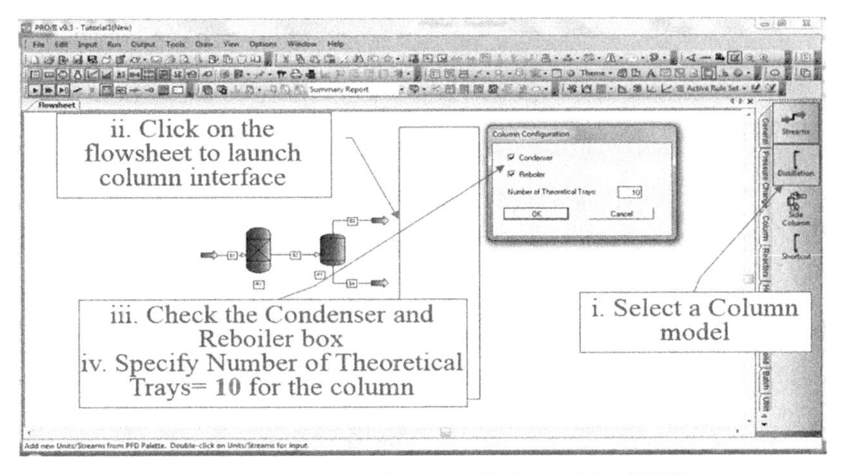

FIGURE 7.11 Launch a rigorous distillation model in PRO/II.

TABLE 7.2 Specification for a Rigorous Distillation Model

Specification		
Pressure	15 psia (top)	20 psia (bottom)
Number of trays	10	
Reflux ratio	10	
Feed tray	4	
Overheads rate, kgmol/h	0.45	
Bottom rate, kgmol/h	0.73	
Ethylene flowrate, kgmol/h	0.359	
Condenser type	Partial	

Eldist algorithm, with the default being inside-out algorithm (PRO/II 9.3.2, Reference Manual, 2015). This method is effective and fast in solving vapor/liquid staged column models. For this case, the "Chemdist" algorithm is selected. Table 7.3 shows suitable systems for different algorithms. The "Eldist" is recommended for electrolyte and equilibrium electrolytic reactions. The "LLEX" can be used for liquid−liquid extractor; chemical reaction; and kinetic, equilibrium (nonelectrolyte), and conversion. For nonpower law kinetics, the "Chemdist" algorithm is more suitable. For hydrocarbon system where water is present, the "Sure" algorithm is recommended. The Sure and Chemdist algorithms also can be used to model vapor-liquid-liquid

TABLE 7.3 Algorithm for Specific Systems (PRO/II 9.3.2, Reference Manual 2015, Schneider Electric)

System	Algorithm
Electrolyte	Eldist
Equilibrium electrolytic reactions	Eldist
Liquid–liquid extractor	LLEX
Chemical reaction	Chemdist, LLEX, RATERFRAC
Kinetic, equilibrium (nonelectrolyte) and conversion reactors	Chemdist (nonideal chemical system)—a Newton–Raphson method, LLEX
Kinetic and equilibrium (nonelectrolyte)	RATERFRAC
Nonpower law kinetics	Chemdist
Hydrocarbon system where water is present Refinery and chemical systems	Sure

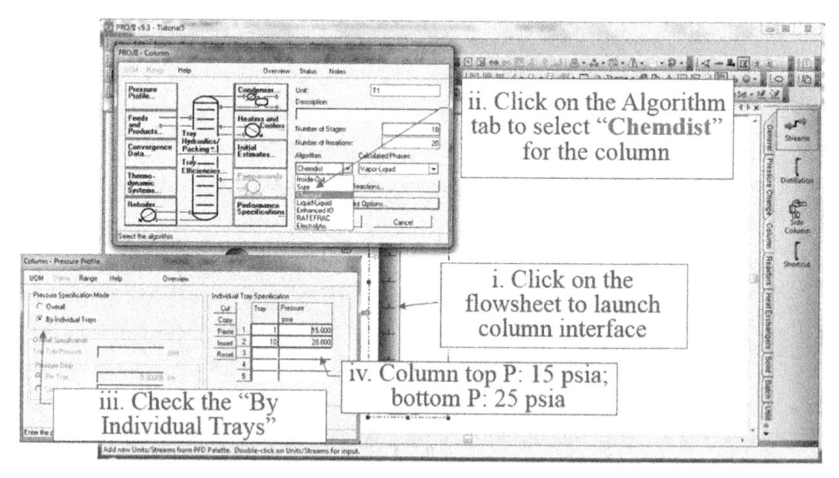

FIGURE 7.12 Specifying algorithm and pressure profile for the distillation column.

equilibrium systems. Detailed steps for launching and specifying algorithm of a rigorous distillation model with pressure profile are given in Fig. 7.12. Next, the condenser type is set to "Partial" (see Fig. 7.13).

For feeds and products specification, refer to Fig. 7.14 to specify feed tray and flowrates for S5 and S6.

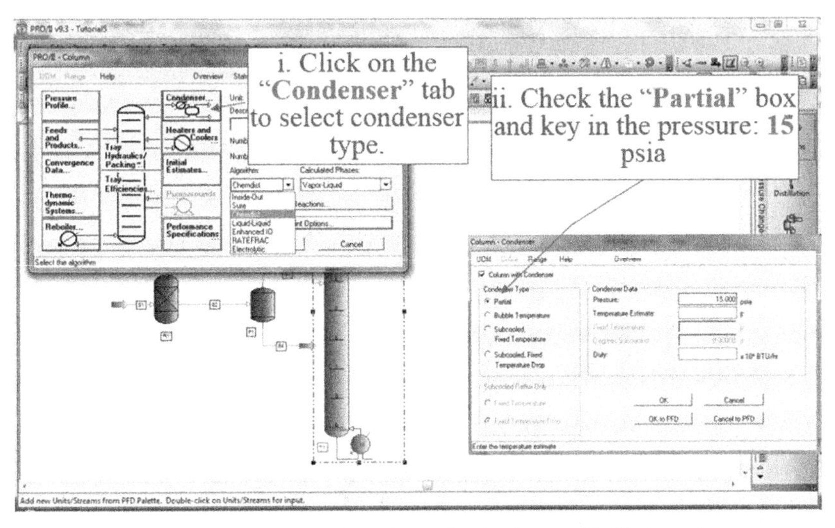

FIGURE 7.13 Select type of condenser for the rigorous distillation column.

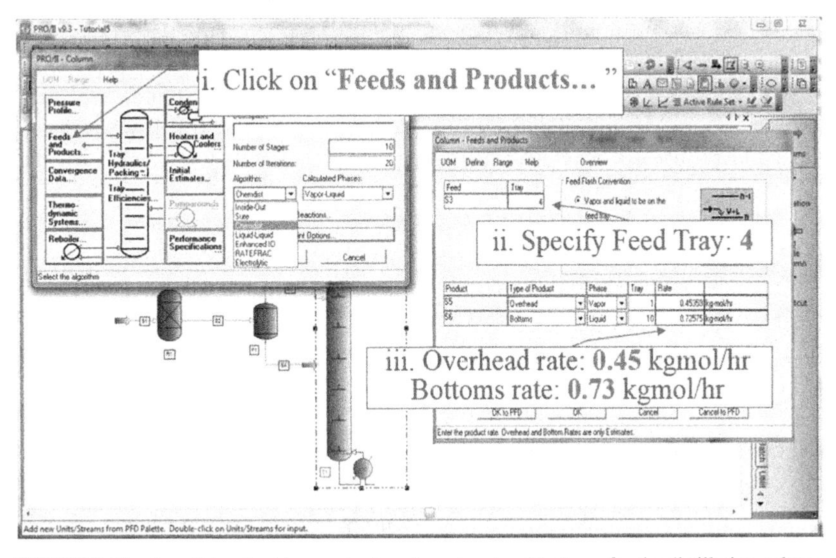

FIGURE 7.14 Specifying feed tray and rates of overhead and bottoms for the distillation column.

Next, the user is required to provide two specifications for the distillation model. It is suggested to use the specific component flowrates and reflux ratio for distillation column simulation. Detailed steps for specifying the variables for the column are shown in Fig. 7.15.

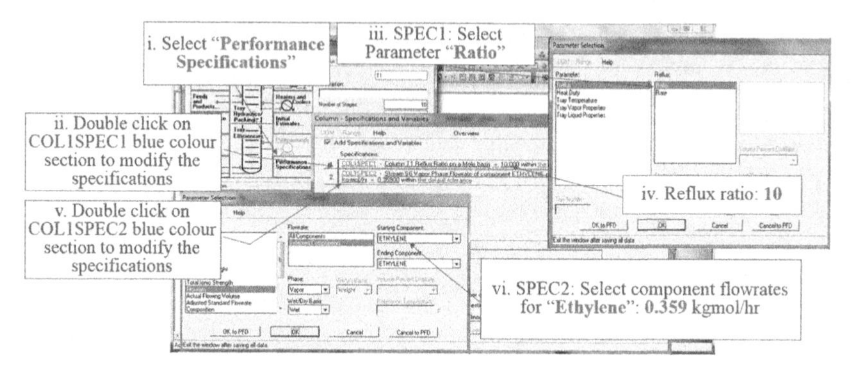

FIGURE 7.15 Column specification and variables for the distillation column.

7.5 STAGE 4: MODELING OF RECYCLE SYSTEMS

In PRO/II software, recycle techniques are divided into three main categories, viz., thermal recycle, mass and energy balance recycles, and purge/makeup systems. The thermal recycle technique is used for heat exchanger networks where only the stream temperature changes. To use this method, users need to specify the output conditions. As long as all data stay within the specified tolerance level, the recycle loop will converge.

Recycle convergence in PRO/II consists of direct-substitution, Wegstein acceleration technique, and Broyden acceleration (Fig. 7.16). The user can select the Problem Recycle Convergence and Acceleration Options from the

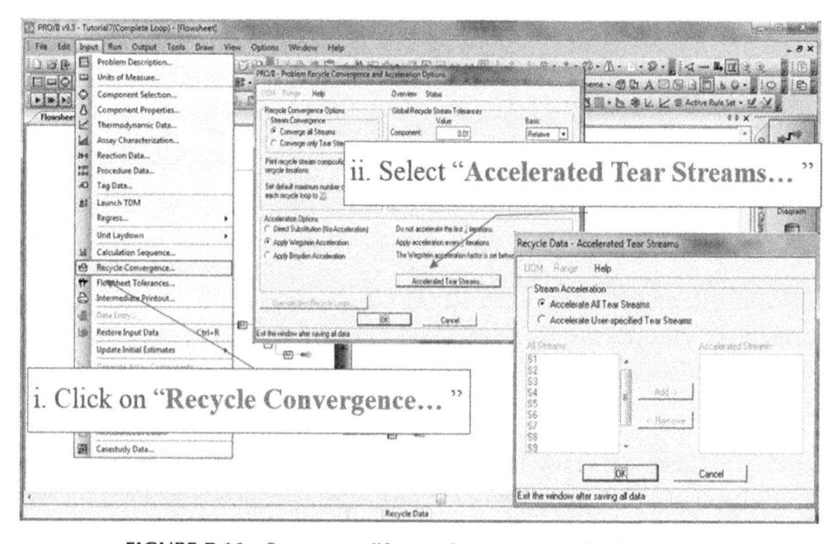

FIGURE 7.16 Steps to modify recycle convergence for the process.

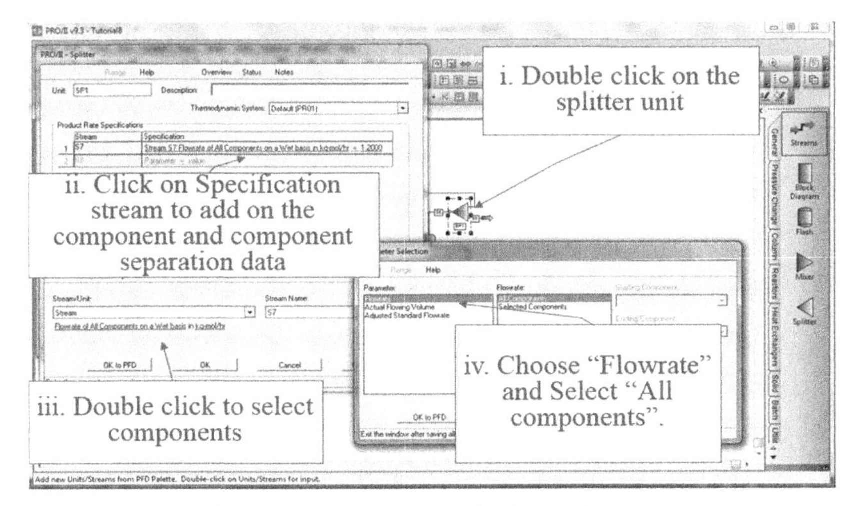

FIGURE 7.17 Splitter specification for recycle process.

input button from the main window. The direct-substitution method used the last computed values for the tear streams for the next trial solution of the recycle loop; hence, it is relatively slower as compared to other methods.[2]

For the *n*-octane case, excessive ethylene is recycled back to the reactor. Prior to recycle, impurities can be purged using a splitter. Remember, the splitter unit must have two or more product streams and it requires specification in terms of rate of a component, recovery of a fraction of the total feed, etc. It is only for identical component composition separation. The splitter calculation is based on the selected outlet pressure parameter and stream specification data. In this case study, a splitter is added to the flowsheet to remove impurities. 95% of all components with flowrate of 1.2 kgmol/h are recycled. Fig. 7.17 shows information required and detailed steps in simulating a component splitter.

To reduce the energy losses, temperature and pressure of the recycle stream are required to be adjusted following the feed stream, S1. Fig. 7.18 shows detailed steps for simulating the compressor and cooler.

To recycle stream, S10, the user is required to reroute stream, S1. Fig. 7.19 shows steps in setting "Breakpoints" prior to create a recycle loop. The reroute command is a specific feature in PRO/II used to recalculate an unobstructed path and correct any problems caused by moving a connection or a unit operation icon and the "Breakpoints" command allows the user to make changes in the input data (PRO/II 9.3.2, Reference Manual, 2015). Fig. 7.20 shows detailed steps in creating a "Breakpoints" for the reactor, R1.

2. See Chapter 4 for more in-depth discussion of this topic.

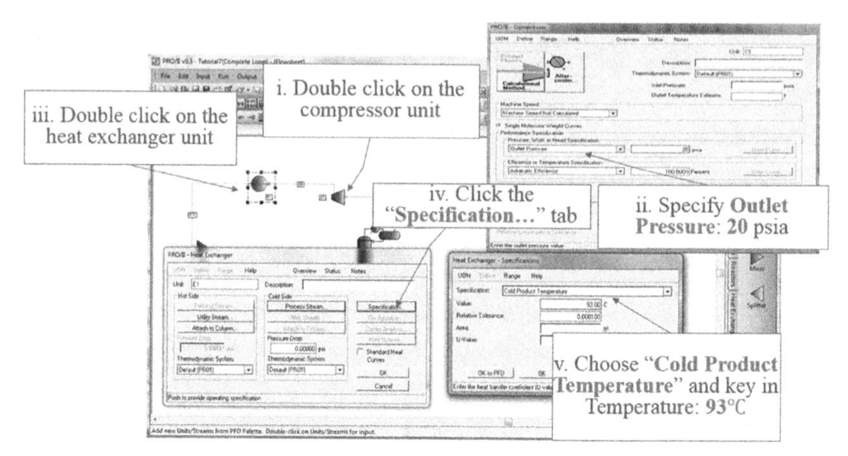

FIGURE 7.18 A compressor and a cooler simulation steps for the process.

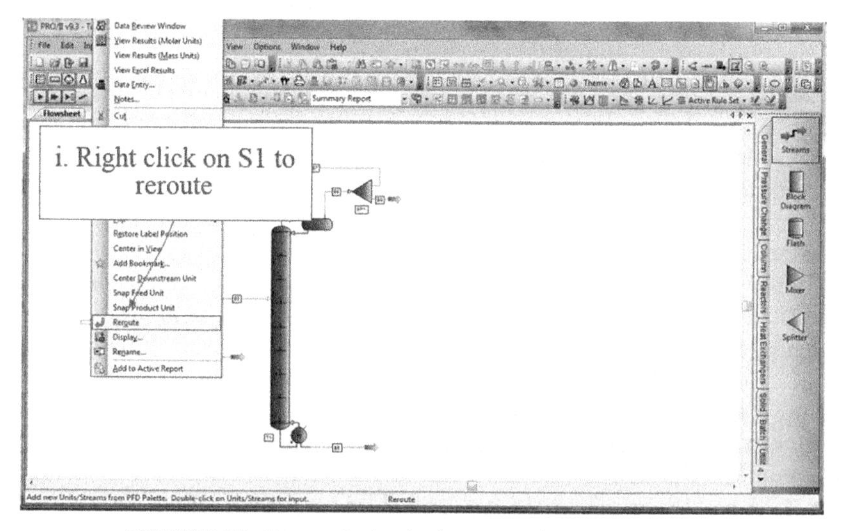

FIGURE 7.19 Reroute the breakpoints stream for recycle purposes.

Furthermore, the user is required to add on a mixer, M1, prior to connecting to the reactor, R1 (see detailed steps in Fig. 7.21) to connect the feed and recycle stream.

Fig. 7.22 shows the steps in simulating heat-integrated flowsheet.

In the final steps, user can add on a worksheet to the process flow diagram (Fig. 7.23).

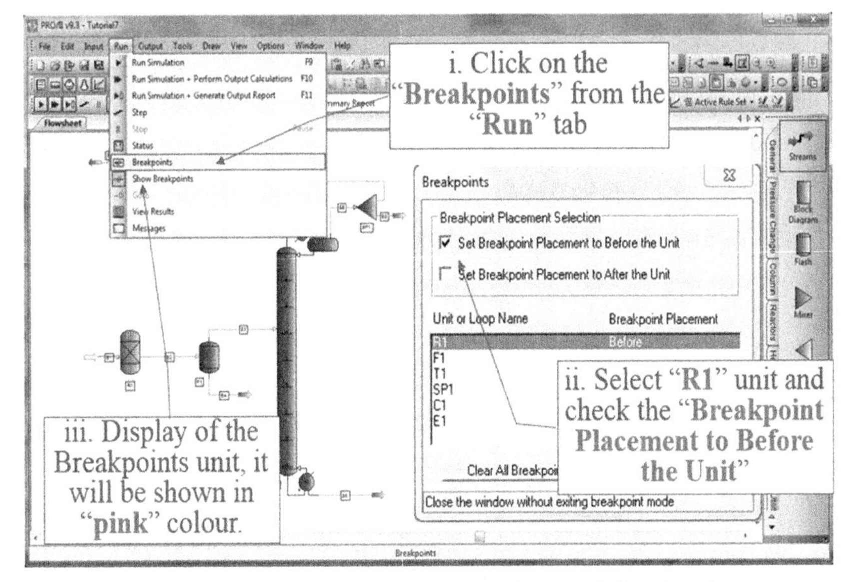

FIGURE 7.20 Specifying the breakpoints placement before the unit.

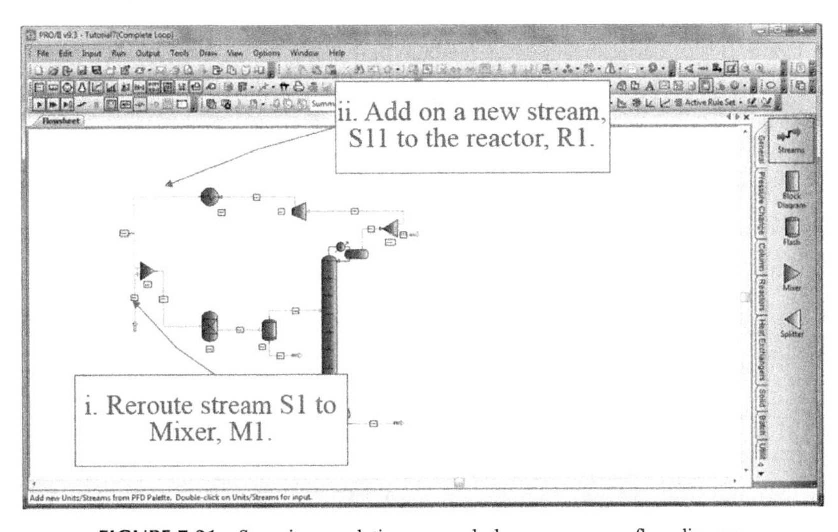

FIGURE 7.21 Steps in completing a recycle loop on process flow diagram.

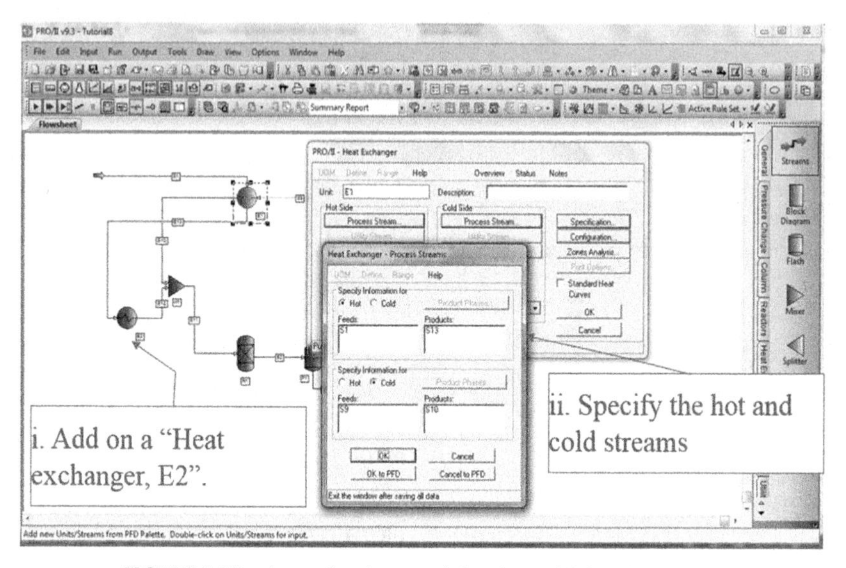

FIGURE 7.22 A complete integrated flowsheet with heat integration.

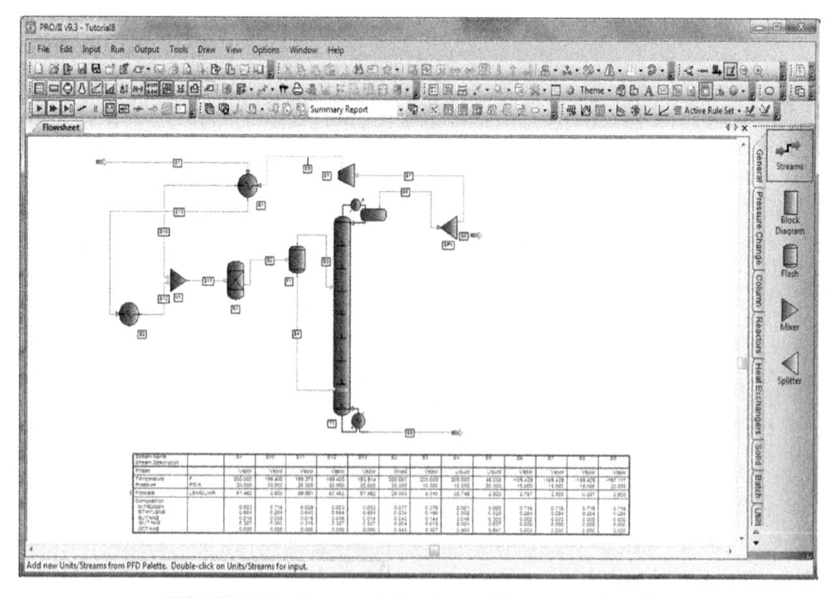

FIGURE 7.23 Integrated flowsheet with a properties table.

7.6 CONCLUSION

A four-stage integrated process flowsheet was developed using SimSci PRO/II to produce *n*-octane (C_8H_{18}). All basic simulation setup starts from *Component Selection window, Thermodynamic window, Reaction Component window* of PRO/II followed by simulation in *flowsheet*. To recycle valuable raw materials, "reroute" and "breakpoints" functions are needed. It is advisable to limit 100 iterations in column simulation. No specification is required for recycle stream for PRO/II.

EXERCISES

1. Simulate the following separation process using data provided in Tables E1 and E2. A single-state expansion of a light hydrocarbon stream. Firstly, feed stream is cooled through a feed chiller (X-1) and any liquid is removed with a vertical separator (F-2). The vapor is expanded (E-4) and liquids are separated using a second vertical separator (F-3).

TABLE E1 Specifications for Process Feed Stream

Components	Flowrate (lb-mol/h)	Condition
Methane	6100	T: 90°F
Ethane	500	P: 994.7 psia
Propane	200	
N-pentane	100	
N-hexane	70	

TABLE E2 Equipment Specifications

Equipment	Tag	Specification
Feed chiller (Model: Exchanger 1)	X-1	Temperature out = −35°F Delta pressure = 10 psi
Two Vertical separators (Model: Flash 1)	F-2 F-4	Adiabatic flash Delta pressure = 0
Expander	E-3	Pressure out = 275 psia Efficiency = 0.80

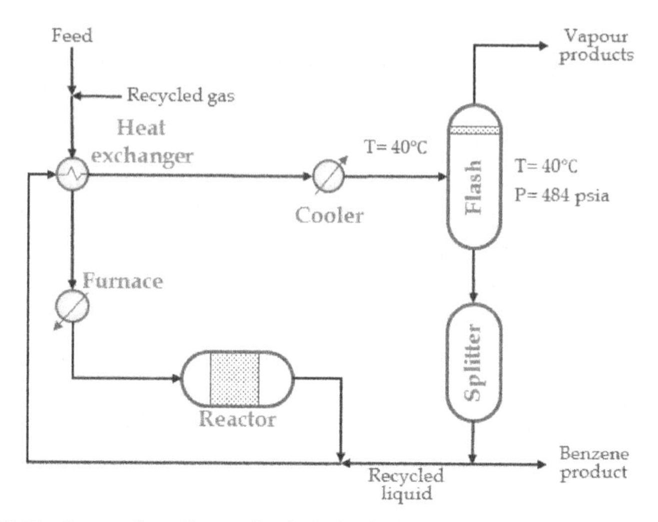

FIGURE E1 Process flow diagram for the hydrodealkylation process (Seider et al., 2004).

2. (Fig. E1) A hydrodealkylation process is adapted from Seider et al. (2004), where toluene is converted to benzene with a side reaction to produce biphenyl (see PFD in Fig. E1). Process feed stream (Table E3) is preheated in a furnace to 500°C along with a recycle stream, before being fed to the reactor. The stream properties are shown in Table E3. The reactor operates with high conversion rate (70%), with effluent temperature of 650°C and pressure drop of 5.0 psi. The reaction stoichiometries are given as follows.

Main reaction (R-1):

$$C_7H_8 + H_2 \rightarrow C_6H_6 + CH_4$$

TABLE E3 Specification for Process Feed Stream

Components	Flowrate (kg-mol/h)	Condition
H_2 (hydrogen)	1200	T = 25°C
CH_4 (methane)	1500	P = 600 psia
C_6H_6 (benzene)	0	
C_7H_8 (toluene)	120	
$C_{12}H_{10}$ (biphenyl)	0	

Side reaction (R-2):

$$2C_6H_6 \rightarrow C_{12}H_{10} + H_2$$

Effluent of the reactor is sent to a Flash unit, where hydrogen and methane (top stream) are separated from toluene, benzene, and biphenyl (bottom stream). The bottom stream will be sent to further purification steps. However, for brevity, the purification steps are approximated using a component splitter (using Splitter model in SimSci PRO/II), where toluene product is withdrawn, while other components are mixed with reactor effluent and recycled to the Flash unit. For the Splitter model, 90% of toluene is set to recycle to the Flash unit, while benzene and biphenyl are withdrawn as product stream. The latter will undergo further purification (excluded from consideration). A simple heat exchanger network can be designed to recover energy from feed stream to the Flash unit, to minimize energy consumption of the furnace. Recommended thermodynamic package for this system is Soave–Redlich–Kwong equation of state.

REFERENCES

Foo, D.C.Y., Manan, Z.A., Selvan, M., McGuire, M.L., October, 2005. Integrate process simulation and process synthesis. Chemical Engineering Progress 101 (10), 25–29.

Schneider Electric, 2015. SimSci PRO/II 9.3.2, Reference Manual 2015. SimSci, USA.

Seider, W.D., Seader, J.D., Lewin, D.R., 2004. Product and Process Design Principles. Synthesis, Analysis, and Evaluation, second ed. John Wiley and Sons, Inc., pp. 136–142

Chapter 8

Modeling for Biomaterial Drying, Extraction, and Purification Technologies

Chien Hwa Chong, Joanne W.R. Chan
Taylor's University, Subang Jaya, Malaysia

Process design involved drying and purification technologies that are essential for herbal processing industry. These techniques are used to retain bioactive compounds of herbs during their processing. In this chapter, herb drying and purification techniques were simulated using process simulation software PRO/II v9.3 (Schneider Electric, 2015). Because major compounds of herbs are not available in PRO/II component libraries, they were developed using properties data through user-defined function of the software, which is shown in this chapter. Furthermore, selection of thermodynamic model is discussed because one of the components exists at its supercritical condition. A herb consisting of some bioactive compounds was selected as the raw material, which can be used to prevent bacteria and viruses. It is important to process this product to retain its valuable compounds because of its short shelf life.

8.1 INTRODUCTION

Strobilanthes crispus (*S. crispus*) or commonly known as "pecah kaca" is a bountiful and low-cost herb that can be obtained from tropical countries. It belongs to the family of *Acanthaceae*. Table 8.1 shows bioactive compounds of *S. crispus*, which can be used for antidiabetic, antilytic, diuretic, and laxative in traditional medicine (Ismail et al., 2000). It has been reported that this plant has high antioxidant effects, viz., antihyperglycemic and hypolipidemic effects, anti-AIDS and anticancer activity (Liza et al., 2010). Koay et al. (2013) reported that dichloromethane extract shows a strong inhibitory effect against the Gram-positive *Staphylococcus aureus* [with *minimum inhibitory concentration* (MIC) of 15.6 μg/mL] and *Bacillus subtilis* (31.0 μg/mL) and moderate activity against Gram-negative bacteria (62.5 μg/mL) using taraxerol. Methanol extract compound 4-acetyl-2,7-dihydroxy-1,4,8-triphenyloctane-3,5-dione is able to fight

Chemical Engineering Process Simulation. http://dx.doi.org/10.1016/B978-0-12-803782-9.00008-X

TABLE 8.1 Bioactive Compound and Medical Usage of the *Strobilanthes crispus*

Herbs (Common Name)	Chemical Constituents (Phytochemical)	Pharmacology	References
Strobilanthes crispus (pecah kaca)	H.E.: 1-heptacosanol, tetracosanoic acid, stigmasterol D.E.: a mixture of four fatty acid esters of β-amyrin, taraxerone, taraxerol M.E.: 4-acetyl-2,7-dihydroxy-1,4, 8-triphenyloctane-3, 5-dione and stigmasterol β-D-glucopyranoside	Antioxidants, which are vitamins C, B_1, and B_2 together with catechin, caffeine and tannin, cytotoxic activity (M.E), treat kidney stones, antidiabetic, antiobesity, antiviral	Koay et al. (2013)

D.E., dichloromethane extract; *H.E.*, hexane extract; *M.E.*, methanol extract.

against *S. aureus* (7.8 μg/mL, 15.6 μg/mL), *Escherichia coli*, *Klebsiella pneumoniae,* and *Salmonella typhimurium* (31–62.5 μg/mL). In general, lower MIC values mean better quality.

8.2 BASIC SIMULATION SETUP

8.2.1 User-Defined and Solid Components

The PRO/II component library consists of SIMSCI, PROCESS, DIPPR, OLI banks, and bankid (PRO/II 9.3.2, Reference Manual, 2015). The SIMSCI and PROCESS banks are pure component databank (default) available in PRO/II, while DIPPR and OLI banks are optional databanks available as add-on. Bankid bank is a user-defined databank, where users can create hypothetical components using the property library manager. The components in PRO/II v9.3 databank are divided into six categories, viz., fixed properties, temperature-dependent properties, solid properties, polymer properties, structure data, and user defined (refer to detail steps to create user-defined component in Fig. 8.1). In this chapter, user-defined and solid properties are discussed in detail. Cellulose and phenolic acids such as gentisic acid, *p*-coumaric acid, *p*-hydroxy benzoic acid, ferulic acid, caffeic acid, vanillic acid, and syringic acid are a solid and user-defined components used in the modeling, respectively.

Properties of newly added components must be supplied in Component Property Section. To supply thermodynamic properties data, the user can refer

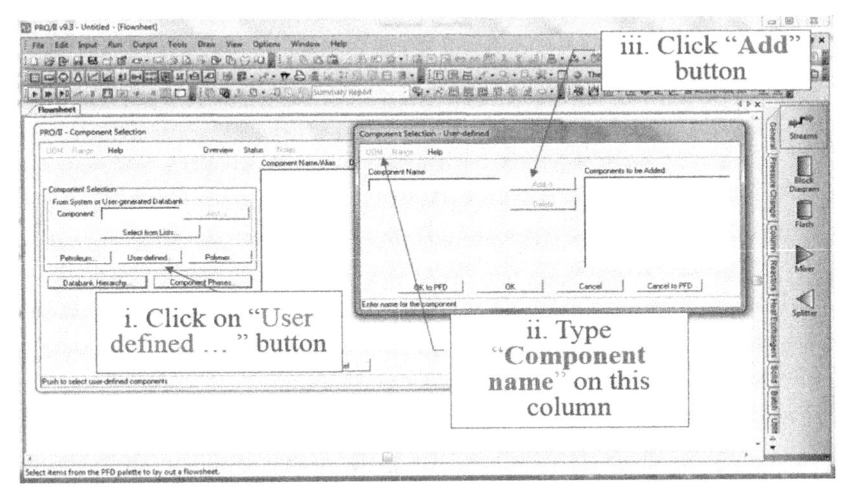

FIGURE 8.1 Create user-defined components for PRO/II.

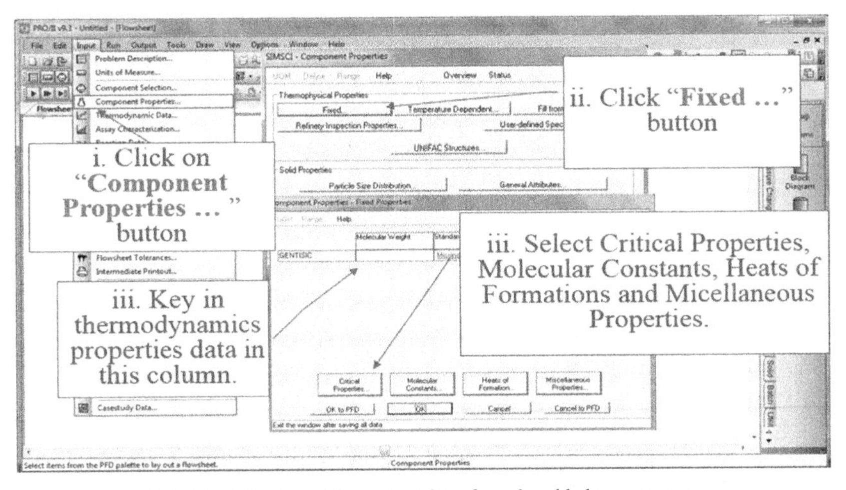

FIGURE 8.2 Specifying properties of newly added components.

detailed steps for thermodynamic data in Fig. 8.2. Table 8.2 shows all physical properties of major acids found in *S. crispus*.

Example of steps for specifying user-defined phenolic acids is shown in Fig. 8.3 using the data stated in Table 8.2.

8.2.2 Specification for Process Feed Stream

Feed stream properties of *S. crispus* are defined. Component flowrates, temperature (T), and pressure (P) of the stream are given in Table 8.3. Detailed

TABLE 8.2 Physical Properties of All Major Phenolic Acids in *Strobilanthes crispus* (Chan et al., 2015)

Components	Normal Boiling Point, NBP[a]	Molecular Weight (g/mol)	Critical Temperature (°C)	Critical Pressure (kPa)
Gentisic acid	405°C	117	650	7280
p-Coumaric acid	380°C	164	614	5037
p-Hydroxy benzoic acid	353°C	138	602	5731
Ferulic acid	410°C	194	679	4233
Caffeic acid	426°C	180	657	6299
Vanillic acid	386°C	168	620	4763
Syringic acid	416°C	198	637	4021

[a]*Detailed calculation of NBP are shown in Appendix A.*

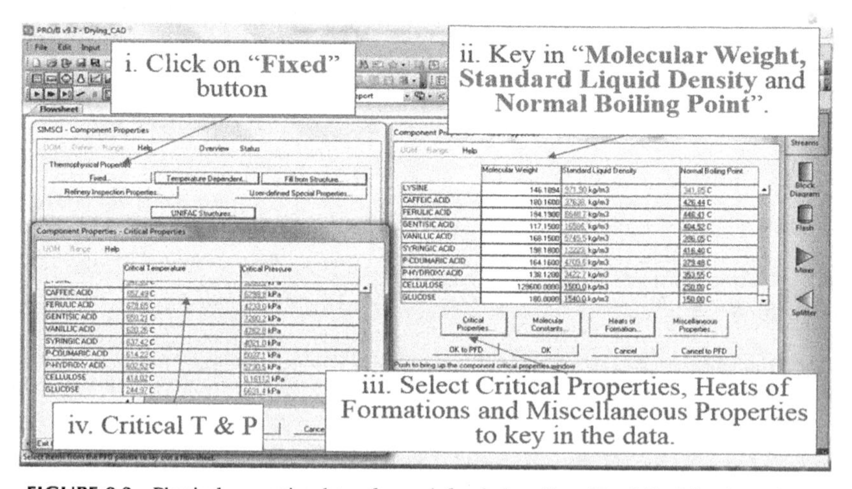

FIGURE 8.3 Physical properties data of user-defined phenolic acids of *Strobilanthes crispus*.

steps to specify components in the feed stream are shown in Fig. 8.4. For the solid component flowrate, see Fig. 8.5 for detailed steps to key in the data.

8.2.3 Thermodynamic Data

Appropriate thermodynamic method is important to model the phase behavior of a process. In this case study, nonrandom two-liquid (NRTL) fluid package

TABLE 8.3 Specifications for Process Feed Stream (Chan et al., 2015)

Components	Flowrate (Kg-mol/h)	Condition
Water	0.609	T: 25°C P: 101.3 kPa
Lysine	0.129	
Gentisic acid	0.005	
p-Coumaric acid	0.004	
p-Hydroxy benzoic acid	0.004	
Ferulic acid	0.004	
Caffeic acid	0.002	
Vanillic acid	0.003	
Syringic acid	0.003	
Glucose	0.24	

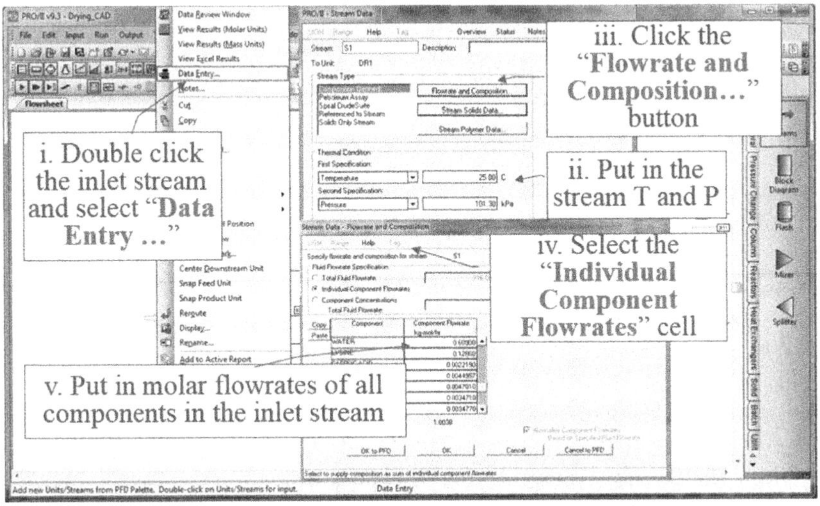

FIGURE 8.4 Defining feed stream properties of *Strobilanthes crispus*.

was selected to model the phase behavior of vapor–liquid extraction and liquid–liquid extraction for polar and electrolyte compounds, while Peng–Robinson package was used to model the supercritical fluid extraction (SFE) process to calculate the solubility of CO_2 (Kondo et al., 2002).

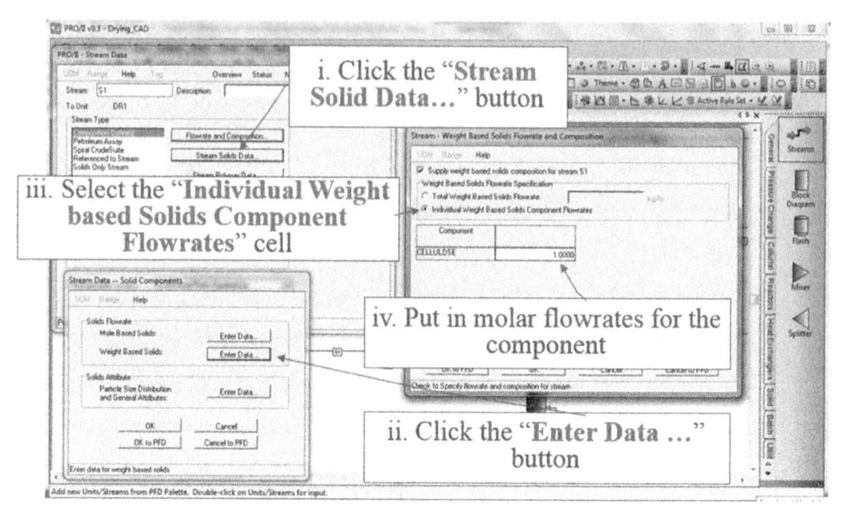

FIGURE 8.5 Specifying component flowrates of a solid component.

8.3 MODELING OF DRYING TECHNOLOGY

Drying is used to preserve bioactive compounds of *S. crispus* from further decomposition due to the growth of bacteria. According to Cerón et al. (2014), drying can increase the extraction rate of antioxidant compounds. This is because dried samples would be more brittle, which allows rapid breakdown of cell wall during grinding and homogenization steps in the extraction process. However, drying process could reduce the retained bioactive compounds because of its operating conditions, such as drying duration, temperature, air velocity, and humidity. All the unit operating conditions can be designed based on existing drying technologies.

Nowadays, new drying technologies such as combined and hybrid dryers are used to remove moisture content of biomaterials, viz., ultrasonic, microwave, heat pump, microwave vacuum, and hot air with low-temperature system. Final moisture content of the dried product using the ultrasonic technique could be reduced to 18% dry basis, while those using the freeze drying (FD), the microwave vacuum drying (MVD), and the convective air drying (CAD) are 6.8%, 9.0%, and 12.7%, respectively (Rodriguez et al., 2013; Schulze et al., 2014). Among other drying techniques, the FD can retain highest amount of bioactive compounds, as water content is removed through the sublimation process. However, the FD is a time-consuming process with at least 48 h of operation, which incurs high operation cost and energy. Therefore, the CAD technique is still the preference in industry with its low operating cost. Note that it is important to avoid drying technologies with high temperature and long drying time, as they will decompose the bioactive ingredient within the products. According to Harbourne et al. (2009), a significant drop in the amount of flavonoid compounds at 70°C and the optimum temperature that yielded highest extracts of phenols is 30°C. Referring to literature, the air velocity for convective air drying of herbs should be ranged from 0.5 to 1.5 m/s (Jałoszyński et al., 2008). Jałoszyński et al. (2008) found that high air velocity would reduce the yield of bioactive compounds.

The modeling of combined microwave vacuum dryer (CMVD) and CAD are next demonstrated.

Example 8.1: Combined Microwave Vacuum Dryer

Solid dryer is used to reduce the moisture content of samples. It can be operated at a fixed temperature, pressure, or at a fixed heat duty requirement. It is recommended to specify either temperature or pressure and placed a design specification, final moisture content on the product stream. The product design specification choice consists of rate, moisture content, ppm, fraction, and vapor fraction. Fig. 8.6 shows a detailed step of simulation for a CMVD drying process of bioactive compounds.

Example 8.2: Convective Air Drying

Fig. 8.7 shows a detailed step of simulation for a CAD drying process of bioactive compounds. Total amount of H_2O removed during the drying process is 3.73 kg-mol/h. Only 1.89 kg-mol/h of H_2O is retained in the process. However, the pressure value is not acceptable as it is below 101.325 kPa. For the CAD, removal of moisture content needs to take into account drying duration as one of the major parameter. The simulated results based only on

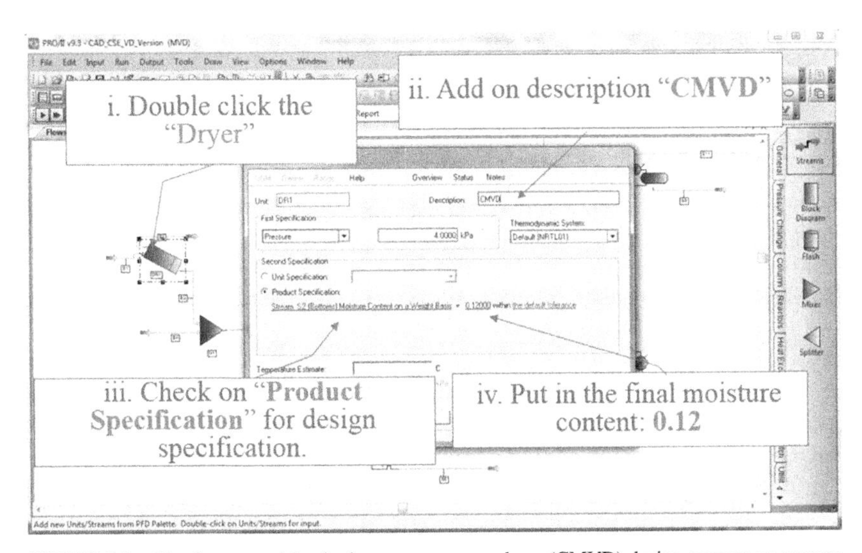

FIGURE 8.6 Simulate a combined microwave vacuum dryer (CMVD) drying process on process flow diagram.

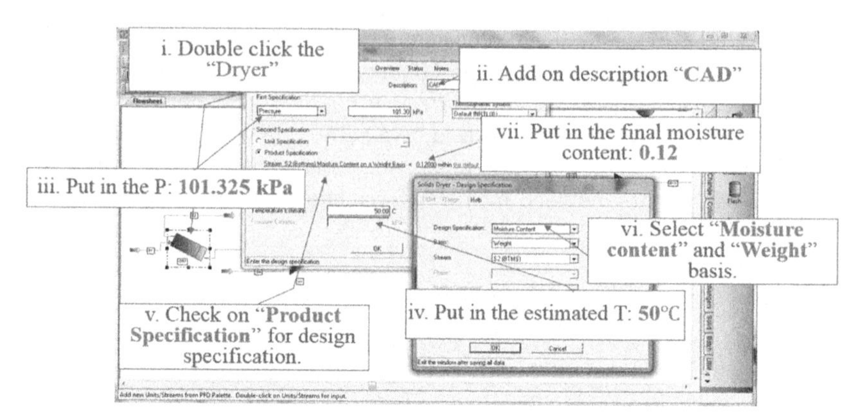

FIGURE 8.7 Simulate a convective air drying (CAD) process on process flow diagram.

the product specification set for moisture content for bottom stream. Therefore, for nonideal condition, some minor reductions should be observed at Stream 3 for all phenolic acids.

8.4 MODELING OF A CONVENTIONAL SOLVENT EXTRACTOR

A typical extraction method applied in industry is a conventional extraction (CE) using different solvents. It deals with transporting solute (bioactive compounds) from one phase (raw materials) into another (solvent) until it reaches the final equilibrium distribution. This method is a mature technology used in industry as it is of low risk and able to cope with most application problems. However, the usage of solvent brings environmental concerns as the effluent may be harmful and hazardous. Therefore, selection of solvent is important to minimize this effect. Solvent is selected by referring to the properties of the solvents, which are mostly organic compounds with lipophilic ("liking fat") and hydrophobic properties. This is because the solute has an opposite properties (hydrophobic and lipophilic) than the organic solvent, so it allows the transfer from aqueous solution into the organic phase. This phenomenon relates to the solubility of the solvent and solute. Researchers are also putting effort in dealing with regeneration of the extractant to achieve sustainability development (Rydberg, 2004).

Within a decade, natural deep eutectic solvents (NaDESs) were developed by researchers. The yield of phenolic compounds extracted using the NaDES as solvent is up to 14% higher than water and ethanol (Dai et al., 2013b). Solvents with this potential are ionic liquid (IL) and deep eutectic solvent (DES). Both solvents have unique properties and are formed when two or more solid crystalline compounds are mixed together. Currently, these solvents are used in purifying biodiesel to remove unwanted substances instead of pharmaceutical industry because of the toxicity characteristics. However, few years ago a researcher found that natural product components such as sugars, organic acids, amino acids, choline, and urea are excellent solvents in the extraction of bioactive compounds (Dai et al., 2013a). These solvents are known as NaDESs. Natural products are ideal source of ILs and DES because of their enormous chemical diversity, high-solubilization power of both polar and nonpolar compounds, biodegradable properties, and pharmaceutically acceptable toxicity profile (Dai et al., 2013b). It has a low melting point of $<100°C$. NaDESs have an advantage over IL because it is easier to prepare them with high purity at low cost (Dai et al., 2013b). The author elucidate the NaDES is a new extraction solvent for phenolic metabolites from *Carthamus tinctorius*. The characteristic of this solvent is beneficial for extraction compared to DES and IL as it has adjustable viscosity, is more sustainable, and exists in liquid state even below 0°C. However, this solvent inherits high viscosity properties, which would reduce the extraction efficiency. Thus far, research related to extraction of herbs using the NaDES as a solvent is rather scarce. It is considered a new technology, and much research is needed before it can be used at industry scale. This solvent is important in extraction processes because of its physicochemical properties i.e., it can dissolve in many kinds of solutes and increase the selectivity of extractions. It can be a potential cosolvent for the CO_2 SFE.

Extraction yield of SFE method is higher than CE using different solvents. The solvent used in SFE technology is CO_2 with organic cosolvents. Organic solvents such as ethanol, methanol, hexane, dichloromethane, and acetate are used in the SFE method. CO_2 is an excellent solvent for extraction of bioactive compounds as it is environmental friendly, inert, nonflammable, nontoxic, and cheap and has a low critical temperature of 31.26 °C and a pressure of 7.38 MPa, where it is effective for extraction of heat-sensitive components. It is also approved as a harmless substance in food and pharmaceutical industry (Kumoro and Hasan, 2007). The supercritical CO_2 has higher diffusivity than other fluids, which enhances mass transfer for higher extraction rate (Chen et al., 2011). In SFE technology, pressure is the primary factor on the solubility of solvent. High pressure is needed to increase the efficiency of extraction. To minimize energy consumption, the pressure was set at 30 MPa in this chapter. Furthermore, a cosolvent is essential to improve the extraction yield. Cerón et al. (2014) recommended 60% of ethanol to be used in the processing of phenolic acids. To prevent the thermal denature of phenolic acids, the temperature of SFE was set at 45°C.

Example 8.3: Conventional Solvent Extraction

The user is required to select a distillation model from the unit operations panel. To simulate an extraction column, pressure profile, feed and products, condenser, and performance specification are required. Table 8.4 shows data

TABLE 8.4 Equipment Specifications

Equipment	Specification
Conventional solvent extraction	Pressure (kPa) = 150 Pressure drop (kPa) = 10 Condenser = bubble temperature Specification 1 = all components in 200 kg/h Reflux ratio = 3 Feed tray = 1 and 5 Bottom flowrate = 1 kgmol/h

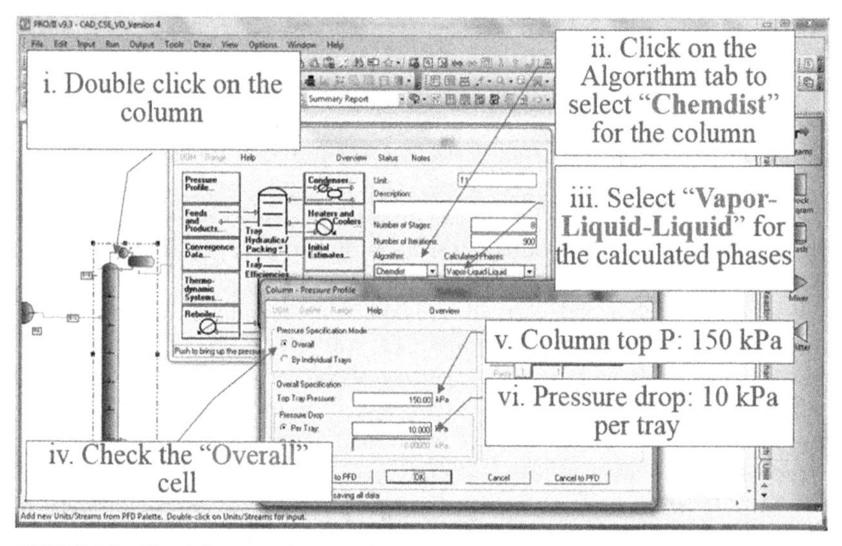

FIGURE 8.8 Simulation steps for extraction process of *Strobilanthes crispus* (Pressure Profile).

required for the simulation. Detailed steps for simulating an extraction process are shown in Figs. 8.8–8.12.

Example 8.4: CO_2 Supercritical Fluid Extraction

Simulation steps for CO_2 SFE are similar to the solvent extraction in Example 8.3. Mainly, the second feed stream is to be replaced by CO_2 (Fig. 8.13). The user can replace with NaDES combined with other solvents as well to evaluate the extraction results following detailed steps stated above. The last stage is the purification process to recycle excessive solvents. Fig. 8.14 shows a complete process flow diagram for a drying, extraction, and purification process of *S. crispus*.

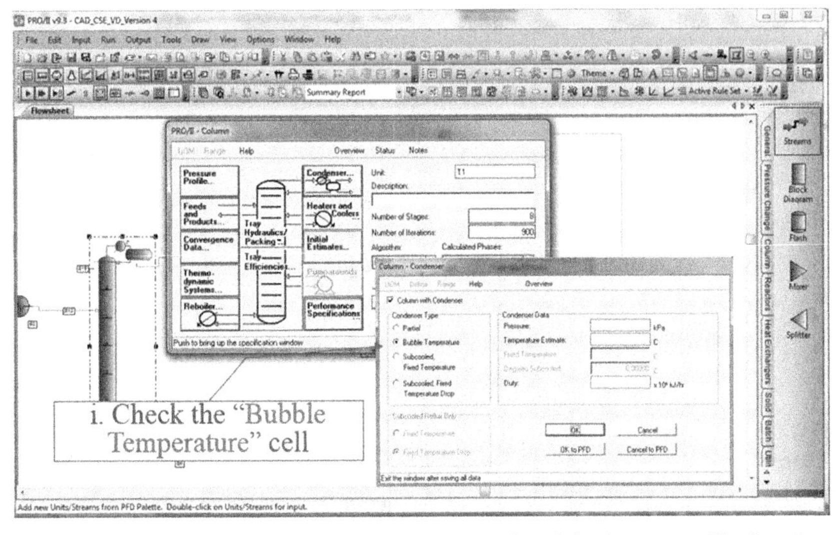

FIGURE 8.9 Simulation steps for extraction process of *Strobilanthes crispus* (Condenser).

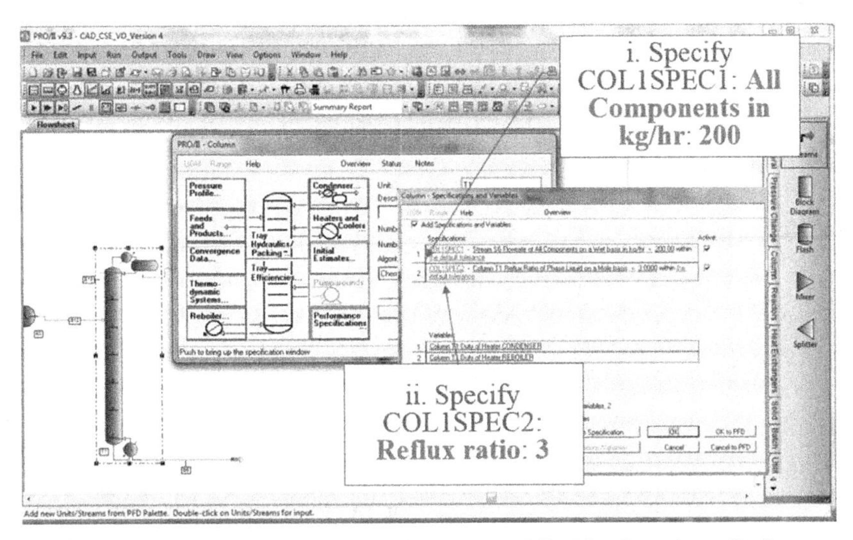

FIGURE 8.10 Simulation steps for extraction process of *Strobilanthes crispus* (Performance Specifications).

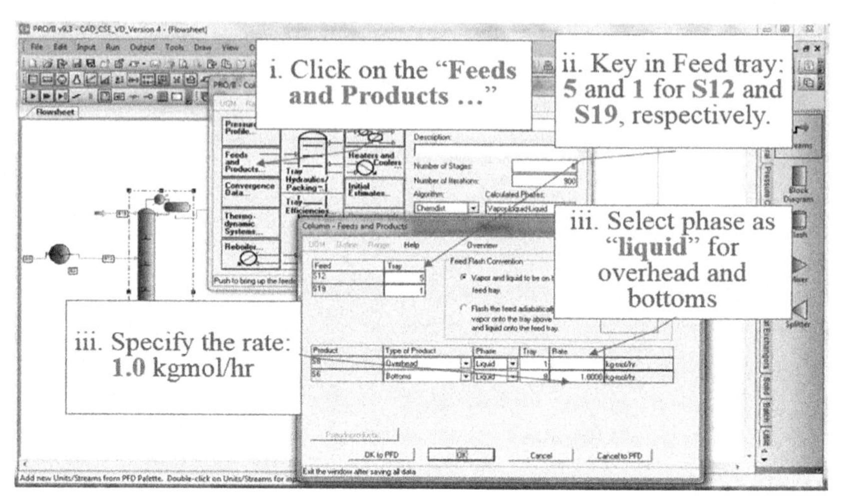

FIGURE 8.11 Simulation steps for extraction process of *Strobilanthes crispus* (Feeds and products).

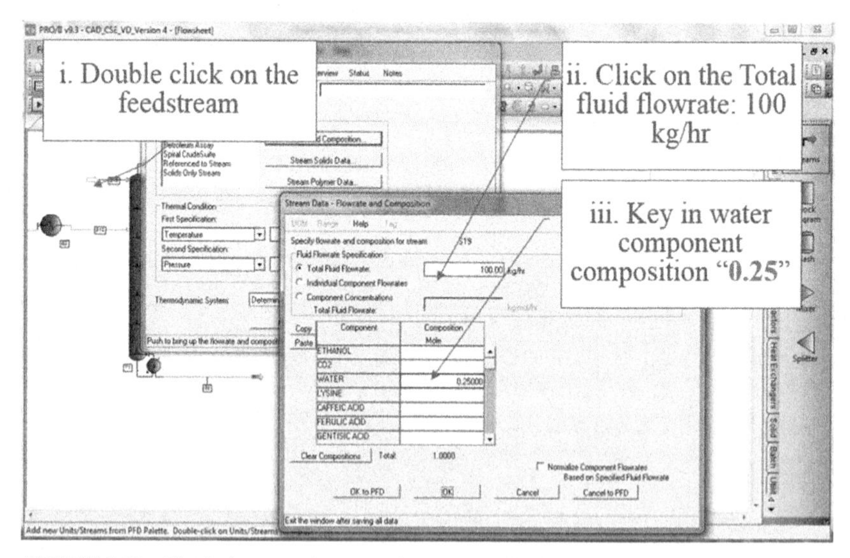

FIGURE 8.12 Simulation steps for extraction process of *Strobilanthes crispus* (Feed stream).

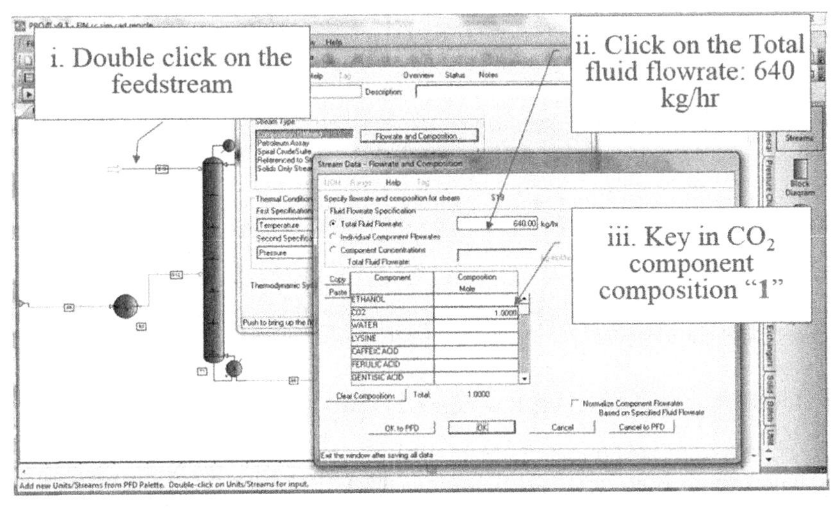

FIGURE 8.13 Simulation steps for CO_2 supercritical fluid extraction process of *Strobilanthes crispus* (Feed stream).

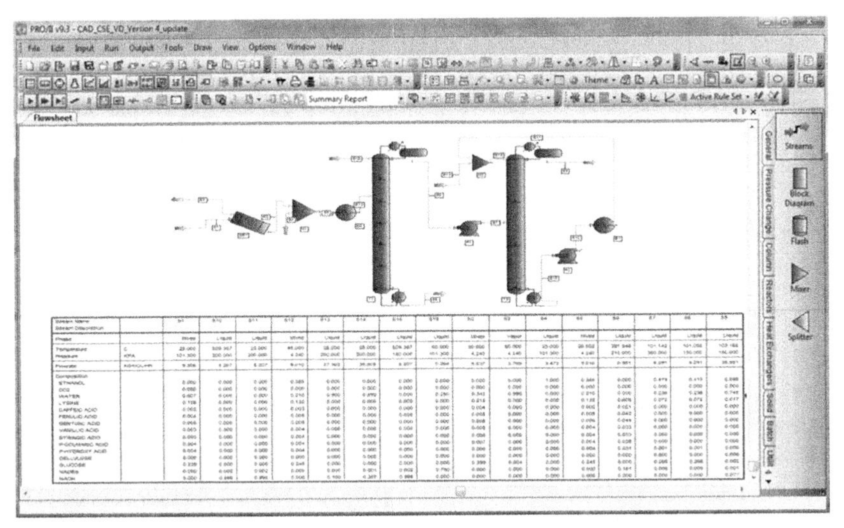

FIGURE 8.14 Process flow diagram for a drying, extraction, and purification process of *Strobilanthes crispus*.

8.5 CONCLUSIONS

Different processing units can be simulated using the PRO/II software to validate the retention of bioactive compounds of biomaterial during processing. Then, the user can use the simulation data related to percentage of yield and cost and can perform optimization for a different processing pathway.

ACKNOWLEDGMENT

The authors wish to acknowledge the support given by Schneider Electric Software, LLC, to Taylor's University, School of Engineering, in sponsoring a LAN copy of steady-state process design and optimization-Bundle #1 for 50 Concurrent Users for 36 months.

EXERCISE

1. A simplified extraction process related to processing of polyphenolic compounds from *Manilkara zapota* is shown in Fig. E1. Samples are kept in a freezer before processing; it is sent to an oven to remove water and ground into powder for extraction purposes. Next, samples are mixed with a solvent to extract compounds and subjected to a vacuum distillation column to recycle CO_2. Detailed composition of process feed compositions are shown in Tables E1 and E2, respectively.

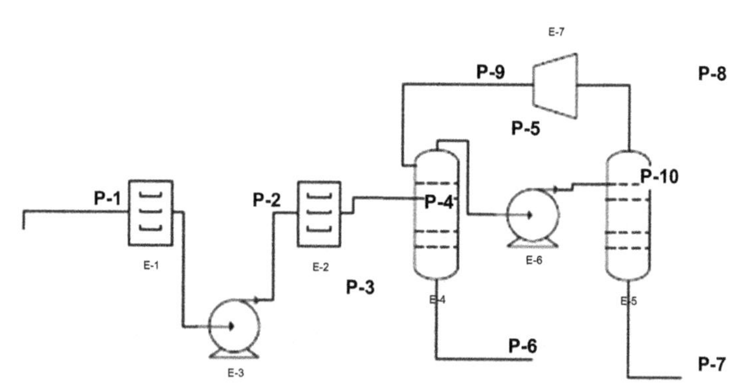

FIGURE E1 Modified extraction process to obtain the highest yield of polyphenolic compounds (Cerón et al., 2014).

TABLE E1 Specification for Process Feed Stream

Components	Composition (kg mol/h)
CO_2	100
H_2O	80
Gallic acid	10
DPPHTH	10

TABLE E2 Equipment Specifications (Cerón et al., 2014)

Equipment	Tag	Specification
Convective air drying	1	Temperature (K) = 333 Air velocity (m/s) = 1.5
Freeze drying	3	Vacuum pressure (mbar) = 400 Temperature (K) = 313
Supercritical CO_2 extractor	8	Pressure (bar) = 300 Temperature (K) = 318
Vacuum distillation	9	Temperature (K) = 318

REFERENCES

Cerón, I., Ng, R., El-Halwagi, M., Cardona, C., April 2014. Process synthesis for antioxidant polyphenolic compounds production from *Matisia cordata Bonpl.*(zapote) pulp. Journal of Food Engineering 134, 5–15.

Chan., J.W.R., Chong, C.H., Denny, K.S. Ng, 2015. Process synthesis and design for extraction of bioactive compounds from *Strobilanthes crispus* Leaves (SC). Journal of Engineering Science and Technology 113–137. (Special issue).

Chen, K., Yin, W., Zhang, W., Pan, J., 2011. Technical optimization of the extraction of andrographolide by supercritical CO_2. Food and Bioproducts Processing 89 (2), 92–97.

Dai, Y., van Spronsen, J., Witkamp, G.-J., Verpoorte, R., Choi, Y.H., 2013a. Natural deep eutectic solvents as new potential media for green technology. Analytica Chimica Acta 766, 61–68.

Dai, Y., Witkamp, G.-J., Verpoorte, R., Choi, Y.H., 2013b. Natural deep eutectic solvents as a new extraction media for phenolic metabolites in *Carthamus tinctorius* L. Analytical Chemistry 85 (13), 6272–6278.

Green, D.W., 2008. Perry's Chemical Engineers' Handbook. McGraw-hill, New York.

Harbourne, N., Marete, E., Jacquier, J.C., O'Riordan, D., 2009. Effect of drying methods on the phenolic constituents of meadowsweet (*Filipendula ulmaria*) and willow (*Salix alba*). Lwt-food Science and Technology 42 (9), 1468–1473.

Ismail, M., Manickam, E., Danial, A.M., Rahmat, A., Yahaya, A., 2000. Chemical composition and antioxidant activity of *Strobilanthes crispus* leaf extract. The Journal of Nutritional Biochemistry 11 (11), 536–542.

Jałoszyński, K., Figiel, A., Wojdyło, A., 2008. Drying kinetics and antioxidant activity of oregano. Acta Agrophysica 11 (1), 81–90.

Koay, Y.C., Wong, K.C., Osman, H., Eldeen, I.M.S., Asmawi, M.Z., 2013. Chemical consituents and biological activities of *Strobilanthes cripus* L. Records of Natural Products 7, 59–64.

Kondo, M., Goto, M., Kodama, A., Hirose, T., 2002. Separation performance of supercritical carbon dioxide extraction column for the citrus oil processing: observation using simulator. Separation Science and Technology 37 (15), 3391–3406.

Kumoro, A.C., Hasan, M., 2007. Supercritical carbon dioxide extraction of Andrographolide from *Andrographis paniculata*: effect of the solvent flow rate, pressure and temperature. Chinese Journal of Chemical Engineering 15, 877–883.

Liza, M.S., Abdul Rahman, R., Mandana, B., Jinap, S., Rahmat, A., Zaidul, I.S.M., Hamid, A., 2010. Supercritical carbon dioxide extraction of bioactive flavonoid from *Strobilanthes Crispus* (Pecah Kaca). Food and Bioproducts Processing 88 (2–3), 319–326.

PRO/II 9.3.2, Reference Manual 2015, SimSci. January 2015. SimSci, Schneider Electric, USA.

Rodriguez, J., Melo, E., Mulet, A., Bon, J., 2013. Optimization of the antioxidant capacity of thyme (*Thymus vulgaris* L.) extracts: management of the convective drying process assisted by power ultrasound. Journal of Food Engineering 119 (4), 793–799.

Rydberg, J., 2004. Solvent Extraction Principles and Practice, Revised and Expanded. CRC Press.

Schulze, B., Hubbermann, E.M., Schwarz, K., 2014. Stability of quercetin derivatives in vacuum impregnated apple slices after drying (microwave vacuum drying, air drying, freeze drying) and storage. LWT-Food Science and Technology 57 (1), 426–433.

APPENDIX A

Normal boiling point (NBP) of phenolic acids is estimated using the Nannoolal method (Green, 2008). It was calculated based on the chemical structure and intramolecular interaction where NBP in Kelvin is obtained from:

$$NBP = \frac{\sum_{i=1}^{N} n_i \times C_i}{n^{0.6583} + 1.6868} + 84.3395 \tag{8.1}$$

where n = number of nonhydrogen atoms, n_i = number of occurrences of group i, N = number of groups, C_i = group contribution.

ID	Gentisic Acid	n_i	C_i	Group Total
15	Aromatic = CH–	3	235.3462	706.0386
17	Aromatic = C< connected to O	2	348.2779	696.5558
16	Aromatic = C< not connected to O	1	315.4128	315.4128
37	OH connected to an aromatic C (phenols)	2	361.4775	722.955
44	COOH connected to C	1	1080.314	1080.314
		9		3521.276
		NBP	K	°C
		Eq. (8.1)	677.6729	404.5229

ID	p-Coumaric Acid	n_i	C_i	Group Total
15	Aromatic = CH−	4	235.3462	941.3848
17	Aromatic = C< connected to O	1	348.2779	348.2779
16	Aromatic = C< not connected to O	1	315.4128	315.4128
37	OH connected to an aromatic C (phenols)	1	361.4775	361.4775
44	COOH connected to C	1	1080.314	1080.314
5	>CH− in a chain	2	249.5809	499.1618
		10		3546.029
		NBP	K	°C
		Eq. (8.1)	652.6293	379.4793

ID	p-Hydroxy Benzoic Acid	n_i	C_i	Group Total
15	Aromatic = CH−	4	235.3462	941.3848
17	Aromatic = C< connected to O	1	348.2779	348.2779
16	Aromatic = C< not connected to O	1	315.4128	315.4128
37	OH connected to an aromatic C (phenols)	1	361.4775	361.4775
44	COOH connected to C	1	1080.314	1080.314
		8		3046.867
		NBP	K	°C
		Eq. (8.1)	626.698	353.548

ID	Ferulic Acid	n_i	C_i	Group Total
15	Aromatic = CH−	3	235.3462	706.0386
17	Aromatic = C< connected to O	2	348.2779	696.5558
16	Aromatic = C< not connected to O	1	315.4128	315.4128
37	OH connected to an aromatic C (phenols)	1	361.4775	361.4775
44	COOH connected to C	1	1080.314	1080.314
5	>CH− in a chain	2	249.5809	499.1618
2	CH_3 connected to C	1	251.8338	251.8338
		11		3910.794
		NBP	K	°C
		Eq. (8.1)	682.8102	409.6602

ID	Syringic Acid	n_i	C_i	Group Total
15	Aromatic = CH−	2	235.3462	470.6924
17	Aromatic = C< connected to O	3	348.2779	1044.834
16	Aromatic = C< not connected to O	1	315.4128	315.4128
37	OH connected to an aromatic C (phenols)	1	361.4775	361.4775
44	COOH connected to C	1	1080.314	1080.314
2	CH_3 connected to C	2	251.8338	503.6676
		10		3776.398
		NBP	K	°C
		Eq. (8.1)	689.5485	416.3985

ID	Vinillic Acid	n_i	C_i	Group Total
15	aromatic = CH−	3	235.3462	706.0386
17	aromatic = C< connected to O	2	348.2779	696.5558
16	aromatic = C< not connected to O	1	315.4128	315.4128
37	OH connected to an aromatic C (phenols)	1	361.4775	361.4775
44	COOH connected to C	1	1080.314	1080.314
2	CH_3 connected to C	1	251.8338	251.8338
		9		3411.632
		NBP	K	°C
		Eq. (8.1)	659.1979	386.0479

ID	Caffeic Acid	n_i	C_i	Group Total
15	Aromatic = CH−	3	235.3462	706.0386
17	Aromatic = C< connected to O	2	348.2779	696.5558
16	Aromatic = C< not connected to O	1	315.4128	315.4128
37	OH connected to an aromatic C (phenols)	2	361.4775	722.955
44	COOH connected to C	1	1080.314	1080.314
5	>CH− in a chain	2	249.5809	499.1618
		11		4020.438
		NBP	K	°C
		Eq. (8.1)	699.589	426.439

Part 4

ProMax

Chapter 9

Basics of Process Simulation With ProMax

René D. Elms
Texas A&M University, College Station, TX, United States

In this chapter, the basics of process simulation with ProMax (BR&E, 2017) will be demonstrated through a step-by-step guide in modeling a simple *n*-octane production case. The concept of simulation is based on a *sequential modular* approach and follows the onion model for flowsheet synthesis (see Chapter 1 for details).

9.1 INTRODUCTION

A simple example that involves the production of *n*-octane (C_8H_{18}; Foo et al., 2005) is demonstrated, with detailed descriptions given in Example 1.1. To model this example in ProMax, the following steps are necessary:

1. Setting the Environment: Selection of the appropriate thermodynamics package and addition of the relevant components to the simulation.
2. Creating and adding the Reaction Set: The Reaction Set that will be used in the Reactor Block is created and added to the Simulation Environment.
3. Flowsheeting:
   a. Addition of Blocks to the flowsheet.
   b. Addition and connection of Process and Energy Streams to the flowsheet.
4. Specification of Blocks and Streams.
5. Determination of a Recycle Block Guess and closing of the recycle loop.
6. Viewing the results.

9.2 SETTING THE ENVIRONMENT

In ProMax, the Environment is where the user defines the thermodynamics package (termed as "Property Package" in ProMax terminology) utilized to solve the flowsheet, as well as to select the components that will be used on the flowsheet. Each flowsheet may have only one Environment. For cases where multiple flowsheets are contained in a project, each flowsheet can have a

Chemical Engineering Process Simulation. http://dx.doi.org/10.1016/B978-0-12-803782-9.00009-1

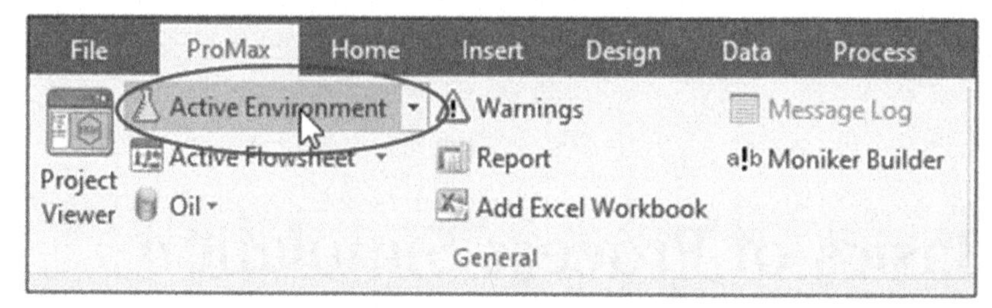

FIGURE 9.1 Selecting the Active Environment option on the General grouping of the ProMax tab.

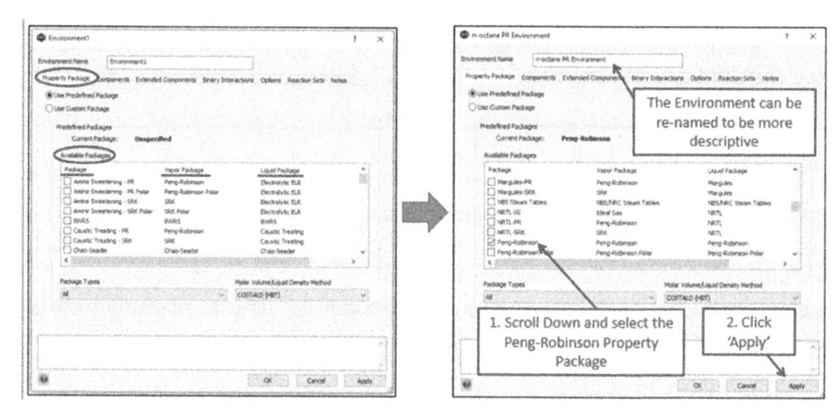

FIGURE 9.2 Selecting the Peng–Robinson Property Package in the Property Package tab of the Environment dialogue.

different Environment. The Environment dialogue is accessed by selecting the "Active Environment" option in the ProMax tab (see Fig. 9.1).

The Environment dialogue (Fig. 9.2) opens by default with the Property Package (thermodynamic models) tab displayed. When the Environment is accessed for the first time, a Property Package must be selected before proceeding to other tabs. Many of the available Property Packages are named for ease of identification for use with certain types of processes. For example, the first package listed is "Amine Sweetening–PR," utilizing Peng–Robinson for the Vapor Package and an Electrolytic Liquid Package. Apart from some commonly encountered packages such as SRK, NRTL, UNIQUAC, and UNIFAC,[1] examples of packages for specific uses include Caustic Treating, Heat Transfer Fluid, National Bureau of Standards (NBS) Steam Tables, and Sulfur. For the current example, the Peng–Robinson Property Package is selected (Fig. 9.2). (Note that the SRK Property Package would also be appropriate.)

Tip: When working with multiple Environments, it is recommended to rename each Environment with a more descriptive name.

---

1. Refer to Chapter 3 for detailed discussion on various physical properties methods.

TABLE 9.1 Components Involved in the *n*-Octane Example

Components	Role in Reaction
Ethylene (C_2H_6)	Reactant
i-Butane (*i*-C_4H_{10})	Reactant
n-Octane (*n*-C_8H_{18})	Product
Nitrogen (N_2)	Inert
n-Butane (*n*-C_4H_{10})	Inert

Once the Property Package is selected, the components involved in the simulation (Table 9.1) are added in the Components tab of the Environment dialogue. Components may be added one at a time by typing their name in the "Pure Components Filtering Criteria" area (see detailed steps in Fig. 9.3). Components may also be added or installed by double-clicking directly on the desired item name in this column.

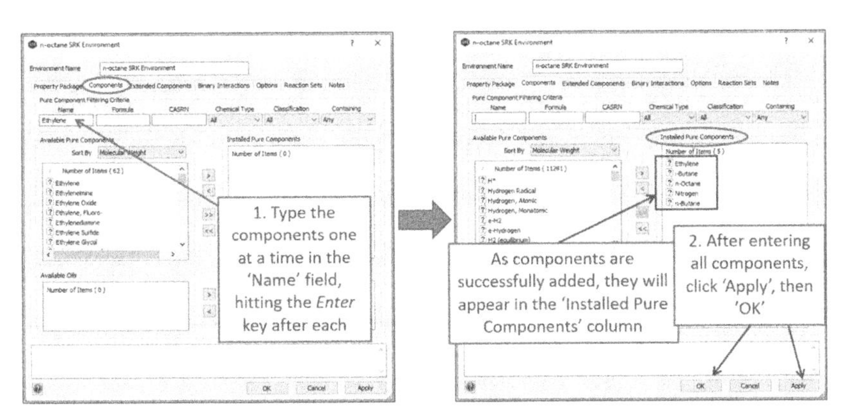

FIGURE 9.3 Adding components to the Environment in the Components tab of the Environment dialogue.

Tip: Some components are also available under alternative names (e.g., "C2" for Ethane). In addition to the name, components may be searched by formula, CASRN, chemical type, classification, and whether they contain certain groups.

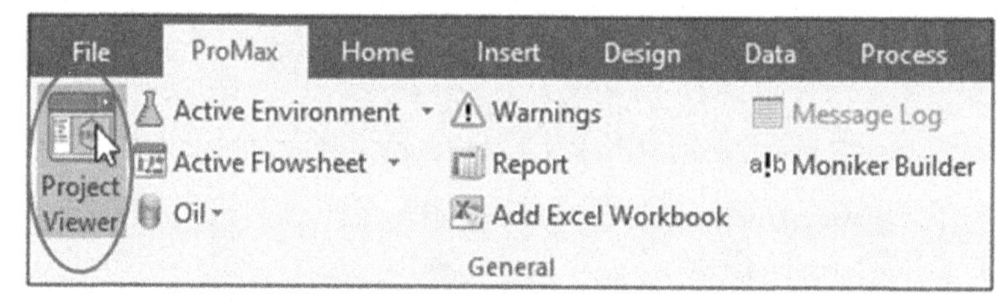

FIGURE 9.4 Selecting the Project Viewer option on the General grouping of the ProMax tab.

9.3 CREATING AND ADDING THE REACTION SET

When modeling a reactor in ProMax with one or more specified reactions, the reaction(s) is first defined or specified in a "Reaction Set." After creating the relevant Reaction Set, it must then be installed or added to the flowsheet Environment before it is available for use.

9.3.1 Creating the Reaction Set

To create a Reaction Set, open the Project Viewer by clicking on the Project Viewer icon in the General grouping of the ProMax tab (Fig. 9.4).

The Project Viewer is the main user interface for entering specifications and retrieving information and results. A Reaction Set is added by following the detailed steps given in Fig. 9.5.

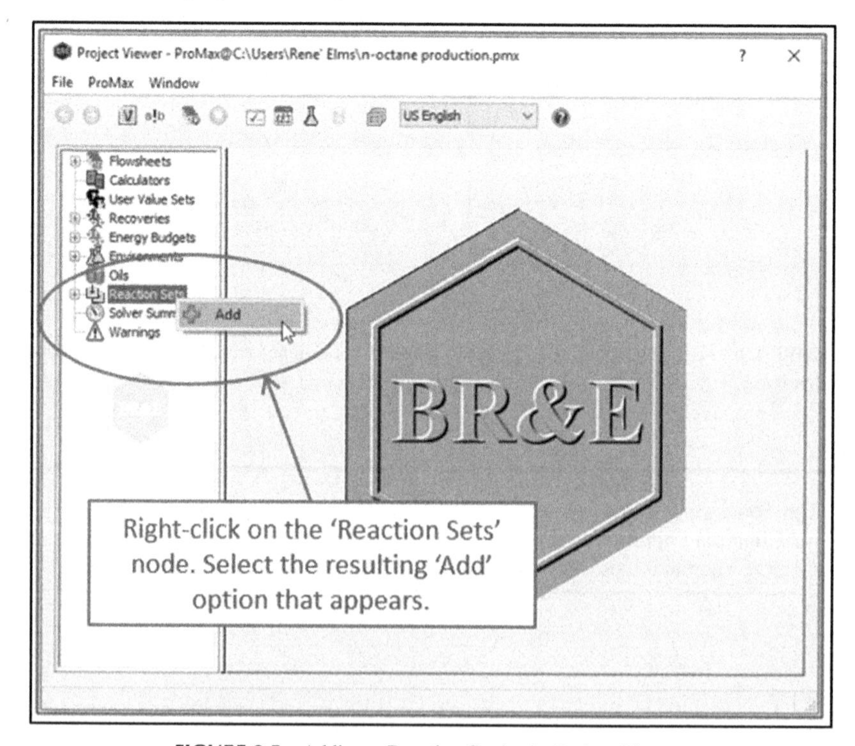

FIGURE 9.5 Adding a Reaction Set in the Project Viewer.

Tip: The Project Viewer can also be opened by double-clicking on any Stream or Block on the flowsheet.

Table 9.2 displays the needed Reaction Set specifications for the current example. The Reaction Set dialogue is where the initial specifications will be made (Fig. 9.6). By default, a "Kinetic" Reactor Type is selected in the General tab of the Reaction Set dialogue. Because it is desired to model a conversion reactor in this example, the "Kinetic" checkbox is to be deselected after the "Conversion" Reactor Type is selected (see Fig. 9.6 for details). Although not needed in this case, multiple Reactor Types may be specified for a given Reaction Set. Although it is not required, it is highly recommended to rename the Reaction Set, especially when there are multiple reactions and Reaction Sets in a simulation exercise.

Fig. 9.7 depicts the detailed steps in specifying the *n*-octane reaction in the Reaction window. As with the Reaction Set, it is recommended to rename the default Reaction Name to be more descriptive. In the Stoichiometry and Reaction Orders table, the reaction components and their stoichiometry are specified. To specify components, either the name can be entered directly or the Species dialogue can be utilized by selecting a cell in the Components column and clicking on the arrow that will appear on the right side of the cell (not shown in Fig. 9.7). Note that when entering the stoichiometry, reactants are represented by negative values and products by positive values.

The long table on the right side of the Reaction window contains various fields to specify different reaction types. Because the current simulation is for a conversion reaction, the relevant specifications of interest are located at the bottom of the table. The conversion basis for this reaction is ethylene, with conversion rate of 98% (hence, the fractional basis for the coefficient of "A" is specified as 0.98). Other detailed steps are found in Fig. 9.7.

TABLE 9.2 Reaction Set Specifications for *n*-Octane Example

Specification	Value	Location of Specification
Reactor Type	Conversion	Reaction Set dialogue, General tab
Conversion Base	Ethylene	Reaction specification dialogue
Conversion Equation Coefficient "A" (Fractional)	0.98	Reaction specification dialogue
Reaction Components, Stoichiometry	Ethylene, -2 *n*-Butane, -1 *n*-Octane, 1	Reaction specification dialogue

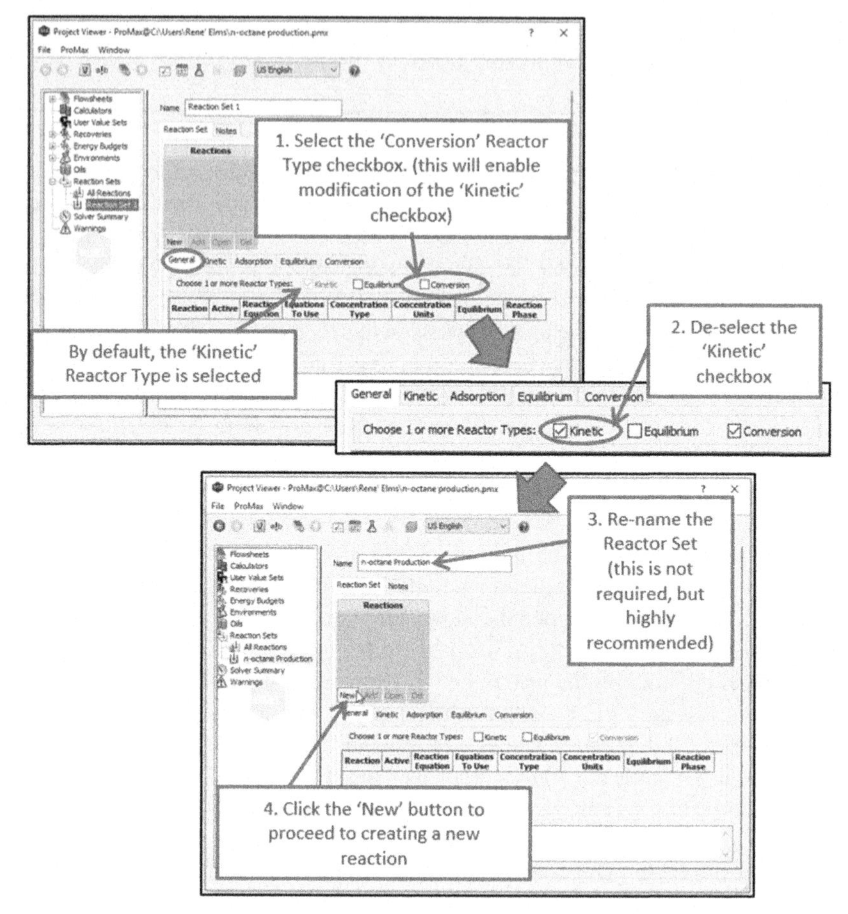

FIGURE 9.6 Initial specification of the Reactor Set in the Reactor Set dialogue.

9.3.2 Adding the Reaction Set to the Environment

Once a Reaction Set is defined, it must be added to the Environment to be used in the simulation. Detailed steps for doing so are found in Fig. 9.8. As shown, the *n*-octane Production Reaction Set will appear in the "Available Reaction Sets" column in the Reaction Sets tab of the Environment dialogue.

9.4 FLOWSHEETING AND SPECIFICATION OF BLOCKS AND STREAMS

ProMax utilizes Microsoft Visio as the simulator interface. The basic structure of ProMax can be seen in Fig. 9.9. Process units and items are segregated into "Shapes," which reside on the left-hand side of the Flowsheeting area. Items are

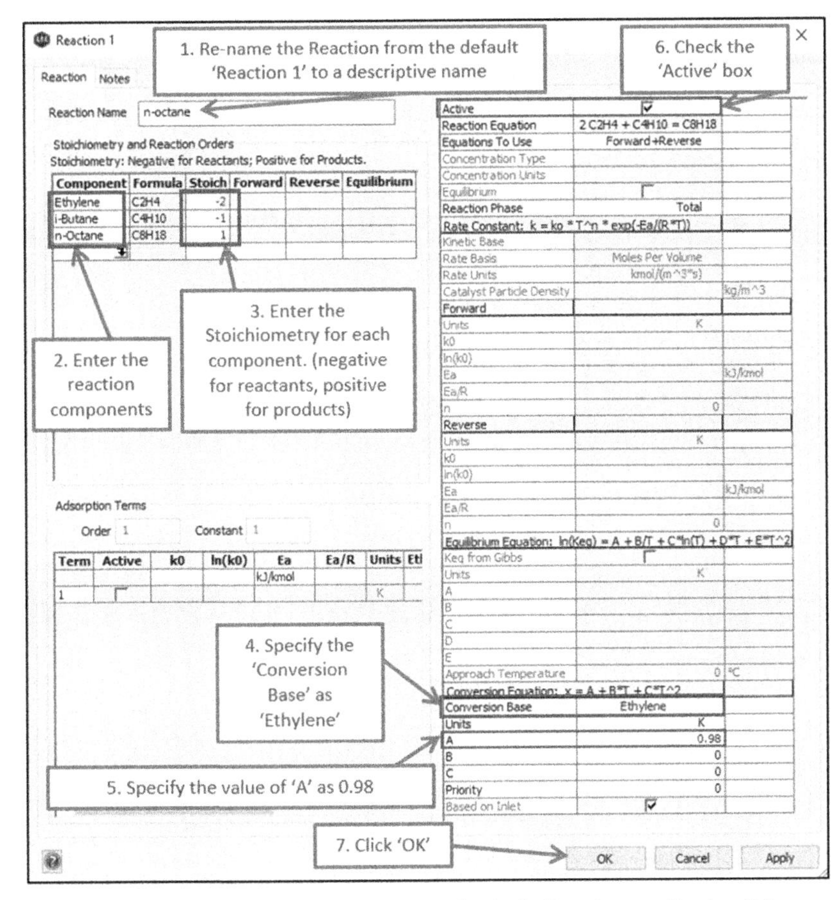

FIGURE 9.7 Specification of the *n*-octane reaction in the Reaction specification dialogue.

added to the flowsheet by dragging and dropping. The message log across the bottom serves as a real-time activity log, displaying and documenting certain changes, errors, warnings, and information pertaining to convergence of Blocks, calculators, and the simulation as a whole. As demonstrated previously, the ProMax tab allows access to various simulator components and functions, such as the Environment, Project Viewer, and "Execute" buttons. Recall the *n*-octane production case in Example 1.1; Table 9.3 provides a summary of the Blocks utilized to model the related process. In addition, the specific Block version used and the ProMax Shape in which each is located are listed.

Construction of the simulation will begin with adding the Reactor, Distillation Column, and Splitter Blocks. In order, each of these Blocks will be specified and solved. Next, each of the components of the recycle loop and inlet preheating will be added, specified, and solved. After a Recycle Guess is obtained, the temporary Reactor Block inlet stream will be removed, the Recycle Loop will be closed, and the full flowsheet will be converged.

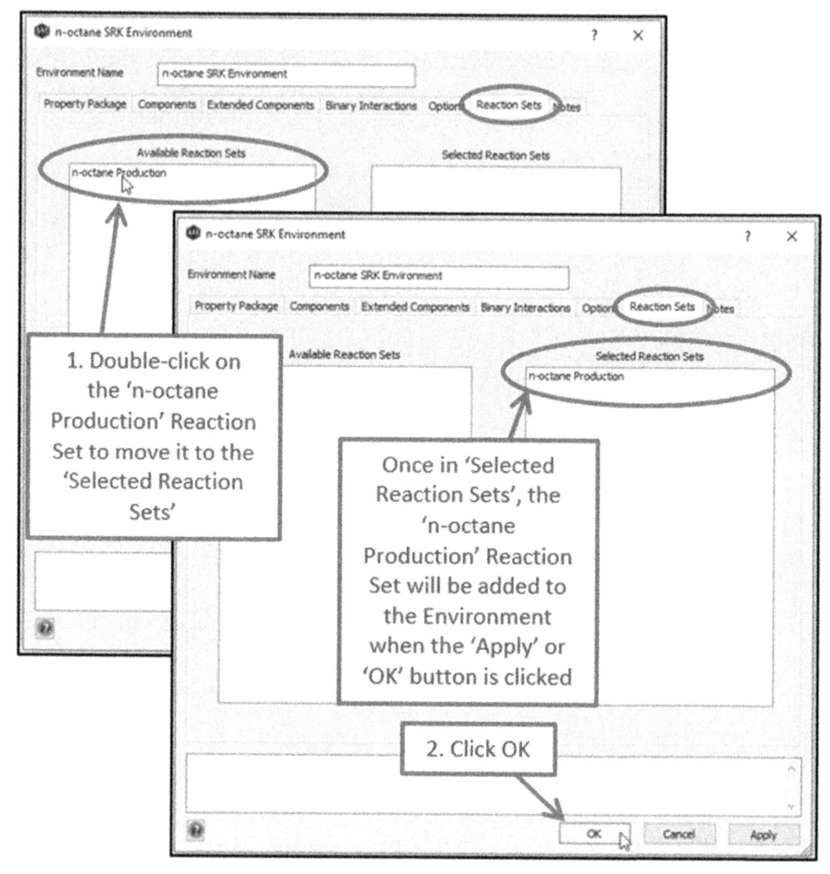

FIGURE 9.8 Adding the "*n*-octane Production" Reaction Set to the "*n*-octane SRK Environment" in the Reaction Sets tab of the Environment dialogue.

9.4.1 Adding and Connecting the Reactor, Distillation Column, and Splitter Blocks

9.4.1.1 Reactor Block

Reactor Blocks in ProMax are found in the "ProMax Reactors" Shape and are available in several configuration types: Single Sided Reactors, Reactive Separators, Cross Reactors, and Multiphase Reactors (see Fig. 9.9). For the current example, a Reactive Separator configuration type is appropriate, specifically either "Reactive Separator 1" or "Reactive Separator 2." Both of these configurations have a single inlet, a vapor outlet, and a liquid outlet. For this example, the "Reactive Separator 1" Block is used and is added to the flowsheet following the detailed procedure as depicted in Fig. 9.9 (Step 1).

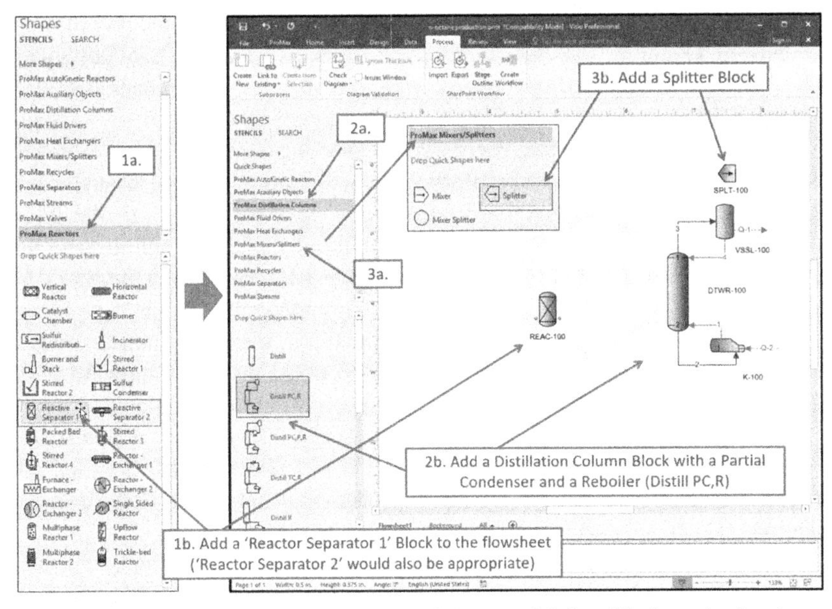

FIGURE 9.9 Adding the Reactor, Distillation Column, and Splitter Blocks to the flowsheet.

TABLE 9.3 Information About the Blocks Used in the *n*-octane Example

Block	Version Used	Related ProMax Shape
Cross Exchanger	Cross Exchanger	Heat Exchangers
Single Sided Exchanger	Single Sided Exchanger	Heat Exchangers
Mixer	Mixer	Mixers/Splitters
Reactor	Reactive Separator 1	Reactors
Distillation Column	Distill PC,R	Distillation Columns
Splitter	Splitter	Mixers/Splitters
Compressor	Compressor 2	Fluid Drivers
Recycle	Recycle	Recycles

9.4.1.2 Distillation Column and Splitter Blocks

Distillation Blocks are found in the "ProMax Distillation Columns" Shape, as can be seen in Fig. 9.9. Although distillation columns can be configured "from scratch" by adding individual Blocks for the column, condenser, and reboiler, it is highly recommended to utilize one of the preconfigured options with various combinations of partial and total condensers, reboilers, and pumps. For the current example, the "Distill PC, R" Block is added to the flowsheet (Fig. 9.9, Step 2). This Block is preconfigured with a partial condenser and reboiler.

To represent the purge that occurs in this process, a Splitter Block is added (Fig. 9.9—Step 3). The Splitter Block splits a stream as defined by the user. The split occurs in the direction of the arrow on the Block, with the inlet attached to the side of the Block with the "tail" of the arrow, and the outlets attached to the side of the Block with the head of the arrow. When the Block is dropped on the flowsheet, by default it will be oriented with the arrow pointing to the right. To rotate it to a better position for this example, click on the rotation handle and rotate to the left so as to point the arrow upward (see Fig. 9.10). This same procedure can be used to reorient other Blocks.

Notice in Fig. 9.10, the Block label is on the top of the Block, which is not a good location. To move Block labels (Fig. 9.11), select the Block, click on the yellow square in the middle of the label, and drag and drop it to the desired location.

To customize a Block name, such as renaming the Splitter Block "Purge," with the Block selected (Fig. 9.12), start typing the desired new Block name. When finished, click away onto a blank part of the flowsheet.

9.4.1.3 Addition and Connection of the Process and Energy Streams

The Process and Energy Streams are located in the "ProMax Streams" Shape. As seen in Fig. 9.13, Process Streams are added starting with the location of

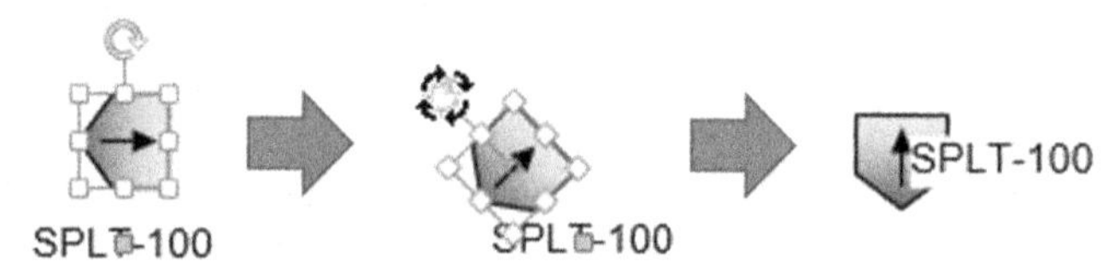

FIGURE 9.10 Rotating the Splitter Block.

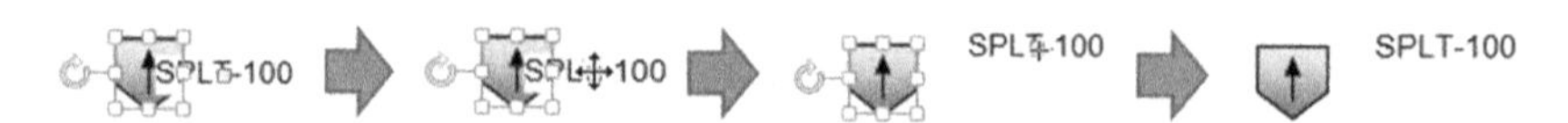

FIGURE 9.11 Moving the Splitter Block label.

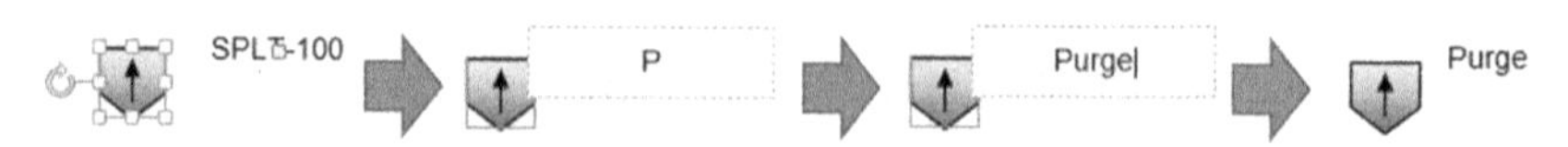

FIGURE 9.12 Renaming the Splitter Block.

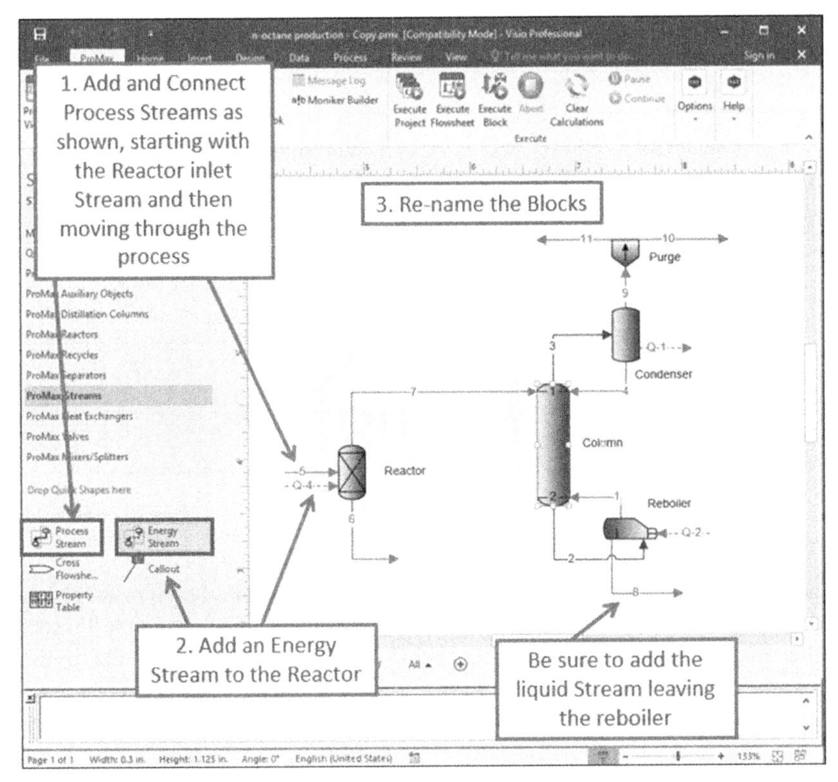

FIGURE 9.13 Adding and connecting the Process and Energy Streams to the Reactor, Distillation, and Splitter Blocks.

the Reactor Inlet and moving through the process to the Distillation Column, then finishing with the Purge Block. Next, an Energy Stream is added to the Reactor at one of the energy connection points, which are denoted by red "Q's." Lastly, the Blocks are renamed. Renaming Blocks is not required but is done here for ease of reference.

Tip: To add and connect Process and Energy Streams more quickly, use the "Connector Tool" in the Tools Grouping of the Home tab. Once finished adding Streams with this tool, be sure to select the "Pointer Tool" in the same location to return the cursor to its default state.

TABLE 9.4 Specifications for the Reactor Inlet Stream

Temperature = 93°C	
Pressure = 20 psia	
Component	**Flowrate (kmol/h)**
Ethylene	20
i-Butane	10
n-Octane	0
Nitrogen	0.1
n-Butane	0.5

9.4.2 Specifying and Executing the Reactor Inlet, Reactor, Distillation Column, and Splitter Blocks

9.4.2.1 Reactor Inlet Process Stream

The specifications for the temporary Reactor inlet stream are shown in Table 9.4. To specify this stream (Stream 5), the Process Stream dialogue in the Project Viewer is first opened by double-clicking on the stream on the flowsheet. In the Stream Properties tab, the temperature (93°C) and pressure (20 psia) are specified in their respective fields in the Total column. Next, the stream composition is specified in the Composition tab (see Fig. 9.14 for details).

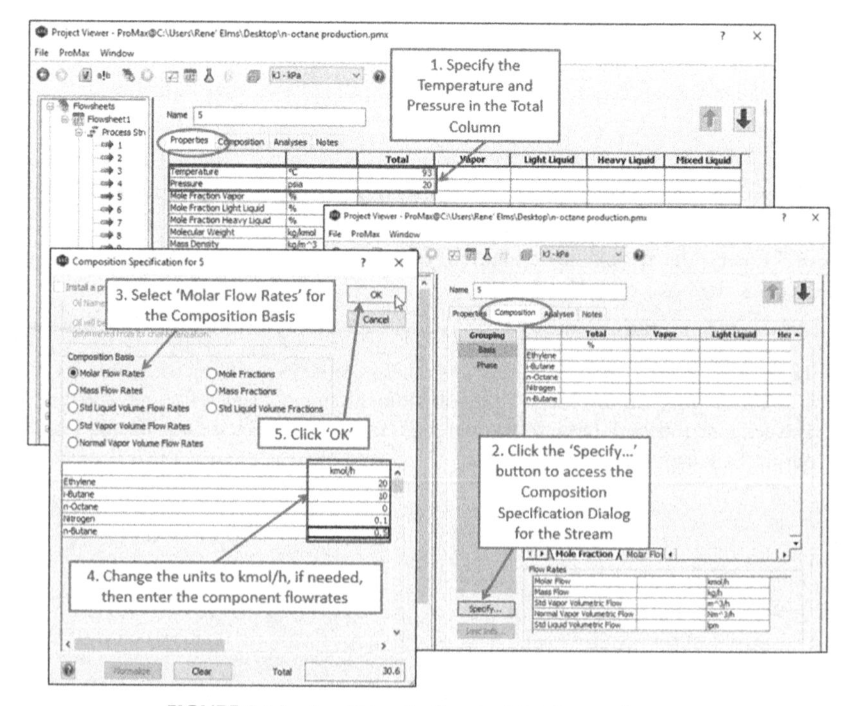

FIGURE 9.14 Specifying the Reactor Inlet Process Stream.

Tip: Values with a blue color in the Project Viewer were specified by a user. Values with a black color were calculated by the simulator.

9.4.2.2 Reactor Block

The needed specifications for the Reactor Block are listed in Table 9.5. There are various options for Reactor Type: Conversion, Equilibrium, Gibbs Minimization, Plug Flow, and Stirred Tank. Because the reaction is being modeled as a simple conversion, the selected Reactor Type is "Conversion." The previously created "n-octane Production" Reaction Set is specified, along with a pressure drop of 5 psi. To account for the isothermal operation of the Reactor, the Reactor Temperature Change is specified as 0°C (see detailed steps in Fig. 9.15). Alternatively, a temperature of 93°C could be specified in either of the Reactor Block outlet streams.

Fig. 9.15 displays the details of completing the Reactor Block specifications.

At this point, the Reactor Block is ready to be executed. There are several ways to execute a Block. One way is to click on the "Execute" button near the top of the Reactor Block dialogue, as seen in Fig. 9.16.

The resulting conversion values for the components in the Reactor Block, as well as the resulting molar flows in the vapor (Stream 7) and liquid (Stream 6) outlet streams, are seen in Fig. 9.17.

TABLE 9.5 Specifications for the Reactor Block

Specification	Value
Type	Conversion
Reaction Set	*n*-octane Production
Pressure Drop	5 psi
Temperature Change	0°C

Location: Process Data tab, Specifications subtab, Reactor Grouping.

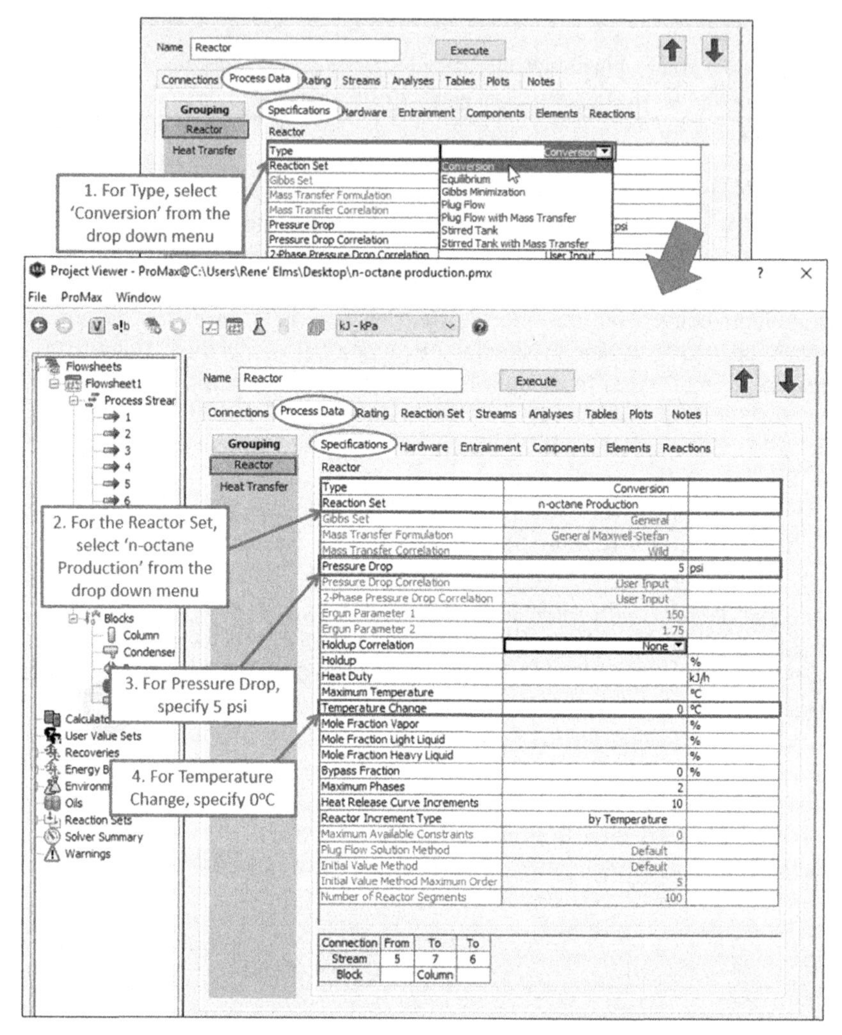

FIGURE 9.15 Specifications for the Reactor Block.

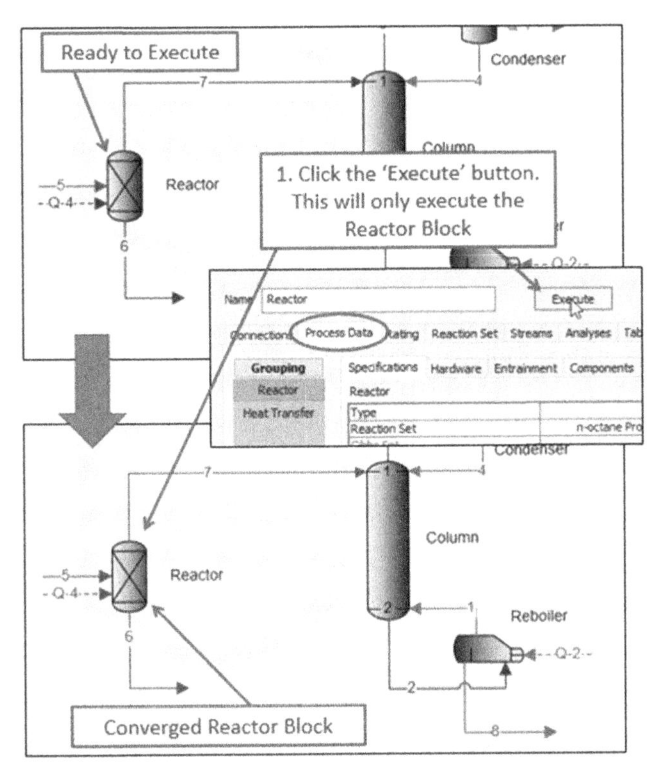

FIGURE 9.16 Executing the Reactor Block via the Block dialogue.

Tip: For Streams and Blocks, green indicates being fully specified or solved. Red indicates one or more specifications (for Streams) or required Streams (for Blocks) are missing. A blue Block indicates the Block has the minimum number of required Streams attached. A brown Stream indicates it is ready for execution.

9.4.2.3 Distillation Column

Several Column Block specifications for the current *n*-octane production example utilize default values. The main default specifications are listed in Table 9.6, along with the location of each within the Block dialogue, as shown in Fig. 9.18. When a column is added to the flowsheet, in any configuration, the default number of stages is 2. The number of stages can be identified by looking at the Column Block on the flowsheet (Figs. 9.9 and 9.13) in the Column Block dialogue Connections tab, or in the Stages dialogue (Properties Grouping) of the Process Data tab. In the latter two locations, the number of stages may be modified (Fig. 9.18).

The remaining default specifications are in the same location within the Column Block dialogue, as seen in Table 9.6. Ideal Stage and Mass Transfer

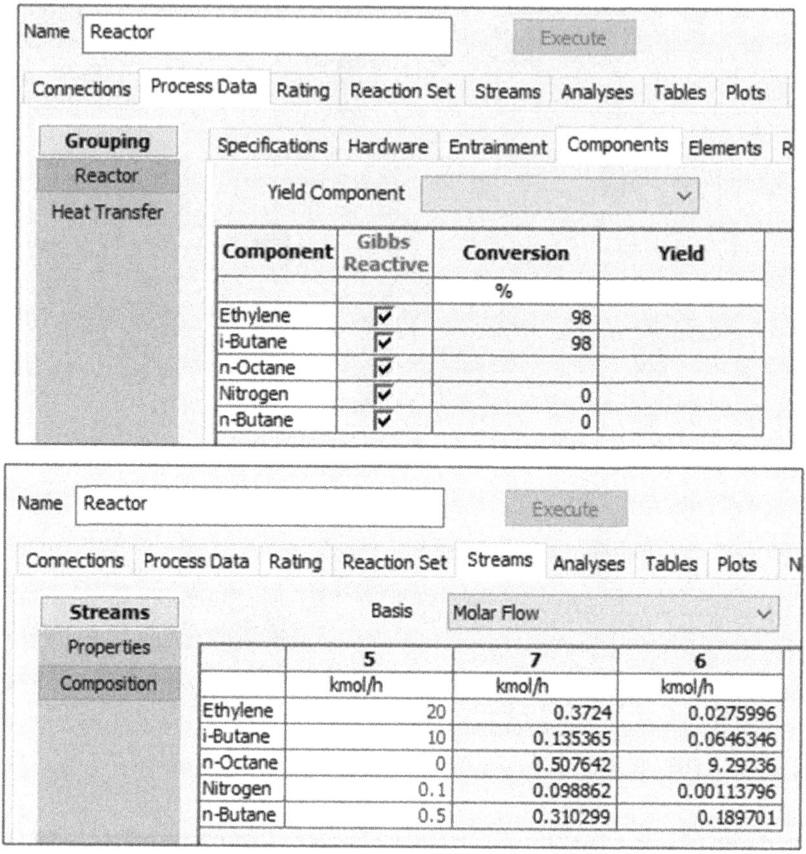

Name: Reactor
Execute

Connections | Process Data | Rating | Reaction Set | Streams | Analyses | Tables | Plots

Grouping: Reactor, Heat Transfer

Specifications | Hardware | Entrainment | Components | Elements

Yield Component

Component	Gibbs Reactive	Conversion	Yield
		%	
Ethylene	☑	98	
i-Butane	☑	98	
n-Octane	☑		
Nitrogen	☑	0	
n-Butane	☑	0	

Name: Reactor
Execute

Connections | Process Data | Rating | Reaction Set | Streams | Analyses | Tables | Plots

Streams: Properties, Composition

Basis: Molar Flow

	5	7	6
	kmol/h	kmol/h	kmol/h
Ethylene	20	0.3724	0.0275996
i-Butane	10	0.135365	0.0646346
n-Octane	0	0.507642	9.29236
Nitrogen	0.1	0.098862	0.00113796
n-Butane	0.5	0.310299	0.189701

FIGURE 9.17 Conversion results in the Reactor Component tab and Outlet Molar Flows in the Composition Grouping of the Reactor Streams tab.

TABLE 9.6 Default Specifications for the Distillation Column Block

Specification	Value	Location of Specification
Number of Stages	2	Connections tab or Process Data tab, Stages dialogue of the Properties grouping
Model Type	Ideal Stage	Process Data tab, Column dialogue of the Properties grouping
Ideal Stage Column Type	General Ideal Stage	
Flash Type	VLE	
Column Add-ons	Partial Condenser with Reboiler	

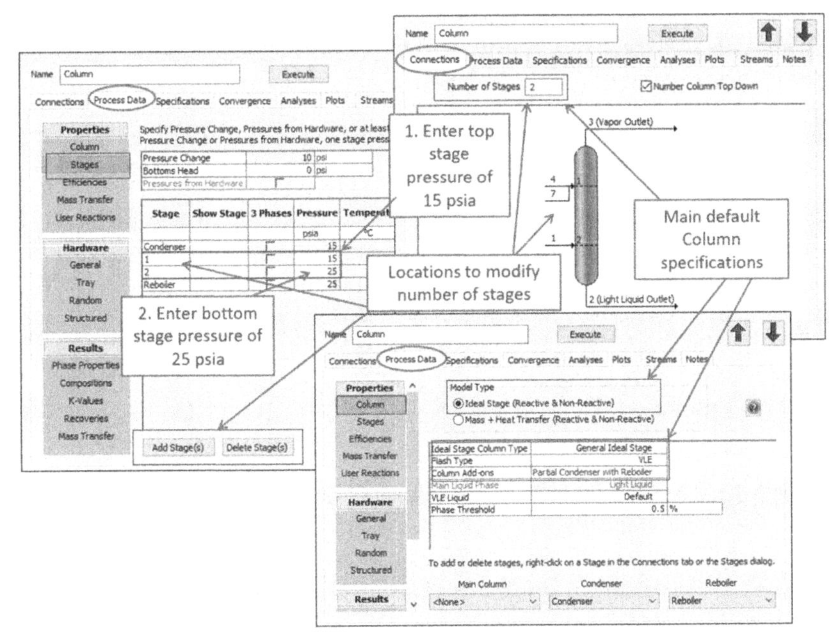

FIGURE 9.18 Main default and Stage specifications for the Distillation Column Block.

Model Types are available, with Ideal Stage being the default. The default Flash Type is VLE, with LLE and VLLE options also available. The Column Add-ons specification is automatically populated depending on the chosen column configuration option placed on the flowsheet. If column add-ons are manually flowsheeted or modified, additional specifications are needed.

TABLE 9.7 Stage Column Block Specifications

Specification	Value
Top Stage Pressure	15 psia
Bottom Stage Pressure	25 psia

Location: Process Data tab, Stages Dialogue of Properties grouping.

Tip: Units can be changed by clicking on the cell containing the unit and using the drop-down menu that appears. In addition, a unit can be specified when entering a property value by typing it directly behind (no space) the numerical value. For example, "100 psia" would be entered into the pressure field to specify a value of 100 psia.

TABLE 9.8 Column Block Specifications—"Active" Specifications and Calculations

Specification Name	Target	Status
Reflux Ratio	1	Specification
Overhead Temperature	85°C	Specification
Bottoms Ethylene Mole Fraction	0.0015	Calculation
Distillate *n*-Octane Mole Fraction	0.3	Calculation
Bottoms *n*-Octane Mole Fraction	0.999	Calculation

Location: Specifications tab (Note: A "Specification" status denotes an item utilized as a criteria for column convergence).

Tables 9.7 and 9.8 outline the remaining Column Block specifications that need to be made manually. The top and bottom stage pressures are entered in the Process Data tab in the Stages dialogue of the Properties grouping, as seen in Fig. 9.18. Alternatively, a Column pressure change may be specified at the top of the same dialogue.

The remaining Column specifications (Table 9.8) are made in the Specifications tab of the Column Block dialogue and are related to selecting appropriate criteria for column convergence and any desired calculations to monitor. The "'Degrees of Freedom" for the Column is determined by its configuration. A reboiler, partial condenser, unspecified Energy Stream, or side draw each add a Degree of Freedom. In the current example, both a reboiler and partial condenser are present. Therefore, the Column has two Degrees of Freedom, as seen in Fig. 9.19. By default, "Reflux Ratio" and "Boilup Ratio" specifications with a "Calculation" status populate the Specifications table because of the connected partial condenser and reboiler, respectively.

The Degrees of Freedom correlate to the number of requisite "active" Column Specifications. As "active" Specifications are made, the value for Degrees of Freedom decreases. When the value is "0," the Column Block is ready for simulation execution (an overspecification option is available, but is beyond the scope of the current discussion). Different categories of Column Specifications are made in the Specifications tab. Specifications having the "Specification" radio button selected are "active" and are utilized as a criterion for column convergence and decrease the available Degrees of Freedom. Specifications with an

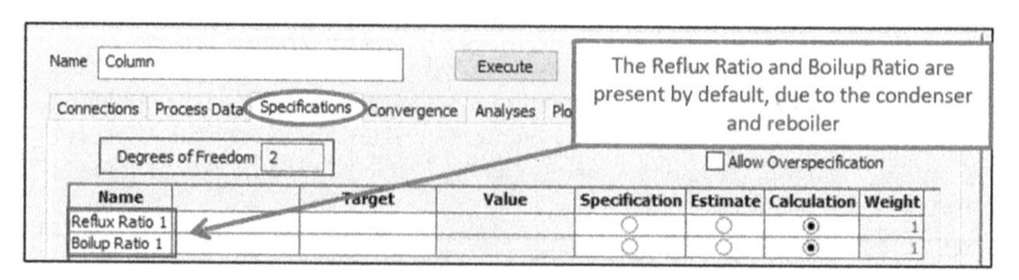

FIGURE 9.19 Default Specifications tab for the Distillation Column Block.

"Estimate" designation or status are used to help the column converge, while those with a "Calculation" status are simply calculated values and have no effect on column convergence. It should be noted that "active" Specifications do not necessarily have to be made in the Specifications tab. Some can be made directly in a column add-on or Streams associated with an add-on.

In the current example, the two "active" Specifications utilized are Reflux Ratio and Overhead Temperature. In addition, three "Calculation" Specifications will be added for relevant informational purposes: the mole fractions of *n*-octane in the distillate and bottoms and the ethylene mole fraction in the bottoms. To utilize the Reflux Ratio as an "active" Specification, a "Target" value of "1" is specified and then the "Specification" radio button is selected. This action will decrease the Degrees of Freedom from a value of "2" to a value of "1." The second Specification to be utilized as "active," the Overhead Temperature, must be created. The related details of this process are shown in Fig. 9.20. Once completed, the Overhead Temperature Specification can be seen in the Specifications table, the Degrees of Freedom decreases from "1" to "0," and the Column simulation is ready for execution.

The remaining three "Calculation" Specifications are created in similar fashion to the Overhead Temperature Specification, with all three having a

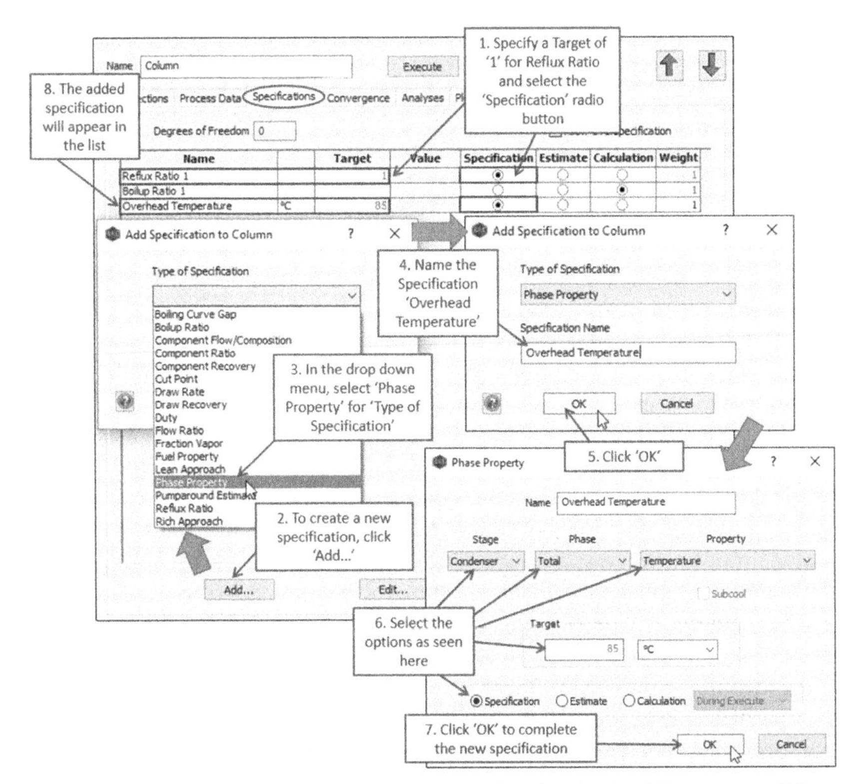

FIGURE 9.20 Specifying the Reflux Ratio and adding the Overhead Temperature Column specification to the Distillation Column Block.

TABLE 9.9 Details for Specifying Column "Calculations"

Specification Name	Stream	Flow Basis	Component To Sum	Target (Fractional)
Bottoms Ethylene Mole Fraction	Bottoms	Mole Fraction	Ethylene	0.0015
Distillate *n*-Octane Mole Fraction	Distillate	Mole Fraction	*n*-Octane	0.3
Bottoms *n*-Octane Mole Fraction	Bottoms	Mole Fraction	*n*-Octane	0.999

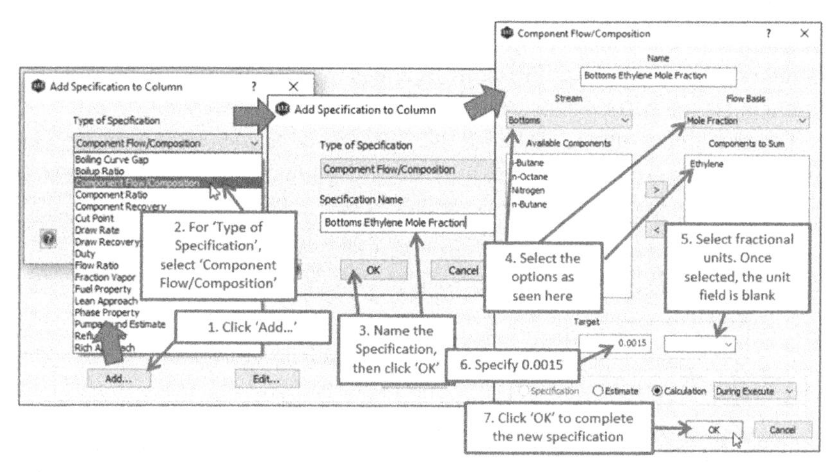

FIGURE 9.21 Adding the Bottoms Ethylene Mole Fraction Calculation to the Distillation Column Block.

"Component Flow/Composition" Specification Type. The needed specifics for creating each are in Table 9.9, with the creation of the "Bottoms Ethylene Mole Fraction" Specification demonstrated in Fig. 9.21, serving as an example for creation of the other two. These calculations are related to product and recycle stream (the overheads) composition and quality. The target value for the bottoms *n*-octane mole fraction represents the desired purity for the *n*-octane product, whereas the target for the bottoms ethylene mole fraction is the desired maximum for ethylene in the *n*-octane product.

After adding all Specifications, the resulting Specifications tab is seen in Fig. 9.22.

As mentioned previously, there are multiple ways to execute a Block. On the flowsheet, select the Column Block. In the ProMax tab "Execute" grouping, click the "Execute Block" button, as seen in Fig. 9.23.

Results displayed in the Specifications tab of the converged Column Block are seen in Fig. 9.24. The Reflux Ratio and Overhead Temperature

Name: Column | Execute

Connections | Process Data | Specifications | Convergence | Analyses | Plots | Streams | Notes

Degrees of Freedom: 0 | ☐ Allow Overspecification

Name		Target	Value	Specification	Estimate	Calculation	Weight
Reflux Ratio 1		1		◉	○	○	1
Boilup Ratio 1				○	○	◉	1
Overhead Temperature	°C	85		◉	○	○	1
Bottoms Ethylene Mole Fraction		0.0015		○	○	◉	1
Distillate n-Octane Mole Fraction		0.3		○	○	◉	1
Bottoms n-Octane Mole Fraction		0.999		○	○	◉	1

FIGURE 9.22 Completely specified Distillation Column Block Specifications tab.

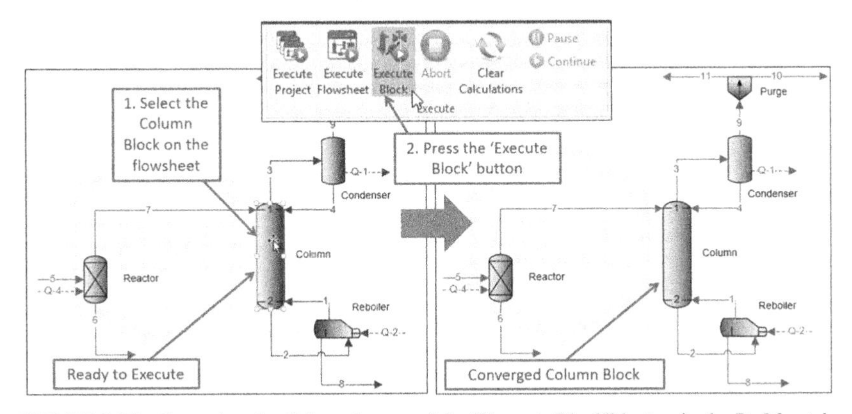

FIGURE 9.23 Executing the Column by use of the "Execute Block" button in the ProMax tab.

Name: Column | Execute

Connections | Process Data | Specifications | Convergence | Analyses | Plots | Streams | Notes

Degrees of Freedom: 0 | ☐ Allow Overspecification

Name		Target	Value	Specification	Estimate	Calculation	Weight
Reflux Ratio 1		1	1	◉	○	○	1
Boilup Ratio 1		1.57447	9.04908	○	○	◉	1
Overhead Temperature	°C	85	85	◉	○	○	1
Bottoms Ethylene Mole Fraction		0.0015	1.21623e-07	○	○	◉	1
Distillate n-Octane Mole Fraction		0.3	0.265341	○	○	◉	1
Bottoms n-Octane Mole Fraction		0.999	0.999884	○	○	◉	1

FIGURE 9.24 Results in the Converged Column specifications tab.

Specifications provided the desired levels of ethylene and *n*-octane in the Bottoms, as well as *n*-octane in the Distillate.

Tip: Hover over any Stream or Block to see a pop-up box with select properties for that item.

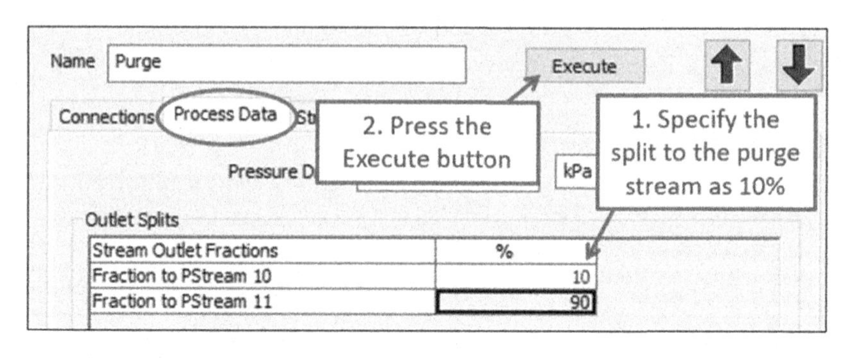

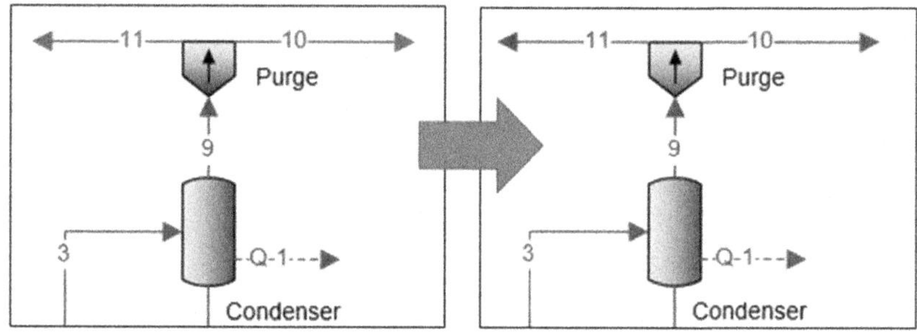

FIGURE 9.25 Specification and execution of the Splitter Block representing the Purge.

9.4.2.4 Splitter (Purge) Block

The Column overhead stream is directed to the Splitter Block representing the purge (Fig. 9.25). In the Process Data tab of the "Purge" Splitter Block dialogue, a 10% purge (Table 9.10) is accomplished by designating a 10% split to Stream 10, which is the purge stream (Fig. 9.25). The remaining 90% will be recycled back to the Reactor via Stream 11.

9.4.3 Recycle Loop and Inlet Preheating: Adding and Connecting the Compressor, Cross Exchanger, Heater, Recycle, and Mixer Blocks

The recycle stream will first be brought up to pressure for subsequent reintroduction to the reactor. After leaving the compressor, it will pass through a cross exchanger at the front end of the process and exchange heat with the

TABLE 9.10 Purge Block Specification

Block	Block Specification
Purge (Splitter)	10% split to purge stream (Stream 10)

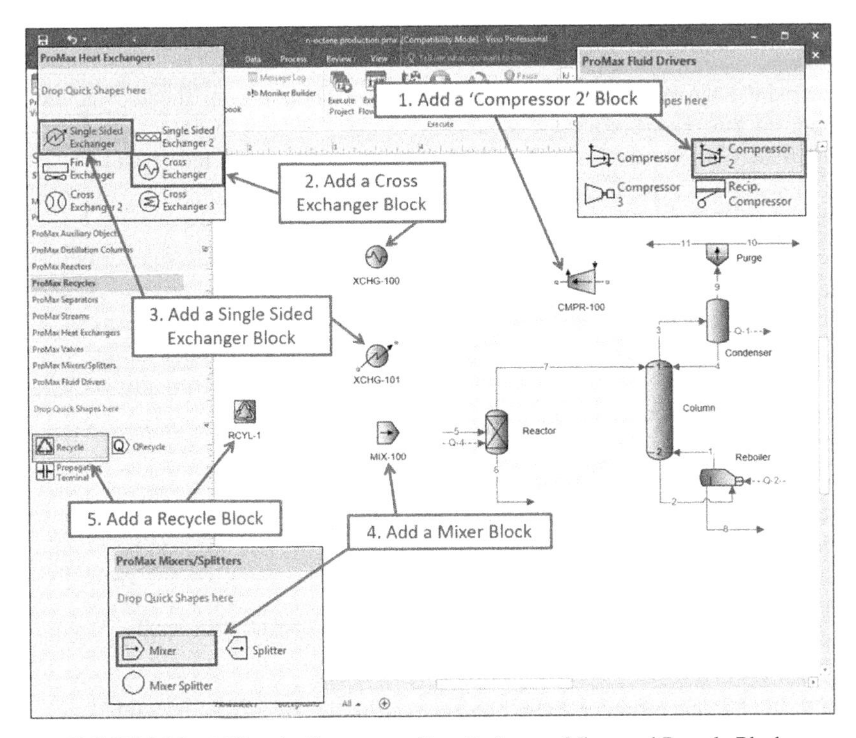

FIGURE 9.26 Adding the Compressor, Heat Exchanger, Mixer, and Recycle Blocks.

fresh feed prior to entering the reactor. The fresh feed inlet first exchanges heat with the recycle stream in the cross exchanger and then is further preheated before entering the reactor. Fig. 9.26 demonstrates addition of the Blocks that will represent the components of the recycle loop and inlet preheating.

9.4.3.1 Compressor Block

Multiple options for Compressor Blocks are found in the "ProMax Fluid Drivers" Shape. "Compressor," "Compressor 2," and "Compressor 3" have the same functionality, only differing by the configuration of the inlet and outlet streams. To best fit with the structure of this particular flowsheet, the "Compressor 2" option is added (Fig. 9.26—Step 1). When added to the flowsheet, the inlet and outlet connections for the Compressor Block are in opposite locations from what is needed. Therefore, the Block needs to be flipped along the vertical axis.

Tip: To "flip" a Block along the vertical axis, select the Block and then hold the CTRL key and hit the "H" key. To flip a Block along the horizontal axis, use CTRL and "J" in the same manner. This can also be used to adjust Process and Energy Streams.

9.4.3.2 Cross Exchanger and Single Sided Exchanger Blocks

The added Cross Exchanger and Single Sided Exchanger Blocks are both found in the "ProMax Heat Exchangers" Shape, as seen in Steps 2 and 3 of Fig. 9.26. The Single Sided Exchanger Block will serve as the Heater.

9.4.3.3 Mixer Block

As with the Splitter Block, the arrow in the Mixer Block indicates the direction of flow. The Mixer Block is needed to combine the fresh feed and the recycle stream because the Reactor Block will only accept one inlet stream. This Block does not perform the mechanical operation of "mixing." Rather, it is utilized to simply combine Streams when needed, whether in reality or as a construct to facilitate the simulation, as in this case.

9.4.3.4 Recycle Block

Because the system contains a loop in which part of the inlet to the reactor is dependent on the reactor solution itself, a Recycle Block is needed to provide a starting point for convergence.

9.4.3.5 Addition and Connection of the Process and Energy Streams

First, the recycle stream leaving the Purge Block (Stream 11) is connected to the inlet of the Compressor. Process Streams are then added and connected as seen in Fig. 9.27. Note, at this point, the outlet of the Mixer and Recycle Blocks are left unconnected. This is to facilitate obtaining a recycle guess, as will be discussed in a subsequent section.

9.4.4 Specification and Execution of the Compressor, Cross Exchanger, Preheater, Recycle, and Mixer Blocks

9.4.4.1 Compressor Block

As seen in Fig. 9.28, in the Process Data tab of the Compressor Block dialogue the Polytropic Efficiency is specified as 75%. In the outlet of the Compressor, Stream 12, a pressure of 22 psia is specified.

Table 9.11 shows the specifications related to the Compressor Block. Fig. 9.29 demonstrates execution of the Compressor Block and viewing a comparison of the inlet (Stream 11) and outlet (Stream 12) streams in the converged Block. Notice that due to the compression, the outlet stream is 98.43°C as compared to 85°C in the inlet. Some of the heat in this stream will be exchanged with the fresh feed in the Cross Exchanger.

9.4.4.2 Inlet Stream

The specifications for the fresh feed Process Stream are shown in Table 9.12. The fresh feed inlet (Stream 15) is specified in the same manner as shown for the Reactor inlet stream in Fig. 9.14. The fresh feed enters the process by first exchanging heat in the Cross Exchanger.

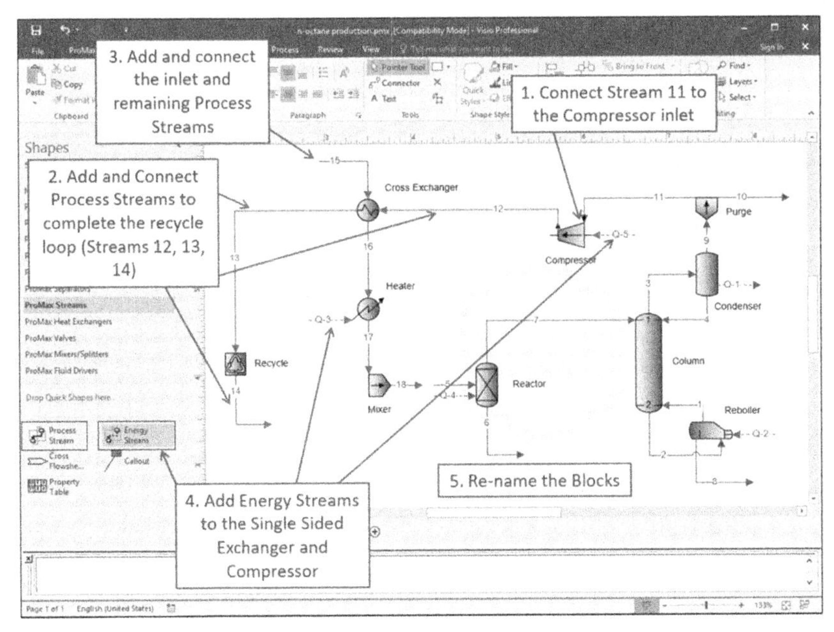

FIGURE 9.27 Adding and connecting the Process and Energy Streams to the Compressor, Heat Exchanger, Mixer, and Recycle Blocks.

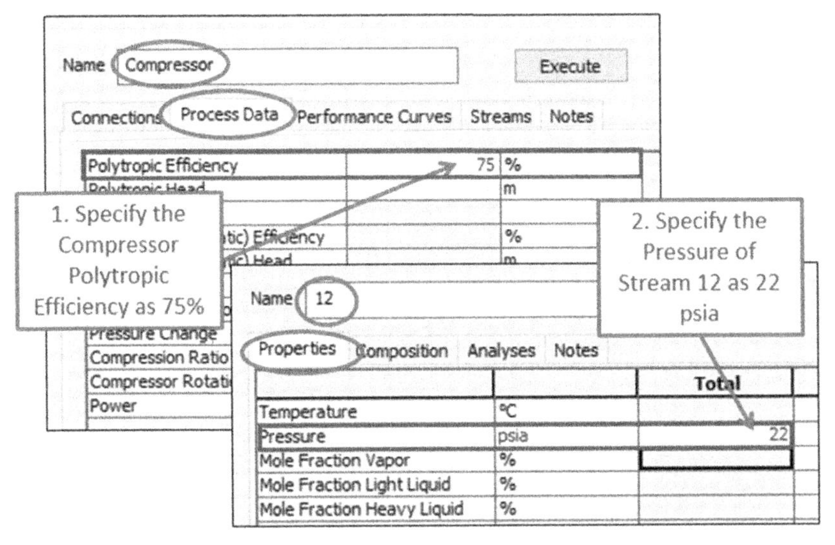

FIGURE 9.28 Specifications related to the Compressor.

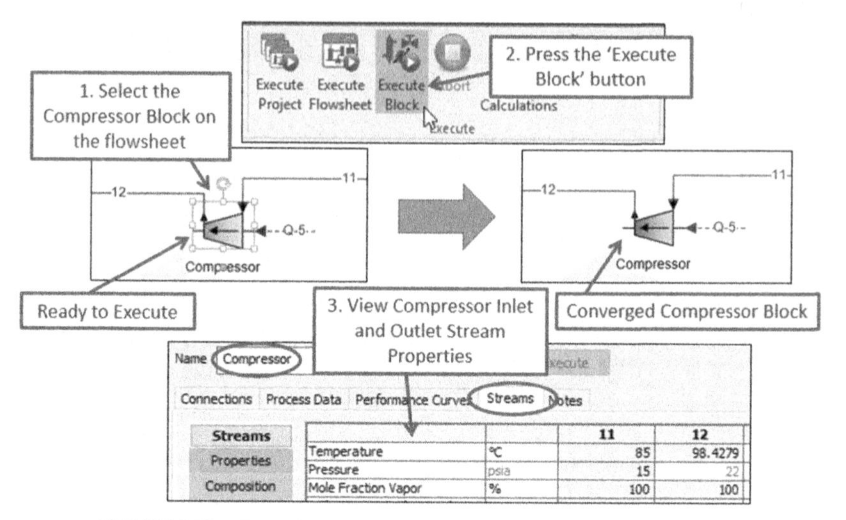

FIGURE 9.29 Execution of the Compressor Block and viewing of results.

TABLE 9.11 Compressor Block Specifications

Block	Block Specifications	Other Specifications
Compressor	75% Polytropic Efficiency	Stream 12 (outlet), Pressure = 22 psia

TABLE 9.12 Specifications for the Fresh Feed Stream

Temperature = 30°C	
Pressure = 24 psia	
Component	**Flowrate (kmol/h)**
Ethylene	20
i-Butane	10
n-Octane	0
Nitrogen	0.1
n-Butane	0.5

TABLE 9.13 Specifications for the Heat Exchanger Blocks

Heat Exchanger Block	Pressure Drop(s)	Other Specifications
Cross Exchanger	Side A: 2 psi Side B: 2 psi	Stream 13 (outlet toward the Recycle Block), temperature = 93°C
Heater	2 psi	Stream 17 (outlet), temperature = 93°C

9.4.4.3 Cross Exchanger and Heater Blocks

Next in the process flow are the Cross Exchanger and Heater Blocks. The related specifications for these Blocks and their connected Process Streams are listed in Table 9.13. The Cross Exchanger facilitates heat exchange from the recycle stream to the fresh feed, to assist in preheating. This exchange is limited by the need for the recycle stream to enter the reactor at 93°C; therefore the recycle stream outlet from the Cross Exchanger (Stream 13) is specified as having a temperature of 93°C.

Double-click on the Cross Exchanger Block on the flowsheet to access the Block dialogue. The pressure drops are specified in the Process Data tab. To switch between sides of the Exchanger, click on the desired side in the Grouping column on the left. The details of specifying the needed pressure drops and outlet stream (Stream 13) temperature are demonstrated in Fig. 9.30.

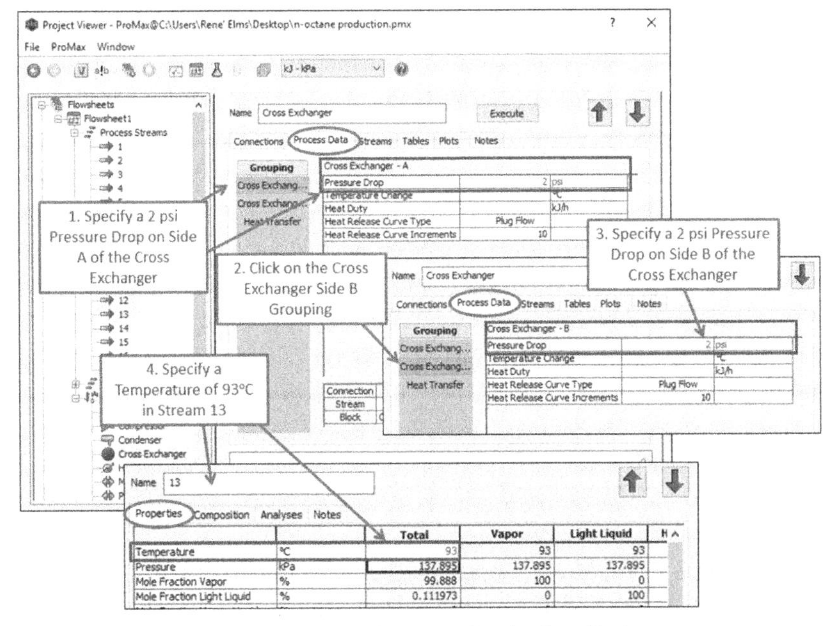

FIGURE 9.30 Specifications related to the Cross Exchanger.

Tip: To prevent several common mistakes and overspecification, for every heat exchanger used in ProMax, specify a pressure drop for each side. For every fluid driver, specify an efficiency.

Once the temperature specification of 93°C is made in Stream 13, notice that the other properties for the Stream are automatically calculated prior to simulation execution, as seen in Fig. 9.30 and by the green color of the Stream in Fig. 9.31. Stream 12, the corresponding inlet for Stream 13, is fully solved, and therefore the temperature, pressure, flowrate, and composition are known. The temperature in Stream 13 is directly specified. The pressure is also already known (the difference between the Stream 12 pressure and the specified pressure drop across that side of the Cross Exchanger). As such, the composition and flowrate information from Stream 12 can be and is propagated into Stream 13.

Fig. 9.31 demonstrates execution of the Cross Exchanger Block. The resulting outlet temperature for the fresh feed (Stream 16) is displayed in a Stream Callout. Callouts can be attached to any Process Stream and customized to display desired Properties. Select information pertaining to the converged Cross Exchanger is viewed (and here displayed) by hovering over the Block.

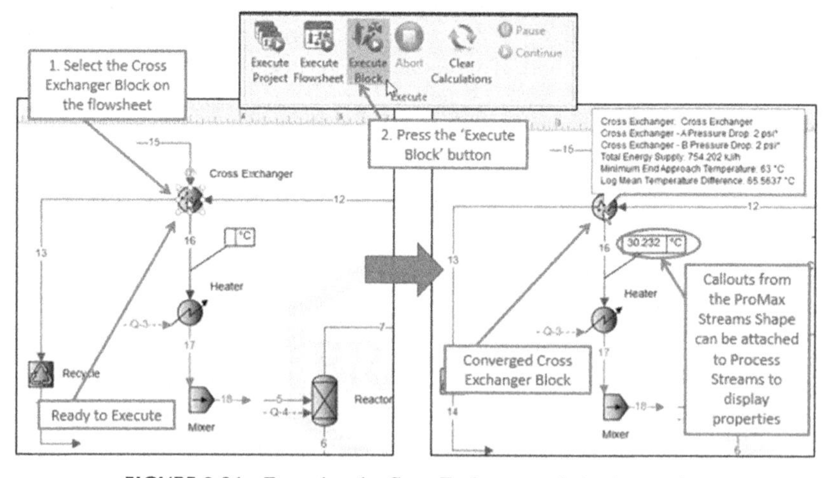

FIGURE 9.31 Executing the Cross Exchanger and viewing results.

Tip: Hover over any Stream or Block to see a pop-up box with select properties for that item.

After the fresh feed inlet passes through the Cross Exchanger, it is further preheated in the Heater Block to a temperature of 93°C. As with the Cross Exchanger, the pressure drop for the single-sided Heater is specified in the Process Data tab of the Block dialogue (Fig. 9.32). When the temperature specification is made in Stream 17, notice that the Block automatically executes and the calculated Stream 17 properties populate the Stream dialogue.

9.5 DETERMINATION OF A RECYCLE BLOCK GUESS AND CLOSING OF THE RECYCLE LOOP

9.5.1 Recycle and Mixer Blocks

As mentioned previously, because the system contains a loop in which part of the inlet to the reactor is dependent on the reactor solution itself, a Recycle Block is needed to provide a starting point for convergence. The Recycle Block requires an inlet stream and a fully specified outlet stream (temperature, pressure, flowrate, composition), which is referred to as the "Recycle Guess." When a reasonable Recycle Guess is known, it can be directly specified in the outlet stream of the Recycle Block and the simulation is executed immediately with the related loop closed. If all or some of the Recycle Guess specifications are not known, a reasonable guess can be obtained as demonstrated in this example, by leaving the Recycle Guess disconnected and the loop open, and then executing the simulation Block-by-Block leading up to the Recycle Block. In Fig. 9.31, it can be seen

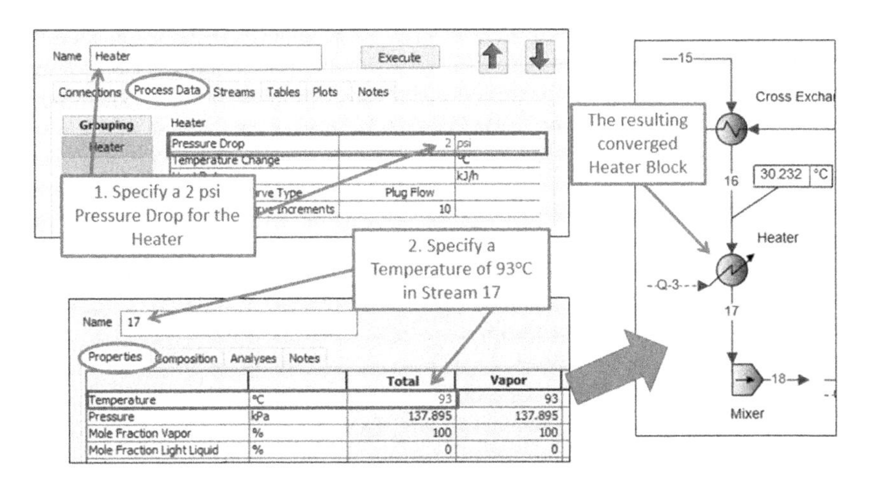

FIGURE 9.32 Specifications related to the Heater and the resulting flowsheet appearance.

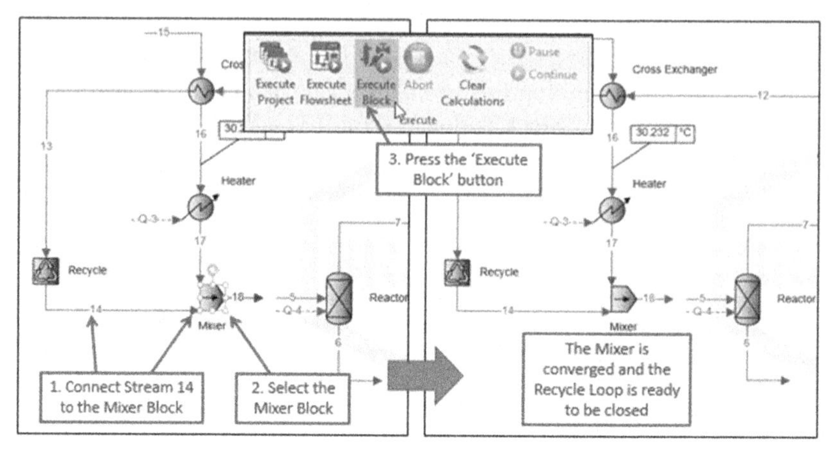

FIGURE 9.33 Connecting and converging the Mixer Block.

that after specifying the Cross Exchanger, the outlet stream of the Recycle Block (Stream 14) is green and hence, fully specified and solved. As such, the Recycle Guess is now provided and the Recycle outlet (Stream 14) is ready to be connected to the Mixer Block. As mentioned previously, the Mixer Block is needed to combine the Recycle Block outlet and the preheated fresh feed stream (Stream 17) because the Reactor Block will only accept one inlet stream. Fig. 9.33 demonstrates connecting the Recycle Block outlet stream to the Mixer Block, and then executing the Mixer. At this point, the recycle loop is now ready to be closed.

9.5.2 Closing the Recycle Loop

Disconnect the temporary Reactor inlet stream (Stream 5) and then connect the Mixer outlet (Stream 18) to the Reactor to close the recycle loop. As shown in Fig. 9.34, press the "Execute Flowsheet" button to converge the flowsheet. (The "Execute Project" button could also be used.)

9.6 VIEWING RESULTS

As demonstrated previously, results throughout the flowsheet can be viewed directly in the Stream or Block of interest. Results can also be displayed directly on the flowsheet in property tables. A converged flowsheet for the *n*-octane simulation is shown in Fig. 9.35. Property tables containing information pertaining to the Process and Energy Streams have been included. Values marked with an "*" indicate a user specified value.

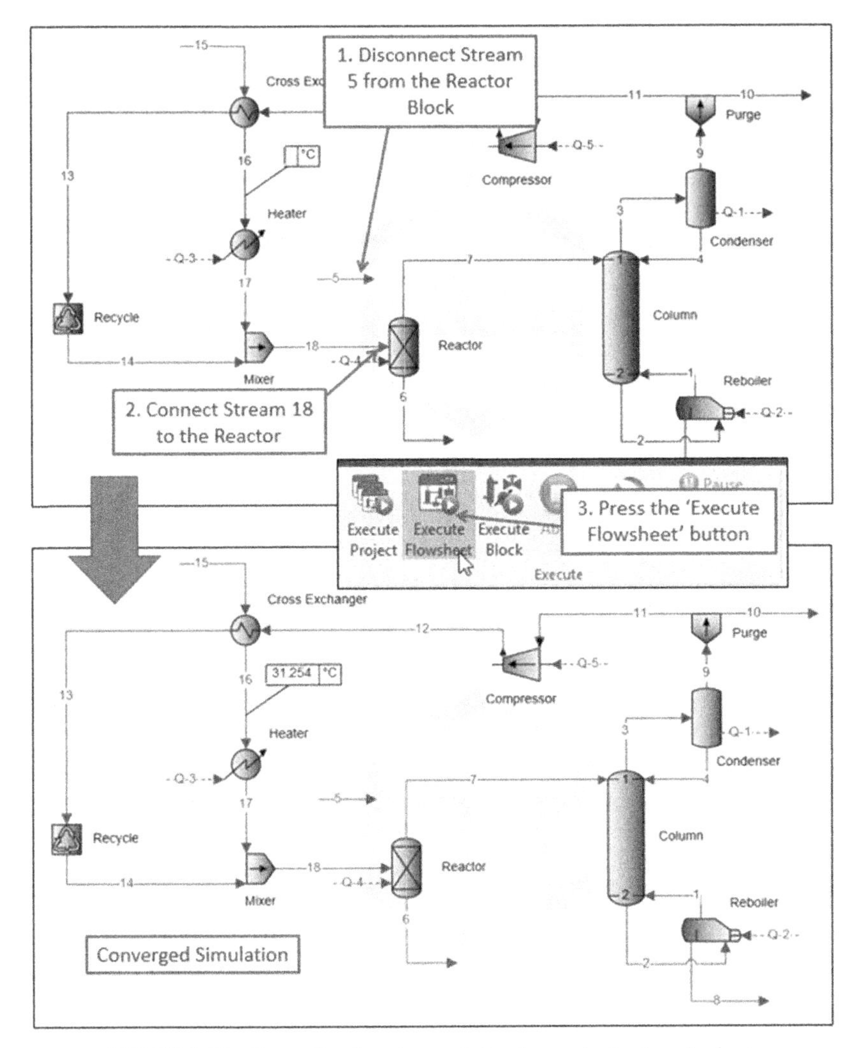

FIGURE 9.34 Connecting the mixer outlet to the reactor to close the loop.

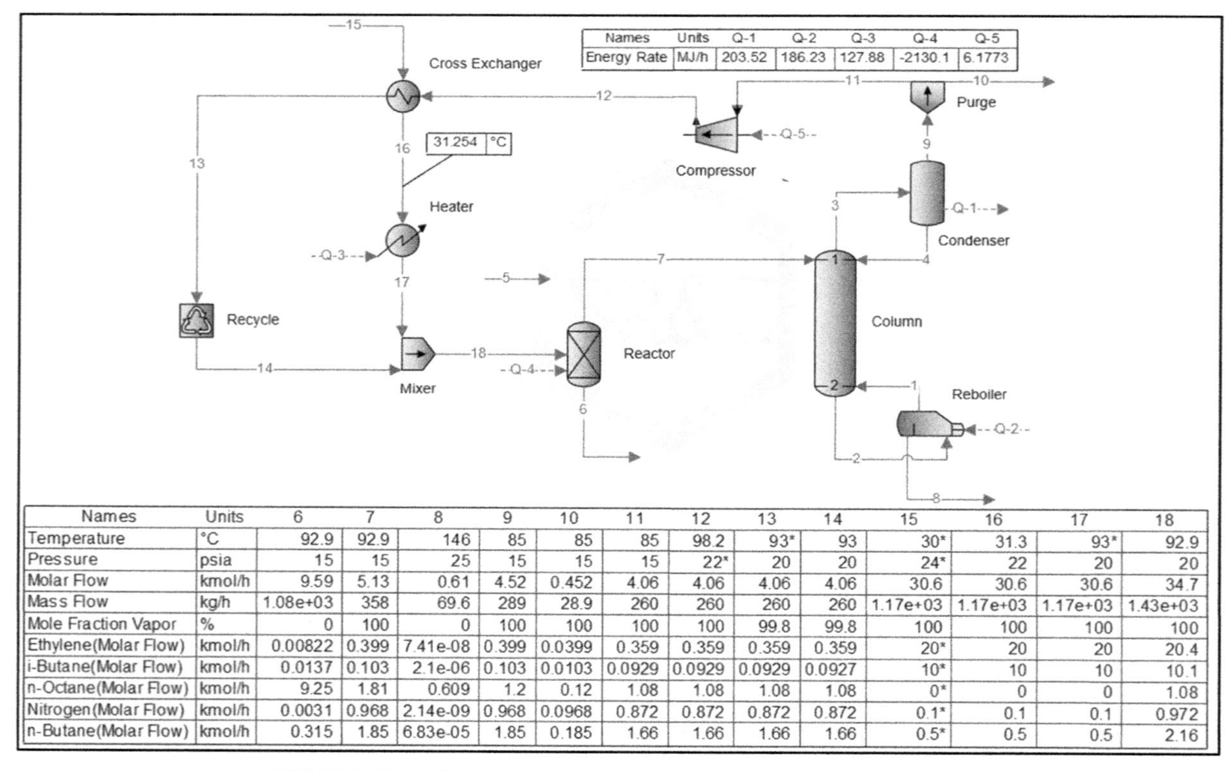

Names	Units	Q-1	Q-2	Q-3	Q-4	Q-5
Energy Rate	MJ/h	203.52	186.23	127.88	-2130.1	6.1773

Names	Units	6	7	8	9	10	11	12	13	14	15	16	17	18
Temperature	°C	92.9	92.9	146	85	85	85	98.2	93*	93	30*	31.3	93*	92.9
Pressure	psia	15	15	25	15	15	15	22*	20	20	24*	22	20	20
Molar Flow	kmol/h	9.59	5.13	0.61	4.52	0.452	4.06	4.06	4.06	4.06	30.6	30.6	30.6	34.7
Mass Flow	kg/h	1.08e+03	358	69.6	289	28.9	260	260	260	260	1.17e+03	1.17e+03	1.17e+03	1.43e+03
Mole Fraction Vapor	%	0	100	0	100	100	100	100	99.8	99.8	100	100	100	100
Ethylene(Molar Flow)	kmol/h	0.00822	0.399	7.41e-08	0.399	0.0399	0.359	0.359	0.359	0.359	20*	20	20	20.4
i-Butane(Molar Flow)	kmol/h	0.0137	0.103	2.1e-06	0.103	0.0103	0.0929	0.0929	0.0929	0.0927	10*	10	10	10.1
n-Octane(Molar Flow)	kmol/h	9.25	1.81	0.609	1.2	0.12	1.08	1.08	1.08	1.08	0*	0	0	1.08
Nitrogen(Molar Flow)	kmol/h	0.0031	0.968	2.14e-09	0.968	0.0968	0.872	0.872	0.872	0.872	0.1*	0.1	0.1	0.972
n-Butane(Molar Flow)	kmol/h	0.315	1.85	6.83e-05	1.85	0.185	1.66	1.66	1.66	1.66	0.5*	0.5	0.5	2.16

FIGURE 9.35 A Converged flowsheet for the *n*-octane production simulation.

9.7 CONCLUSION

The use of ProMax to simulatc a simple *n*-octane production process was demonstrated. A conversion reactor was utilized to represent the 98% conversion of a feed containing 65.4 mol% ethylene and 32.7 mol% *i*-butane to *n*-octane. The product was further purified in a distillation column, with the distillate recycled back to the reactor by use of a Recycle Block.

EXERCISES

1. For the converged flowsheet in Fig. 9.35, determine and discuss the effects of changing the Compressor efficiency from 75% to (a) 85% and (b) 65%.
2. Determine and discuss the effects of changing the Overhead Temperature Specification in the Column Block from 85°C to 82°C.

REFERENCES

Bryan Research & Engineering (BR&E), 2017. ProMax. www.bre.com.

Foo, D.C.Y., Manan, Z.A., Selvan, M., McGuire, M.L., October 2005. Integrate process simulation and process synthesis. Chemical Engineering Progress 101 (10), 25–29.

Chapter 10

Modeling of Sour Gas Sweetening With MDEA

René D. Elms
Texas A&M University, College Station, TX, United States

In this chapter, modeling of a methyldiethanolamine (MDEA) sweetening unit is demonstrated in ProMax (BR&E, 2017). The use of Saturator and Makeup/Blowdown blocks is introduced and described, as well as the appropriate column specifications for an amine absorber and stripper.

10.1 INTRODUCTION

Following collection and gathering, natural gas is treated to remove various contaminants prior to transportation and/or processing. The majority of these contaminants contribute to either solids formation or pipeline corrosion. The presence of acid gases, specifically carbon dioxide (CO_2) and hydrogen sulfide (H_2S), can contribute to pipeline and equipment corrosion. Various processes can be utilized to remove CO_2 and H_2S, including the use of chemical solvents, physical solvents, scavengers, and membranes (GPSA, 2012). The use of *alkanolamines* for acid gas removal, commonly referred to as "amine sweetening," is a well-established process. Alkanolamines (or simply "amines") are the most commonly used solvent type for removal of acid gases (GPSA, 2004).

Fig. 10.1 shows a typical amine sweetening unit, which consists of an absorber or contactor column through which the sour gas is passed upward coming into contact with the amine solution that is flowing downward. The rich solution (rich amine) exits the bottom of the absorber. For high-pressure systems, a flash drum is installed downstream of the absorber to assist with hydrocarbon removal from the rich amine prior to treatment in the amine stripper column. The rich amine is heated prior to entering the stripper column by cross-exchange of heat in the lean/rich exchanger with the lean amine solution exiting the stripper column. The regenerated lean amine is then pumped to match the absorber column pressure and cooled to an appropriate temperature prior to being reintroduced into the absorber column (Kohl and Riesenfeld, 1979; GPSA, 2004).

Chemical Engineering Process Simulation. http://dx.doi.org/10.1016/B978-0-12-803782-9.00010-8

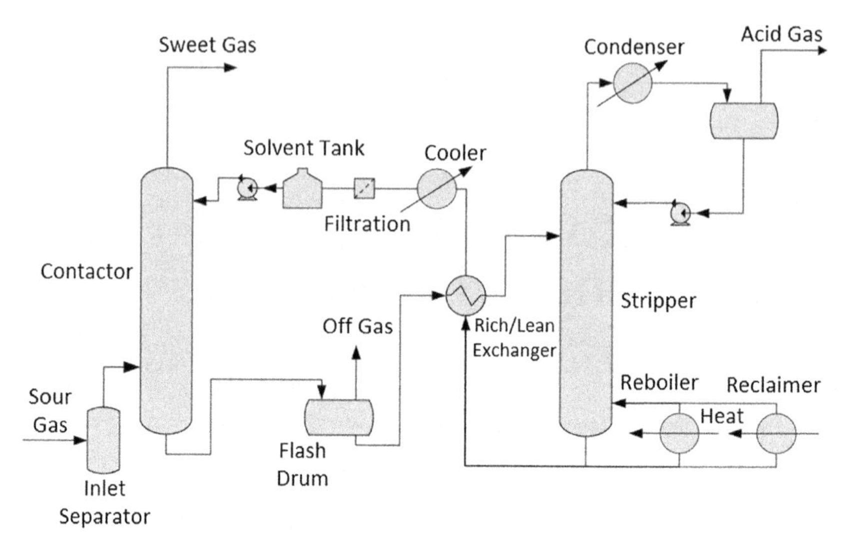

FIGURE 10.1 A typical amine gas sweetening unit (GPSA, 2012).

Various amines are available for use. Selection of an appropriate amine depends on various factors such as the sour gas composition, pressure and temperature, and the desired acid gas content of the treated gas, commonly referred to as "sweet gas" (Kohl and Riesenfeld, 1979). The amine to be utilized in this example is MDEA. MDEA is a tertiary amine and can be utilized to selectively remove H_2S in the presence of CO_2 at moderate to high pressures (GPSA, 2012).

10.1.1 Background—MDEA Sweetening Example

A sour gas stream is being treated in an MDEA sweetening unit to remove its H_2S and CO_2 content. The stream conditions are given in Table 10.1. The resulting sweet gas should contain less than 2 mol% CO_2 and 5 ppm H_2S. A 50 wt% MDEA/50 wt% H_2O solution is circulated at a flowrate of 275 sgpm (62.5 m^3/h). The reboiler duty is equivalent to 1.1 lb steam/gallon of circulating amine.

10.2 PROCESS SIMULATION

The MDEA sweetening simulation will be developed with the following steps:

1. Setting the Environment
2. Adding Blocks to the flowsheet
3. Adding and connecting Process and Energy Streams

TABLE 10.1 Sour Gas Inlet Stream Conditions and Composition [Temperature = 100°F (38°C), Pressure = 1000 psia (69 bar), Flow Rate = 50 MMSCFD (59,000 Nm^3/h)]

Components	Mole % (Dry Basis)
H_2S	1.5
CO_2	3.5
Methane	87.5
Ethane	5
Propane	1.49
n-Butane	0.5
n-Pentane	0.5
Benzene	0.007
Toluene	0.001
Ethylbenzene	0.001
o-Xylene	0.001

4. Specification of Blocks and Streams
5. Viewing the results

10.3 SETTING THE ENVIRONMENT

First, the "Amine Sweetening–PR" Property Package is selected in the Environment, as demonstrated in Chapter 9. Note that "Amine Sweetening–SRK" is also an appropriate Property Package for this application. Both of these Property Packages are electrolytic, which is required to obtain appropriate and accurate results when modeling an amine sweetening unit. Next, the components involved in the simulation are added to the Environment. Apart from those listed in Table 10.1, MDEA and water are also added.

Tip: The order in which components are listed in the Installed Pure Components column of the Components tab of the Environment will be the order in which they appear elsewhere in the simulation. The components can be reordered in the Installed Pure Components column by selecting a component and dragging and dropping to a new location in the list.

10.4 ADDING BLOCKS TO THE FLOWSHEET

The blocks utilized in this simulation, the related process unit and/or name, the specific version used, and the related ProMax Shape in which each block is located are summarized in Table 10.2.

10.4.1 Adding Stages to the Columns

Before attaching Process Streams, the requisite number of stages will be specified for the Absorber and Stripper Column Blocks. To represent a typical amine absorber, seven ideal stages are needed, while a typical amine stripper is represented by 10 ideal stages. (More details will follow in a subsequent section.) In the Connections tab of the block dialogs for each Column Block, change the number of stages from the default of 2 to 7 for the Absorber and 10 for the Stripper, as shown in Fig. 10.2.

TABLE 10.2 Information About the Blocks Used in the MDEA Sweetening Simulation

Block	Process Unit/ Name	Version Used	Related ProMax Shape
Saturator	Saturator	Saturator	Auxiliary Objects
Distillation Column	Absorber	Distill	Distillation Columns
Separator	Rich Flash	Two Phase Separator Horizontal	Separators
Cross Exchanger	Lean/Rich Exchanger	Cross Exchanger	Heat Exchangers
Distillation Column	Stripper	Distill PC,R	Distillation Columns
Recycle	Recycle	Recycle	Recycles
Makeup/ Blowdown	MDEA Makeup	Makeup/Blowdown	Auxiliary Objects
Pump	Circulation Pump	Centrifugal Pump	Fluid Drivers
Heat Exchanger	Trim Cooler	Fin Fan Exchanger	Heat Exchangers

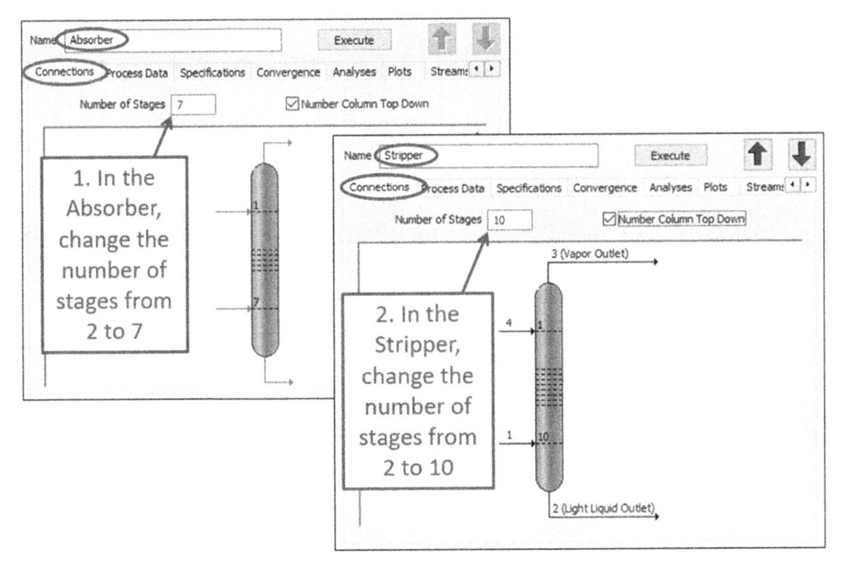

FIGURE 10.2 Adding stages to the Absorber and Stripper Column Blocks.

Tip: The Connections tab of each Block will indicate missing required Streams by flashing red arrows. Connected Streams will be represented by fixed (not flashing) black arrows having the respective Stream number or name.

10.4.2 Showing Stages in the Stripper Column Block

To connect a Stream to a Column Block stage, the stage must be visible on the Column Block icon on the flowsheet. By default, only the top and bottom stages for Column Blocks are visible. The process of selecting additional stages to be displayed on the Column Block is shown in Fig. 10.3.

10.5 ADDITION AND CONNECTION OF PROCESS AND ENERGY STREAMS

Process and Energy Streams are added, connected, and renamed as demonstrated in Fig. 10.4.

Tip: The location and position of connected Streams can be customized. Select the relevant Stream and locate the blue diamonds, or "handles", on the Stream. Click on a "handle" to resize and/or reposition that portion of the Stream. Sometimes manipulation of multiple "handles" is necessary to make desired changes.

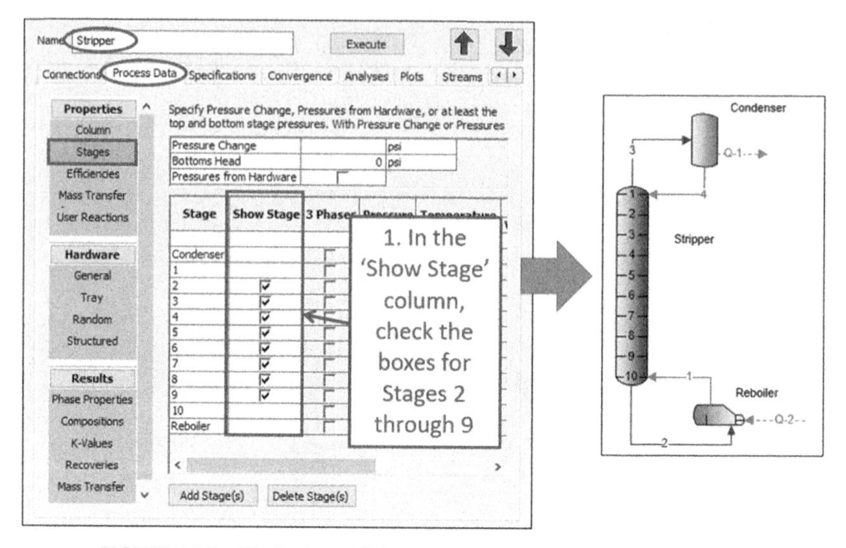

FIGURE 10.3 Displaying additional stages on the Stripper Column Block.

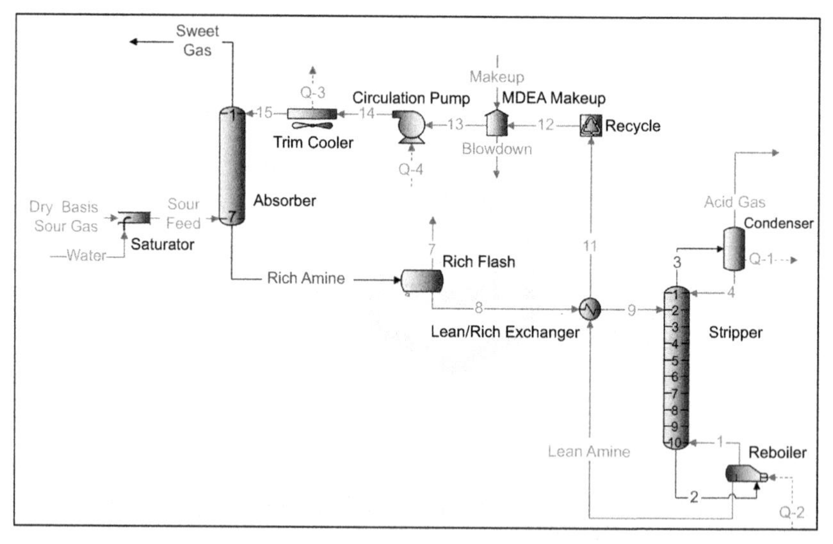

FIGURE 10.4 Process flow diagram for MDEA sweetening simulation.

Tip: When selected on the flowsheet, Streams and Blocks can be rotated in increments of 90 degrees by holding the CTRL key then hitting the "L" key (to rotate to the left) or by holding the CTRL key then hitting the "R" key (to rotate to the right).

10.6 SPECIFICATION OF BLOCKS AND STREAMS

10.6.1 Dry Basis Sour Gas Process Stream and the Saturator Block

The sour gas inlet conditions and composition on a dry basis are shown in Table 10.1. Often, natural gas compositional analyses are provided on a dry basis. The Saturator Block facilitates the saturation of the Dry Basis Sour Gas Stream, providing the "wet" sour gas feed that will enter the Absorber. (The Saturator Block can also be used for other saturation-related purposes.)

The Dry Basis Sour Gas Process Stream is specified, as seen in Fig. 10.5. In Chapter 9, the Stream Composition Specification dialog was accessed by clicking on the "Specify..." button located on the Composition tab. Here, the alternative of double-clicking on the Composition Table is demonstrated. In Fig. 10.5, no values are entered for MDEA and water. By default, if values are missing in the Composition Specification dialog, the window shown in Step 4 is generated, asking whether the user would like to assign zeros to all missing specifications. Alternatively, zeros could have been specified directly in Step 3.

Typically, when defining inlet Process Streams, temperature, pressure, composition, and flowrate must be specified. The "Water" Process Stream that is connected to the bottom of the Saturator Block is a special case. This Stream

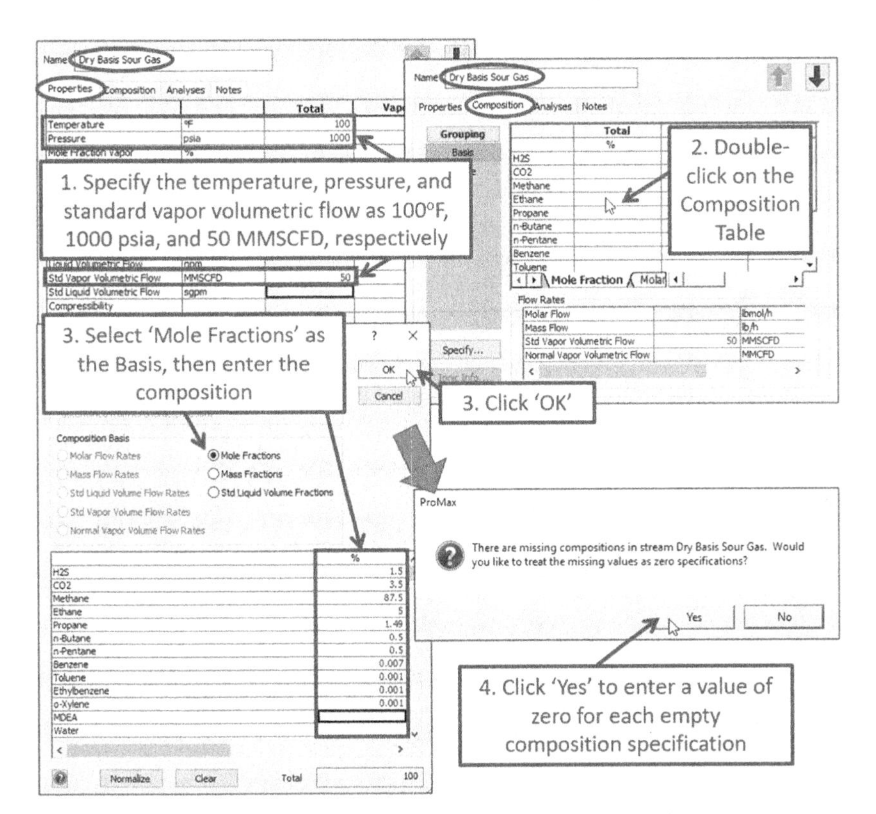

FIGURE 10.5 Specifying the Dry Basis Sour Gas Process Stream.

TABLE 10.3 Specifications Related to the Saturator Block

Block or Stream	Specification
"Water" Process Stream	Composition: 100% water
Saturator Block	Fraction of Saturation: 100%

serves as the saturant for the saturation that is facilitated by the Saturator Block. Because the Saturator Block calculates the necessary temperature, pressure, and flowrate of the saturant to accomplish the specified level of saturation, only the composition is specified in the Process Stream serving as the saturant (here the "Water" Process Stream).

Specifications for the Water Process Stream and the Saturator Block are shown in Table 10.3. The composition of the Water Process Stream is specified using the steps demonstrated for other Streams in Chapter 9 and Fig. 10.5. The default value for "Fraction of Saturation" is 100%. Because the Saturator Block inlets are fully specified and the desired saturation level is the default, the Saturator Block is ready to be executed, as demonstrated in Fig. 10.6.

The composition of the saturated inlet sour gas (Sour Feed Process Stream) is shown in Fig. 10.7. This saturated stream is the sour gas feed that will enter the amine absorber for removal of H_2S and CO_2.

10.6.2 Amine Absorber

The specifications for the Absorber and their respective locations in the Absorber Block dialog are outlined in Table 10.4.

The specifications in Table 10.4 are required to model the amine absorber. With respect to the first four specifications in Table 10.4, general information and their locations in the Column dialog have been shown previously in Chapter 9. Please note that the TSWEET Kinetics Column Type is specific for modeling

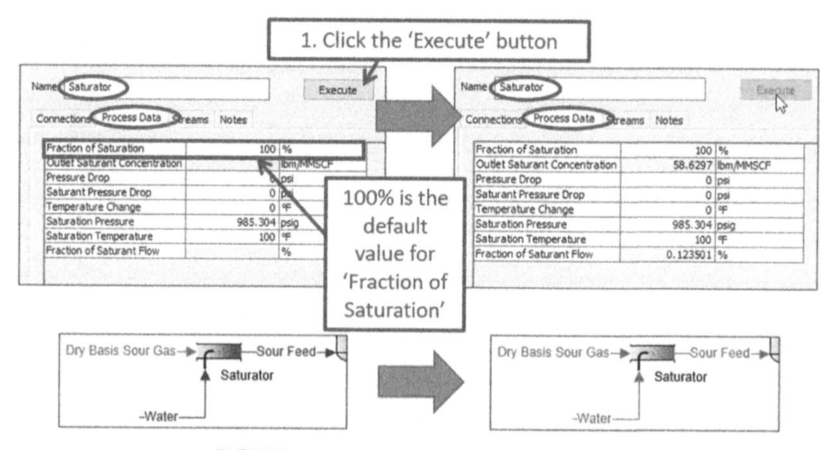

FIGURE 10.6 Executing the Saturator Block.

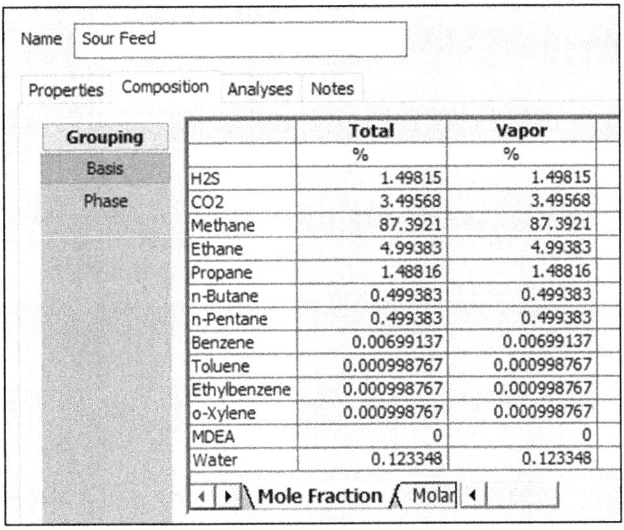

Name: Sour Feed

Properties | Composition | Analyses | Notes

Grouping: Basis, Phase

	Total %	Vapor %
H2S	1.49815	1.49815
CO2	3.49568	3.49568
Methane	87.3921	87.3921
Ethane	4.99383	4.99383
Propane	1.48816	1.48816
n-Butane	0.499383	0.499383
n-Pentane	0.499383	0.499383
Benzene	0.00699137	0.00699137
Toluene	0.000998767	0.000998767
Ethylbenzene	0.000998767	0.000998767
o-Xylene	0.000998767	0.000998767
MDEA	0	0
Water	0.123348	0.123348

Mole Fraction | Molar

FIGURE 10.7 The composition of the saturated inlet sour gas (Sour Feed Process Stream).

TABLE 10.4 Specifications Related to the Absorber Column Block

Specification	Value	Location of Specification
Model Type	Ideal Stage (default)	Process Data tab, Column Dialog of the Properties Grouping
Ideal Stage Column Type	TSWEET Kinetics	
Flash Type	VLE (default)	
Pressure Change	3 psi (0.2 bar)	Process Data tab, Stages Dialog of the Properties Grouping
Fraction Flooding	70%	Process Data tab, General Dialog of the Hardware Grouping
Real/Ideal Stage Ratio	3	
System Factor	0.8	
Tray Spacing	2 feet (0.61 m)	Process Data tab, Specifications subtab of the Tray Dialog of the Hardware Grouping
Weir Height	3 inches (7.62 cm)	

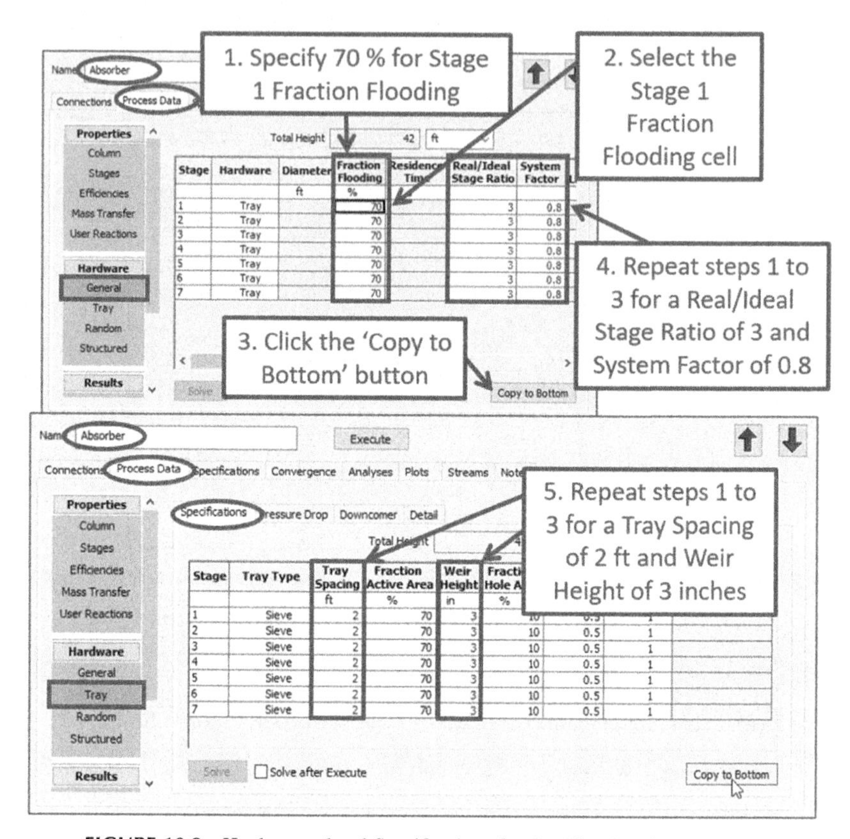

FIGURE 10.8 Hardware-related Specifications for the Absorber Column Block.

amine sweetening. Completion of the remaining specifications is shown in Fig. 10.8. Although use of the "Copy to Bottom" button is demonstrated, values for each of these specifications can be manually entered for all stages.

It should be noted, if the diameter of the column is known, it can be specified instead of Fraction Flooding. The typical amine absorber has ~20 trays. In ProMax, this is modeled by using 7 ideal stages and a Real/Ideal Stage Ratio of 3 (7 ideal stages * 3 real stages/ideal stages = 21 real stages). Please note that the Real/Ideal Stage Ratio is dependent on the application. The system factor compensates for foaming, with 0.8 being the recommended value for amine absorbers. The values for Tray Spacing and Weir Height utilized here are common values and can be used when modeling this type of system if these values are unknown.

10.6.3 Rich Flash and the Lean/Rich Exchanger

Specifications related to the Rich Flash and Lean/Rich Exchanger are shown in Table 10.5. The Rich Flash operates at 85 psia (5.9 bar). This is represented by

TABLE 10.5 Rich Flash and Lean/Rich Exchanger—Related Specifications

Block	Block Specification	Other Specifications
Rich Flash	None	Stream 7 (vapor outlet) Pressure = 85 psia (5.9 bar)
Lean/Rich Exchanger	Side A Pressure Drop: 5 psi (0.34 bar) Side B Pressure Drop: 5 psi (0.34 bar)	Stream 9 (outlet to Stripper) Temperature = 210°F (99°C)

specifying the Rich Flash outlet pressure in Stream 7 as 85 psia (5.9 bar). This specification could also be made in the liquid outlet (Stream 8). The pressure drops for each side of the Lean/Rich Exchanger are specified, as well as a typical temperature of 210°F (99°C) for the rich amine entering the Stripper (Stream 9).

10.6.4 Amine Stripper

The minimum required specifications for the Stripper Column Block are shown in Table 10.6 (nonconvergence-related) and Table 10.7 (convergence-related). If it is desired to size the Stripper Column, additional specifications are required.

TABLE 10.6 Stripper Column Specifications

Specification	Value	Location of Specification
Model Type	Ideal Stage (default)	Process Data tab, Column Dialog of the Properties Grouping
Ideal Stage Column Type	TSWEET Stripper	
Flash Type	VLE (default)	
Stage 1 (Top) Pressure Stage 10 (Bottom) Pressure	12 psig (0.83 barg) 16 psig (1.1 barg)	Process Data tab, Specifications subtab of the Tray Dialog of the Hardware Grouping

TABLE 10.7 Convergence-Related Stripper Column Specifications

Specification	Target/Value	Location of Specification
Condenser Temperature	120°F (49°C) (Status: Specification)	Stripper Block Specifications tab
Reboiler Duty	16.6 MMBtu/h (4900 kW)	Reboiler Energy Stream (Q-2)

As discussed with the Column in Chapter 9, each add-on provides a degree of freedom. The Stripper Column Block has 2 add-ons and, therefore, two degrees of freedom. The "active" specifications that will serve as criteria for column convergence are listed in Table 10.7.

As with the Column Block in Chapter 9, a Phase Property Specification will be added (Condenser Temperature) in the Specifications tab of the Stripper Column Block. The details of the added Condenser Temperature specification are seen in Fig. 10.9. See Chapter 9 for a review of adding this type of specification.

The second "active" specification, the reboiler duty, is specified directly in the Energy Stream of the Reboiler (Q-2), as shown in Fig. 10.10. The Energy Rate of 16.6 MMBtu/h (4900 kW) is derived from the industry standard rule of

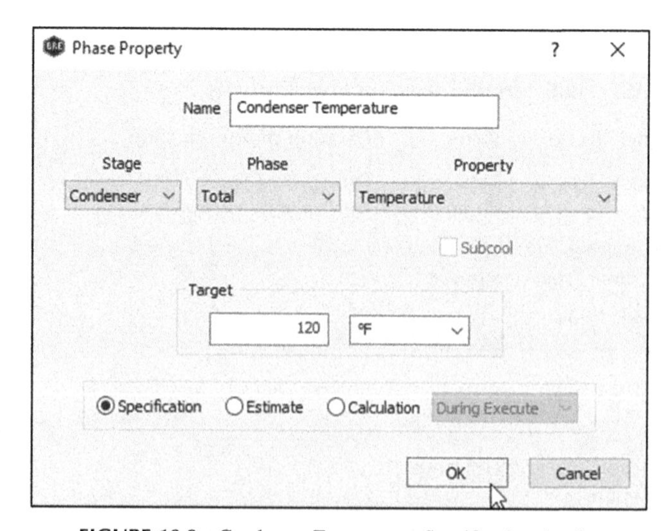

FIGURE 10.9 Condenser Temperature Specification details.

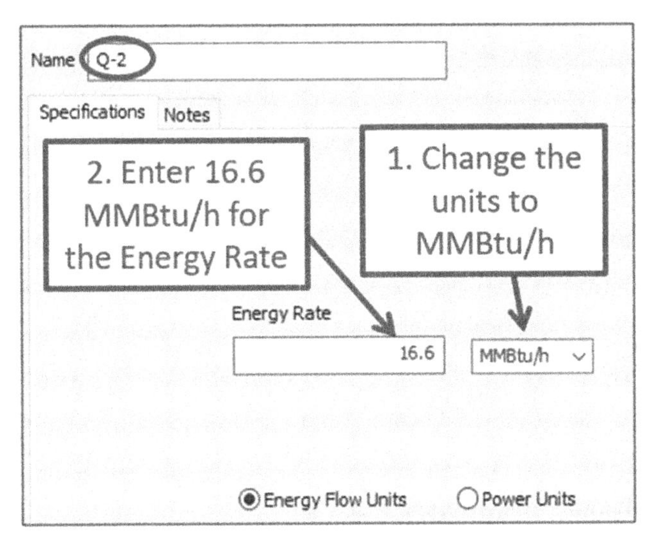

FIGURE 10.10 Condenser Temperature Specification details.

thumb for amine stripper reboiler duty, approximately 1 lb of steam per gallon of circulating amine. Here, the equivalent of a steam rate of 1.1 lb of steam/gal amine is utilized (275 gallons of amine will be circulated). [At a rate of 1.1 lb/min of 50 psig (3.45 barg) steam, the equivalent reboiler energy rate is ~0.0605 MMBtu/h or 17.7 kW.]

10.6.5 Recycle Block

The Recycle Guess does not have to be accurate in this case, especially in terms of the flowrate and composition, because the Makeup/Blowdown Block immediately following sets the composition and flowrate of the lean amine entering the Absorber Block. Aside from this, the stripped or lean amine leaving the Stripper Column Block should be very close in concentration to the original fresh amine concentration, and the circulating amine concentration is known. In addition, a pressure guess can be derived from knowledge of the pressure of the lean amine leaving the Stripper Column and the pressure drop across the Lean/Rich Exchanger. If the user is not familiar with typical temperature ranges of the lean amine at this point in the process, a logical guess would be a temperature lower than the rich amine inlet to the Stripper (210°F or 99°C) and higher than the Trim Cooler outlet temperature. Regardless, again, the Recycle Guess does not need to be accurate in this particular case. The Recycle Guess that is utilized here in Process Stream 12 is shown in Table 10.8. Please note that when working with amine concentrations they are expressed as a mass fraction (wt%).

TABLE 10.8 Initial Recycle Guess (Stream 12)

Specification	Value
Temperature	170°F (77°C)
Pressure	11 psig (0.8 barg)
Flowrate	275 sgpm (62.5 m^3/h)
Component	**Composition (wt%)**
MDEA	50
Water	50

MDEA, methyldiethanolamine.

10.6.6 Makeup/Blowdown Block

In an actual sweetening unit, addition of amine occurs at singular points in time, not continually. But, to provide flow assurance in this steady-state model, the Makeup/Blowdown Block is used to represent this process on a continual basis. The block calculates and provides any needed amine and water makeup and/or blowdown to maintain the specified amine concentration and flowrate. Table 10.9 displays the specifications related to the Makeup/ Blowdown Block.

The Makeup Stream temperature is typically specified approximately as the ambient temperature or expected temperature in the tank holding the amine. The pressure needs to be higher than the lean amine temperature entering the Makeup/Blowdown Block. Fig. 10.11 shows the steps to specify

TABLE 10.9 Specifications Related to the Makeup/Blowdown Block

Block/Stream	Specification(s)
Makeup Stream	Temperature = 80°F(27°C) Pressure = 20 psig (1.4 barg)
Makeup/Blowdown Block	Target Outlet Composition = 50 wt % MDEA Makeup Bulk Composition = 100 wt% water
Stream 13 (outlet of Makeup/ Blowdown Block)	Standard Liquid Volumetric Flow = 275 sgpm (62.5 m^3/h)

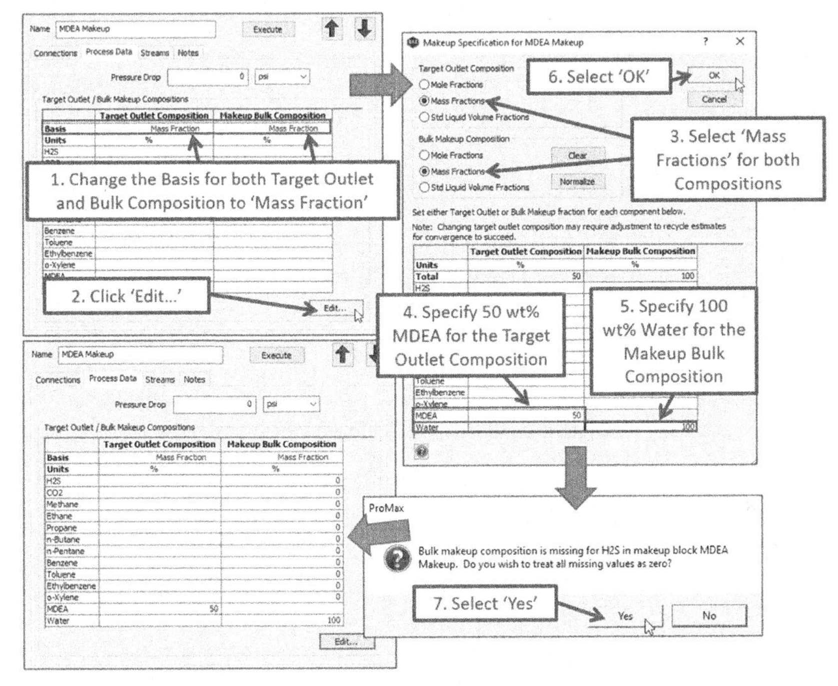

FIGURE 10.11 Specifying the Makeup/Blowdown Block.

the Makeup/Blowdown Block to maintain circulation of 50 wt% MDEA. The outlet of the Makeup/Blowdown Block (Stream 13) is where the desired amine flowrate is specified.

10.6.7 Circulation Pump and Trim Cooler

Table 10.10 outlines the needed specifications for the Circulation Pump and Trim Cooler Blocks. It should be noted that the lean amine entering the

TABLE 10.10 Specifications for the Circulation Pump and Trim Cooler Blocks

Block	Specification(s)
Circulation Pump	Overall Efficiency = 65% Stream 14 (outlet) pressure = 1010 psia (70 bar)
Trim Cooler	Pressure Drop = 5 psi (0.34 bar) Stream 15 (outlet) temperature = 110°F (44°C)

Absorber needs to be at least 10°F (5.5°C) higher than the sour gas inlet to prevent hydrocarbon condensation.

Once the specifications in Table 10.10 are completed, the flowsheet is ready for execution.

10.7 VIEWING RESULTS

Fig. 10.12 demonstrates how to view the resulting Sweet Gas CO_2 and H_2S content on the flowsheet in a callout. From the resulting callout, the Sweet Gas H_2S is ~3.6 ppm, which meets the desired specification of < 5 ppm. The Sweet Gas CO_2 is ~1.32 mol%, which also meets the CO_2 specification of <2 mol%.

The Rich and Lean Amine Loadings are also important parameters to determine when modeling an amine unit. These values provide information

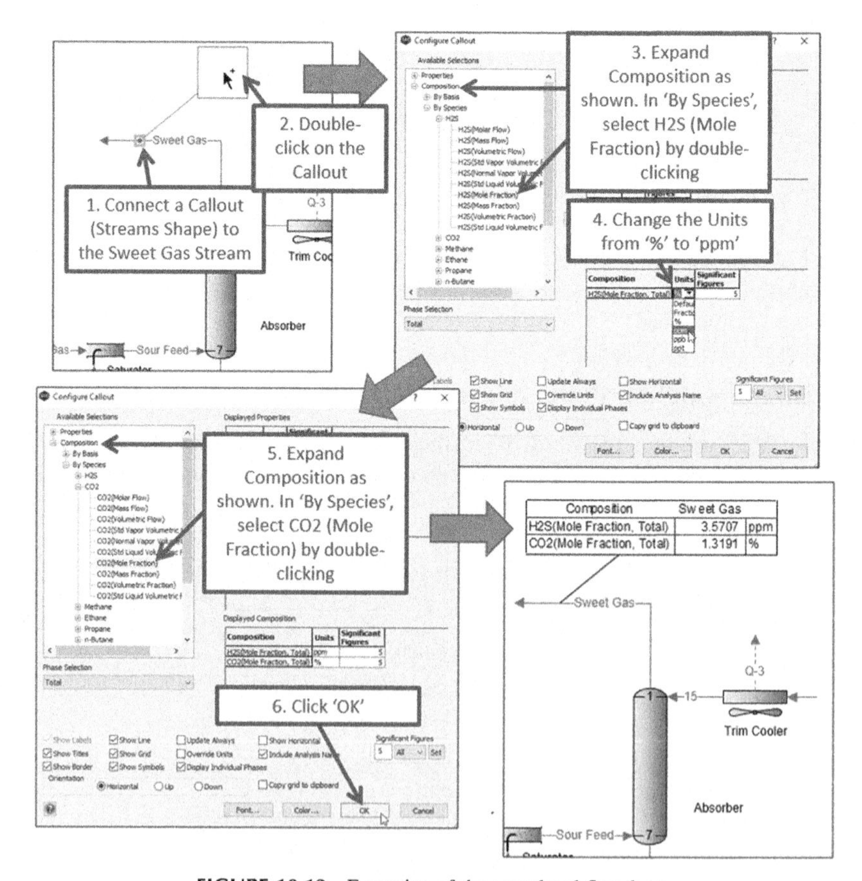

FIGURE 10.12 Execution of the completed flowsheet.

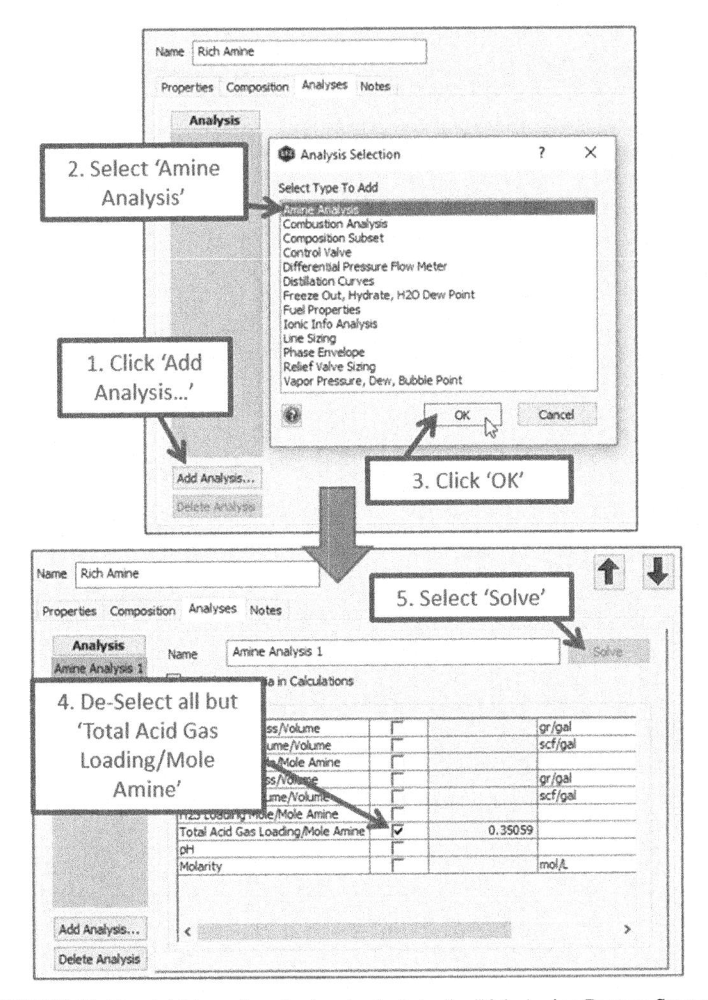

FIGURE 10.13 Addition of an Amine Analysis to the Rich Amine Process Stream.

about the performance of the unit, as well as the risk of corrosion. Loading values can be determined for any Process Stream by the addition of an Amine Analysis. Fig. 10.13 demonstrates this process for the Rich Amine Stream.

The same process outlined in Fig. 10.13 can be used to add an Amine Analysis to the Lean Amine Stream to determine the Lean Loading. Both Loading values are displayed in a Property Table in Fig. 10.14. The Rich Loading is 0.351, which is in a safe range in terms of corrosion. The Lean Loading is 0.003, indicating sufficient stripping. Appropriate Rich and Lean Loading values will vary by amine type.

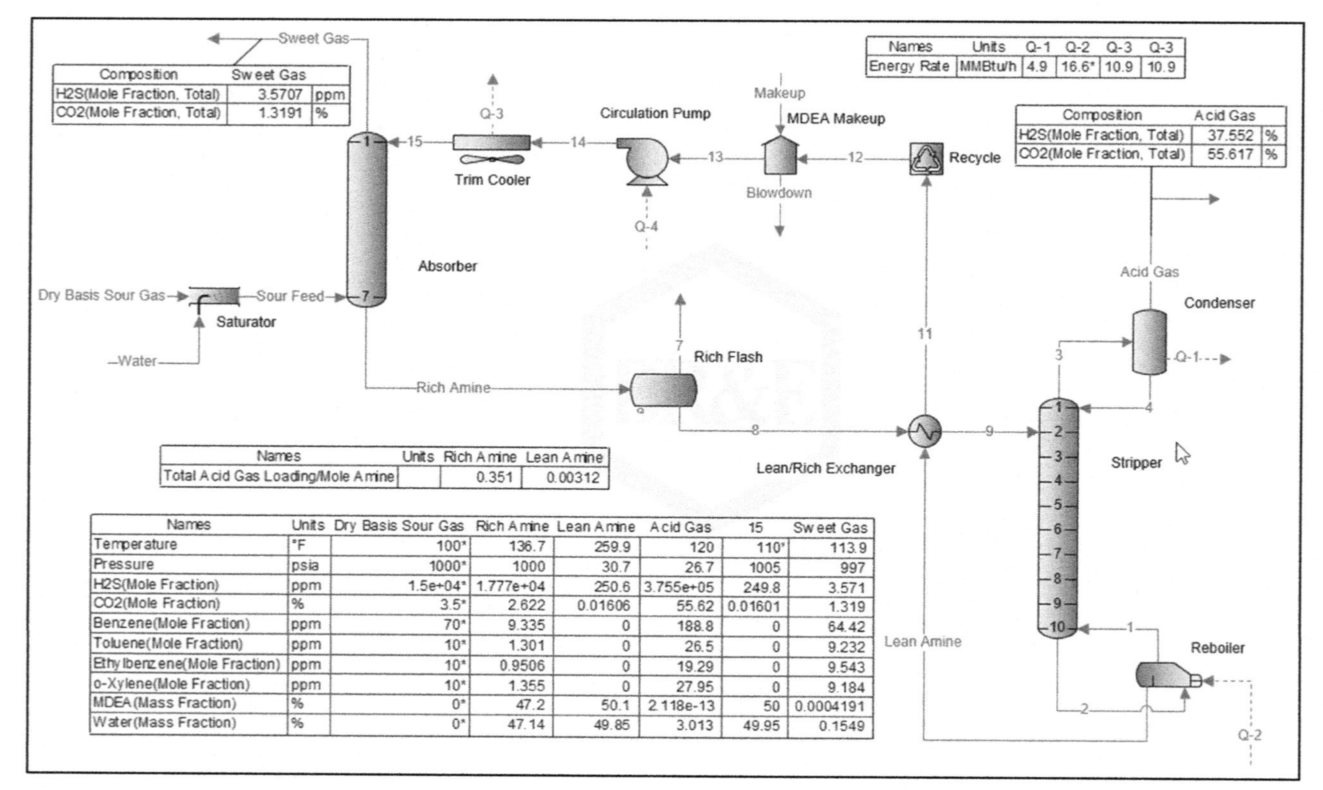

Composition	Sweet Gas	
H2S(Mole Fraction, Total)	3.5707	ppm
CO2(Mole Fraction, Total)	1.3191	%

Names	Units	Q-1	Q-2	Q-3	Q-3
Energy Rate	MMBtu/h	4.9	16.6*	10.9	10.9

Composition	Acid Gas	
H2S(Mole Fraction, Total)	37.552	%
CO2(Mole Fraction, Total)	55.617	%

Names	Units	Rich Amine	Lean Amine
Total Acid Gas Loading/Mole Amine		0.351	0.00312

Names	Units	Dry Basis Sour Gas	Rich Amine	Lean Amine	Acid Gas	15	Sweet Gas
Temperature	°F	100*	136.7	259.9	120	110*	113.9
Pressure	psia	1000*	1000	30.7	26.7	1005	997
H2S(Mole Fraction)	ppm	1.5e+04*	1.777e+04	250.6	3.755e+05	249.8	3.571
CO2(Mole Fraction)	%	3.5*	2.622	0.01606	55.62	0.01601	1.319
Benzene(Mole Fraction)	ppm	70*	9.335	0	188.8	0	64.42
Toluene(Mole Fraction)	ppm	10*	1.301	0	26.5	0	9.232
Ethylbenzene(Mole Fraction)	ppm	10*	0.9506	0	19.29	0	9.543
o-Xylene(Mole Fraction)	ppm	10*	1.355	0	27.95	0	9.184
MDEA(Mass Fraction)	%	0*	47.2	50.1	2.118e-13	50	0.0004191
Water(Mass Fraction)	%	0*	47.14	49.85	3.013	49.95	0.1549

FIGURE 10.14 A Converged MDEA sweetening simulation.

REFERENCES

Bryan Research & Engineering (BR&E), 2017. ProMax. www.bre.com.

Gas Processors Suppliers Association (GPSA), 2012. Engineering Data Book, thirteenth ed. FPS. GPSA, Tulsa, OK, USA.

Gas Processors Suppliers Association (GPSA), 2004. Engineering Data Book, twelfth ed. FPS. GPSA, Tulsa, OK, USA.

Kohl, A., Riesenfeld, F., 1979. Gas Purification, third ed. Gulf Coast Publishing, Houston, TX, USA.

EXERCISE

1. Determine the effect of different Steam Rates (ranging between 0.8 and 1.3 lb steam/gal circulating amine) on sweet gas composition and the lean amine and rich amine loadings.

Part 5

aspenONE Engineering

Chapter 11

Basics of Process Simulation With Aspen HYSYS

Nishanth Chemmangattuvalappil, Siewhui Chong
University of Nottingham Malaysia Campus, Semenyih, Malaysia

In this chapter, a step-by-step guide is provided for the simulation of an integrated process flowsheet using Aspen HYSYS. The concept of simulation is based on *sequential modular* approach and follows the onion model for flowsheet synthesis (see Chapter 1 for details). The case study on *n*-octane production is used for illustration throughout the chapter.

11.1 EXAMPLE ON *N*-OCTANE PRODUCTION

A simple example that involves the production of *n*-octane (C_8H_{18}) (Foo et al., 2005) is demonstrated, with detailed descriptions given in Example 1.1. The basic simulation setup involving registration of components, thermodynamic models, and reaction stoichiometry is to be carried out in the *Basis Environments* of Aspen HYSYS, while the modeling of reactor, separation, and recycle system are to be carried out in the *Simulation Environments*. The individual steps are discussed in the following subsections.

Step 1: Basic Simulation Setup

The first step in simulation using Aspen HYSYS is the definition of components. All components involved in the process are entered into the flowsheet by selecting from the component database, as shown in Fig. 11.1.

Once the components are chosen, the appropriate thermodynamic model (this is known as "fluid package" in Aspen HYSYS terminoogy) for the system is chosen.[1] Because this process involves the hydrocarbons at high pressure, "Peng–Robinson" equation of state has been chosen as the fluid package, with steps shown in Fig. 11.2.

1. See Chapter 3 for guidelines on thermodynamic package section.

Chemical Engineering Process Simulation. http://dx.doi.org/10.1016/B978-0-12-803782-9.00011-X

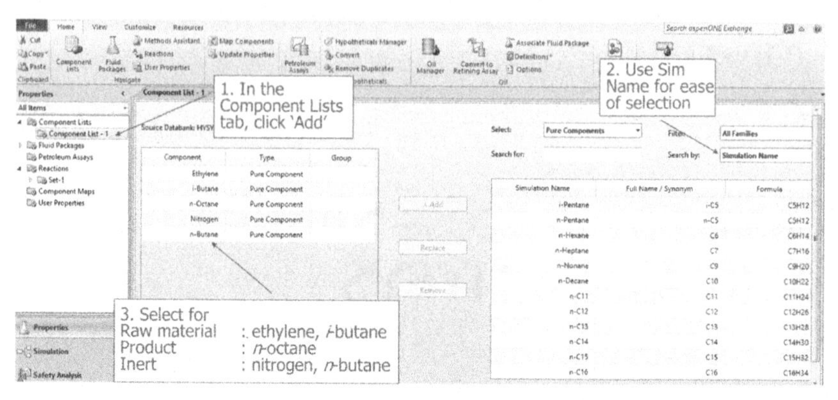

FIGURE 11.1 Defining components.

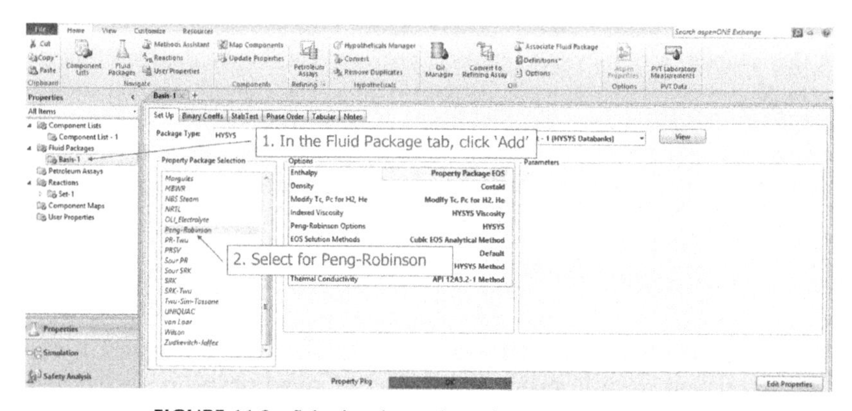

FIGURE 11.2 Selecting thermodynamic model (fluid package).

In the next step, the reactions involved in the process must be defined at the "Reaction" tab (Fig. 11.3). The first step is the selection of reactor model. In this example, the reaction is modeled as a "conversion reactor." The conversion reactor model will treat the reactor as a stoichiometric problem and solve the mass balance based on the specified conversion (see steps in Fig. 11.3A).

Once the reactor model is chosen, all components taking part in the reaction are selected accordingly and their stoichiometric coefficients are entered. The limiting component must be chosen along with the conversion. It is to be noted that, in Aspen HYSYS, the conversion must be specified in percentage (see Fig. 11.3B). Once all information is entered, a thermodynamic package must be assigned to this reaction to estimate the conditions after the reaction. In this example, the reaction is linked to the Peng–Robinson equation of state model (see steps in Fig. 11.3C).

These are the basic information required for creating the simulation flowsheet. Now we may proceed to the *Simulation Environments* to perform

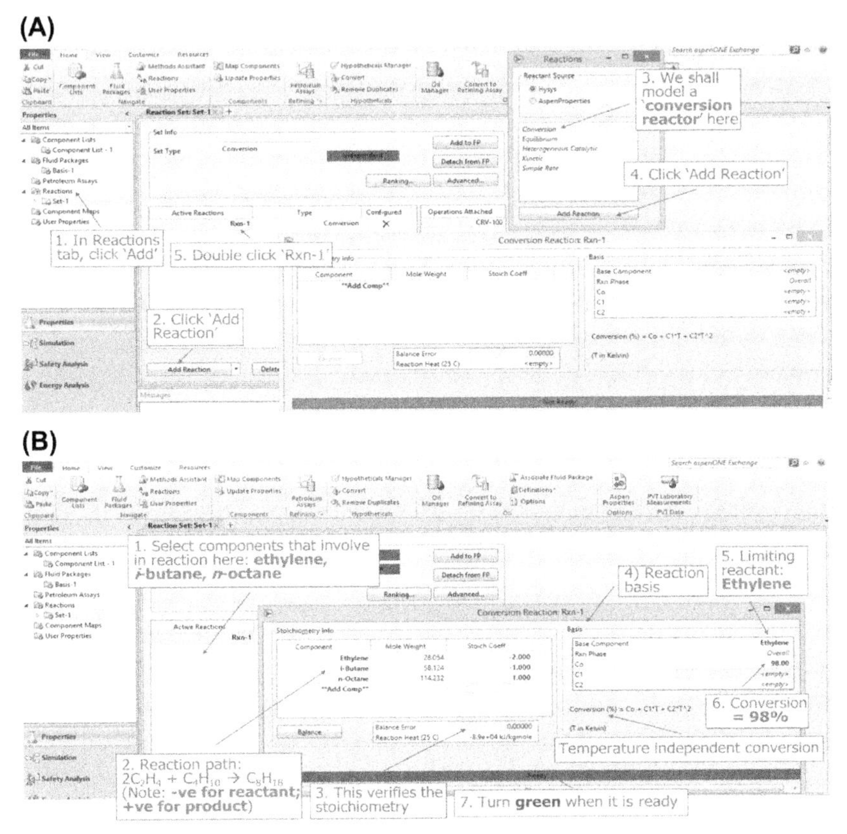

FIGURE 11.3 Steps in specifying reaction sets. (A) Select reactor type, (B) enter reaction details, (C) assign a fluid package, (D) enter the Simulation Environments.

modeling of the unit operations. To enter the *Simulation Environments*, we shall press the "Enter Simulation Environment" button on the *Simulation Basis Manager* (Fig. 11.3D).

Step 2: Modeling of Reactor

The *Simulation Environments* of Aspen HYSYS consist of main flowsheet, subflowsheet, and column subflowsheet environments. For the *n*-octane production example, only the main flowsheet is used. In this stage, the topology of the flowsheet must first be defined to identify the sequence of unit operations. Now, according to the onion model,[2] the reactor system is simulated in the first step. It is necessary to define the reactions in a flowsheet before entering the Simulation Environment. At this stage, the type of reactor, information on conversion,

2. See Example 1.1 for details.

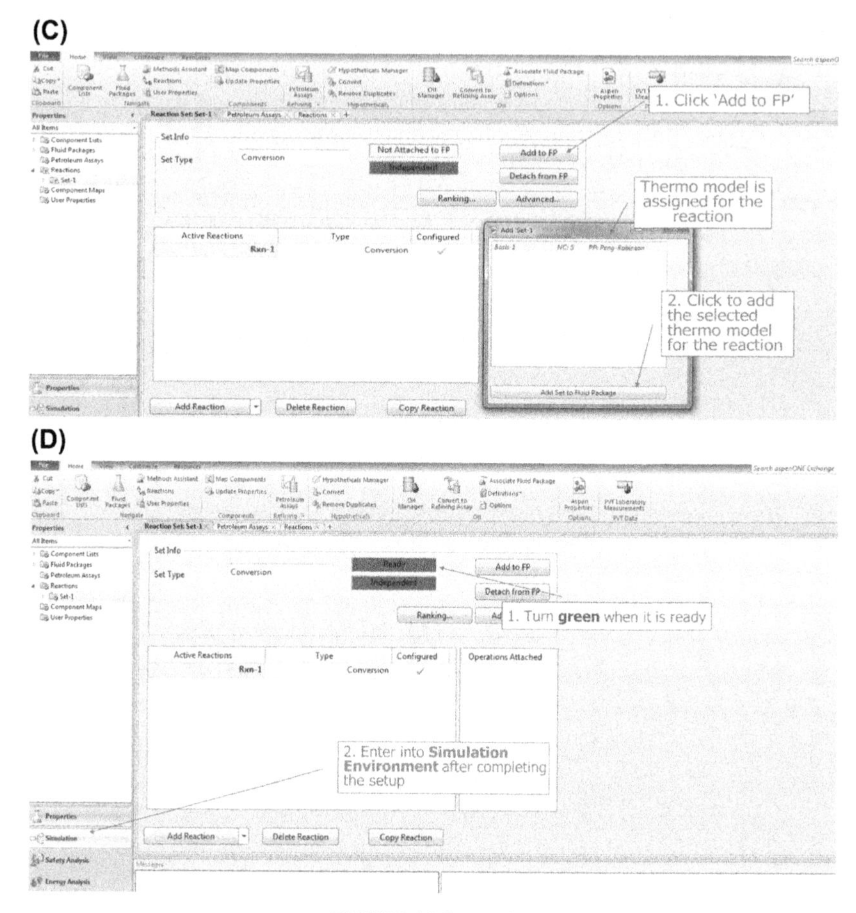

FIGURE 11.3 cont'd

kinetics, equilibrium, etc., (depend on the available information) have to be specified as well. Although there are different types of reactor models available in Aspen HYSYS such as continuous stirred tank reactor and plug flow reactor, a conversion model is chosen for this example. The kinetic reaction models require the information on the reaction kinetics. For the *n*-octane production example, we shall utilize the "conversion reactor" model. It is to be noted that the conversion reactor model can only perform mass and energy calculations based on the stoichiometry.

Detailed steps to draw the flowsheet on the process flow diagram (PFD) are shown in Fig. 11.4, where the conversion reactor consists of two inlet (Streams 1 and Q-101) and two outlet streams (Streams 2 and 3). Note that Streams 1, 2, and 3 are the actual process streams that consist of material (termed as Material Stream in HYSYS), while Stream Q-101 is actually virtual stream that is used for performing heat balances.

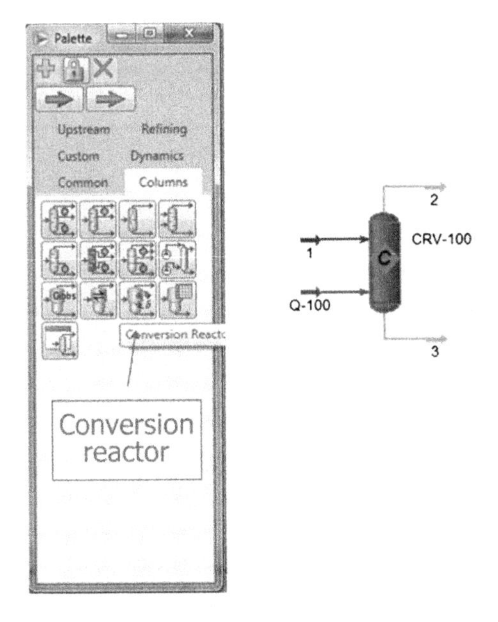

FIGURE 11.4 Setting up feed.

After the connections are made, the conditions need to be specified in the incoming stream according to Table 11.1. It is advisable to conduct a degree of freedom analysis before designing the equipment to make sure that there are enough process parameters for the design of equipment and the variables are not overspecified. To do a degree of freedom analysis, we list all the variables involved in the process units. These variables can be operating conditions, such as temperature and pressure, flowrates, and compositions. Once sufficient conditions, which are molar flowrates, temperature, and pressure in this case, are entered, the incoming stream is completely defined.

TABLE 11.1 Feed Condition

Component	Flowrate (kmol/h)	Condition
Nitrogen, N_2	0.1	$T = 93°C$ $P = 20$ psia
Ethylene, C_2H_4	20	
n-Butane, C_4H_{10}	0.5	
i-Butane, C_4H_{10}	10	
n-Octane, C_8H_{18}	0	

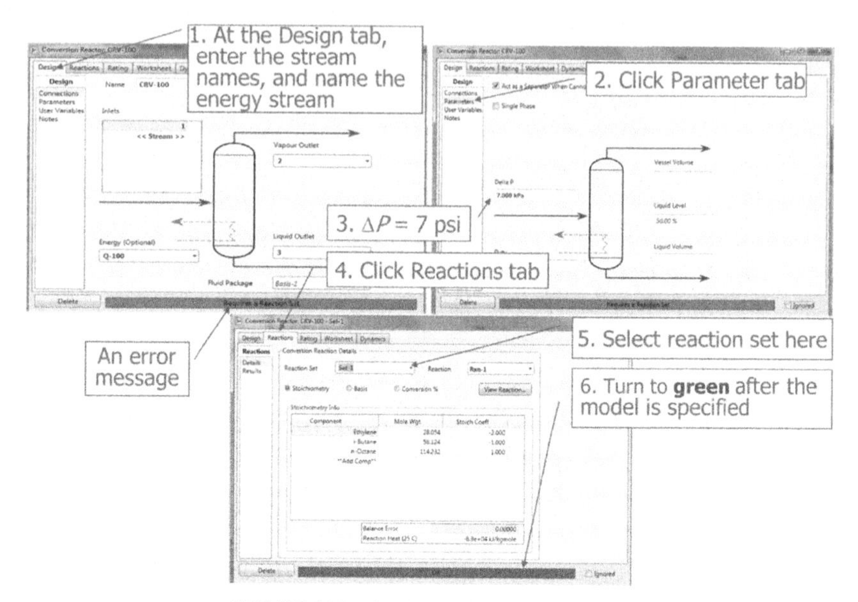

FIGURE 11.5 Setting up reactor connection.

In the next step, the reactor model must be completely specified. We may enter the pressure drop and also define the reaction set to completely define the reaction (see detailed steps in Fig. 11.5). In this example, there is only one reaction set that is chosen. Because the reaction is conducted at isothermal mode, the outlet stream temperature should be selected and set to the same temperature as the incoming stream. We may notice that the energy stream now indicates a heat flow to maintain the reactor at isothermal conditions. In case of adiabatic reactors, no energy stream must be connected to the reactor. At this stage, we may notice that the reactor model is converged because all variables have been specified.

Aspen HYSYS has been set to solve once the necessary data are sufficient. A convenient way of displaying the simulation results is via the *Workbook*. Fig. 11.6 shows the detailed steps in displaying molar flowrates of all components on the Workbook. One may also insert the *Workbook Table* within the PFD. This is illustrated with Fig. 11.7, where material and energy streams are displayed (one may also choose to display the stream compositions).

Note that the energy stream in the converged reactor model has a negative value, which indicates that energy must be removed to maintain the isothermal conditions. This indicates an exothermic reaction in the reactor. In the final step, the basic mass balance calculations can be performed to see that the outlet composition from the reactor is in agreement with the stoichiometry. However, in more complex flowsheets, it may not be easily verified through hand calculations.

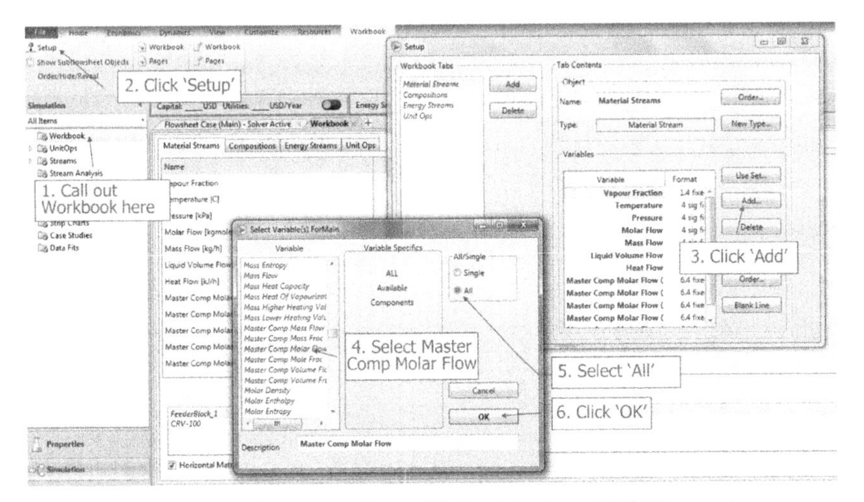

FIGURE 11.6 Setting up Workbook in Aspen HYSYS.

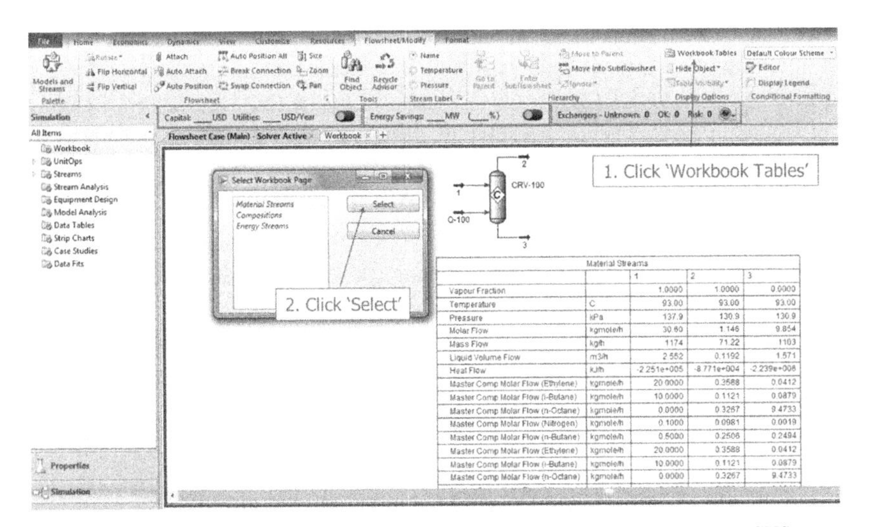

FIGURE 11.7 Adding Workbook Table to process flow diagram in Aspen HYSYS.

Step 3: Separation Units

In this process, the only separation system is a distillation column. The preliminary design of the distillation column can be done using the "shortcut distillation" model in HYSYS. This model is based on the Fenske–Underwood–Gilliland model, which is useful for conducting the preliminary design of a distillation column. The parameters obtained from shortcut distillation model can be used as initial estimates in the rigorous distillation model, which performs stage-by-stage calculations. To continue in building

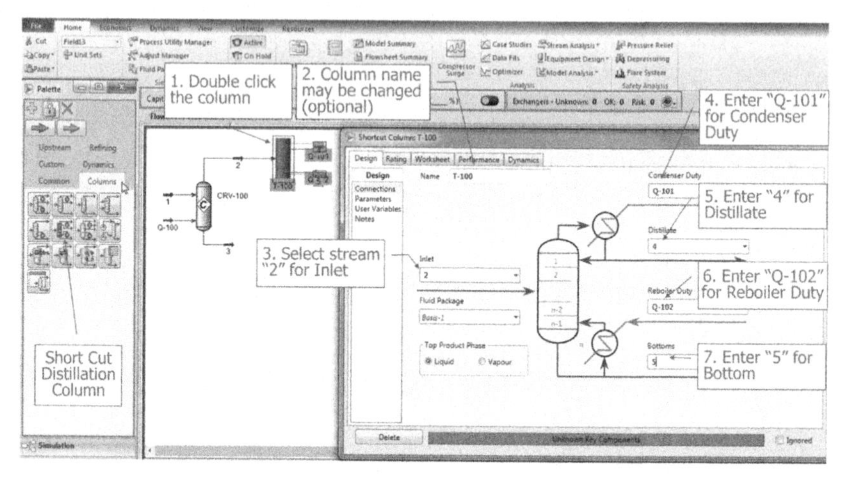

FIGURE 11.8 Adding shortcut distillation model to the process flow diagram.

the topology of the flowsheet, one may refer to Fig. 11.8 for the alternative steps in connecting the material and energy streams. Fig. 11.8 also shows that the column is set to operate with partial condenser, where the column top stream exists in vapor form.

The following design parameters are defined for this separation unit (Fig. 11.9):

- Condenser type: partial condenser
- Light key and mole fraction: ethylene (0.0015 in bottom stream)
- Heavy key and mole fraction: *n*-octane (0.28 in distillate stream)
- Column pressure: 25 psia in condenser
- Pressure drop: 10 psia

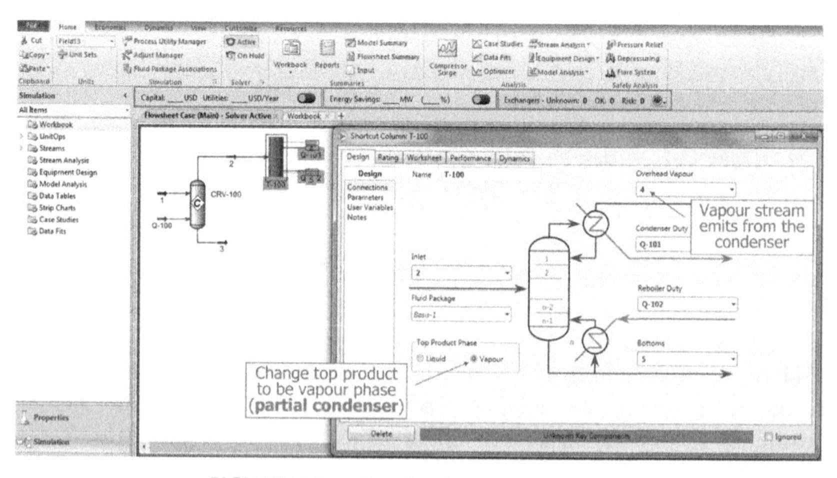

FIGURE 11.9 Changing the type of condenser.

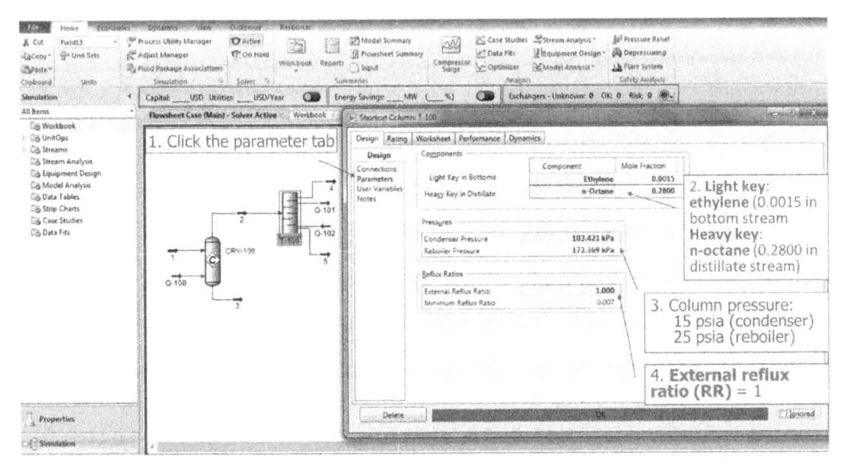

FIGURE 11.10 Specifications for the shortcut distillation model.

In the next step, the specifications of final product stream are added (mole fractions of the key component), and the pressure of condenser and reboiler is defined as shown in Fig. 11.10. After entering the given specifications, it can be noticed that the minimum reflux ratio is automatically calculated. In this case it is 0.007. So, an actual reflux ratio must be chosen to estimate the number of trays required to obtain the desired separation. In this case, we use an actual reflux ratio of 1. Once this information is entered, the column simulation converges. The calculated number of trays and column conditions can be seen on the tab "Performance" as shown in Fig. 11.11.

We next move on to display the simulation results using the *Workbook Table*, as shown in Fig. 11.12.

Step 4: Recycle System (Materials)

Example 1.1 shows that the *n*-octane case contains both material and heat recycle streams. In this step, the convergence of these recycle streams is illustrated. Fig. 11.12 indicates that the distillate stream (Stream 4) contains some unconverted raw material that can be recycled to the reactor. Hence, we need to insert a recycle loop into the simulation to model the material recycle stream, which involves a purge stream unit, compressor, and cooler. Purging unit is necessary to avoid the trapping of materials inside the system. In this case, it is assumed that 10% of the distillate is purged. The compressor and a cooler are then used to adjust the pressure and temperature of the recycle stream to match those of the reactor. Specifications for these units are given in Table 11.2.

To model the purge unit, a stream splitting model (called the "Tee" in Aspen HYSYS) is introduced in PFD. Detailed steps to connect the Tee model (to distillation top stream) and to provide its model specifications are shown in

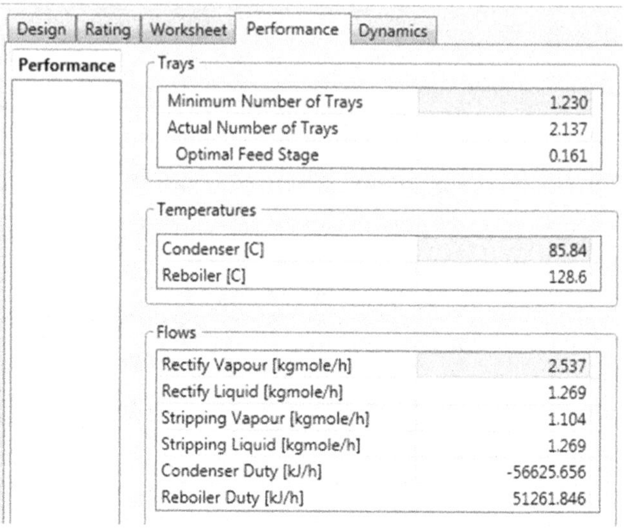
Design
Rating
Worksheet
Performance
Dynamics
Performance
Trays
Minimum Number of Trays 1.230
Actual Number of Trays 2.137
Optimal Feed Stage 0.161
Temperatures
Condenser [C] 85.84
Reboiler [C] 128.6
Flows
Rectify Vapour [kgmole/h] 2.537
Rectify Liquid [kgmole/h] 1.269
Stripping Vapour [kgmole/h] 1.104
Stripping Liquid [kgmole/h] 1.269
Condenser Duty [kJ/h] -56625.656
Reboiler Duty [kJ/h] 51261.846

FIGURE 11.11 Results of shortcut distillation column design.

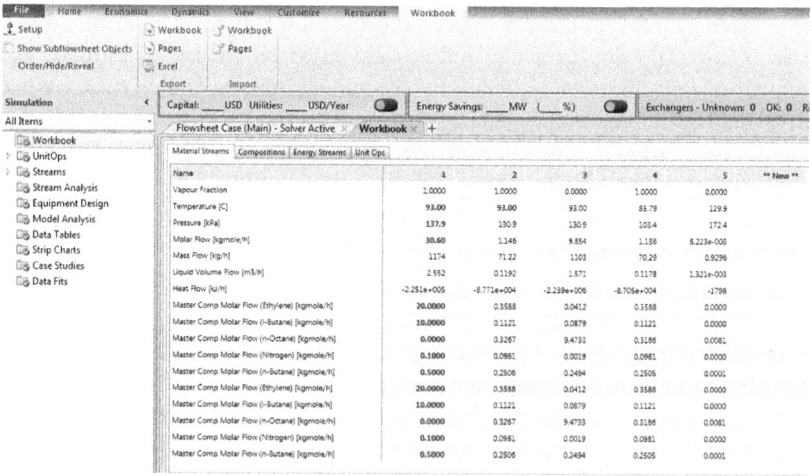
Workbook
Setup
Show Subflowsheet Objects
Order/Hide/Reveal
Workbook
Pages
Excel
Export
Import
Simulation
All Items
Workbook
UnitOps
Streams
Stream Analysis
Equipment Design
Model Analysis
Data Tables
Strip Charts
Case Studies
Data Fits
Capital: ____USD Utilities: ____USD/Year
Energy Savings: ____MW (____%)
Exchangers - Unknown: 0 OK: 0
Flowsheet Case (Main) - Solver Active
Workbook
Material Streams
Compositions
Energy Streams
Unit Ops
Name 1 2 3 4 5 ** New **
Vapour Fraction 1.0000 1.0000 0.0000 1.0000 0.0000
Temperature [C] 93.00 93.00 93.00 85.79 129.9
Pressure [kPa] 137.9 130.9 130.9 103.4 172.4
Molar Flow [kgmole/h] 30.60 1.146 9.854 1.138 8.223e-003
Mass Flow [kg/h] 1174 71.22 1103 70.29 0.9296
Liquid Volume Flow [m3/h] 2.552 0.1192 1.571 0.1178 1.321e-003
Heat Flow [kJ/h] -2.251e+005 -8.771e+004 -2.239e+006 -8.705e+004 -1798
Master Comp Molar Flow (Ethylene) [kgmole/h] 20.0000 0.3588 0.0412 0.3588 0.0000
Master Comp Molar Flow (i-Butane) [kgmole/h] 10.0000 0.1121 0.0879 0.1121 0.0000
Master Comp Molar Flow (n-Octane) [kgmole/h] 0.0000 0.3267 9.4733 0.3186 0.0081
Master Comp Molar Flow (Nitrogen) [kgmole/h] 0.1000 0.0981 0.0019 0.0981 0.0000
Master Comp Molar Flow (n-Butane) [kgmole/h] 0.5000 0.2506 0.2494 0.2505 0.0001
Master Comp Molar Flow (Ethylene) [kgmole/h] 20.0000 0.3588 0.0412 0.3588 0.0000
Master Comp Molar Flow (i-Butane) [kgmole/h] 10.0000 0.1121 0.0879 0.1121 0.0000
Master Comp Molar Flow (n-Octane) [kgmole/h] 0.0000 0.3267 9.4733 0.3186 0.0081
Master Comp Molar Flow (Nitrogen) [kgmole/h] 0.1000 0.0981 0.0019 0.0981 0.0000
Master Comp Molar Flow (n-Butane) [kgmole/h] 0.5000 0.2506 0.2494 0.2505 0.0001

FIGURE 11.12 Simulation results.

Fig. 11.13. Because 10% of the distillate is purged, the flow ratio for recycle stream is set to 0.9.

The distillate is collected at a pressure of 15 psia, whereas the reactor operates at 20 psia. To match the pressure of the distillate steam to that of reactor, a compressor is added to raise its pressure to 20 psia. Detailed steps to connect the compressor model and to provide its specifications are shown in Fig. 11.14.

TABLE 11.2 Specifications for Units in the Material Recycle System

Equipment	Specifications
Purge unit	Flow ratio for recycle stream: 0.9
Compressor	Outlet P: 22 psia
Cooler	Outlet T: 93°C
	Delta P: 2 psi

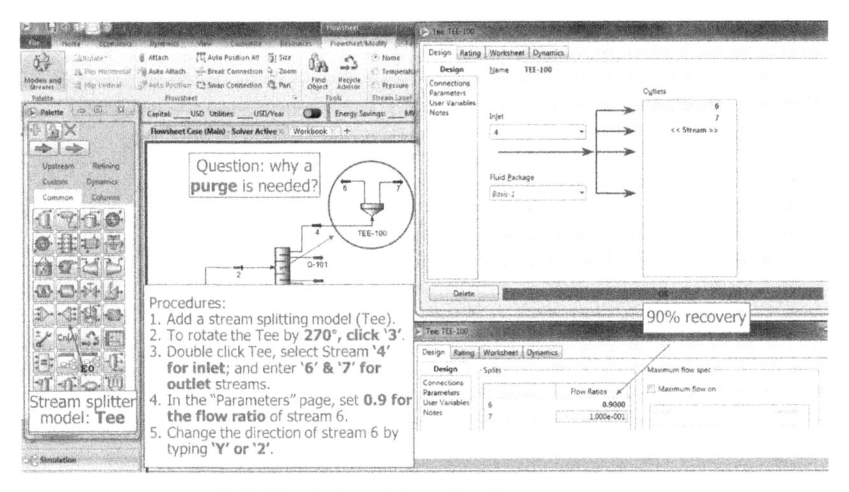

FIGURE 11.13 Adding a Tee for purge stream.

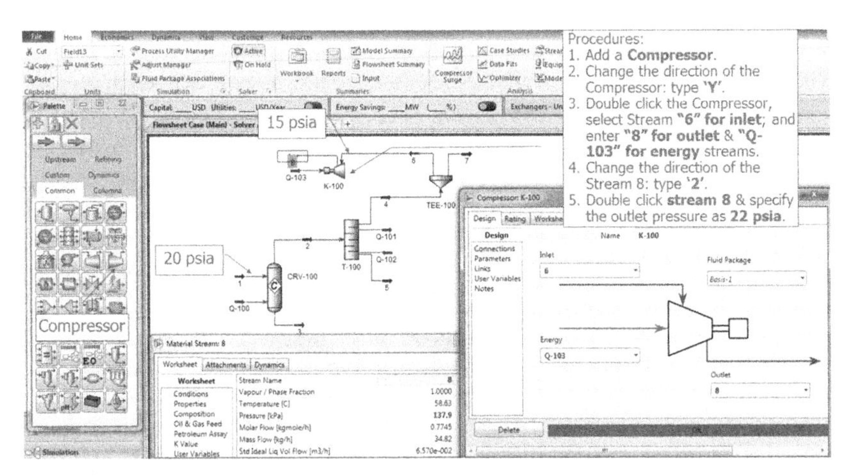

FIGURE 11.14 Adjusting the stream pressure.

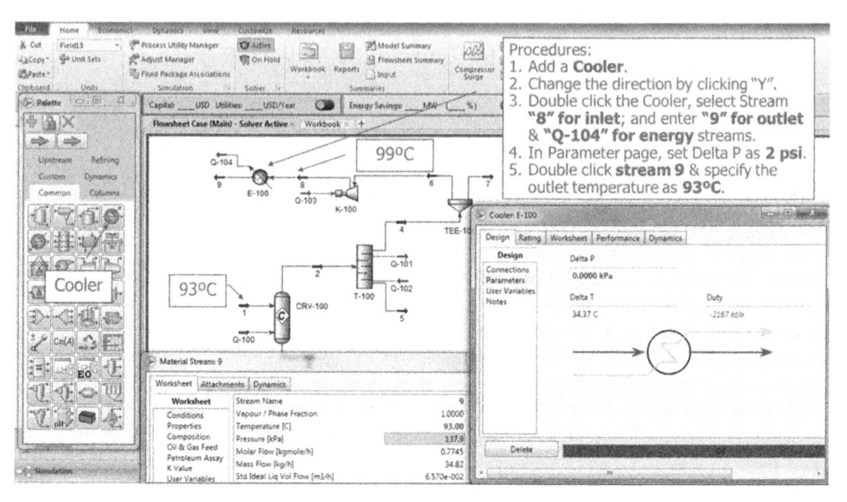

FIGURE 11.15 Adjusting the stream temperature.

Compression will increase the temperature of streams and in this case, the temperature of recycle stream after compression is raised to 99°C, which is higher than the reactor operating temperature. A cooler unit is added to reduce the temperature to the reactor temperature of 93°C. Fig. 11.15 shows the detailed steps to connect the cooler model and to provide its specifications.

Because both pressure and temperature of the recycle stream are now adjusted to match those of the reactor, we can connect the recycle stream to the reactor. To converge this material recycle stream, we can make use of the "Recycle" unit. The latter facilitates the convergence of a recycle loop following the "tear stream" concept.[3] Detailed steps to converge the recycle stream to the reactor with the Recycle unit are shown in Figs. 11.16–11.19. Note that the Recycle unit shows a yellow outline when the recycle stream is first connected to the reactor. This means that some parameters are not converged after 20 rounds of iteration (default setting in Aspen HYSYS). Hence more iteration is needed to ensure all parameters are converged completely (by pressing the "Continue" button in its Connections page). The simulation results are also displayed in the *Workbook Table* in Fig. 11.20.

After the material recycle system is converged, we next proceed to converge the energy recycle stream. This will be done using the "tear stream" concept, i.e., without the use of the Recycle unit. Specifications for heat exchanger and heater in the energy recycle system are given in Table 11.3.

3. The Recycle unit in Aspen HYSYS performs the tear-stream calculation (see Chapter 4 for details) to converge the recycle stream automatically.

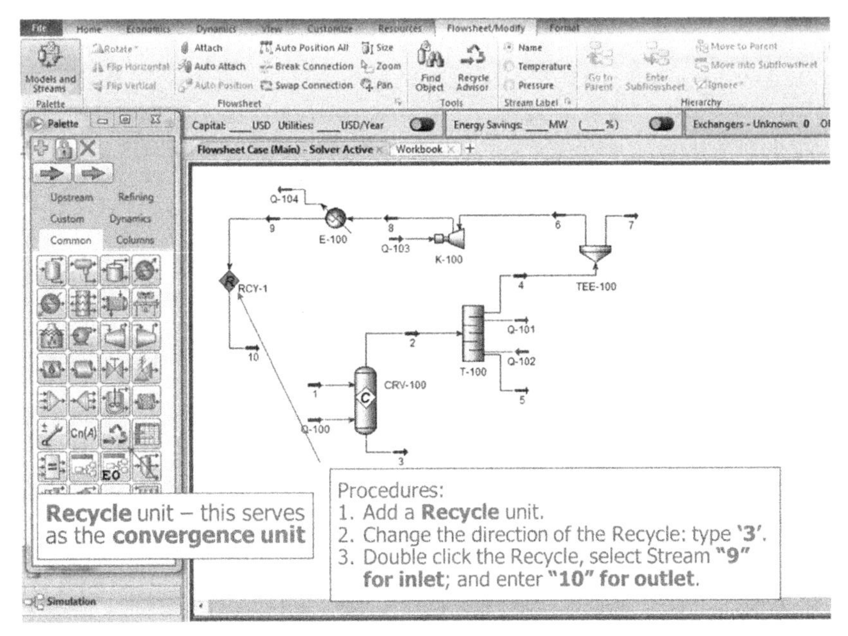

FIGURE 11.16 Adding the Recycle unit.

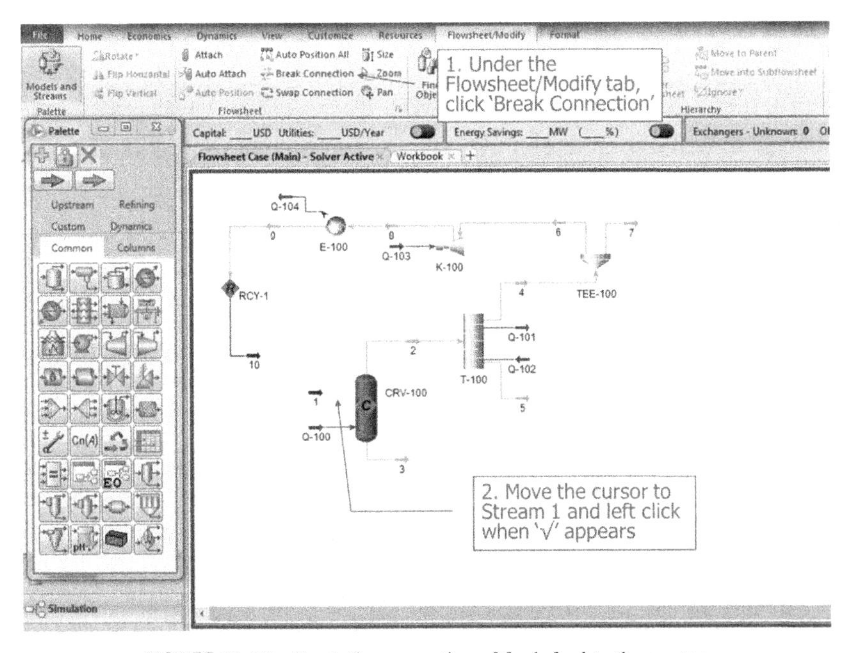

FIGURE 11.17 Break the connection of fresh feed to the reactor.

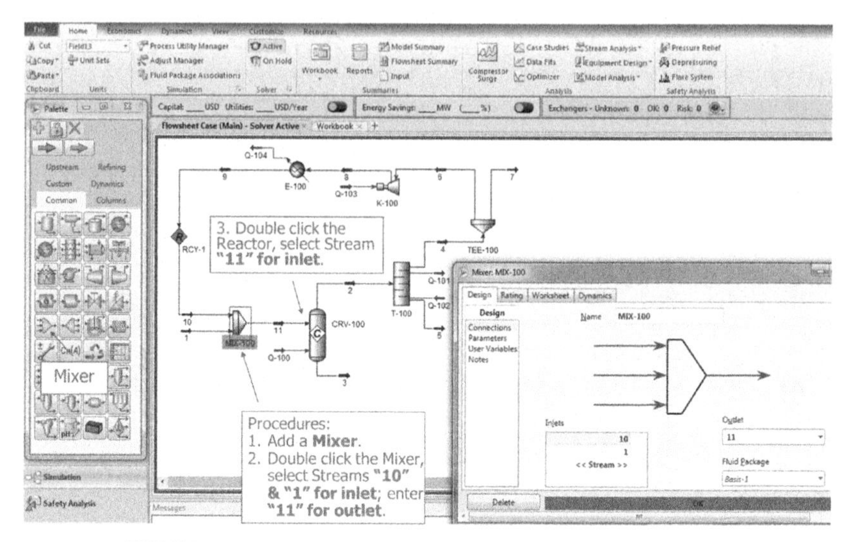

FIGURE 11.18 Add the fresh feed and recycle stream to the reactor.

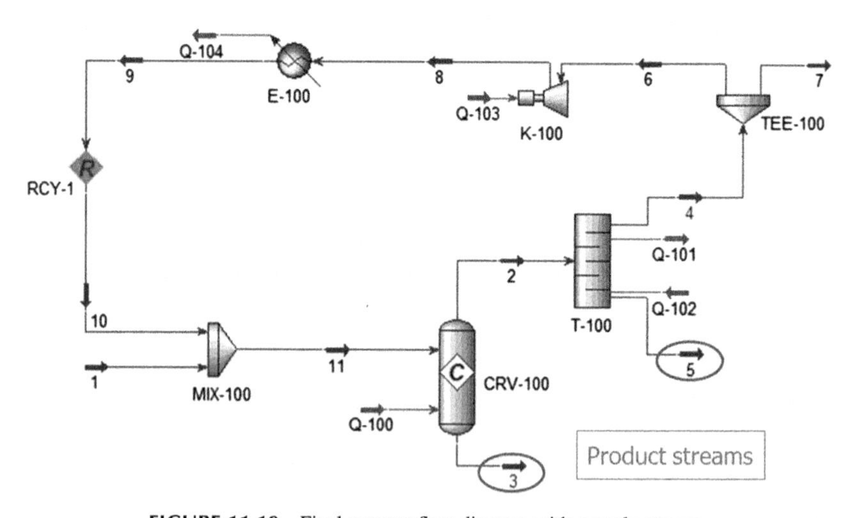

FIGURE 11.19 Final process flow diagram with recycle stream.

In earlier steps, it has been assumed that the fresh feed stream is available at 93°C (see Table 11.2). This assumption is now relaxed. A heater is added to raise the temperature of the fresh feed stream from 30°C. Detailed steps to do so are given in Fig. 11.21. The simulated results indicate that the heater requires a total heating duty of 131 MJ/h (indicated by energy stream Q-105); while 5.4 MJ/h of energy needs to be removed by the cooler (indicated by energy stream Q-104).

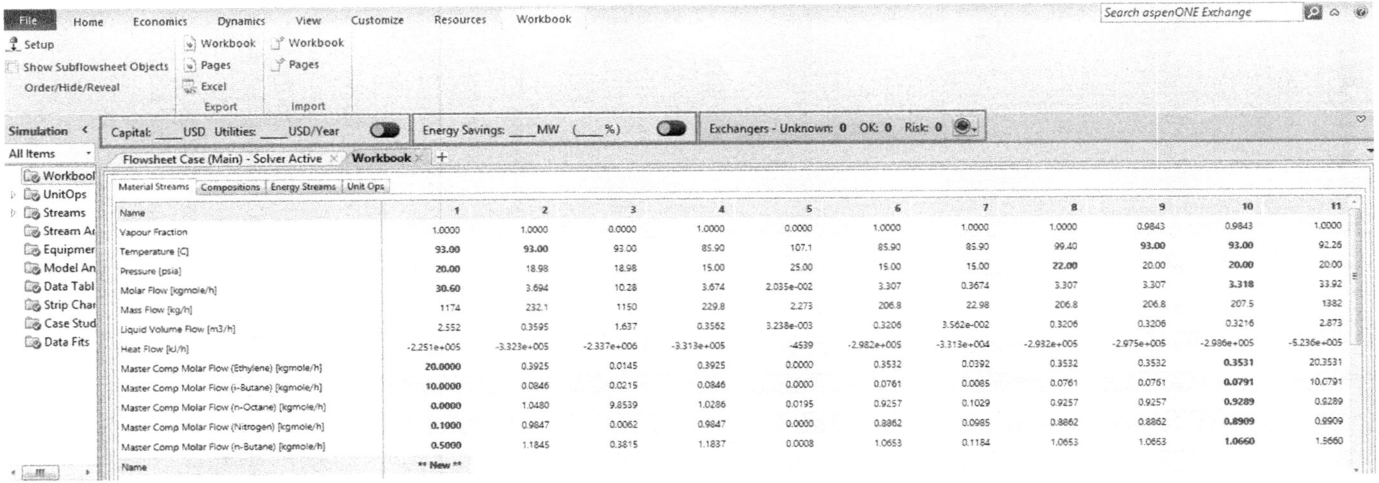

Name	1	2	3	4	5	6	7	8	9	10	11
Vapour Fraction	1.0000	1.0000	0.0000	1.0000	0.0000	1.0000	1.0000	1.0000	0.9843	0.9843	1.0000
Temperature [C]	**93.00**	**93.00**	93.00	85.90	107.1	85.90	85.90	99.40	**93.00**	**93.00**	92.26
Pressure [psia]	**20.00**	18.98	18.98	15.00	25.00	15.00	15.00	**22.00**	20.00	**20.00**	20.00
Molar Flow [kgmole/h]	**30.60**	3.694	10.28	3.674	2.035e-002	3.307	0.3674	3.307	3.307	**3.318**	33.92
Mass Flow [kg/h]	1174	232.1	1150	229.8	2.273	206.8	22.98	206.8	206.8	207.5	1382
Liquid Volume Flow [m3/h]	2.552	0.3595	1.637	0.3562	3.238e-003	0.3206	3.562e-002	0.3206	0.3206	0.3216	2.873
Heat Flow [kJ/h]	-2.251e+005	-3.323e+005	-2.337e+006	-3.313e+005	-4539	-2.982e+005	-3.313e+004	-2.932e+005	-2.975e+005	-2.986e+005	-5.236e+005
Master Comp Molar Flow (Ethylene) [kgmole/h]	**20.0000**	0.3925	0.0145	0.3925	0.0000	0.3532	0.0392	0.3532	0.3532	**0.3531**	20.3531
Master Comp Molar Flow (i-Butane) [kgmole/h]	**10.0000**	0.0846	0.0215	0.0846	0.0000	0.0761	0.0085	0.0761	0.0761	**0.0791**	10.0791
Master Comp Molar Flow (n-Octane) [kgmole/h]	**0.0000**	1.0480	9.8539	1.0286	0.0195	0.9257	0.1029	0.9257	0.9257	**0.9289**	0.9289
Master Comp Molar Flow (Nitrogen) [kgmole/h]	**0.1000**	0.9847	0.0062	0.9847	0.0000	0.8862	0.0985	0.8862	0.8862	**0.8909**	0.9909
Master Comp Molar Flow (n-Butane) [kgmole/h]	**0.5000**	1.1845	0.3815	1.1837	0.0008	1.0653	0.1184	1.0653	1.0653	**1.0660**	1.5660
Name	** New **										

FIGURE 11.20 Simulation results after material recycle.

TABLE 11.3 Specifications for Units in the Energy Recycle System

Equipment	Specifications
Heat exchanger	Delta P: 2 psi (tube side)
	Delta P: 2 psi (shell side)
Heater	Outlet T: 93°C
	Delta P: 2 psi

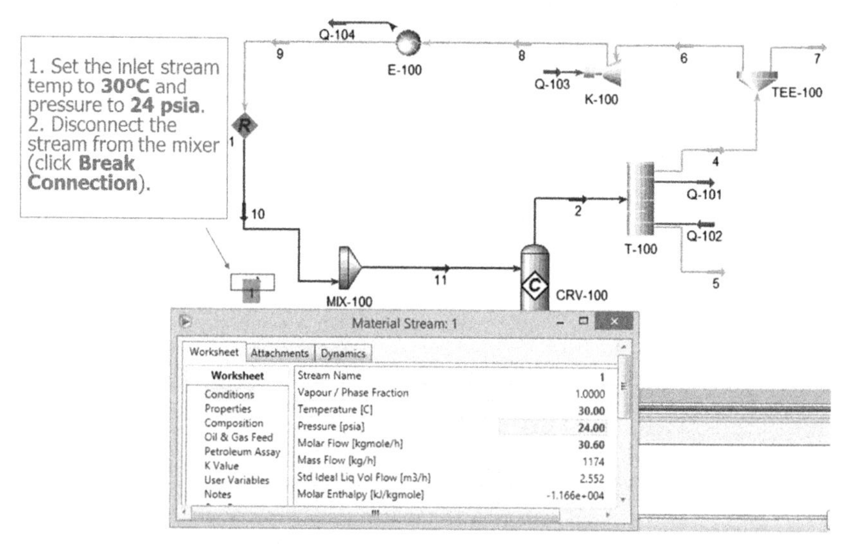

FIGURE 11.21 Set the inlet condition of feed.

Fig. 11.22 shows a *temperature–enthalpy plot*[4] for the streams undergoing heating and cooling in the heaters and cooler. As shown, the temperature profiles of the cooler (Q-104—the material recycle stream) are higher than those of the heater (Q-105—fresh feed). Hence, energy released from the heater can be completely recovered to the cold stream. In other words, part of the heating requirement of the heater (5.4 MJ/h) is to be fulfilled by the cooling duty of the cooler, through a process-to-process heat exchanger. The remaining heating duty of the cold stream (Q_H = 125.6 MJ/h) is to be supplied by the heater, as shown in Fig. 11.22.

4. This is the most basic form of *heat transfer composite curves* in *process integration*; see Linnhoff et al. (1982) or Smith (2016) for more details.

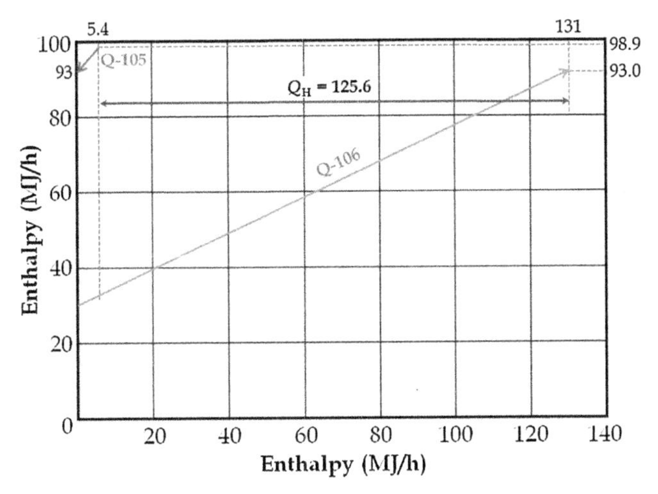

FIGURE 11.22 Temperature–enthalpy plot for heat recovery system.

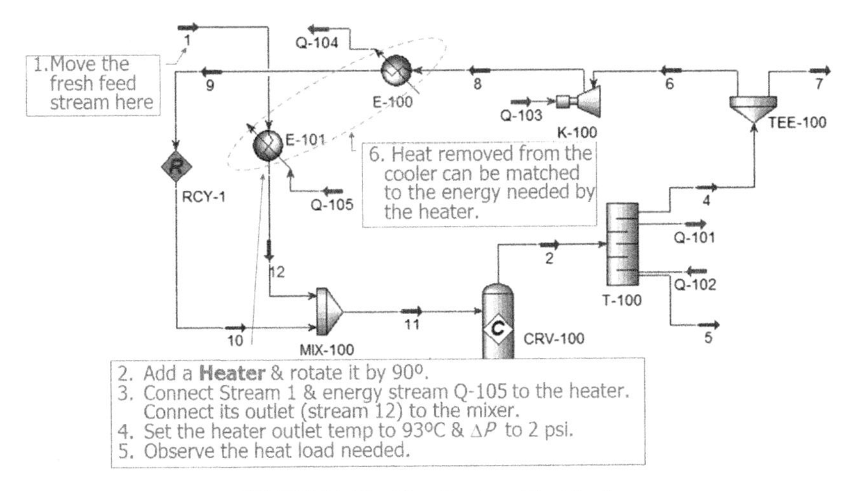

FIGURE 11.23 Identification of heat load.

With the heating and cooling requirements identified, we can now move on to simulate the process-to-process heat exchanger in the original PFD (Fig. 11.23). The "heat exchanger" model is utilized and added to replace the cooler model. Because the recycle model is not utilized in this case, we shall create a *tear stream*[5] for the energy recycle system, for the stream connecting the heat exchanger and heater. Detailed steps for doing so are shown in Fig. 11.24. Note that the flowsheet is unconverged at this stage.

5. See Chapters 1 and 4 for detailed discussion on the use of tear stream for recycle simulation.

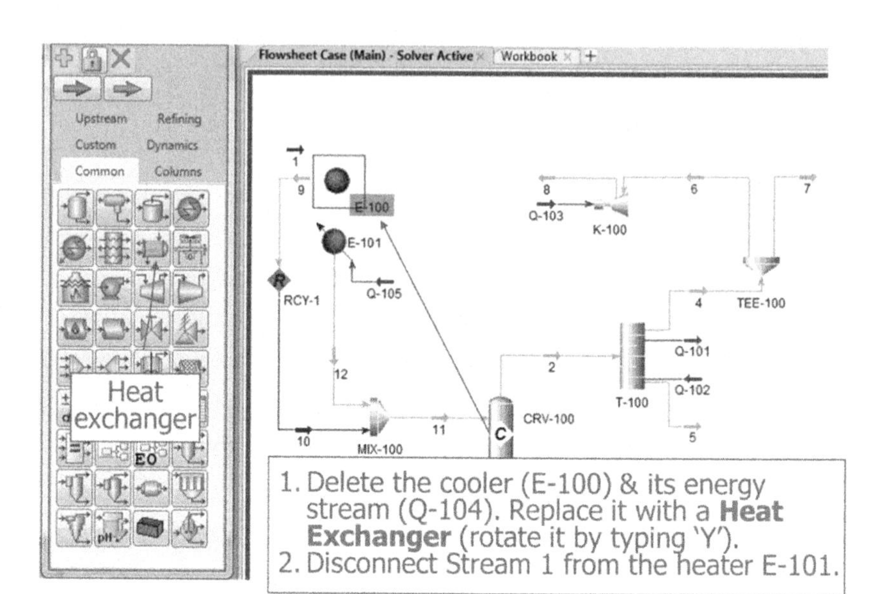

FIGURE 11.24 Delete cooler to bring heat exchanger.

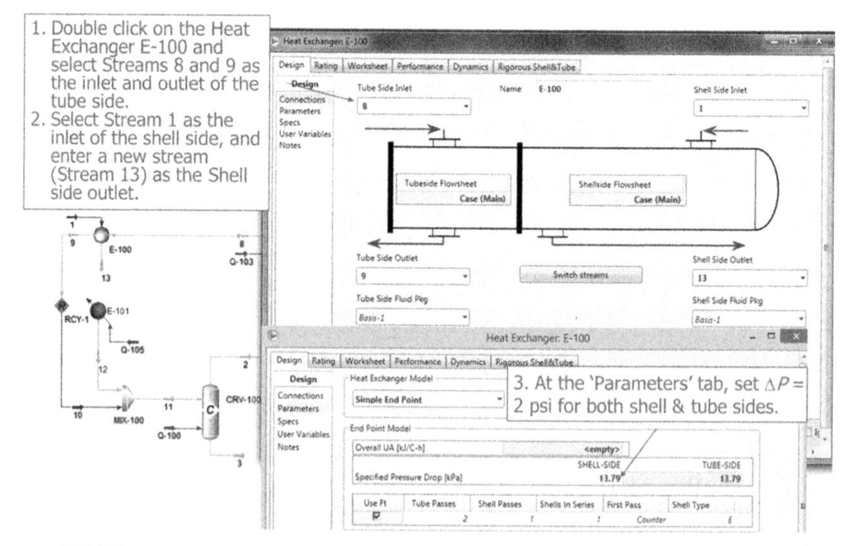

FIGURE 11.25 Bring heat exchanger to make use of the process stream temperature.

We next proceed to provide the missing parameters to converge the flowsheet. These include the estimation of values for the tear stream. Because this stream is essentially the same fresh feed stream that enters the heat exchanger at the shell side, its condition should be very similar (except that

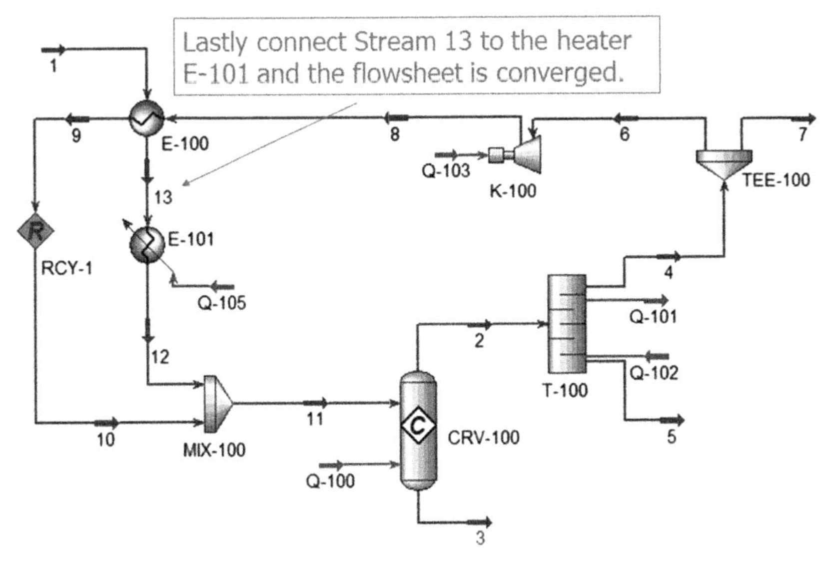

FIGURE 11.26 Final process flow diagram.

Name	1	2	3	4	5	6	7
Vapour Fraction	1.0000	1.0000	0.0000	1.0000	0.0000	1.0000	1.0000
Temperature [C]	30.00	93.00	93.00	85.90	107.1	85.90	85.90
Pressure [psia]	24.00	18.98	18.98	15.00	25.00	15.00	15.00
Molar Flow [kgmole/h]	30.60	3.694	10.28	3.674	2.035e-002	3.307	0.3674
Mass Flow [kg/h]	1174	232.1	1150	229.8	2.273	206.8	22.98
Liquid Volume Flow [m3/h]	2.552	0.3595	1.637	0.3562	3.238e-003	0.3206	3.562e-002
Heat Flow [kJ/h]	-3.568e+005	-3.323e+005	-2.337e+006	-3.313e+005	-4539	-2.982e+005	-3.313e+004
Master Comp Molar Flow (Ethylene) [kgmole/h]	20.0000	0.3925	0.0145	0.3925	0.0000	0.3532	0.0392
Master Comp Molar Flow (i-Butane) [kgmole/h]	10.0000	0.0846	0.0215	0.0846	0.0000	0.0761	0.0085
Master Comp Molar Flow (n-Octane) [kgmole/h]	0.0000	1.0480	9.8539	1.0286	0.0195	0.9257	0.1029
Master Comp Molar Flow (Nitrogen) [kgmole/h]	0.1000	0.9847	0.0062	0.9847	0.0000	0.8862	0.0985
Master Comp Molar Flow (n-Butane) [kgmole/h]	0.5000	1.1845	0.3815	1.1837	0.0008	1.0653	0.1184

Name	8	9	10	11	13	12	** New **
Vapour Fraction	1.0000	0.9843	0.9843	1.0000	1.0000	1.0000	
Temperature [C]	99.40	93.00	93.00	92.26	32.03	93.00	
Pressure [psia]	22.00	20.00	20.00	20.00	22.00	20.00	
Molar Flow [kgmole/h]	3.307	3.307	3.318	33.92	30.60	30.60	
Mass Flow [kg/h]	206.8	206.8	207.5	1382	1174	1174	
Liquid Volume Flow [m3/h]	0.3206	0.3206	0.3216	2.873	2.552	2.552	
Heat Flow [kJ/h]	-2.932e+005	-2.975e+005	-2.986e+005	-5.236e+005	-3.525e+005	-2.251e+005	
Master Comp Molar Flow (Ethylene) [kgmole/h]	0.3532	0.3532	0.3531	20.3531	20.0000	20.0000	
Master Comp Molar Flow (i-Butane) [kgmole/h]	0.0761	0.0761	0.0791	10.0791	10.0000	10.0000	
Master Comp Molar Flow (n-Octane) [kgmole/h]	0.9257	0.9257	0.9289	0.9289	0.0000	0.0000	
Master Comp Molar Flow (Nitrogen) [kgmole/h]	0.8862	0.8862	0.8909	0.9909	0.1000	0.1000	
Master Comp Molar Flow (n-Butane) [kgmole/h]	1.0653	1.0653	1.0660	1.5660	0.5000	0.5000	

FIGURE 11.27 Workbook table of final process flow diagram.

with different temperature). Once the tear stream is specified, the "open loop" flowsheet is converged (Fig. 11.25). In the final step, the tear stream is removed and the outlet stream from the heat exchanger is connected to the heater. A converged "close loop" flowsheet is resulted and is shown in Fig. 11.26. The material stream conditions are shown using the *Workbook Table* in Fig. 11.27.

REFERENCES

Foo, D.C.Y., Manan, Z.A., Selvan, M., McGuire, M.L., October, 2005. Integrate process simulation and process synthesis. Chemical Engineering Progress 101 (10), 25–29.

Linnhoff, B., Townsend, D.W., Boland, D., Hewitt, G.F., Thomas, B.E.A., Guy, A.R., Marshall, R.H., 1982. A User Guide on Process Integration for the Efficient Use of Energy. IChemE, Rugby, UK.

Smith, R., 2016. Chemical Process: Design and Integration, second ed. John Wiley and Sons, New York.

FURTHER READING

Foo, D.C.Y., Manan, Z.A., Selvan, M., McGuire, M.L., October, 2005. Integrate process simulation and process synthesis. Chemical Engineering Progress 101 (10), 25–29.

Chapter 12

Process Simulation for VCM Production

Siewhui Chong
University of Nottingham Malaysia Campus, Semenyih, Malaysia

In this chapter, the application of Aspen HYSYS is demonstrated for an industrial case study on the production of vinyl chloride monomer (VCM). Some advanced simulation skills, e.g., kinetic reaction, heterogeneous catalytic reaction, rigorous distillation, and logical operation tools are also demonstrated. Solving guide is provided to aid users in converging the simulation.

12.1 INTRODUCTION

Vinyl chloride (C_2H_3Cl) is one of the world's most important commodity chemicals. It is a colorless gas with sweet odor, flammable at room temperature, and carcinogenic (EPA, 2000). It is used as the precursor for the world's third most widely produced commodity plastics, i.e., polyvinyl chloride by polymerizing VCMs. For this reason, vinyl chloride is commonly abbreviated as VCM. As outlined in Table 12.1, VCM can be produced via several reaction pathways with the balanced process and thermal cracking of ethylene dichloride (EDC) from chlorination of ethylene being most commonly used (Pathways A and B). In order for the design engineers to perform evaluation on various pathways, process simulations are usually carried out at conceptual design stage. The following sections will demonstrate how Aspen HYSYS can be utilized to perform this task. The flowsheet for reaction pathway A will be simulated with Aspen HYSYS. The case study considers an annual production of ~360,000 ton of VCM, assuming an annual operating time of 330 days. This is equivalent to a production rate of 45,500 kg/h, or 728 kmol/h.

12.2 PROCESS SIMULATION

12.2.1 The Balanced Process

VCM is exclusively produced via the integrated balanced ethylene (C_2H_4) route. In the balanced process, 1,2-dichloroethane (commonly known as ethylene dichloride) is synthesized by direct chlorination of ethylene, which is

Chemical Engineering Process Simulation. http://dx.doi.org/10.1016/B978-0-12-803782-9.00012-1

TABLE 12.1 The Reaction Paths for Vinyl Chloride Monomer (VCM) Manufacture

Path #	Name	Features (Seider et al., 2009)
A	Balanced process for chlorination of ethylene ($2C_2H_4 + Cl_2 + ½\ O_2 \rightarrow 2C_2H_3Cl + H_2O$)	Converts the expensive chlorine atoms with HCl by-product consumed.
B	Thermal cracking of EDC from chlorination of ethylene ($C_2H_4 + Cl_2 \rightarrow C_2H_3Cl + HCl$)	Produces HCl by-product; high yield of EDC (98%) in the chlorination process of ethylene, which is then consumed in the dehydrochlorination process that follows thus EDC intermediate not in significant quantities.
C	Thermal cracking of EDC from oxychlorination of ethylene ($C_2H_4 + HCl + ½\ O_2 \rightarrow C_2H_3Cl + H_2O$)	HCl as the source of Cl thus works well when HCl cost is low; high yield of EDC (95%) in the oxychlorination process of ethylene, which is then consumed in the dehydrochlorination process that follows thus EDC intermediate not in significant quantities.
D	Hydrochlorination of acetylene ($C_2H_2 + HCl \rightarrow C_2H_3Cl$)	Requires HCl; high yield of VCM at moderate reaction conditions.
E	Direct chlorination of ethylene ($C_2H_4 + Cl_2 \rightarrow C_2H_3Cl + HCl$)	Requires Cl_2 and low yield of VCM; produces large amounts of EDC and HCl by-products.

EDC, ethylene dichloride; *HCL*, hydrogen chloride.

then dehydrochlorinated resulting in the formation of VCM. The hydrogen chloride (HCl) produced is then utilized in the oxychlorination to produce more EDC. Thus overall, this process offers the advantages of consuming HCl, the by-product, and converts both chlorine atoms, the expensive reactants, to VCM, ensuring a tight closure of material balance to consisting only VCM as the final product.

Direct chlorination (60°C, 1.5 atm):

$$C_2H_4 + Cl_2 \rightarrow C_2H_4Cl_2 \tag{12.1}$$

Oxychlorination (225°C, 4 atm):

$$C_2H_4 + 2HCl + 1/2O_2 \rightarrow C_2H_4Cl_2 + H_2O \tag{12.2}$$

EDC Pyrolysis (550°C, 26 atm):

$$2C_2H_4Cl_2 \leftrightarrow 2C_2H_3Cl + 2HCl \tag{12.3}$$

Overall reaction:

$$\mathbf{2C_2H_4 + Cl_2 + 1/2O_2 \rightarrow 2C_2H_3Cl + H_2O} \tag{12.4}$$

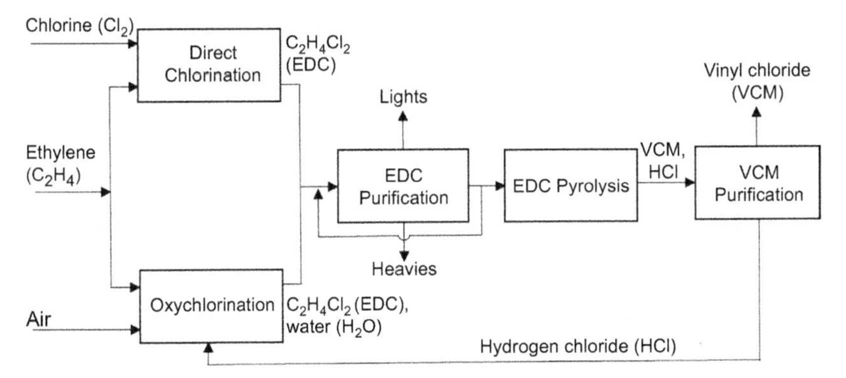

FIGURE 12.1 Block diagram for Pathway A, the balanced vinyl chloride monomer process.

The balanced process consists of four main steps (Dimian and Bildea, 2008; Hess et al., 1995): (1) direct chlorination and oxychlorination, (2) EDC purification, (3) EDC pyrolysis, and (4) VCM purification. The process block diagram with overall material balance is shown in Fig. 12.1. C_2H_4 is fed to both direct chlorination reactor and oxychlorination reactor. The former is supplied with chlorine (Cl_2) and the latter with air (or oxygen). The outputs from the two reactors are combined and fed to an EDC purification process to obtain EDC, which then undergoes pyrolysis (thermal cracking) to produce VCM and HCl. The mixture is then fed to a VCM purification system to separate VCM with others. EDC is returned to the EDC purification process while HCl is recycled back to the oxychlorination reactor with no net production/consumption.

Using the sequence of input steps outlined in Chapter 1, all chemical components are first entered into the component list, including C_2H_4, Cl_2, 1,2-dichloroethane, C_2H_3Cl, HCl, H_2O, O_2, and N_2 (assuming air consists of 21% O_2 and 79% N_2). Next, a suitable thermodynamic model (termed as "fluid package" in Aspen HYSYS) is selected. In this case, because H_2O is involved, Peng–Robinson–Stryjek–Vera fluid package is chosen to account for the moderately nonlinear systems before a flowsheet is built. The subsequent sections entail the simulation procedure for the flowsheet defined in Fig. 12.1.

Part I: Direct Chlorination and Oxychlorination Reactors System

Direct chlorination of ethylene to EDC (Eq. 12.1) can be conducted at low temperatures of 50–70°C and 0–2 atm, or high temperatures of 90–150°C and 1.5–5 atm, on 0.1–0.5 wt% ferric chloride ($FeCl_3$) catalyst. The process is very exothermic with $\Delta H^0_{298} = -218$ kJ/mol (Dimian and Bildea, 2008). The low-temperature chlorination process has a higher adoption rate of 70% in the industries because of its high EDC selectivity up to 99%, but it requires catalyst removal and external temperature control; the high-temperature process offers efficient heat recovery at the expense of lower selectivity. In this case study, we adopt the low-temperature chlorination process using 60°C and

1.5 atm as the simulation condition and the kinetics from Orejas (2001), with the activation energy in kJ/kmol and concentrations in kmol/m^3.

$$\begin{aligned} R_1 &= k_1 C_{C_2H_4} C_{Cl_2}\ \text{kmol/m}^3/\text{s}; \\ k_1 &= 11{,}493\ \exp(-17{,}929.81/RT)\ \text{m}^3/\text{kmol/s} \end{aligned} \tag{12.5}$$

Therefore, as shown in Fig. 12.2, in the "Reactions" properties of the Aspen HYSYS, Eq. (12.1) is entered as a *kinetic* equation, with kinetics parameters in Eq. (12.5).

Oxychlorination of ethylene to EDC (Eq. 12.2) is usually conducted in a fixed bed (230–300°C, 1.5–14 atm gauge pressure) or a fluidized bed (220–235°C, 1.5–5 atm gauge pressure) plug flow reactor (PFR) on copper II chloride ($CuCl_2$)/alumina catalyst (Lakshmanan and Biegler, 1997). This process is also highly exothermic with $\Delta H^0_{298} = -295$ kJ/mol (Dimian and Bildea, 2008) and therefore requires good temperature control. The kinetic model from Wachi and Asai (1994) is adopted, who reported that the oxychlorination reaction exhibited first-order kinetics on $CuCl_2$ catalyst, and the dependency on ethylene concentration can be interpreted by a Langmuir–Hinshelwood mechanism (Eq. 12.6), with the activation energy in kJ/kmol and concentrations in kmol/m^3.

$$R_2 = \frac{269\ K_a C_{C_2H_4} C_{Cl_2} \exp(-37{,}800/RT)}{1 + K_a C_{C_2H_4}}\ \text{kmol/m}^3/\text{s} \tag{12.6}$$

From Wachi and Asai (1994), K_a was found to be 630 m^3/kmol, the particle density of the γ-alumina powder was 1369 kg/m^3, and the solid density was

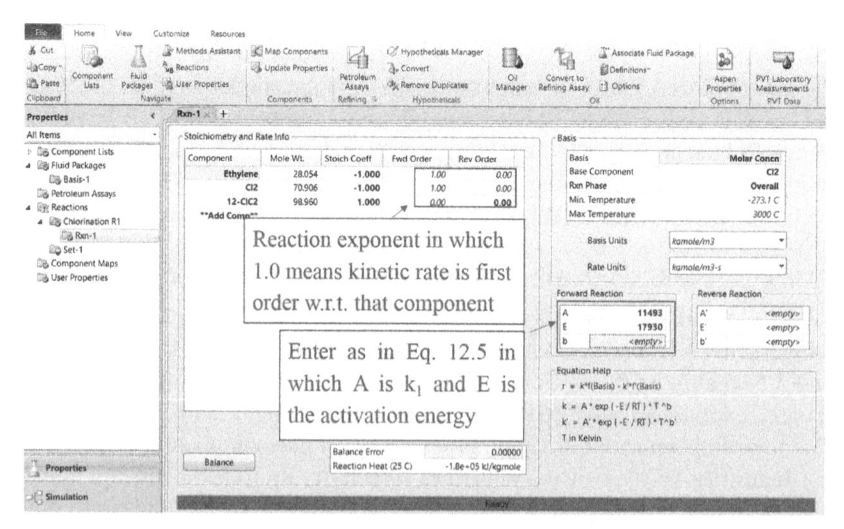

FIGURE 12.2 Entering direct chlorination reaction (Eqs. 12.1 and 12.5) in Aspen HYSYS.

3075 kg/m^3. The particles had an internal surface area of 221,000 m^2/kg and the amount of copper loaded into the alumina carrier was 4.23 wt%, equivalent to C_{Cl_2} of 0.993 kmol/m^3 (analyzed by an atomic spectrometer). Therefore, Eq. (12.6) becomes:

$$R_2 = \frac{168,284\ C_{C_2H_4}\exp(-37,800/RT)}{1\ +\ 630\ C_{C_2H_4}}\ \mathrm{kmol/m^3/s} \qquad (12.7)$$

Using an operating condition of 225°C and 4 atm, Eq. (12.2) is modeled as a *heterogeneous catalytic* reaction with kinetics parameters entered correspondingly as shown in Table 12.2.

TABLE 12.2 Entering Oxychlorination Reaction (Eqs. 12.2 and 12.7) in Aspen HYSYS

<table>
<tr><th>Page</th><th>Tab</th><th>Essential Setting (Default for Unspecified Parameters)</th></tr>
<tr><td>Stoichiometry</td><td>-</td><td>Stoichiometry:
<table>
<tr><td>Component</td><td>Stoich Coeff</td></tr>
<tr><td>Ethylene</td><td>-1</td></tr>
<tr><td>HCl</td><td>-2</td></tr>
<tr><td>Oxygen</td><td>-0.5</td></tr>
<tr><td>12-ClC2</td><td>1</td></tr>
<tr><td>H2O</td><td>1</td></tr>
</table>
Basis:
<table>
<tr><td>Basis</td><td>Molar Conc</td></tr>
<tr><td>Base Component</td><td>Ethylene</td></tr>
<tr><td>Rxn Phase</td><td>Overall</td></tr>
</table>
Basis Units: kgmole/m^3
Rate Units: kgmol/m^3-s</td></tr>
<tr><td>Reaction Rate</td><td>-</td><td>Numerator:
Forward Reaction
<table>
<tr><td>A</td><td>1.6828e+05</td></tr>
<tr><td>B</td><td>36800</td></tr>
<tr><td>ß</td><td><empty></td></tr>
</table>
<table>
<tr><td>Components</td><td>Forward Order</td><td>Reverse Order</td></tr>
<tr><td>Ethylene</td><td>1</td><td>0</td></tr>
<tr><td>HCl</td><td>0</td><td>0</td></tr>
<tr><td>Oxygen</td><td>0</td><td>0</td></tr>
<tr><td>12-ClC2</td><td>0</td><td>0</td></tr>
<tr><td>H2O</td><td>0</td><td>0</td></tr>
</table>
Denominator:
<table>
<tr><td>A</td><td>E [kJ/kg mole]</td><td>Ethylene</td><td>HCl</td><td>Oxygen</td><td>12-ClC2</td><td>H2O</td></tr>
<tr><td>630.00</td><td>0</td><td>1</td><td>0</td><td>0</td><td>0</td><td>0</td></tr>
</table>
Note: 1.0 entered for the component Ethylene only, as K_a corresponds to the adsorption constant on ethylene.</td></tr>
</table>

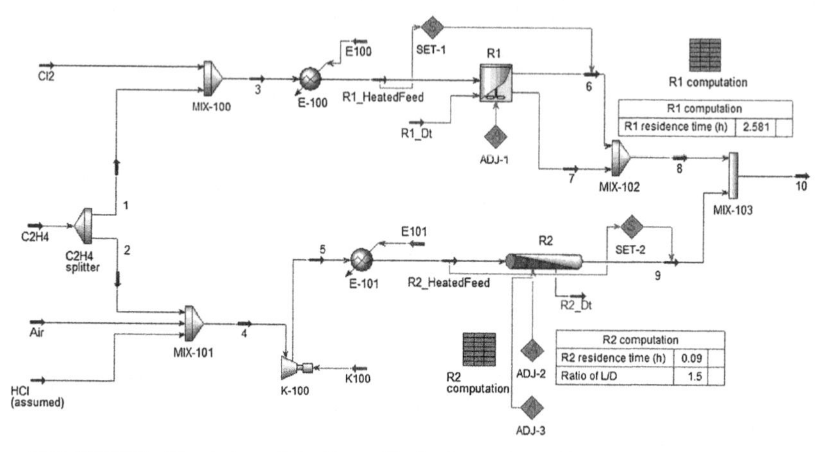

FIGURE 12.3 Simulation of the direct chlorination and oxychlorination reactors system.

After adding all reactions into fluid package, we can enter the simulation environment to build the process flowsheet as outlined in Fig. 12.3. All streams and unit operations are defined according to the data given in Table 12.3, with the following assumptions:

- C_2H_4 stream is split equally to R1 and R2.
- There is no pressure drop throughout unless specified.
- The reactors are controlled at the specified temperatures by external heat removal/addition, and thus modeled as isothermal.

TABLE 12.3 Feed Streams and Reactor Feed Data (Shaded Boxes Show the Parameters to be Entered by User)

Stream Names	C_2H_4	Cl_2	Air	HCl (Assumed)	R1_HeatedFeed	R2_HeatedFeed
Temperature (°C)	25	25	25	25	60	225
Pressure (atm)	1.5	1.5	1.5	1.5	1.5	4
Component molar flow (kmol/h):						
C_2H_4	757[a]	0	0	0	378.56	378.56
Cl_2	0	364	0	0	364	0
$C_2H_4Cl_2$	0	0	0	0	0	0
C_2H_3Cl	0	0	0	0	0	0
HCl	0	0	0	800	0	800
H_2O	0	0	0	0	0	0
O_2	0	0	210	0	0	210
N_2	0	0	790	0	0	790
Total molar flow (kmol/h)	757	364	1000	800	742.56	2178.56
Actual volume flow (m^3/h)	12,230	5,827	16,300	12,930	13,390	22,230

TABLE 12.4 Computation for R1 Reactor Volume

<table>
<tr><th>Page</th><th>Tab</th><th>Essential Setting (Default for Unspecified Parameters)</th></tr>
<tr><td colspan="3">1. In Chlorination R1</td></tr>
<tr><td>Design</td><td>Parameters</td><td>Volume: 30000 (Assume a value)</td></tr>
<tr><td colspan="3">2. Add a Spreadsheet: Name it as "R1 Computation"</td></tr>
<tr><td>Spreadsheet</td><td>-</td><td>Type in/import values to respective cells:
<table>
<tr><th></th><th>A</th><th>B</th></tr>
<tr><td>1</td><td>Type "R1_HeatedFeed"</td><td></td></tr>
<tr><td>2</td><td>Type
"Actual volume flow"</td><td>Right click and import the Actual Volume Flow of R1HeatedFeed (the value 1.339e+0.004 m3/h will automatically appear)</td></tr>
<tr><td>3</td><td>Type
"R1 volume"</td><td>Right click and import the R1 Tank Volume (the value 3.000e+004 m3 will automatically appear)</td></tr>
<tr><td>4</td><td>Type
"Residence time (h)"</td><td>Type "=B3/B2" and the value will be automatically calculated (in this case, 2.240)</td></tr>
</table></td></tr>
<tr><td colspan="3">3. Add an Adjust Tool: the Default Name is ADJ-1</td></tr>
<tr><td>Connections</td><td>Connections</td><td>Select the Adjusted Variable:
Object: R1
Variable: Tank Volume

Select the Target Variable:
Object: R1 computation@B4
Variable: B4:

Specify the Target Value:
Source: User Specified
Specified Target Value: 2.5814</td></tr>
</table>

1. First Assume a Reactor Volume (Keeping the Default Liquid Level Percentage of 50%).
2. Create a "Spreadsheet" and Compute the Residence Time; and
3. Use an "Adjust" to Meet the Desired Residence Time of 2.5814 h.

- C_2H_4 feed is supplied in slight excess (1.04 in excess of the stoichiometric amount) with Cl_2 and HCl as the limiting agents in R1 and R2 respectively.
- A *pseudo* HCl is created as one of the feeds for R2 and will be updated later by recycle stream.

R1 is modeled as a continuous-stirred-tank-reactor (CSTR) with residence time 2.5814 h (Lakshmanan et al., 1999). An "Adjust" logical operation tool[1] along with a "Spreadsheet" are used to achieve this residence time based on the volumetric flowrate. As shown in Table 12.4, an initial tank volume is first

1. Tip: To use the "Adjust" tool, an initial value must first be given to the adjusted variable.

assumed and the "Actual volume flow" of the stream "R1_HeatedFeed" is extracted to the spreadsheet. The residence time is then computed as:

$$\text{Residence time(h)} = \text{Reactor volume}(\text{m}^3)/\text{Volumetric flowrate}(\text{m}^3/\text{h}) \quad (12.8)$$

The "Adjust" tool then adjusts the "Adjusted Variable," which is the tank volume of R1, to meet the specified residence time of 2.5814 h. At this residence time, R1 volume is found to be $3.458 \times 10^4\ \text{m}^3$.

R2 is modeled as a PFR with residence time 0.09 h (Lakshmanan et al., 1999). Similarly, as shown in Table 12.5, an initial PFR volume and its diameter are assumed, and a "Spreadsheet" is created to compute the residence time as well as the PFR tube length-to-diameter ratio (L/D). "Adjust" tools are then used to adjust the adjusted variables to meet the desired residence time of 0.09 h and L/D ratio of 1.5. The PFR volume value is found to be 2001 m^3, with L = 17.90 m, and D = 11.93 m. To define isothermal reactor, a "Set" tool can be used so that the reactor outlet temperature is equal to the inlet, as displayed in Table 12.6. This applies to both R1 and R2. The liquid and vapor product from R1 are then mixed with the product stream from R2.

Part II: Ethylene Dichloride Purification

Part II consists of the EDC purification section as shown in Fig. 12.4. The products from the direct chlorination, oxychlorination, and other recycle streams (which will be added later) contain not only EDC but also a large amount of water and small amounts of C_2H_4, HCl, and air (in this case study we have not considered other by-products such as trichloroethane). An EDC purity of at least 99.5% must be met to prevent coking and fouling of the pyrolysis reactor. Water easily forms minimal azeotrope with not only EDC but also with most of the organic compounds present in the process (Cordeiro et al., 2013). Therefore, the industrial EDC purification process consists of a wash tower to recover the catalyst and a reboiler type thermosiphon. For better representation of the real system, Cordeiro et al. (2013) proposed using a reboiled absorption, followed by a condenser and a decanter. Similar configuration is adopted in this case study, in which a flash is first placed to separate the light gas components, followed by a liquid–liquid extractor to combine aqueous washing with some characteristics of azeotropic separation, and finally a reboiled absorber to achieve the purification target.

Flash column is commonly used in the industry to reduce the load on separation followed on. The temperature of the flash column is selected in such a way that the amount of water is minimal in the vapor product, at the same time economic in terms of the heating/cooling utilities and equipment costs. Here, a temperature of 15°C is selected yielding an EDC purity of 65.65% in the liquid product while removing most of the C_2H_4 and air components for further gas treatment. Following that, a five-stage liquid–liquid extractor is set

TABLE 12.5 Computation for R2 Reactor Volume and Dimensions

<table>
<tr><th>Page</th><th>Tab</th><th>Essential Setting (Default for Unspecified Parameters)</th></tr>
<tr><td colspan="3">In Oxychlorination R2</td></tr>
<tr><td>Rating</td><td>Sizing</td><td>Total Volume: 2000 (Assume)
Diameter: 11 (Assume)</td></tr>
<tr><td colspan="3">Add a Spreadsheet: Name it as "R2 Computation"</td></tr>
<tr><td>Spreadsheet</td><td>-</td><td>Type in:
<table>
<tr><th></th><th>A</th><th>B</th></tr>
<tr><td>1</td><td>Type "R2_HeatedFeed"</td><td></td></tr>
<tr><td>2</td><td>Type "Actual volume flow"</td><td>Right click and import the Actual Volume Flow of R2HeatedFeed</td></tr>
<tr><td>3</td><td>Type "R2 volume"</td><td>Right click and import the R2 Tank Volume</td></tr>
<tr><td>4</td><td>Type "Residence time (h)"</td><td>Type "=B3/B2" and the value will be automatically calculated</td></tr>
<tr><td>5</td><td></td><td></td></tr>
<tr><td>6</td><td>Type "Tube diameter"</td><td>Right click and import the R2 Tube Diameter</td></tr>
<tr><td>7</td><td>Type "Tube length"</td><td>Right click and import the R2 Tube Length</td></tr>
<tr><td>8</td><td>Type "Ratio of L/D"</td><td>Type "=B7/B6"</td></tr>
</table></td></tr>
<tr><td colspan="3">Add an Adjust Tool: the Default Name is ADJ-2</td></tr>
<tr><td>Connections</td><td>Connections</td><td>Select the Adjusted Variable
Object: R2
Variable: Reactor Volume

Select the Target Variable
Object: R2 computation@B4
Variable: B4

Specify the Target Value:
Source: User Specified
Specified Target Value: 0.09</td></tr>
<tr><td colspan="3">Add an Adjust Tool: the Default Name is ADJ-3</td></tr>
<tr><td>Connections</td><td>Connections</td><td>Select the Adjusted Variable
Object: R2
Variable: Tube Diameter

Select the Target Variable
Object: R2 computation@B8
Variable: B8:

Specify the Target Value:
Source: User Specified
Specified Target Value: 1.50</td></tr>
</table>

1. First Assume a Reactor Volume and Tube Diameter;
2. Create a "Spreadsheet" to Compute for the Residence Time and the L/D Ratio; and
3. Use an "Adjust" to Converge to the Desired Residence Time of 0.09 h, and Another "Adjust" to Converge to the Desired L/D Ratio of 1.5.

TABLE 12.6 Example of Using a "Set" Tool to Set up an Isothermal Reactor (With the y = mx + C Relation)

Page	Tab	Essential Setting (Default for Unspecified Parameters)
Add a Set tool: the default name is Set-1		
Connections	-	Target Variable: Object: 6 Variable: Temperature Source: Object: R1_HeatedFeed
Parameters	-	Parameters Multiplier: 1.0000 Offset [C]: 0 C

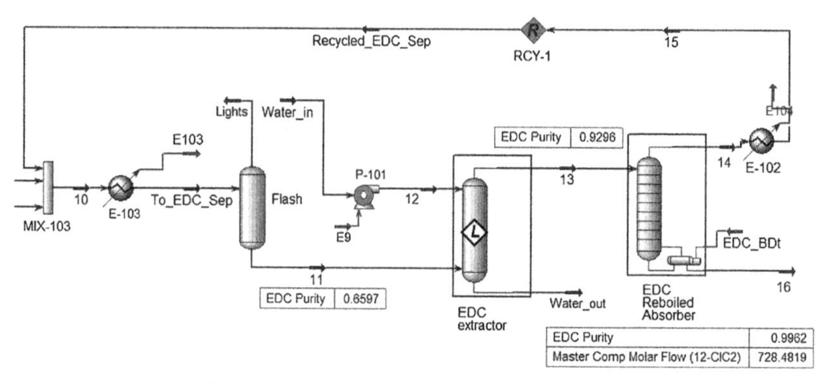

FIGURE 12.4 Ethylene dichloride purification flowsheet.

up, using water as an extractor and an operating temperature and pressure of 15°C and 4 atm. The second purification stage increases the EDC purity to 93.02%.

In the final-stage EDC purification, a 17-stage reboiled absorber is set up according to Fig. 12.5, using the same operating temperature and pressure of 15°C and 4 atm throughout. This last stage of purification results in an EDC purity of 99.62%, which meets the requirement of at least 99.5%. Purified EDC exits as bottom liquid product (Stream 16) with ~13.75% EDC loss to the vapor product (Stream 14), which is then cooled to 90°C (to avoid evaporation when mixing) and recycled back to the Flash column[2] for repeating the purification

2. Tip: Do not forget to use the "Recycle" tool for the recycle streams for Aspen HYSYS to use the tear-stream solving approach.

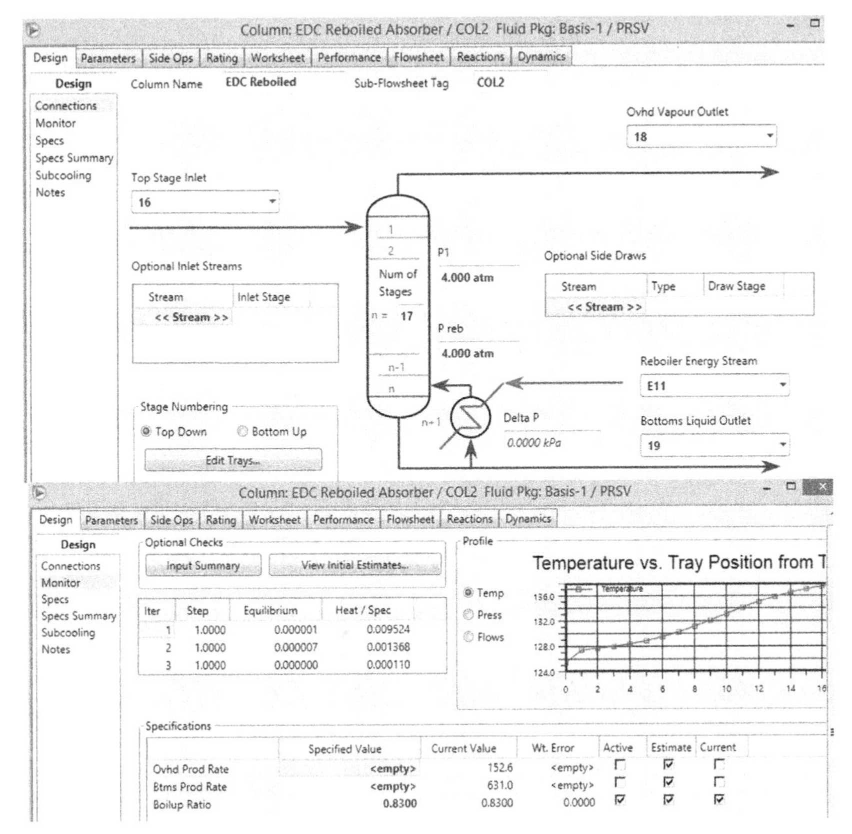

FIGURE 12.5 Setting up the reboiled absorber in the final-stage ethylene dichloride purification (assuming a boilup ratio of 0.83).

process, as demonstrated in Fig. 12.4. The recycled stream has apparently resulted in 16% increase in the amount of purified EDC produced, as signaled by the increased EDC flow from 628.6 to 728.5 kmol/h in Stream 16.

Part III: Ethylene Dichloride Pyrolysis

The purified EDC now enters the endothermic pyrolysis section, which usually occurs at a temperature range of 480–530°C and gauge pressures of 6–35 atm (Ranzi et al., 1993) in a long tubular coiled placed in a furnace, producing VCM with other by-products such as acetylene, ethylene, butadiene, trichloroethane, and vinyl acetylene (Lakshmanan et al., 1999; Dimian and Bildea, 2008). Here, we consider only the main reaction of EDC to form VCM (Eq. 12.3), with kinetics parameters from Lakshmanan et al. (1999):

$$R_3 = k_3 C_{EDC}\, \text{kmol}/\text{m}^3/\text{s} \tag{12.9}$$

with

$$k_3 = 10^{13.6} \exp(-242,672/RT) s^{-1} \quad (12.10)$$

To bring the result closer to the industrial data, Eq. (12.10) is modified to become (Dimian and Bildea, 2008):

$$k_3 = 1.14 \times 10^{14} \exp(-242,672/RT)\ s^{-1} \quad (12.11)$$

These kinetic data are entered as a new *kinetic* equation in the "Reactions" properties of the Properties mode. After adding this reaction into fluid package, a PFR is used to model the pyrolysis reactor in the simulation flowsheet using an operating condition of 550°C and 26 atm. The residence time is taken as 30 s (0.0083 h), and the tube length is fixed at 250 m (Dimian and Bildea, 2008). Similar to R1, a "Spreadsheet" is constructed for computational purpose, and an "Adjust" tool is used for achieving the desired residence time of 30 s. A "Set" tool is also added to equalize the outlet temperature with the inlet.

Part IV: Vinyl Chloride Monomer Purification

To minimize the formation of coke and also to prevent further reaction, a quench system or a transfer line exchanger is essential to rapidly quench or cool the pyrolysis product. Following the quench system is the VCM purification section to separate the final VCM product from a mixture consisting mainly of VCM, HCl, and EDC; thus, at least one piece of separation unit will be required. Apart from deciding the number of separation units, the operating temperatures and pressures also need to be identified. For economical operations, the separation devices are usually controlled to operate at a reboiler temperature below 186°C (366°F) and a condenser temperature in the range of 25–49°C (80–120°F), so that the low- or medium-pressure steam and air or cooling (river) water from the cooling tower can be utilized to avoid, if possible, extra expenditure on expensive heating and cooling media.

To aid justification, a graph can be plotted to see the dependence of boiling points on pressure for each species by utilizing the "Case Studies" tool in Aspen HYSYS; one example as shown in Fig. 12.6. First, right click the VCM purification feed stream (Stream 18), and select "Clone Selected Objects" key in the vapor fraction as 0 to indicate bubble point, the same pressure and component flowrates, and name it as 18-2. However, because this feed stream to the VCM purification section is a high-pressure gaseous mixture, we will have to add a new fluid package as Basis 2 in the "Properties" mode and apply it as the basis for the cloned streams. In this case, a Soave–Redlich–Kwong fluid package is suitable for the hydrocarbon mixtures at elevated pressure (Seider et al., 2009). Clone several streams from 18-2 if needed, and retain only one (or two) species each time. Next, from the left menu, click on "Case

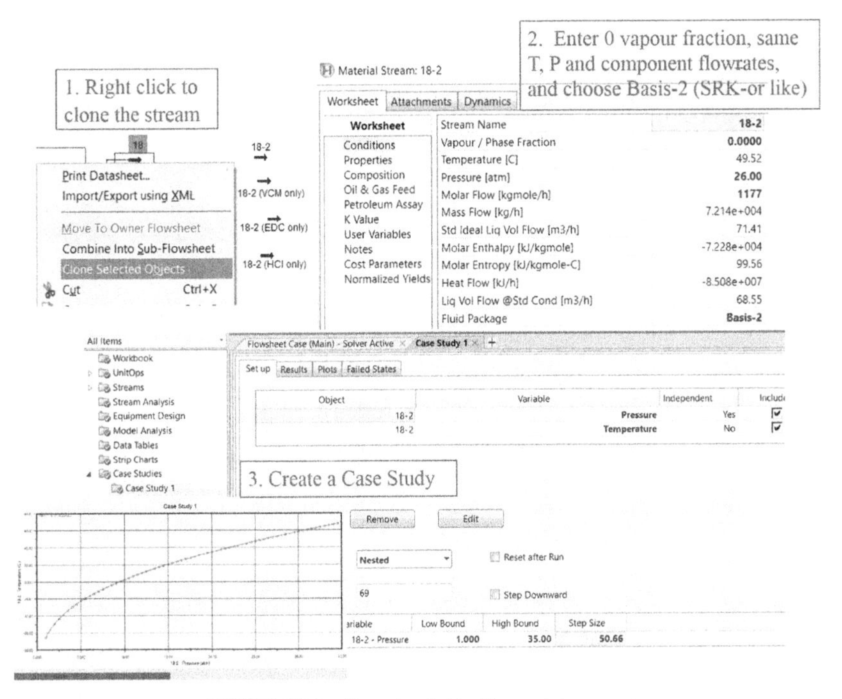

FIGURE 12.6 Steps to plot boiling point curves.

Studies," and for each case study, add in the respective independent and dependent variables, and specify the low bound, high bound, and step size. Lastly, click on "Run" and under the "Plots" tab; you can then see the plot. To determine the critical properties, make use of the "Create Stream Analysis"—"Critical Properties" tool as shown in Fig. 12.7. Alternative way to see the correlation of pressure-temperature is to use the "Envelope" tool, similar to determining the critical properties.

Fig. 12.8 compiles all the relevant boiling point curves in the pressure range of 1–35 atm. The critical temperatures and pressures of each species are indicated on the top left corner, which are to be avoided as operating in the critical region will result in difficult separation. Because of the rather large difference in the boiling points of different species, use of distillation columns could be the most economical route. At the first glance, HCl is a good option to be removed first due to its relatively lowest boiling points than EDC and VCM, and also due to the high operating pressure (26 atm) from the previous pyrolysis section. For these reasons, Goodrich (1963) selected an operating pressure of 12 atm for the first column and 4.8 atm for the second column. A general guide for setting up a rigorous distillation column in Aspen HYSYS is provided below.

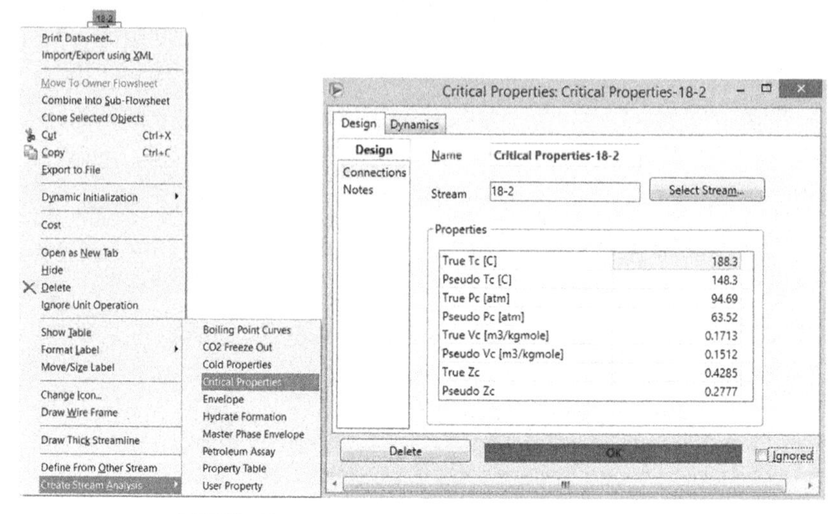

FIGURE 12.7 Steps to determine the critical properties.

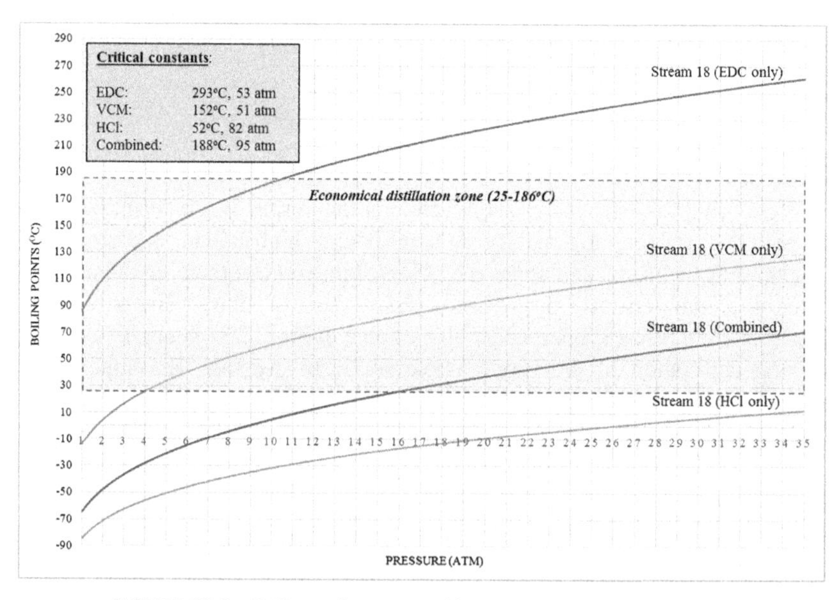

FIGURE 12.8 Boiling point curves with respect to changes in pressure.

Simulation of Rigorous Distillation Column

Step 1: Justification of Operating Pressure With Splitter

Saturated liquid feed is initially supplied at 10.32°C and 12 atm. A suggested way to evaluate the pressure effect is to make use of the splitter. As depicted in

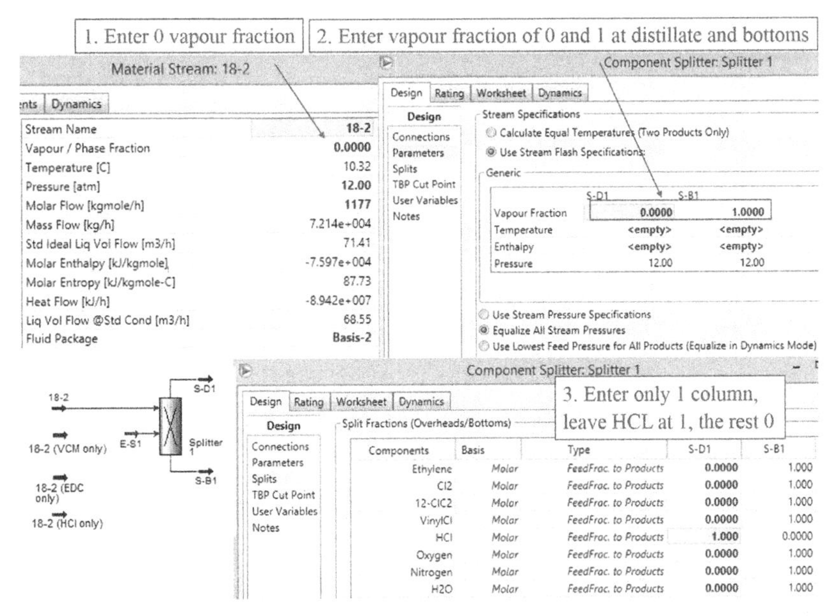

FIGURE 12.9 Setting up a splitter to evaluate pressure effect on top and bottoms temperatures.

Fig. 12.9, a splitter is connected to the cloned feed, with the distillate at its bubble point (vapor fraction equal to 0), and the bottoms at its dew point (vapor fraction equal to 1). All stream pressures are equalized with the assumption that the pressure drop across the column is negligible. Next, the split fractions are specified allowing HCl to leave as distillate (S-D1), and EDC and VCM mixture as bottoms (S−B1). The splitter is now converged, resulting in a distillate temperature of −26°C and a bottoms temperature of 147°C. A "case study" can also be created to aid the selection of a suitable operating pressure. As shown in Fig. 12.10, decreasing pressure will result in reduced reboiler heating load, but at the same time increased condenser refrigerant load.

On taking the considerations of the heating/cooling utility costs and the critical points for the first column, operating pressure of 16 atm is used. This in turn gives an approximate feed temperature of 24°C and condenser temperature of −17°C for HCl condensation, corresponding to a reboiler temperature of 160°C. Thus liquid refrigerant such as refrigerated brine and medium-pressure steam will be required.

Step 2: Estimation of Reflux Ratio, Number of Trays, and Feed Stage With Shortcut Distillation Model

After the operating pressure is selected, a shortcut distillation column is constructed to estimate the reflux ratio, the number of trays, and the optimal

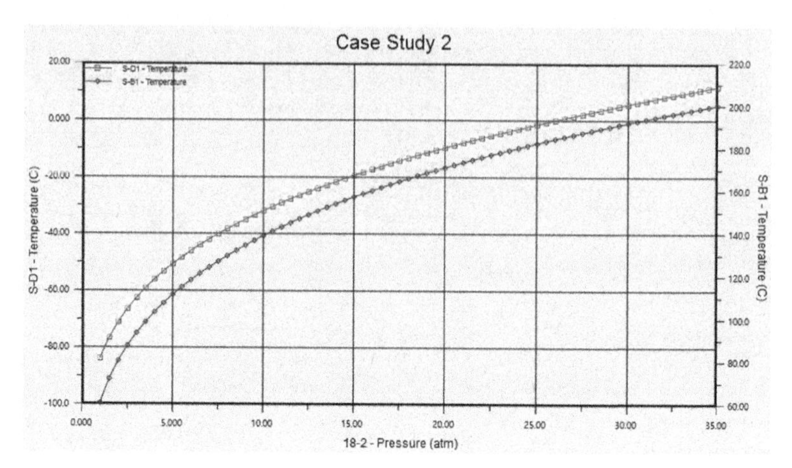

FIGURE 12.10 The effect of operating pressure on the distillate and bottoms temperatures generated by Aspen HYSYS with a splitter.

feed stage. HCl, the light key, will recover as liquid in D1 with mole fraction 0.0001, and the rest as bottoms in B1 bottoms, with VCM as the heavy key in the distillate (with mole fraction 0.0001). By taking a reflux ratio 1.75 times of the minimum reflux generated by Aspen HYSYS, the column is converged, providing us the required information to proceed.

Step 3: Modeling With Rigorous Distillation Column

Before distillation process, as depicted in Fig. 12.11, a cooler is used to model the quench tower to bring the pyrolysis outlet vapor stream (Stream 18) to its

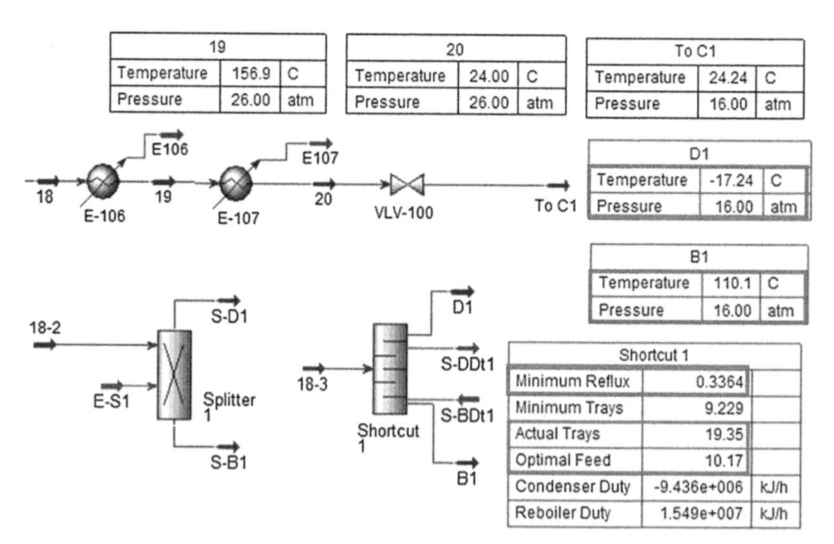

FIGURE 12.11 Preparation of feed to vinyl chloride monomer purification.

dew point, which is 157°C at 26 atm (Stream 19). Having decided an operating pressure of 16 atm for the first column, a condenser (modeled as a cooler) is used to produce a saturated liquid feed at 24°C followed by a control valve to reduce the pressure from 26 atm to 16 atm. With all the information obtained from Steps 1 and 2, the stream (To C1) is now ready to enter the distillation column for purification.

When setting up the rigorous distillation column, the circled information in Fig. 12.11 is entered by following the sequence as displayed in Table 12.7. After the "Done" button is clicked, the column will not converge unless we have specified two variables, one of which can be the reflux ratio, taken as 1.75 times the minimum reflux, the other can be the column component recovery parameter. The target component recovery of HCl from the feed is taken as 99.99%. Now with zero degree of freedom, we hit on the "Run" button and the column is converged.

TABLE 12.7 Setting up a Rigorous Distillation Column (Assuming Once-Through Reboiler Type, and Reflux Ratio 1.75 Times of the Minimum Reflux)

Page	Tab	Essential Setting (Default for Unspecified Parameters)
Distillation Column Input Expert		
1. Connect streams	-	Inlet Streams: Select "To C1", 10_Main Tower # Stages: n = 20 Condenser: Total Ovhd Liquid Outlet: Type "HCl" Condenser Energy Stream: Type "C1_DDt" Reboiler Energy Stream: Type "C1_BDt" Bottoms Liquid Outlet: C1 Bottoms Then click *Next*.
2. Specify pressures	-	Condenser Pressure: 16 atm Reboiler Pressure: 16 atm Then click *Next*.
3. Specify temperatures	-	Optional Condenser Temperature Estimate: -17.25 C Optional Reboiler Temperature Estimate: 110.1 C Then click *Next*.
4. Specify reflux ratio	-	Reflux Ratio: 0.5887 Flow Basis: Molar Then click *Done*.
In Column: Column 1 / COL3		
Design	Monitor	Click on "Add Spec" and select "Column Component Recovery", then click "Add Spec" after selection. In the pop-up window: Name: Comp Recovery (default) Draw: Select "HCl @COL3" Spec Value: Enter "0.9999" Components: Select "HCl" Back to the Monitor screen, select the checkboxes of Reflux Ratio and Comp Recovery as the "Active" parameters. (For distillation column, two variables must be specified and selected) Finally, click on the "Run" button to simulate.

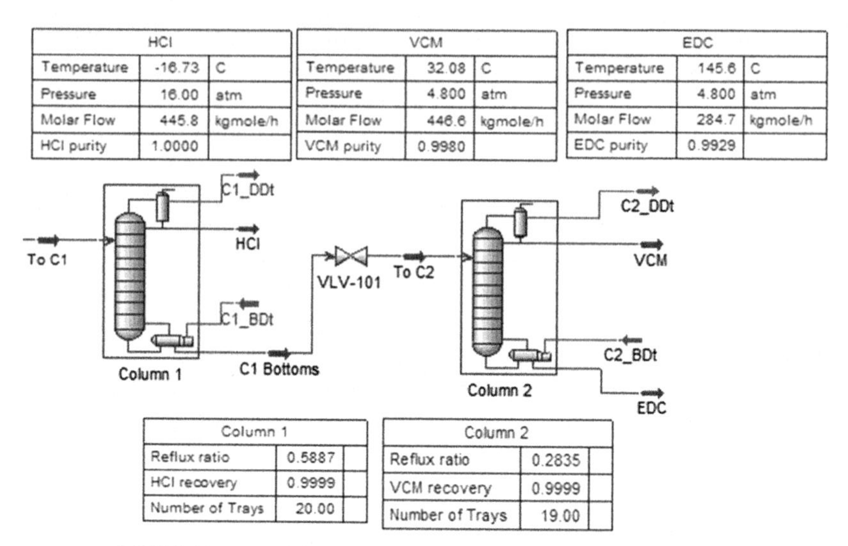

FIGURE 12.12 Results of vinyl chloride monomer purification process.

For the second column, we have chosen an operating pressure as recommended by Goodrich (1963), which is 4.8 atm, as it falls within the economical distillation zone, thus requiring only the use of cooling water at the condenser and the low-pressure steam at the reboiler. A control valve is placed at the feed to Column 2 to reduce the pressure from 16 atm to 4.8 atm. The same procedure follows to set up the second rigorous distillation column. As shown in Fig. 12.12, the purified VCM product has a purity of 99.8% at a production rate of 445.7 kmol/h.

Flowsheet Integration

Last but not least, the entire flowsheet is augmented by the addition of recycle streams and logical operation tools to achieve the desired production demand. The simulation so far yields a VCM production rate of 445.7 kmol/h. To increase the production, the following approaches are taken, corresponding to the final flowsheet shown in Fig. 12.13.

1. Recycle the recovered HCl from Column 1 back to the oxychlorination reactor R2. The HCl stream is heated to 25°C before replacing the *pseudo* HCl stream located earlier.
2. Recycle the recovered EDC from Column 2 back to the EDC purification section. The EDC from the bottoms of Column 2 is first cooled to 90°C to avoid vaporization before being mixed with the reactors effluent.
3. A "Set" tool (depicted as "SET-2" in Fig. 12.13) is used to always maintain the C_2H_4 supply at 1.04 times in excess of the Cl_2 feed with the following relation;

$$C_2H_4\ (\text{kmol/h}) = 2.08 \times \text{Molar flow of } Cl_2 \tag{12.12}$$

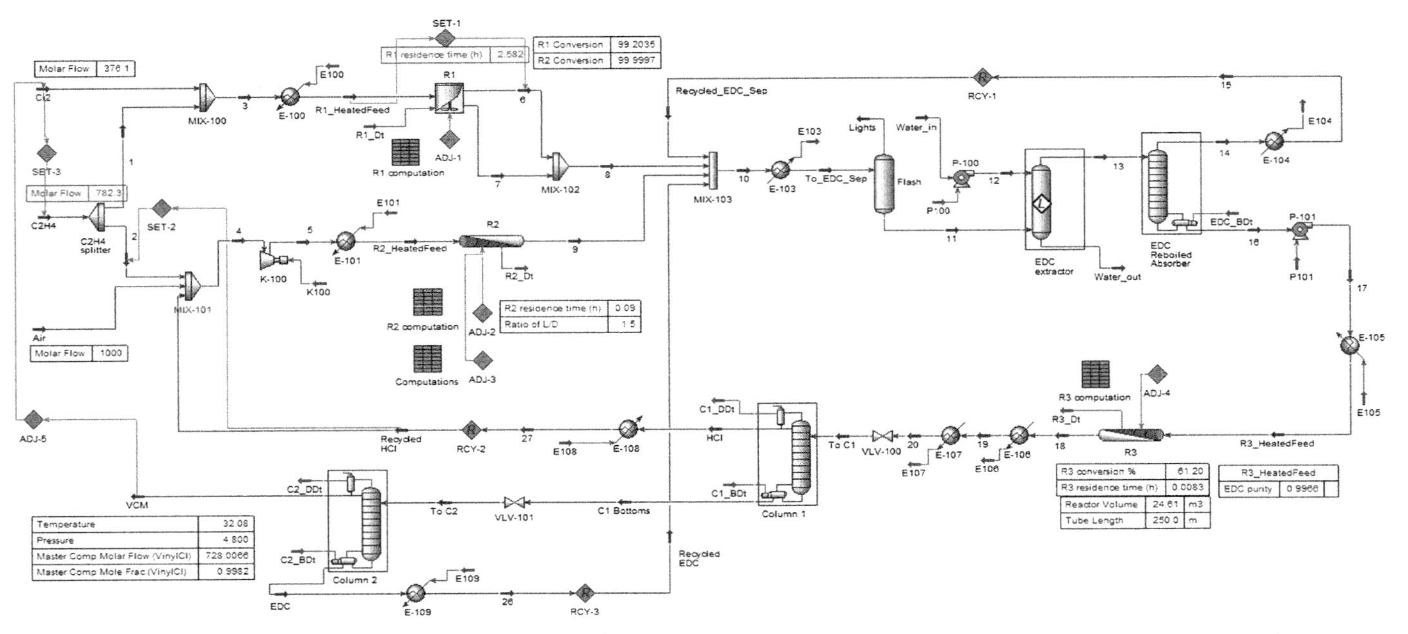

FIGURE 12.13 The final simulation flowsheet for vinyl chloride monomer manufacture (flowrates in kmol/h, T in °C, and P in atm).

4. A "Set" tool (depicted as "SET-3") is used to control the split fraction of C_2H_4 splitter, by adjusting the C_2H_4 supply to the oxychlorination reactor at a stoichiometric relation (Eq. 12.2) with the recycled HCl from Column 1:

$$\begin{aligned} &C_2H_4 \text{ supply to R2 (Stream 2) (kmol/h)} \\ &= 0.5 \times \text{Molar flow of the recycled HCl} \end{aligned} \tag{12.13}$$

5. An "Adjust" tool (depicted as "ADJ-5") is added to ensure VCM produced always meets the target of 728 kmol/h (100,000 lb/h) by adjusting the Cl_2 feed (which is the limiting reactant).

As presented in Fig. 12.13, to produce 728 kmol/h of VCM with a purity higher than 99.8%, the amount of Cl_2 supply needs to be at least 376 kmol/h. This balanced route of VCM manufacture is to be designed in a way that all Cl_2 and HCl are consumed in the process thereby leaving far lesser harmful by-products when compared to other alternatives, for instance, Reaction Pathway B, which produces a huge amount of HCl. The process simulation can also be continued to carry out heat integration for minimizing the cost of heating/cooling equipment and utilities.[3] Side reactions can also be added to simulate a closer process condition. The use of process simulators enables the design team to carry out a number of important design duties, including process optimization, process control (dynamics simulation), equipment design, heat exchange network design, cost analysis, comparison studies with different process alternatives, and also validation with the pilot-plant data for parametric studies.

12.3 CONCLUSION

Having practised this chapter, the reader should be able to design a process flowsheet with the ability to justify the use of recycle streams, different types of reactors and their kinetics, and the logical operation tools including Adjust, Set, and Spreadsheet. The reader should be sufficiently familiar with the Splitter-Shortcut Distillation-Rigorous Distillation approach of designing a distillation column. In addition, the reader should know how to determine some physical and chemical properties from Aspen HYSYS by utilizing the Case Study, Envelope, and Critical Properties features, and to independently explore more advanced features and options in Aspen HYSYS and other software in order to tackle some other chemical engineering problems.

3. Refer to Chapter 14 for the use of Aspen Energy Analyzer for heat integration.

EXERCISES

1. Simulate Reaction Pathway B. Find out the current price of the relevant chemicals and compare your result with the current case study in terms of gross profit.
2. Simulate the case study by replacing all reactors with conversion reactors. What is the difference and which approach is more appropriate?

REFERENCES

Cordeiro, G.M., Dantas, S.R., Vasconcelos, L.G.S., Brito, R.P., 2013. Effect of Two Liquid Phases on the Separation Efficiency of Distillation Columns.

Dimian, A.C., Bildea, C.S., 2008. Chemical Process Design: Computer-aided Case Studies. John Wiley & Sons.

Epa, 2000. Vinyl Chloride [Online]. United States Environmental Protection Agency. Available from: https://www.epa.gov/sites/production/files/2016-09/documents/vinyl-chloride.pdf.

Goodrich, B.F., 1963. Preparation of Vinyl Chloride. North Carolina (1975) Patent Application Instrument Society of America Standard Isa-s5-1.

Hess, W.T., Kurtz, A., Stanton, D., 1995. In: Kirk-Othmer (Ed.), Kirk-Othmer Encyclopedia of Chemical Technology. John Wiley & Sons Ltd., New York.

Lakshmanan, A., Biegler, L.T., 1997. A case study for reactor network synthesis: the vinyl chloride process. Computers & Chemical Engineering 21 (Suppl.), S785–S790.

Lakshmanan, A., Rooney, W.C., Biegler, L.T., 1999. A case study for reactor network synthesis: the vinyl chloride process. Computers & Chemical Engineering 23, 479–495.

Orejas, J.N.A., 2001. Model evaluation for an industrial process of direct chlorination of ethylene in a bubble-column reactor with external recirculation loop. Chemical Engineering Science 56, 513–522.

Ranzi, E., Grottoli, M., Bussani, G., Che, S.C., 1993. A new simulation program predicts EDC furnace performances. Chimica El Industria-Milano 75, 261–269.

Seider, W.D., Seader, J.D., Lewin, D.R., 2009. Product & Process Design Principles: Synthesis, Analysis and Evaluation (With Cd). John Wiley & Sons.

Wachi, S., Asai, Y., 1994. Kinetics of 1, 2-dichloroethane formation from ethylene and cupric chloride. Industrial & Engineering Chemistry Research 33, 259–264.

Chapter 13

Process Simulation and Design of Acrylic Acid Production

I-Lung Chien[1], Bor-Yih Yu[1], Hao-Yeh Lee[2]
[1]*National Taiwan University, Taipei, Taiwan;* [2]*National Taiwan University of Science and Technology, Taipei, Taiwan*

13.1 INTRODUCTION

Acrylic acid, and its salt and esters, is widely used in polymeric flocculants, dispersants, coatings, paints, adhesives, and binders for leather, paper, and textile (Straathof et al., 2005). There are several alternative processes to produce this important commodity chemical, but the most common way, nowadays, is by partial oxidation of propylene (Lin, 2001).

The usual mechanism for producing acrylic acid utilizes a two-step process in which propylene is first oxidized to acrolein and then further oxidized to acrylic acid. Some side reactions will also occur in the reactor, which include partial oxidation reaction to produce acetic acid as a by-product and fully oxidation reaction to form carbon dioxide and water.

In Turton et al. (2009) flowsheet, a fluidized-bed reactor is used. It is assumed that the bed of catalyst behaves as a well-mixed tank and operated at isothermal condition with reaction heat removed by molten salt. Steam is also fed into the reactor. The purpose of the steam is twofold. First, the steam acts as a heat carrier to reduce the temperature excursion inside the reactor. Second, coking reactions that can deactivate the catalyst are suppressed when steam is present.

In Suo et al. (2015), the original flowsheet in Turton et al. (2009) was modified by reducing the steam fed into the reactor. A different separation method utilizing heterogeneous azeotropic distillation to save energy was also proposed in this modified flowsheet.

13.2 PROCESS OVERVIEW

There are two further modifications from the flowsheet of Suo et al. (2015) in this chapter. First, a more realistic operating condition is assumed in this study

Chemical Engineering Process Simulation. http://dx.doi.org/10.1016/B978-0-12-803782-9.00013-3

because the steam feed rate in the flowsheet of Suo et al. (2015) was somewhat arbitrarily cut from 1017.6 kmol/h in the original flowsheet of Turton et al. (2009) to 148.2 kmol/h. Because the purpose of the steam feed is to suppress the coking reaction that deactivates the catalyst. It is not known if this significant reduction of the steam feed is feasible. Additional information from the book of Turton et al. (2009) stated that a new catalyst was considered to improve the process so that the requirement of steam can be reduced to 1 kg steam per kilogram of propylene in the reactor feed. This information is used in the simulation to calculate the required steam feed as 296.3 kmol/h.

Because the steam feed rate is doubled than that of the flowsheet of Suo et al. (2015), a promising energy-saving design by utilizing extraction to remove large amounts of water in a diluted mixture is used in the current design, instead of using heterogeneous azeotropic distillation as in the flowsheet of Suo et al. (2015). The detailed integrated flowsheet is explained in the following section.

13.2.1 Reaction Kinetics

The reaction mechanism for producing acrylic acid utilizes a two-step partial oxidation process in which propylene is first oxidized to acrolein and then further oxidized to acrylic acid as below.

$$\underset{\text{Acrolein}}{C_3H_6 + O_2 \rightarrow C_3H_4O + H_2O} \tag{13.1}$$

$$\underset{\text{Acrylic Acid}}{C_3H_4O + \frac{1}{2}O_2 \rightarrow C_3H_4O_2} \tag{13.2}$$

Each reaction step as above actually takes place over a separate catalyst and at different operating conditions. In this simulation, as in the design flowsheets in Turton et al. (2009) and Suo et al. (2015), the reaction mechanism is simplified to express as an overall reaction in Eq. (13.3). A side reaction to produce acetic acid as a by-product is also included in the simulation as in reaction Eq. (13.4). Another fully oxidation reaction as in reaction Eq. (13.5) is also considered.

$$\underset{\text{Acrylic Acid}}{C_3H_6 + \frac{3}{2}O_2 \rightarrow C_3H_4O_2 + H_2O} \tag{13.3}$$

$$\underset{\text{Acetic Acid}}{C_3H_6 + \frac{5}{2}O_2 \rightarrow C_2H_4O_2 + CO_2 + H_2O} \tag{13.4}$$

TABLE 13.1 Reaction Kinetic Parameters of the Studied System

Reaction	E_i kcal/kmol	$k_{0,i}$ kmol/m^3 reactor·h/(kPa)^2
3	15,000	1.59×10^5
4	20,000	8.83×10^5
5	25,000	1.81×10^8

$$C_3H_6 + \frac{9}{2}O_2 \rightarrow 3CO_2 + 3H_2O \quad (13.5)$$

The kinetics for the reactions presented above is described by the following rate expression:

$$-r_i = k_{o,i} \exp\left[\frac{-E_i}{RT}\right] p_{propylene}\, p_{oxygen} \quad (13.6)$$

In the above expression, partial pressures are in kpa. The kinetic expressions are in power-law form and can easily be implemented in Aspen Plus. The kinetic parameters for the above three reactions are summarized in Table 13.1.

13.2.2 Phase Equilibrium

For the above reactions, oxygen is provided by air; thus for the reaction section there are a total of eight components in the simulation. They are propylene, oxygen, nitrogen (come with air), water, acrylic acid, acetic acid, carbon dioxide, and diisopropyl ether (DIPE). In the separation section of the overall design flowsheet, DIPE will be introduced into the system as a solvent for extraction.

In the Aspen Plus simulation, NRTL-HOC thermodynamic model is selected to describe the phase equilibrium behavior of this process with NRTL to account for nonideal liquid behavior and HOC to account for possible associate behavior of acrylic acid and acetic acid in vapor phase. Table 13.2 shows the NRTL binary parameters used in the simulation. For the binary pairs not shown in Table 13.2, ideal phase equilibrium behavior is assumed. Also, Henry's law is applied for describing the solubility of gas into liquid.

13.2.3 Upstream Process Flowsheet

The conceptual design flowsheet of this process is presented in Fig. 13.1. There are three feed streams, including propylene, air (assumed containing 79 mol% of nitrogen and 21 mol% of oxygen), and steam. As mentioned, low-pressure

TABLE 13.2 NRTL Model Parameters for the Studied System

Component *i*	AA	ACE	AA	DIPE	DIPE	DIPE
Component *j*	Water	Water	ACE	Water	AA	ACE
Source	Aspen VLE-HOC	Aspen VLE-HOC	Regression	Aspen LLE	Aspen VLE-HOC	Aspen VLE-HOC
a_{ij}	0	−1.9763	0.539728	0.0035	−0.1958	0
a_{ji}	0	3.3293	−1.79443	8.0209	1.6953	0
b_{ij} (K)	−293.649	609.889	−32.7296	422.978	750.656	344.4
b_{ji} (K)	919.456	−723.888	914.333	−766.417	−840.241	82.134
c_{ij}	0.3	0.3	0.3	0.2	0.3	0.3

AA, acrylic acid; *ACE*, acetic acid; *DIPE*, diisopropyl ether.

Aspen Plus NRTL:

$$\ln \gamma_i = \frac{\sum_j x_j \tau_{ji} G_{ji}}{\sum_k x_k G_{ki}} + \sum_j \frac{x_j G_{ij}}{\sum_k x_k G_{kj}} \left[\tau_{ij} - \frac{\sum_m x_m \tau_{mj} G_{mj}}{\sum_k x_k G_{kj}} \right] \quad \text{Where} : G_{ij} = \exp(-\alpha_{ij} \tau_{ij})$$

$$\tau_{ij} = a_{ij} + \frac{b_{ij}}{T}$$

$$\alpha_{ij} = c_{ij}, \tau_{ii} = 0, G_{ii} = 1$$

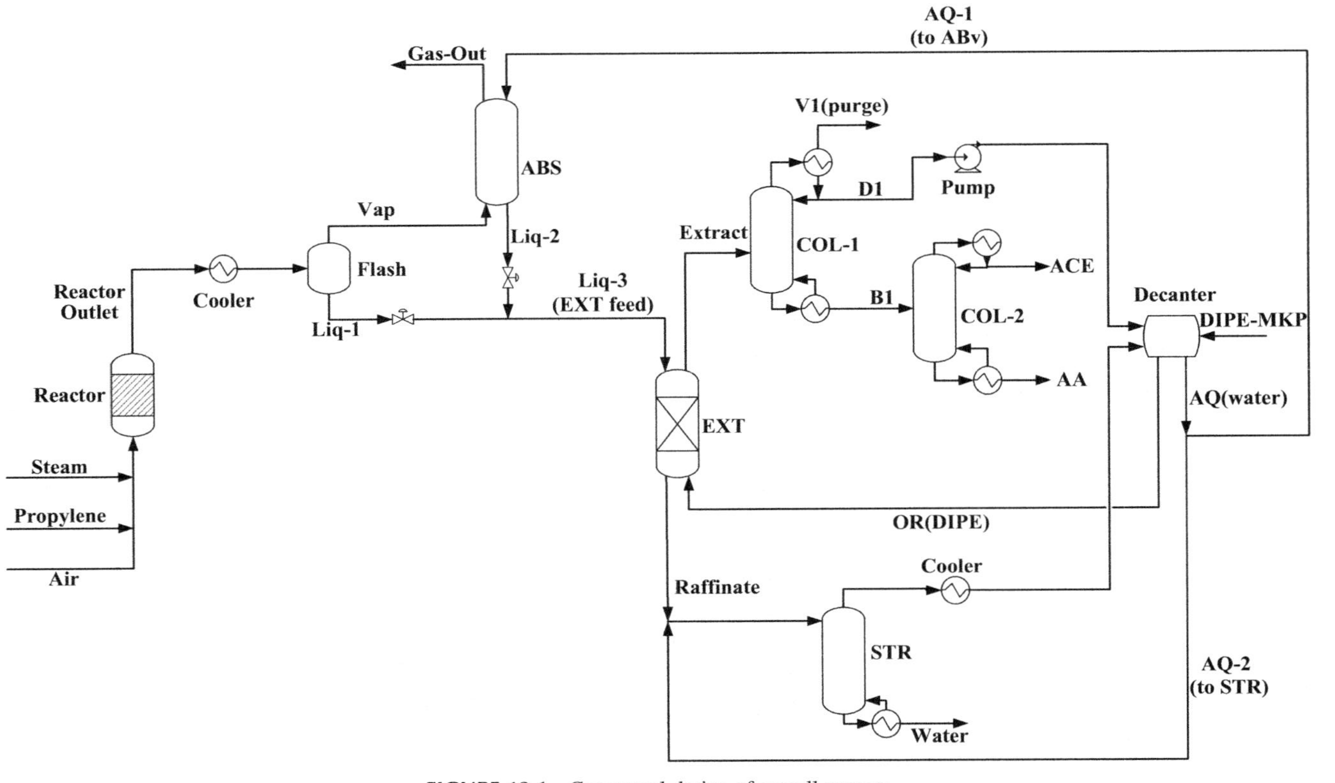

FIGURE 13.1 Conceptual design of overall process.

steam as heat carrier is necessary to slow down the increasing rate of temperature in the reactor.

A new catalyst was considered to improve this process, which requires only 1 kg steam per 1 kg of propylene in the feed to the reactor. With fresh propylene feed rate set at 127 kmol/h to produce about 50,000 ton per year of 99.5 mol% acrylic acid, the steam feed rate is calculated to be 296.3 kmol/h. From the main reaction in Eq. (13.3), the stoichiometric ratio of propylene to oxygen is 1:1.5. However, excess oxygen is required to increase the conversion of propylene in the reactor. A ratio of 1:2.2 as in the original flowsheet in Turton et al. (2009) was selected. This will make the air feed rate to be 1337.6 kmol/h. In Fig. 13.1, the combined vapor feed rate is shown with total feed rate at 1760.9 kmol/h and feed temperature at 191°C, and feed pressure at 4.3 bar.

In reality, in any reaction involving partial oxidation of a fuel-like gas (e.g., propylene in this case), considerable attention must be paid to the composition of hydrocarbons and oxygen in the feed stream. The feed condition needs to be either below the lower explosion limit so that it is too lean to burn or higher than the upper explosion limit so that it is too rich to burn. However, in this simulation, the purpose was to demonstrate using of Aspen Plus to simulate a typical integrated process; those explosion limits were not considered.

The reaction section in this process flowsheet was simplified to contain only one fluidized-bed reactor as in the flowsheets of Turton et al. (2009). Because the reactions in Eqs. (13.3)–(13.5) are highly exothermic, there will be heat transfer between the molten salt and the fluidized-bed to remove the reaction heat so that the catalyst sintering in the reaction tubes will not occur. Because of the nature of circulating inside a fluidized-bed reactor, the reactor is assumed to be operated isothermally with the operating temperature as an important design variable. The upper limit of the operating temperature is 603K following the suggestion of Turton et al. (2009) to avoid damage to the catalyst. Besides, the operating temperature should not be too low for maintaining acceptable reaction rate.

Inside Aspen Plus, the fluidized-bed reactor is simulated as an isothermal plug-flow reactor (PFR) so that the composition profile along the inlet to outlet of the fluidized-bed reactor can be described in the simulation. The length of the reactor is assumed to be 10 m, while the diameter is determined based on the target reaction conversion and selectivity of products. The void fraction of the catalyst in the tube is assumed to be 0.5 with the solid catalyst density of 1600 kg/m^3. The reactor diameter is also another important design variable, which will be investigated in Section 13.3.3. In the simulation, it is assumed that 10% of the gas fed to the reactor bypasses the catalyst. This accounts for the channeling phenomenon that occurs in a real fluidized-bed reactor. With the above reactor conditions specified, the reactor effluent composition can be predicted.

For the pressure drop of this tubular reactor, Eq. (13.7) is used based on the fact that the fluidized-bed reactor is operated in the fluidized or bubbling region in which the gravity force of solid equals to the drag force:

$$\Delta p = g\left(\rho_s - \rho_g\right)(1 - \varepsilon)h \tag{13.7}$$

where Δp is pressure drop in pascal, g is gravitational acceleration in m^2/s, ρ_s is the density of solid particle in kg/m^3, ρ_g is the density of gas in kg/m^3, ε is the void fraction inside the bed (−), and h is the bed height in m. More detailed information about the simulation of the fluidized-bed reactor is addressed in Section 13.4.

The reactor effluent is immediately cooled to 47°C, in a cooler to avoid further oxidation reactions that may occur in the downstream equipment. The reactor outlet temperature is high (250°C); thus heat can be recovered through heat integration. The first part of the heat recovery (from 250 to 170°C) can be performed to produce industrial low-pressure steam (6 bar, 160°C). Another portion of it (<170°C) can be used to exchange heat with other sections inside the integrated process. The last cooling part to the final temperature of 47°C can be achieved using cooling water. For simplicity in this work, it is assumed that all the cooling of this hot stream to 47°C is achieved by cooling water, with 0.3 bar pressure drop in this cooler.

The stream after cooling is fed into an adiabatic flash drum operated at the same pressure as cooled reactor outlet stream. The liquid stream of the flash drum containing mostly acrylic acid, acetic acid, and water will go to separation system for further separation. All the lighter components (such as nitrogen, unreacted oxygen, and carbon dioxide) should go to vapor stream of the flash drum. However, significant amounts of the product will lose through this vapor stream if no recovery equipment is present.

In the design flowsheet, a gas absorber is designed to recover products (acrylic acid and acetic acid) from this vapor stream by adding water into this absorber. The water feed rate to the absorber is set so that only 0.01% of the main product (acrylic acid) is allowed to be lost from the vapor outlet stream of the absorber. Note that in the design flowsheet, the liquid outlet stream of flash drum and absorber liquid outlet stream are combined to go to further separation section described in next section. The design flowsheet of the upstream process is illustrated in Fig. 13.2.

13.2.4 Downstream Further Separation Flowsheet

From Fig. 13.2, it is noted that the feed mixture into the further separation section contains large amounts of water. There are three major sources of this water component. First, it was added into the reactor to act as a heat carrier to reduce temperature excursion inside the reactor. Second, water was added into the gas absorber so as to reduce the loss of main product from a gas stream. Third, water was formed from the side reaction of propylene partial

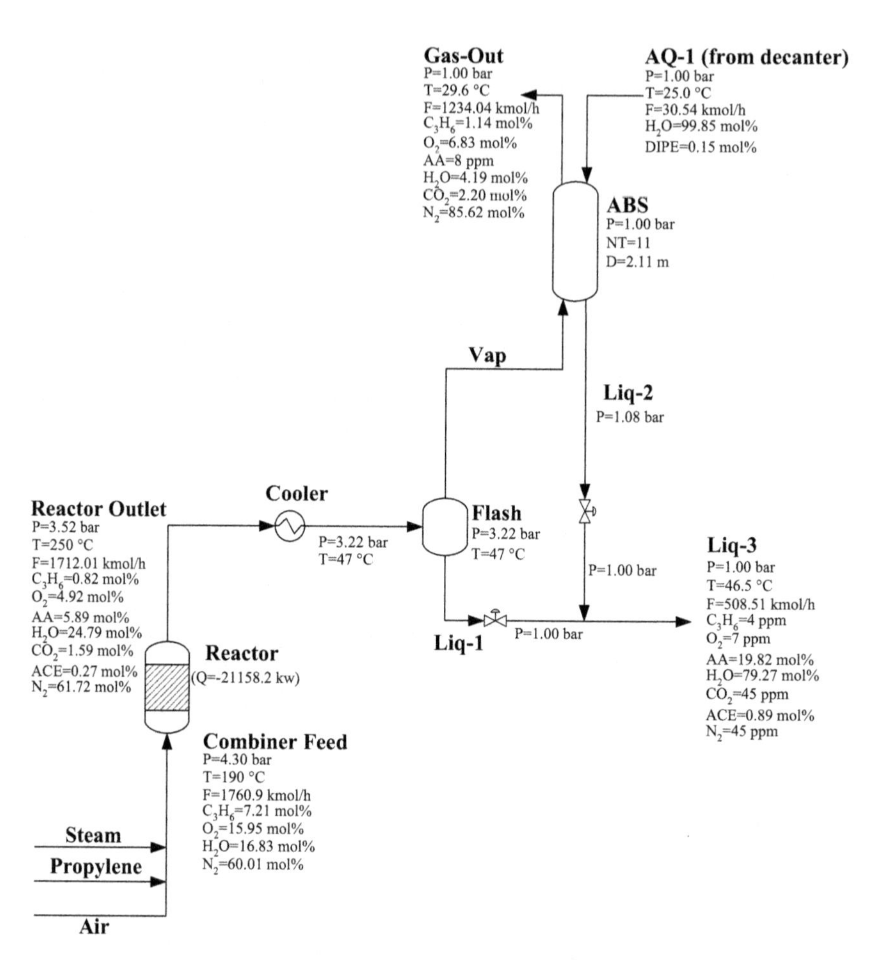

FIGURE 13.2 Design flowsheet of upstream process.

oxidation. The other two major components are acrylic acid (product) and acetic acid (by-product).

The specifications for the outlet stream of this further separation section are set to be 99.5 mol% acrylic acid and 96.0 mol% acetic acid. As for the water outlet stream, the impurity specification in this water outlet stream is to have acids below 1000 ppm (weight-based). There is also another high-temperature operating constraint to prevent acrylic acid to become polymerized. This temperature constraint is set at 110°C following the suggestion of Suo et al. (2015).

From T-xy plots of Fig. 13.3, it is illustrated that a mixture of acrylic acid and water and also a mixture of acetic acid and water both do not contain azeotrope. However, there is a tangent pinch behavior toward pure water end for the mixture of acetic acid and water. This means that to obtain a water product at stringent purity specification, many stages are required. Also, water

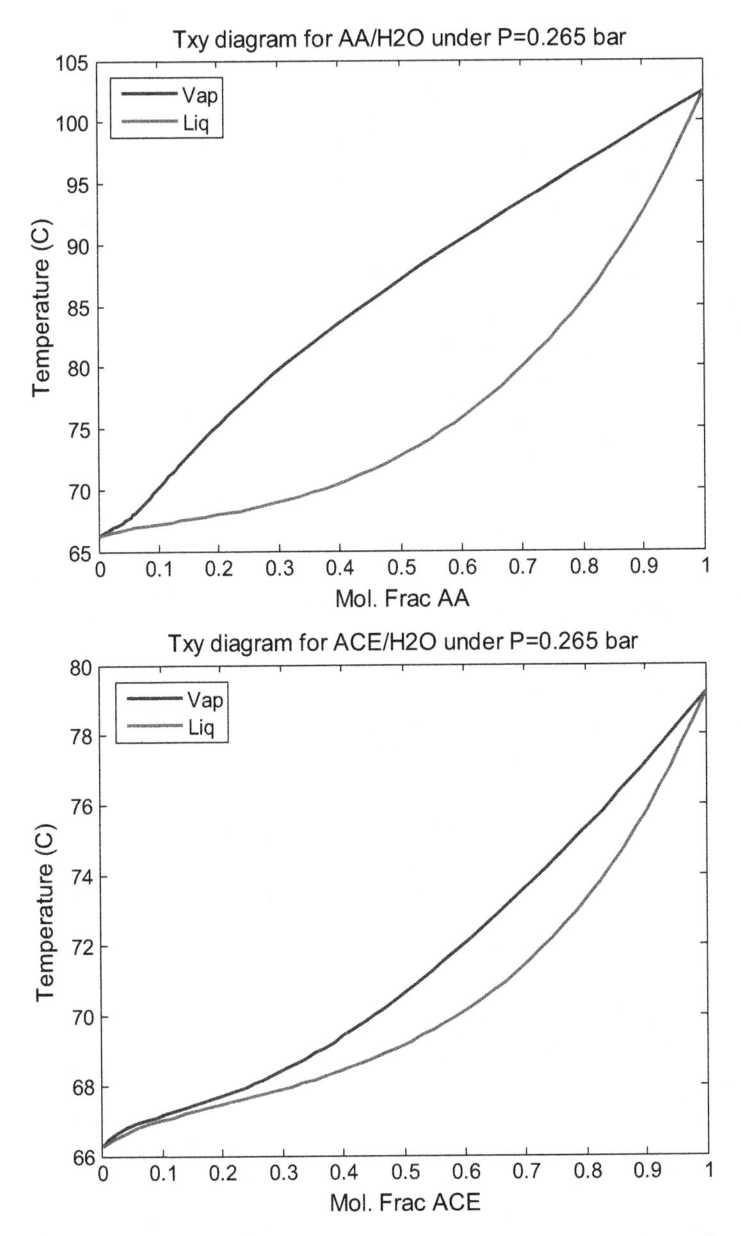

FIGURE 13.3 T-xy diagrams. (A) T-xy diagram for AA/H_2O, and (B) T-xy diagram for ACE/H_2O. *AA*, acrylic acid; *ACE*, acetic acid.

is the lightest component of this ternary mixture, which should go out as distillate in a regular distillation column. This means large energy will be required to boil up water into column top.

An energy-saving design for separating this ternary system with large amounts of water is to utilize extraction in this process. The concept is to use a solvent to extract acrylic acid and acetic acid out from this ternary mixture as an extract phase. The raffinate phase is desirable to contain mostly water. Selecting a proper solvent to enhance the selectivity of this extraction separation is most important. In this work, DIPE, one of the solvents suggested by Kürüm et al. (1995) in their acetic acid dehydration study, is selected as the solvent.

Following the idea in Chen et al. (2015), the conceptual design of this hybrid extraction-distillation system can be seen in Fig. 13.4. Here in this work, several modifications are included. In the work by Chen et al., two strippers are designed to purify both outlet streams (extract phase and raffinate phase) of the extractor. But in this work, gaseous species are presented in the system and are dissolved in liquid. Although the amount of dissolving is slight, it still cannot be neglected if not taking out from a vapor purge stream of a partial-vapor condenser. Thus, a stripper and a regular column with partial-vapor condenser are used for the purpose of separation.

The overall design is described as follows. The raffinate (rich in water) is sent to a stripper (STR) to further purify the water outlet stream, with impurity specification of 1000 ppm acids and high water recovery (99.9%). Because the impurities inside the raffinate contains some DIPE, which is the lighter component inside the system, DIPE will come out from the top of STR, while water and acids are from the bottom.

Also note that there is a tangent pinch of water/ACE in T-xy diagram, which means that ACE can barely be separated from water in STR. From these features, it can be concluded that enough extraction trays will be required so that the acids go with the raffinate phase would not violate the impurity specification of 1000 ppm acids at stripper bottom stream. Here in this part, reboiler duty of STR and the number of stages of EXT are adjusted to satisfy the water purity and recovery requirements mentioned above. The detailed information of related specifications is illustrated in Table 13.3.

The extract phase from the extractor is sent to a regular column (COL-1) with a partial vapor–liquid condenser. Overall, three product streams can be designed by this column. One is the vapor purge stream, another one is the liquid distillate rich in DIPE and water. The remaining one is the mixture of acrylic acid and acetic acid, which is sent to another column (COL-2) for further separation to obtain the purity requirement of both products (99.5 mol % AA and 96 mol% ACE).

As mentioned in the beginning of this section, there is a high-temperature constraint at 110°C in the separation section to prevent acrylic acid polymerization. Thus, both COL-1 and COL-2 should be operated at vacuum pressure. In this work, the pressure of COL-1 is set at 0.265 bar. For the partial

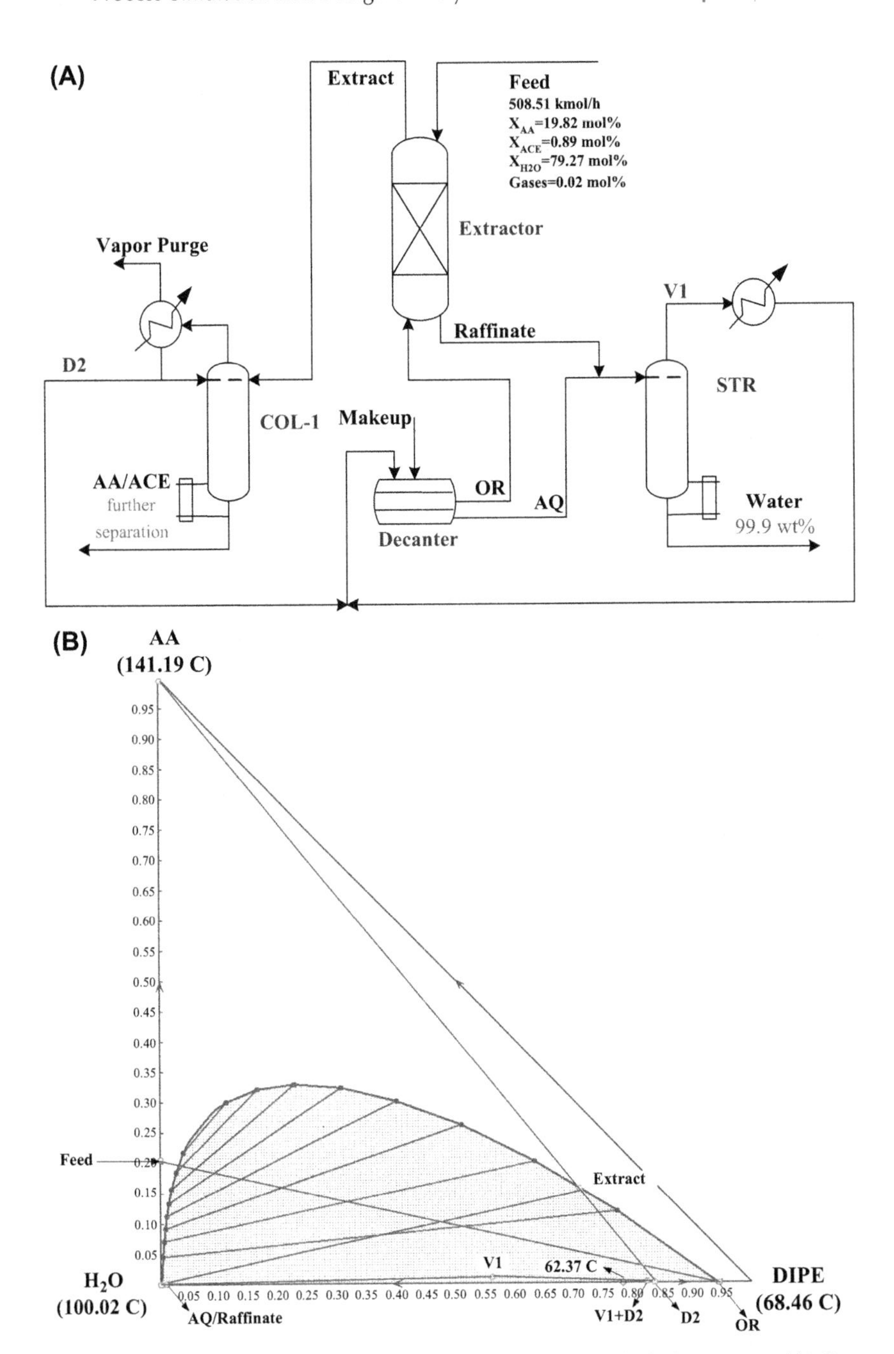

FIGURE 13.4 Conceptual design flowsheet of hybrid extraction-distillation process. (A) Conceptual design flowsheet. (B) Conceptual design through ternary diagram for AA/Water/DIPE. *AA*, acrylic acid; *ACE*, acetic acid; *DIPE*, diisopropyl ether.

TABLE 13.3 Specification of Variables in Further Separation Process

Column	Specification	Vary
Hybrid Extraction-Distillation Process		
EXT	Water purity = 99.9 wt% at STR bottom	NT-EXT (NT-STR is fixed at three stages)
STR	Water recovery = 0.999 (bot)	Reboiler duty
COL-1	ACE recovery = 0.99 (top)	Reflux ratio
	Vapor fraction = 0.1% in partial vapor-liquid condenser	(–)
	AA + ACE = 99.5 mol% (bot)	Reboiler duty
COL-2	ACE purity = 96 mol%	Reflux ratio
	AA purity = 99.5 mol%	Reboiler duty
Sequenced-Separation Process		
COL-3	Water purity = 99.9 wt% (top)	Reflux ratio
	Vapor fraction = 0.1% in partial vapor-liquid condenser	(–)
	AA + ACE = 99.5 mol% (bot)	Reboiler duty
COL-4	ACE purity = 96 mol%	Reflux ratio
	AA purity = 99.5 mol%	Reboiler duty

AA, acrylic acid; *ACE*, acetic acid.

vapor–liquid condenser in COL-1, it is assumed that 0.1% of distillate is condensed to vapor phase and is purged out from the process to avoid accumulation. Under this pressure and the specified purge ratio, the temperature at the column top is around 25°C and thus chilled water can be used at the condenser. The pressure of COL-2 is set at 0.08 bar, and this setting makes the use of cooling water allowable in this condenser. The detailed specifications and the related adjusted variables are listed in Table 13.3.

The top vapor from the STR is rich in DIPE and is cooled to 25°C, where temperature is the same as the liquid distillate from COL-1. COL-1 liquid distillate is pumped back to 1 bar and is sent to a decanter together with the cooled vapor stream from STR, and also the small makeup DIPE to perform liquid–liquid separation.

The decanter is operated at 1 bar and 25°C. The organic phase of the decanter is rich in DIPE and is then recycled back to the extractor (EXT). The aqueous outlet is divided into two streams: the first one is recycled to

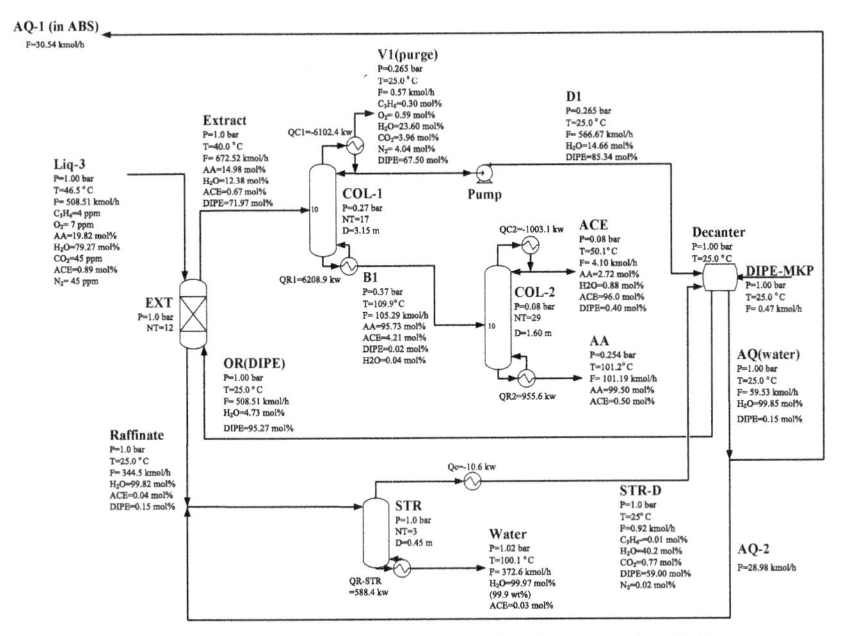

FIGURE 13.5 Design flowsheet of downstream process in hybrid extraction-distillation process.

the absorber (ABS) for recovering the acids and the remaining portion is recycled to STR for further purification of water. The overall design flowsheet of this further separation section can be seen in Fig. 13.5.

13.2.5 Sequenced-Separation Process

Hybrid extraction-distillation process is a design configuration for energy saving. Apart from this design configuration, the reactor outlet can also be directly separated by a series of columns because there is no azeotrope between AA, ACE, and water. As mentioned in Section 13.2.4, there is a tangent pinch in T-xy plot of ACE/water mixture, which may need many theoretical trays and large amount of energy for stringent separation to meet the requirement in wastewater stream. Through simulation using Aspen Plus, a quantitative and qualitative comparison between these two separation methods can be easily obtained. The overall design flowsheet for sequenced-separation process is illustrated in Fig. 13.6.

In the sequenced-separation process, the combined reactor outlet is first sent to a column (COL-3) to separate water from AA and ACE. The temperature limit of 110°C also holds in this design method to avoid acids from polymerizing, and thus the column is also operated under vacuum condition. Because many trays are needed for stringent separation, the pressure drop inside the column may be large. Thus the column pressure is set at 0.13 bar at the top.

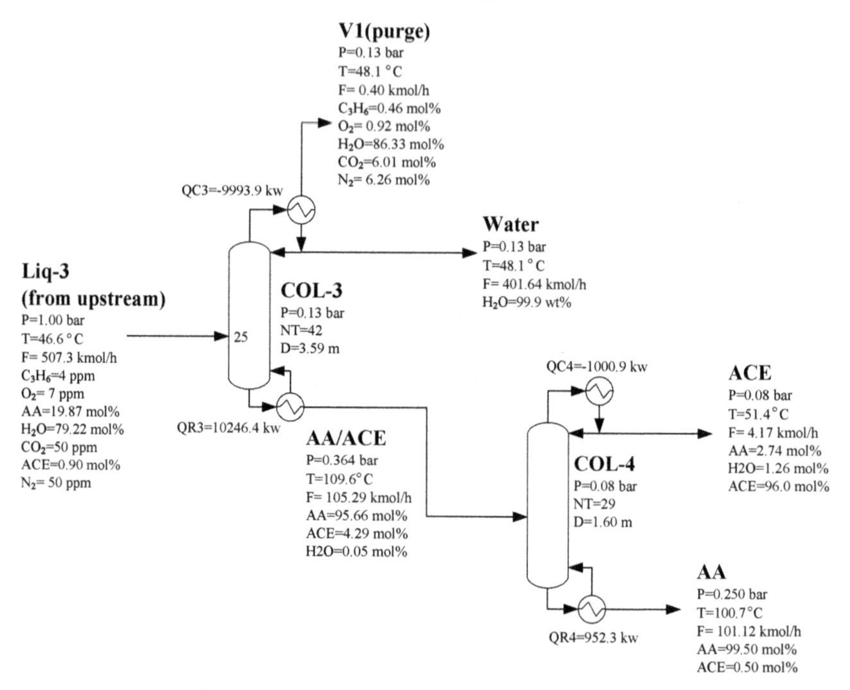

FIGURE 13.6 Design flowsheet of sequenced-separation process.

Similar to the hybrid extraction-distillation process, the condenser is also a partial vapor-liquid one to remove the slightly dissolved gaseous species in liquid. In COL-3, water comes out as the liquid distillate, and the mixture of AA and ACE comes out at the bottom. The target of separation will be 99.9 wt % water purity at the top, and 99.95 mol% for combined AA and ACE at the bottom. Reflux ratio and reboiler duty can be adjusted to reach the separation targets. For the partial vapor—liquid condenser in COL-3, it is assumed that 0.1% of distillate is in vapor phase and is purged from the process. With this assumption, the corresponding temperature at the outlet of partial vapor—liquid condenser is around 48°C, and cooling water can be used.

Note that in hybrid extraction-distillation process, DIPE is an extra lighter component in COL-1 to be separated into distillate along with water. While in sequenced-separation process, water is the only lighter component in COL-3 to be separated into liquid distillate. This difference leads to the different cooling sources used in these processes.

The acid mixture obtained at COL-3 bottom is then sent to COL-4 for further separation. In order for fair comparison, the pressure of COL-4 is also set at 0.08 bar, with two product specifications the same as in hybrid extraction-distillation process (ACE = 96 mol%, AA = 99.5 mol%). The optimization results of sequenced-separation process and the comparison between these two processes are illustrated in Section 13.3.4.

13.3 EFFECT OF IMPORTANT DESIGN VARIABLES AND EXAMPLES OF OPTIMIZATION WORKS

In this section, we will first discuss the effect of some important design variables for the upstream process (in Fig. 13.2) and also the downstream further separation section (in Fig. 13.5) for the hybrid extraction-distillation process. Then a demonstrative optimization work for minimizing total annual cost (TAC) is performed on the hybrid extraction-distillation process. Finally, the optimized hybrid process can be compared with the optimized sequenced-separation process.

13.3.1 Reactor Temperature and Its Size

Reactor temperature and its size are two important design variables in the upstream process, and the determination of these two variables are based on the targeted conversion and selectivity. Overall, the reactions in this study are in parallel, thus the temperature will influence the reaction rate. The higher the temperature, the greater the rates for all the reactions will be. Thus it also affects the selectivity between AA and ACE. The reactor size has a direct impact on the conversion. With different reactor temperature, the required reactor size to reach the targeted conversion will also be different. Note that the kinetic expressions are assumed to be first order to partial pressure of both propylene and oxygen for all three reactions. Thus reactor pressure will solely influence the reaction rate without altering the selectivity of products.

Fig. 13.7 illustrates the sensitivity test of this reactor with varying temperature and diameter. Because the length is assumed to be 10 m in this study, varying diameter has the same meaning of varying size. In Fig. 13.7A, it illustrates peak in AA production rate. As the reactor size becomes larger, the peak shifts to the right. This is due to the competition of oxygen among these three reactions. The left side of the peak is the region that reaction Eq. (13.3) dominates, and AA still has the trend to form under the given temperature and size. The right side of the peak is the region that the undesired side reaction, reaction Eq. (13.5), starts to dominate. In Fig. 13.7B, it illustrates the profile of ACE production that it becomes greater as the temperature rises. Thus the ACE reaction, reaction Eq. (13.4), does not have the dominating performance in this reactor.

Generally, the selection of reactor temperature and size should be based on the following three goals: (1) The AA selectivity should be high. (2) The overall conversion should be decent. (3) The reactor should be designed to be economically effective. In this work, our targeting reaction conversion is at around 99% (exclude the 10% bypass). From Fig. 13.7A and C, it is illustrated that the enhancement in the reaction performance is not obvious with the diameter greater than 1.5 m. Thus, the reactor diameter is set to be at this value

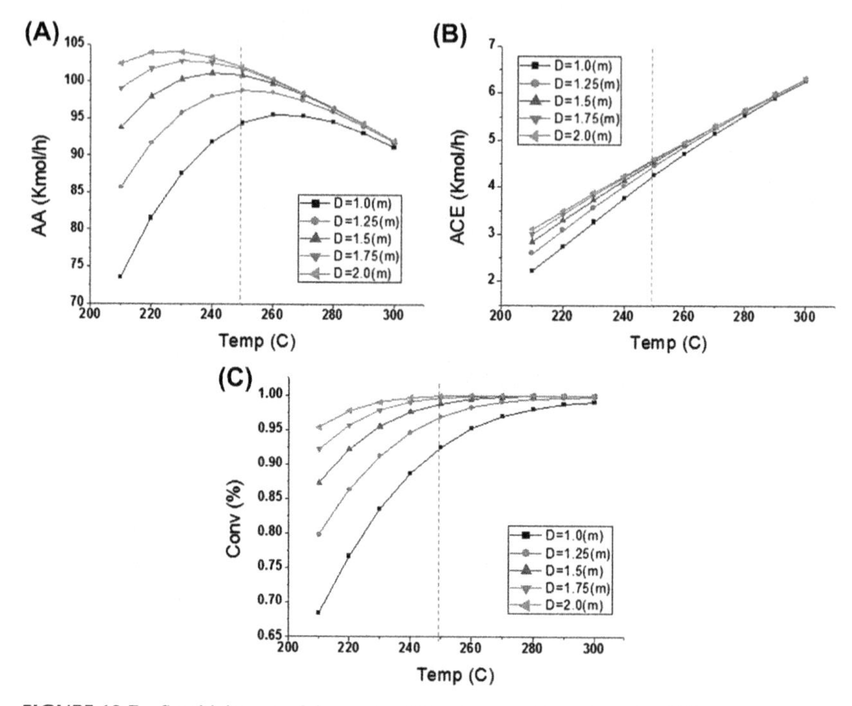

FIGURE 13.7 Sensitivity test of the reactor. (A) AA production rate under different temperature and reactor size. (B) ACE production rate under different temperature and reactor size. (C) Overall conversion under different temperature and reactor size. *AA*, acrylic acid; *ACE*, acetic acid.

as larger reactor size leads to greater equipment and catalyst costs. From Fig. 13.7C, it illustrated that the operating temperature at 250°C under this reactor size leads to conversion around 99%. Also, 250°C of temperature and 1.5 m of diameter lead to a high production rate of AA, which is around the peak of the profile as illustrated in Fig. 13.7A. Thus, temperature at 250°C and diameter equals to 1.5 will be reasonable choices for our design purposes. Following this set of condition, the reactor conversion will be 98.8% for the portion that undergoes reaction and will be 88.9% overall with consideration of 10% bypass.

13.3.2 Water Rate Into Absorber

In this work, the absorber is operated so that 99.99% of AA is recovered from the vapor stream of the flash unit, with the required amount of water feeding in. The water rate is dependent on the number of trays in absorber. As this number increases, the required water rate decreases under the same specification, but with its decrement becoming not obvious. More importantly, the water amount will have a great influence on the downstream separation, especially in the performance of extraction. Thus the determination of water feed rate, or in other words, the number of trays in absorber, should be discussed with the variables in downstream process in next subsection.

13.3.3 Design Variables in Further Separation Section

There are six important design variables in the hybrid extraction-distillation process, including ratio of solvent/feed (FS/FF); total stages of the extractor (NT-EXT); total stages of STR (NT-STR), and total stage and feed location of COL-1 (NT-COL-1 and NF-COL-1). Another one is the total number of stages of absorber (NT-ABS) mentioned in Section 13.3.2.

The optimal design flowsheet of this further separation section is obtained by minimizing TAC with these design variables. TAC includes annual total operating cost and total capital cost divided by a payback period, which is assumed to be 3 years. Operating cost includes the cost of utilities (such as steam, cooling water, and chilled water) and solvent makeup; capital cost includes the costs of columns, decanter, heat exchangers, and vacuum system. The formulae for calculating the total annual cost were mainly adopted from Luyben (2011). The sizing calculations of the extraction column (its diameter and height) are adopted from page 583 of Seider et al. (2009). The calculation of vacuum cost is also adopted from page 589–590 of Seider et al. (2009), and one-staged steam ejector is assumed to be applicable in both hybrid and sequenced-separation process. Table 13.4 summarizes the formulae for all the TAC calculations.

Some preliminary studies can be done before optimization. These works can be helpful in eliminating those variables that are insensitive to TAC change and for better usage of the degree of freedoms in the process. In this work, the acids in the raffinate phase of extractor should be less enough to ensure the production of water with required impurities (<1000 ppm) at stripper (STR). Thus the NT-EXT is set to be the minimized value to produce the qualified water stream (with 99.9% water recovery at STR bottom as mentioned in Section 13.2.4) and is removed from the optimization works.

It is noted that in typical situation, the tray efficiency of liquid–liquid phase separation is not high, which is often in the range of 20–30%. Thus the theoretical stages for extraction should not be too many. In this optimization work, it is assumed that the upper limit for NT-EXT is 15.

Also, the purpose of STR is to separate the slight amount of DIPE in the raffinate phase of EXT; thus its equipment size does not have great influence on TAC, either. Thus, NT-STR is set at three trays in the following study and is also removed from optimization works. After the preliminary study, four optimization variables remained, which are FS/FF, NT-COL-1, NF-COL-1, and NT-ABS.

The sequential iterative optimization procedure is outlined in Fig. 13.8. The ratio of extraction solvent and fresh feed (FS/FF) is the most important

TABLE 13.4 Basis of Economics and Equipment Sizing

Column diameter (D): Aspen tray sizing Column length (L): NT trays with 2-feet spacing plus 20% extra length Column and other vessel (D and L are in meters) Capital cost $= 17{,}640(D)^{1.066}(L)^{0.802}$
Condensers (area in m^2) Heat-transfer coefficient = 0.852 kW/°C-m^2 Differential temperature = reflux-drum temperature −42°C Capital cost $= 7296(\text{area})^{0.65}$
Reboilers (area in m^2) Heat-transfer coefficient = 0.568 kW/C-m^2 Differential temperature = steam temperature – base temperature ($\Delta T > 20°C$) Capital cost $= 7296(\text{area})^{0.65}$
Extraction tower Length: NT trays with 4 feet HETP plus additional 3 feet at the top and 3 feet at the bottom. Diameter: Following calculations on page 583 of Seider et al. (2009) with maximum total liquid throughput = 120 ft^3/h ft^2 and safety factor f = 0.6 Capital cost $= 17{,}640(D)^{1.066}(L)^{0.802}$ (D and L are in meters)
Decanter A tank with aspect ratio of L/D = 2 and 40 min holdup time with liquid half full Capital cost $= 17{,}640(D)^{1.066}(L)^{0.802}$ (D and L are in meters)
Energy cost HP steam = $9.88/GJ (41 barg, 243°C) MP steam = $8.22/GJ (10 barg, 184°C) LP steam = $7.78/GJ (5 barg, 160°C) Cooling water = $0.354/GJ Chilled water = $4.43/GJ (5°C in and 15°C out) Electricity = $16.9/GJ
TAC = (capital cost/payback period) + energy cost Payback period = 3 years

design variable and should be assigned as the outer-most one. The number of trays in absorber (NT-ABS) affects the required water feed rate to recover the acid and then influences the performance in a later process. Thus NT-ABS is assigned in the second loop. The other two variables (NT-COL-1 and NF-COL-1) majorly affect the energy and equipment cost of COL-1, and they are set in the inner loop. In each simulation run, the specifications of products should be achieved. The detailed information of specifications and the variables being adjusted to reach the requirement are listed in Table 13.3.

Fig. 13.9A–D shows the summary of the TAC plots at FS/FF = 1.00 under different NT-ABS, NT-COL-1, and NF-COL-1. TAC plots at other FS/FF ratio

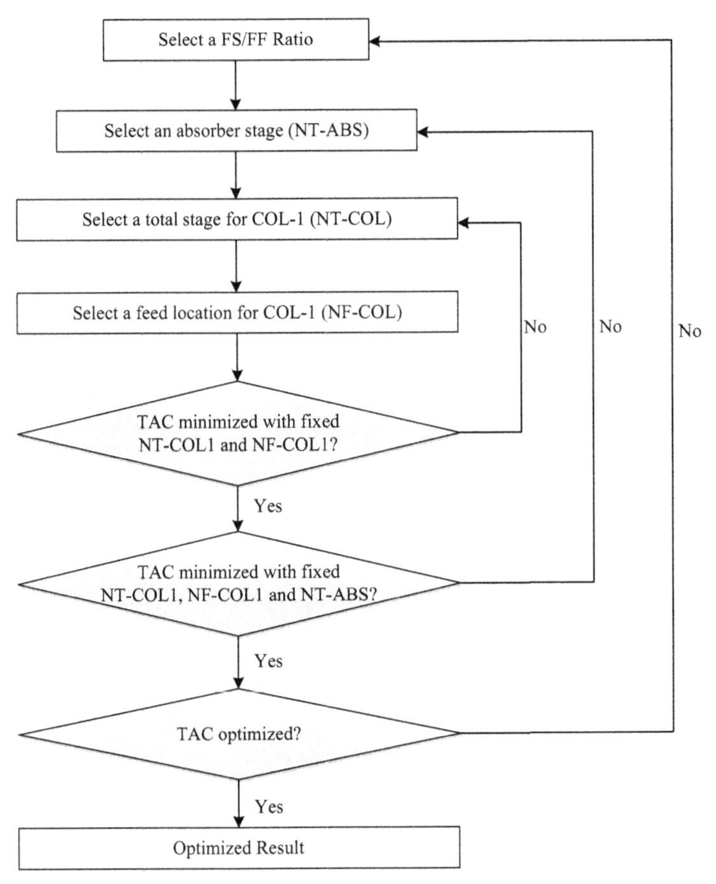

FIGURE 13.8 Iterative algorithm for total annual cost (TAC) optimization of hybrid extraction-distillation process.

are also available but not shown in this figure. From these subfigures, it is illustrated that NT-ABS = 11 yields the lowest TAC under FS/FF = 1.00. Besides, there is a temperature limit of 110°C to avoid polymerization of AA, thus the bottom temperature of COL-1 should be carefully investigated. The bottom temperature of COL-1 is illustrated in Fig. 13.9E. As NT-COL-1 is larger than 17, the pressure drop inside the column causes the bottom temperature to rise over the limitation. From these subfigures, although the TAC decreases as NT-COL-1 increases, the bottom temperature exceeds the limit when NT-COL-1 is larger than 17. Also note that it requires at least seven stages below the feed location, otherwise the separation targets at bottom cannot be met. This phenomenon can be clearly observed from Fig. 13.9A–C when TAC reaches a constraint with increased NF-COL-1. From the results mentioned above, NT-COL-1 is set at 17, and the optimal NF-COL-1 is at 10th tray.

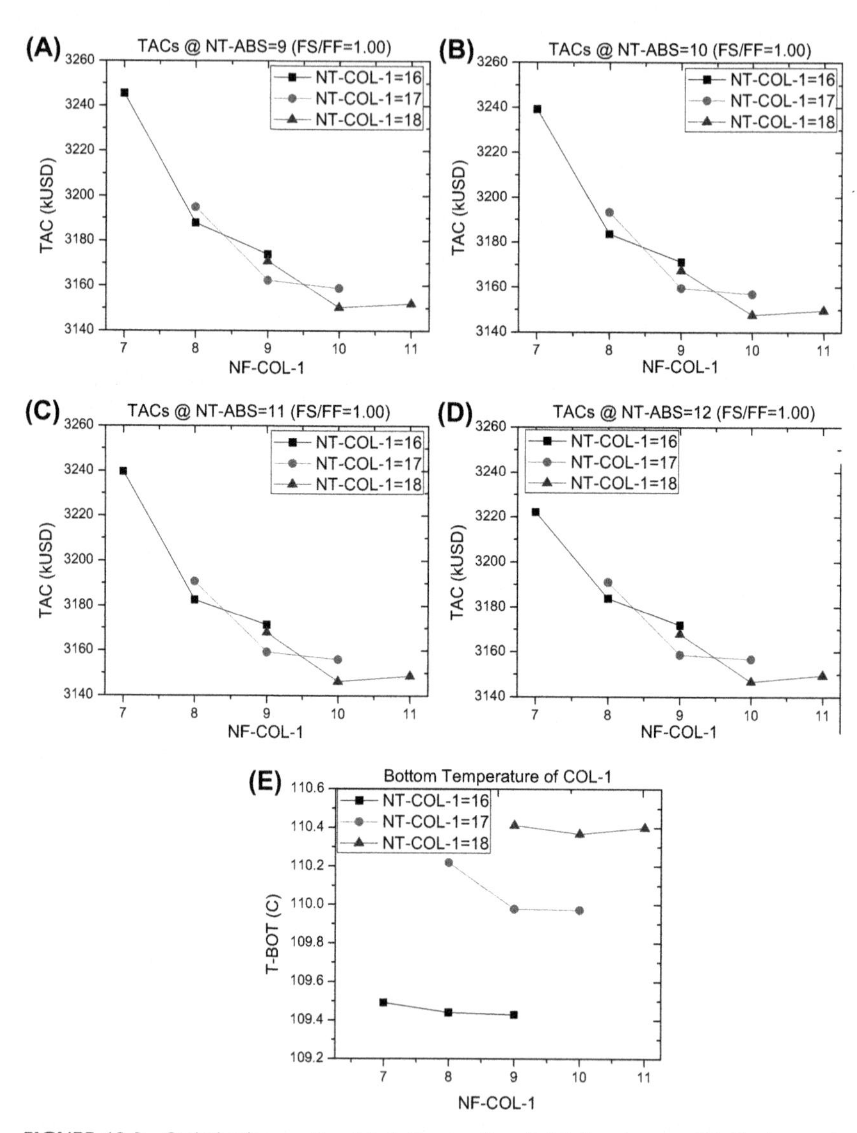

FIGURE 13.9 Optimization results of hybrid extraction-distillation process. (A) Total annual cost (TAC) under NT-ABS = 9, FS/FF = 1.00. (B) TAC under NT-ABS = 10, FS/FF = 1.00. (C) TAC under NT-ABS = 11, FS/FF = 1.00. (D) TAC under NT-ABS = 12, FS/FF = 1.00. (E) Bottom temperature of COL-1 under different NT-COL-1 and NF-COL-1.

Fig. 13.10A and B shows the relationship between NT-ABS, TAC, and the required water flowrate under FS/FF = 1.00. As it is illustrated, a higher NT-ABS leads to less water feed rate required, and less TAC. This is because less water leads to less energy required in separation of water in further separation process, and the optimal NT-ABS is 11 stages. Fig. 13.10C shows the TAC

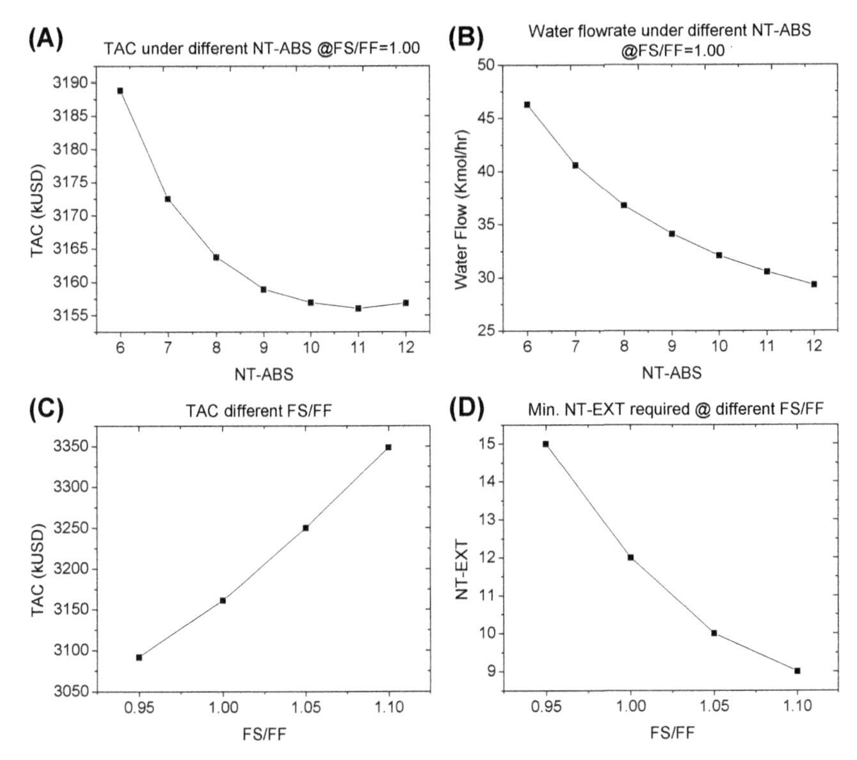

FIGURE 13.10 Optimization results of hybrid extraction-distillation process (A) Total annual cost (TAC) under different NT-ABS and FS/FF = 1.00. (B) Required water flow rate under different NT-ABS and FS/FF = 1.00. (C) TAC under different FS/FF. (D) Minimum NT-EXT required for reaching the water purity under different FS/FF.

results under different FS/FF ratio, and Fig. 13.10D shows the minimum NT-EXT required for reaching the water purity at STR bottom.

By summarizing Fig. 13.10C and D, it is found that the optimal FS/FF is at 0.95. Also, as FS/FF increases, it leads to less NT-EXT but resulting in higher TAC. Note that from Fig. 13.10C, the optimal case reaches a low constraint for the system, which means that when using less amount of solvent, the specification of water product cannot be achieved. This is because a minimum amount of solvent is necessary to achieve the required liquid−liquid separation in the extraction column. Thus, this optimal case may not be suitable to handle feed composition disturbance in dynamic simulation. There will be an economic trade-off between design and control in this case. A similar discussion of trade-off between design and control can be seen from Chen et al. (2015). Thus, to allow handling feed composition disturbances, the feed ratio is set at FS/FF = 1.00, with 12 extractive stages applied to reach the specification of water product at STR bottom.

After this hybrid extraction-distillation process is obtained, the optimal design variables in the last product column (COL-2) can easily be determined by varying total stages and feed location to minimize TAC of this column. The resulting optimal total stages are 29 and feed from bottoms of COL-1 is entered into this column at 10th stage. The detailed information of the design flowsheet is shown in Fig. 13.5.

13.3.4 Comparison Between Hybrid Extraction-Distillation Process and Sequenced-Separation Process

In the sequenced-separation process illustrated in Fig. 13.6, several important variables are also needed to be discussed. The first one is the absorber stage (NT-ABS). Similar to the hybrid process, NT-ABS affects the water flowrate input to the absorber under a given specification (0.01% AA loss) and then further influences the separation performance in COL-3. Because in this part the separation performance of two processes are compared, it is reasonable to assume that the NT-ABS in sequenced-separation process is the same as in hybrid process (NT-ABS = 11). The optimization of COL-3 and COL-4 can be performed independently with each other once the separation targets for each column are specified. The specifications are also listed in Table 13.3.

In this section, TAC for separation of water and acid mixture in two processes are compared. In hybrid process, the included items are ABS, STR, COL-1, HX, Decanter, and vacuum system. While in sequenced-separation process, only ABS, COL-3, and vacuum system are included. Because of the same specification for AA/ACE mixture from COL-1 bottoms in the hybrid process and COL-3 bottoms in the sequenced-separation process, the further separation of them is considered to have minor difference in TACs. It is assumed that the previously studied optimal NT and NF from COL-2 in the hybrid process hold in COL-4 of the sequenced-separation process. Thus, the further AA/ACE separation part is scratched from this comparison.

Under given column pressure of COL-3 (0.13 bar), two design variables are left to be determined, namely, total number of trays (NT) and feed location (NF). The optimal NT-COL-3 is 42 and NF-COL-3 is 25. The plot of TAC results can be easily obtained and are not shown here. Also note that the AA losses in two processes should be taken into consideration when comparing TACs. In this work, the loss rate of AA in sequenced-separation process is 0.06 (kmol/h) more than in the hybrid extraction-distillation process, which corresponds to about 80.5 (kUSD) loss annually. Thus this loss is added to the operating cost in sequenced-separation process. The unit price of AA is set at 1.95 USD/kg and can be obtained from ICIS, 2014.

The comparison of TAC between the optimized hybrid extraction-distillation process and the sequenced-separation process is illustrated in Table 13.5. From the result, the hybrid process saves about 10.94% in TAC. From the results, it clearly shows that the addition of solvent can enhance the performance of separation. Although the design configuration is more

TABLE 13.5 Total annual Cost (TAC) Comparison in Between Hybrid Extraction-Distillation Process and Sequenced-Separation Process

	Hybrid Extraction-Distillation Process		Sequenced-Separation Process	
Capital Cost (kUSD)				
ABS	193.066		193.088	
STR	Column	10.362	–	
	Reboiler	46.498	–	
	Cooler	3.615	–	
COL-1/ COL-3	Column	431.289	Column	1057.383
	Condenser	373.291	Condenser	980.043
	Reboiler	240.875	Reboiler	97.499
EXT	389.179		–	
Decanter	452.989		–	
Vacuum system	3.523		3.544	
Operating Cost (kUSD/year)				
LP-steam	1525.495		2313.735	
Cooling water	–		102.688	
Chilled water	781.685		–	
Vacuum steam	128.014		269.653	
Solvent makeup	11.301		–	
AA loss	–		80.504	
Total Cost				
Capital (kUSD)	2144.687		2331.557	
Operating (kUSD/y)	2446.495		2766.580	
TAC (kUSD/y)	3156.039 (−10.94%)		3543.765	

AA, acrylic acid.

complicated and higher-ranked cooling medium (chilled water) is needed in hybrid process, it is still more economically attractive.

The author would also like to mention that the advantage of using hybrid extraction-distillation is much more obvious when there is more water inside the system. In this work, the steam added is cut from original value to 1 kg steam per 1 kg propylene. Thus the cost for separation is already greatly reduced comparing to the process in Turton et al. (2009).

13.4 FURTHER COMMENTS ON ASPEN PLUS SIMULATION

In this section, helpful comments will be provided to simulate this integrated process using Aspen Plus.

13.4.1 Reaction Section

In this process, a fluidized-bed reactor is used for the AA production reaction. Because of its nature of circulating, it often provides an environment with homogeneous temperature distribution. There is also effective heat removal around inside the fluidized-bed reactor by the arrangement of a tube bundle (Fogler, 2006). Thus, this is especially suitable for a largely exothermic reaction. Also, in a real fluidized-bed reactor, the gas reactants enter from the bottom of reactor and go upward; thus the composition profile can be obtained inside of the reactor. Based on these backgrounds, it is reasonable to simulate the fluidized-bed reactor as an isothermal PFR. In this work, a 10% channeling of reactants inside the reactor is also adopted based on work by Turton et al. (2009). This can be simulated by a bypass stream using Aspen built-in FSPLIT unit. In this unit, the author assumed that 10% of the reactant gases bypasses from the reactor, but with the temperature and pressure to be the same as the reactor outlet. Thus a HEATER model and a valve are also needed for this purpose.

Note that the reactor pressure drop can be calculated based on Eq. (13.7) mentioned in Section 13.2.3, and this can be calculated manually. However, there is another smarter way to specify it through the built-in function "Flowsheeting Design-spec" or "Calculator." Here the author takes the "Flowsheeting Design-spec" function for example. In this function, the user needs to first import the variable mentioned in this equation (such as ΔP, ρ_s, ρ_g, and h) in the "Define" page. After that, the user can specify the pressure drop based on Eq. (13.7) in the "Spec" page. Next in "Vary" page, access the pressure drop input in the RPLUG module. Briefly, the pressure drop is calculated and transferred to RPLUG as an input through this function. This function is useful and effective for getting a converge run, especially in the cases that some variables are changing (or need to be changed) in simulation runs.

With these setting inside Aspen Plus mentioned above, the readers only need to assign the size and temperature of the reactor. A heat removal rate is calculated by Aspen Plus to maintain the reactor operated at the designated temperature. The summation of this heat removal rate and the one calculated by the HEATER mentioned above for the bypass stream will be the actually total heat removal rate in the fluidized-bed reactor. Although the assumptions greatly simplify the simulation works, it still can provide a qualitative simulation result of a fluidized-bed reactor.

There is one more thing to be carefully investigated, the unit of reaction rate and concentration of reactant. Sometimes large amount of efforts are needed for unit conversion because the units of reaction rates and concentration properties are mandatory to be specific in Aspen Plus. For Aspen built-in reactor models (including RCSTR, RPLUG, and RADFRAC for reactive distillation), the accepted units for reactions are kmol/m^3 s for expressing in volume basis, and kmol/kgcat s for expressing in catalyst weight basis. The unit of concentration is also important, and there is mandatory unit for each concentration property. For example, the partial pressure should be expressed in pascal like the one in this work. The complete information for the required units in kinetic expression can be found in Aspen built-in "Help" function.

13.4.2 Further Separation Section

In performing simulation work, to get a converging run as a base case study is of top priority. Once the base case simulation is obtained, the later optimization works and dynamic studies can be performed. However, in a system containing recycle stream, especially with liquid–liquid separation, convergence is often difficult. Here the author provides two tips to help converging.

Firstly, a good initial guess as inputs in Aspen Plus can be helpful in getting a converge run. Thus, how to get a good one(s) is important. Take the example from this work. The outputs from decanter are important in the process because they are recycled back to the former parts of the process (organic phase to extraction tower, and aqueous phase to STR). To obtain a good initial guess, it is better to start from the ternary diagram. The prediction from tie-line can be a good starting point.

Secondly, instead of being the last step, the process makeup stream is suggested to be set in simulation at the beginning stage. Take another example in this process. The solvent to feed flow ratio is specified for a base case. When the recycle streams are tried to be connected, iteration of composition and temperature between the initially broken recycle streams are needed. If the makeup flow is not specified (or randomly specified), it might cause a problem in getting a converge run. The solvent flowrate is likely to be constantly decreased or increased as iteration goes on. As the ternary map illustrated, the prediction of organic phase and aqueous phase compositions depends on the tie-line. Thus a little change in decanter input may result in large variation in decanter output. This is how the problem affects the simulation works and is the reason why the author suggested specifying makeup flowrate at the beginning stage of simulation. For setting the makeup flowrates, Aspen built-in "Calculator Block" or "Flowsheeting Design Specs" can be helpful. The makeup flowrate can be specified as the summation of solvent flowrate that is lost from the product streams. The detailed settings of these calculations are left for readers to investigate.

13.5 CONCLUSIONS

In this chapter, the simulation of acrylic acid process was studied, with the relevant optimization work and the general concepts of determining the design and operating variables also illustrated.

Industrial-relevant integrated process for the production of acrylic acid was simulated using process simulation via Aspen Plus. Effect of important design variables for this process can easily be revealed once the simulation of the integrated process is established. Because of the difficulty in separation of acrylic acid and acetic acid with large amounts of water, a hybrid extraction-distillation process is introduced in this work to show its ability to be an energy-saving alternative design. From the simulation results, the proposed design performs better than the traditional sequenced-separation process by saving of 10.94% TAC.

Through these demonstration works, authors hope that the readers can have better understanding of the process design concept using Aspen Plus software.

EXERCISES

Exercise 1: Furfuryl Alcohol Production Process

Furfuryl alcohol (FOL) is an industrial chemical used primarily in binders for foundry sands in the production of cores and molds in metalworking. Furfuryl alcohol can be produced by hydrogenation of furfural (FAL) in the gas phase over a copper chromite catalyst.

Reactions and Kinetics

The desired reaction is the hydrogenation of furfural:

$$\text{Furfural(FAL)} + \text{H}_2 \rightarrow \text{Furfuryl Alcohol(FOL)}$$

Further hydrogenation of furfuryl alcohol produces the undesired by-product 2-methyl furan:

$$\text{Furfural alcohol(FOL)} + \text{H}_2 \rightarrow 2 - \text{methyl furan(2-MF)} + \text{H}_2\text{O}$$

The kinetics reported by Borts et al. (1986) using a copper chromite catalyst is employed:

$$r_{FOL} = 1.57 \times 10^{14} \exp\left(-\frac{10,740}{T}\right) C_{FAL} C_{H_2}^2$$

The second reaction is:

$$r_{2-MF} = 7.05 \times 10^{17} \exp\left(-\frac{19,000}{T}\right) C_{FOL} C_{H_2}$$

where the concentration of reactants $C_{FAL}, C_{FOL},$ and C_{H_2} are in mol/L and the temperature T is in Kelvin.

TABLE E1 NRTL Binary Parameters

Component i	FAL	FAL	FOL	2-MF	2-MF	2-MF
Component j	FOL	H_2O	H_2O	FOL	FAL	H_2O
a_{ij}	0	−4.7563	0	1.6859	0.007942	0
a_{ji}	0	4.2362	0	0.1097	−0.001577	0
b_{ij} (K)	69.0160	1911.4222	60.3941	−154.3	1319.0490	852.5402
B_{ji} (K)	24.0213	−262.2408	845.5429	−126.8	−279.9706	1915.0555
α_{ij}	0.3	0.3	0.3	0.3	0.3	0.3

FAL, furfural; *FOL*, furfuryl alcohol; *2-MF*, 2-methy-furan.

Thermodynamics

The NRTL model with the Hayden-O'Connell correlation was used to model the vapor–liquid equilibrium in the simulation. The numerical values of the NRTL binary interaction parameters are given in Table E1.

Process Design

The process flowsheet is shown in Figure E1. The main unit is the PFR, which is to be modeled using RPlug block in Aspen Plus. The material and energy balance equations were solved assuming perfect mixing in the radial direction and no mixing in the axial direction. Coolant temperature was assumed to be constant and heat transfer between the reactor tubes and the coolant was described by an overall heat-transfer coefficient. The reaction rate was converted to a reactor volume basis using an assumed catalyst density of 1.2 g/cm^3.

The feed to the reactor is assumed to be hydrogen that is nearly saturated with furfural vapor at the reactor inlet temperature and pressure (135 °C and 1.1 bar). The feed ratio of hydrogen to furfural at the reactor inlet was considered as a design variable.

The procedure for the design of the reactor then is as follows:

1. The coolant temperature was set to 140°C.
2. The tube diameter was set to 3 cm, the smallest value that was thought to be reasonable, to provide maximum area for heat transfer.
3. A value for the catalyst dilution factor was specified.
4. The number of tubes and the length of each tube were determined to achieve 99% conversion of furfural with a reasonable overall reactor aspect ratio.
5. The maximum temperature difference between the reactor and jacket was determined. If the value was greater than 10 °C, the catalyst dilution factor was increased and the procedure was repeated from step 3.

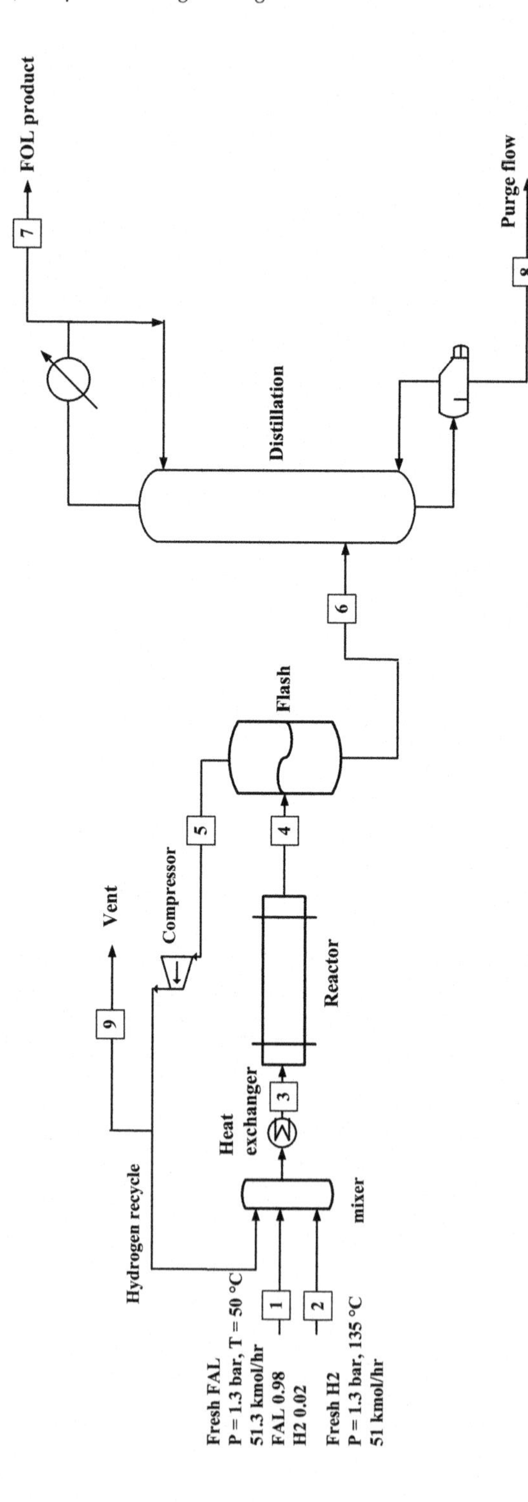

FIGURE E1 Flowsheet of furfuryl alcohol production process.

The process requires a single distillation column to separate the product furfural alcohol from the unconsumed furfural and the by-products water and 2-methylfuran. It is advantageous to operate the column at reduced pressure because the relative volatility between furfural and furfural alcohol is larger at lower pressure. The top column pressure was set at 0.2 atm so that cooling water could be used to cool the distillate. The column bottom composition was set at 99.2% furfural alcohol by adjusting the bottom flowrate, and the mole fraction of furfural alcohol in the top product was set to 1% by adjusting the reflux ratio.

Please simulate the process flowsheet using Aspen Plus to determine the optimal result with minimum TAC. (Hint: the setting of the reactor and distillation column can be seen in Tables E2 and E3. The equations of TAC calculation can be found in Appendix.)

Exercise 2: 2-Methylfuran and Furfuryl Alcohol Coproduction Process

2-Methylfuran is used as a solvent and as a precursor in the manufacture of various specialty chemicals. Because the demand for 2-methylfuran is considerably less than the demand for furfuryl alcohol, and because 2-methylfuran

TABLE E2 Reactor Configuration for Furfuryl Alcohol Process

Reactor Property	Value
Heat-transfer coefficient	90.85 W/m^2K
Catalyst dilution factor	200
Number of tubes	6000
Length of each tube	4.66 m
Diameter of each tube	0.03 m
Inlet pressure	1.1 bar
Pressure drop	0.027 bar
Inlet temperature	135°C
Bed voidage	0.5
Catalyst bulk density	1.2 g/cm^3
Catalyst pellet size	0.01 m
Critical temperature difference	17°C
Residence time	9.45 s

TABLE E3 Distillation Column Configuration

Property	Value
Total number of stages	12
Feed stage	9
Condenser	Partial vapor
Tray diameter	0.66 m
Top stage pressure	0.2 atm
Bottom stream specification	99.2 mol% of FOL
Bottom stream flowrate	50 kmol/h
Reboiler duty	375.05 kW
Reflux ratio	5.27

FOL, furfuryl alcohol.

is produced as a by-product (coproduct) in the production of furfuryl alcohol, it is not economical to produce 2-methylfuran with a dedicated process. Instead, furfuryl alcohol and 2-methylfuran are produced in the same process by the hydrogenation of furfural.

Kinetic and Thermodynamic Models

Furfuryl alcohol and 2-methylfuran can be produced by the hydrogenation of furfural in the gas phase. The reactions are:

$$\text{Furfural(FAL)} + H_2 \rightarrow \text{Furfuryl Alcohol(FOL)}$$

$$\text{Furfural alcohol(FOL)} + H_2 \rightarrow \text{2-methyl furan(2-MF)} + H_2O$$

A number of researchers have studied the hydrogenation of furfural over various catalysts; unfortunately, few of them also considered the reaction that produces 2-methylfuran. Brown and Hixon (1949) reported a set of data for the yield of furfuryl alcohol and 2-methylfuran in a reactor packed with another copper chromite catalyst at temperatures between 100 and 164 °C. Based on their data, the following kinetic model was shown:

$$r_{FOL} = 9.98 \times 10^7 \exp\left(-\frac{4166}{T}\right) C_{FAL}\, C_{H_2}^2 \tag{E1}$$

$$r_{2-MF} = 4.09 \times 10^{15} \exp\left(-\frac{14838}{T}\right) C_{FOL}\, C_{H_2} \tag{E2}$$

where concentrations are in mol/L (kmol/m^3) and the reaction rates are kmol/s m^3.

Vapor–liquid and vapor–liquid–liquid equilibria were modeled with the NRTL equation in Aspen Plus. Values of the NRTL parameters used in this work can be taken from Table E1 of Exercise 1.

The Guideline of Process Design

The design flowsheet for this process is shown in Figure E2. The major difference between the process considered in this work and compared to that in Exercise 1 is that, the reactor is operated at higher temperature, so useful quantities of 2-methylfuran are produced in addition to furfuryl alcohol.

In the process, fresh furfural and hydrogen are fed together with recycled hydrogen to a reactor in a molar ratio such that the partial pressure of furfural is less than the saturation pressure. A high conversion of furfural >99% is achieved in the reaction. The product distribution is determined primarily by the reactor temperature, with higher temperature favoring the production of 2-methylfuran. The reactor effluent is cooled and unconsumed hydrogen is recycled to the reactor. The liquid phase containing furfuryl alcohol, 2-methylfuran, and water is fed to a vacuum distillation column (0.1 bar). Furfuryl alcohol is collected at the bottom, while 2-methylfuran and water are collected at the top. Because water and 2-methylfuran are almost completely immiscible, a single decanter is sufficient to separate them with high purity.

The process is designed to produce furfuryl alcohol and 2-methylfuran with a total production rate of 50 kmol/h. The conversion of furfural was 99.5% when the product distribution was 10% 2-methylfuran. This guarantees that the conversion of furfural will be at least 99.5%, since a higher yield of 2-methylfuran is achieved by increasing the reactor temperature. Reactor tube diameter was specified to be 3 cm, and the number of tubes is set to 2400 and the tube length was determined to get 99.5 mole percent conversion of furfural entering into the reactor. Properties of the reactor are shown in Table E4.

The distillation column is to be designed by adjusting the total number of trays and the feed tray location to minimize the total annual cost, while adjusting the reboiler and condenser duties to maintain the desired composition of furfural at the top and bottom of the column. A single decanter, operating at 20 °C and 1 bar, could produce high-purity 2-methylfuran. Please simulate this process by using the flowsheet in Figure E2. Both furfuryl alcohol and 2-methylfuran products should have a concentration of to 99.2 mol %. The equations of TAC calculation is given in Appendix.

APPENDIX A: COST EQUATIONS OF TWO EXERCISES

This appendix lists equations used in determining the capital and operating costs of two exercises. The value of the M&S index used in the calculations is 1500.

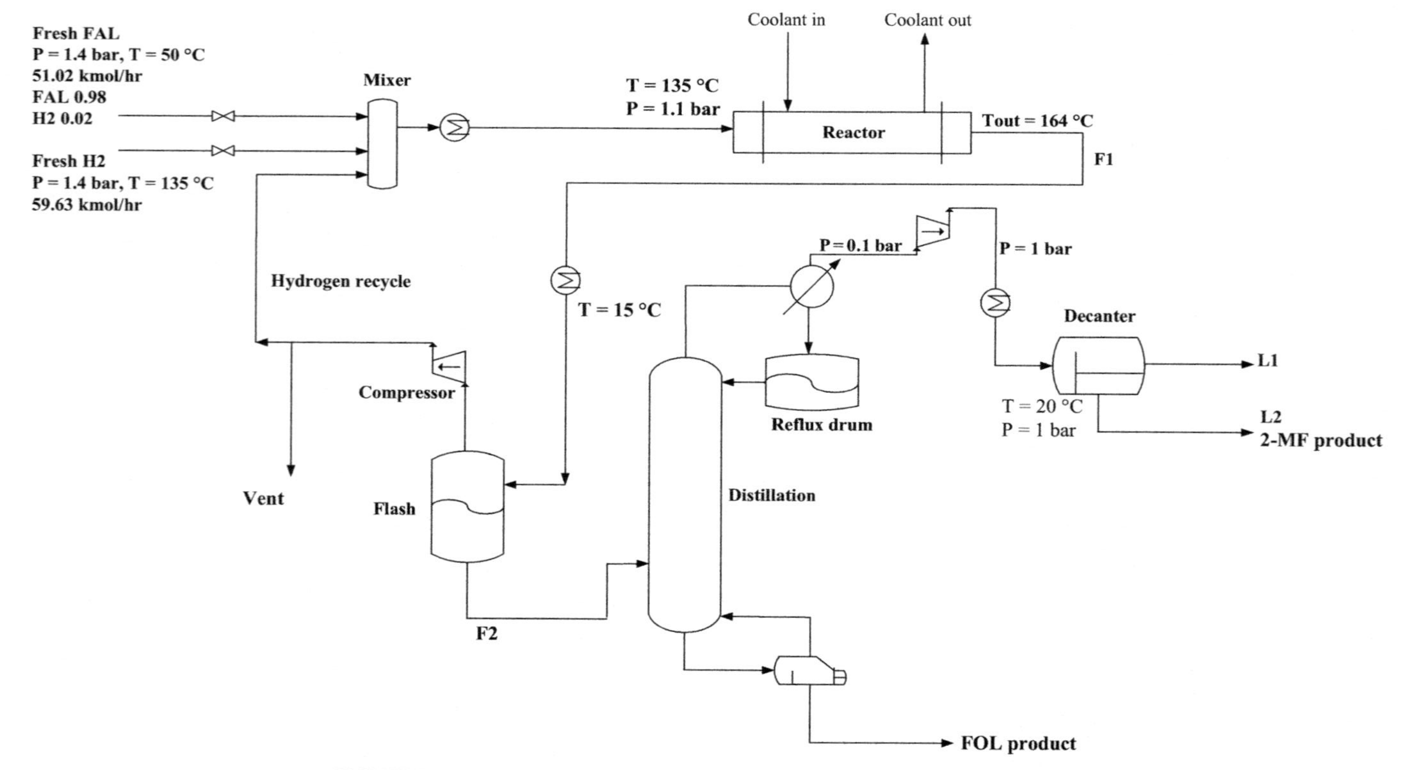

FIGURE E2 Flowsheet of 2-methylfuran and furfuryl alcohol coproduction process.

TABLE E4 Reactor Properties of 2-Methylfuran and Furfuryl Alcohol Coproduction Process

Reactor Property	Value
Heat-transfer coefficient	140 W/m^{2}K
Number of tubes	2400
Length of each tube	3.89 m
Diameter of each tube	0.03 m
Inlet pressure	1.1 bar
Pressure drop	0.030 bar
Inlet temperature	135 °C
Bed voidage	0.5
Catalyst bulk density	1.2 g/cm^3
Catalyst pellet size	0.01 m
Residence time	2.06 s

A1. Reactor and Column Shell Cost

$$\text{Reactor and column shell cost [\$]} = \frac{\text{MS}}{280} \times 101.9 \times \text{D}^{1.066}\text{H}^{0.802} \times (2.18 + (1.0))$$

$$\text{Catalyst cost}\left[\frac{\$}{\text{kg}}\right] = 100$$

$$\text{H} = \text{height, ft}$$

$$\text{D} = \text{diameter, ft}$$

A2. Heat Exchanger Cost

$$\text{Heat exchanger cost [\$]} = \frac{\text{MS}}{280} \times 101.3 \times \left(\text{A}^{0.65}\right) \times (1.35 \times 1.0)$$

$$\text{A} = \text{area, ft}^2$$

A3. Cooling Water Cost

$$\text{Cooling water cost}\left[\frac{\$}{\text{year}}\right]=\frac{\$0.354}{\text{GJ}}(Q_c)\left(8000\frac{\text{h}}{\text{year}}\right)$$

$$Q_c = \text{cooling duty, GJ/h}$$

A4. Flash Cost

$$\text{Flash cost [\$]}=\frac{\text{MS}}{280}\times 101.9\times D^{1.066}H^{0.802}\times(2.18+(1.35))$$

$$H = \text{height, ft}$$
$$D = \text{diameter, ft}$$

A5. Compressor Cost

$$\text{Compressor installed cost [\$]}=\frac{\text{MS}}{280}\times 517.5\times \text{bhp}^{0.82}\times(2.11+1)$$

$$\text{Compressor operating cost}\left[\frac{\$}{\text{year}}\right]=\frac{\$0.06}{\text{kWh}}(E)\left(8000\frac{\text{h}}{\text{year}}\right)$$

$$E = \text{electricity, kW}$$

A6. Pump Cost

$$\text{Pump operating cost}\left[\frac{\$}{\text{year}}\right]=\frac{\$0.06}{\text{kWh}}(E)\left(8000\frac{\text{h}}{\text{year}}\right)$$

$$E = \text{electricity, kW}$$

A7. Furnace Cost

$$\text{Furnace installed cost [\$]}=\frac{\text{MS}}{280}\times 5.52\times 10^3\times Q^{0.85}\times 2.27$$

$$\text{Furnace operating cost}\left[\frac{\$}{\text{year}}\right]=\frac{\$6.88}{\text{GJ}}(Q)\left(8000\frac{\text{h}}{\text{year}}\right)$$

$$Q = \text{heat duty, GJ/h}$$

A8. Column Tray and Tower Internals Cost

$$\text{Cost } \$ = \frac{\text{MS}}{280} \times 4.7 \times \text{D}^{1.55}\text{H} \times 4.5$$

$$\text{H} = \text{height, ft}$$

$$\text{D} = \text{diameter, ft}$$

REFERENCES

Borts, M.S., Gilchenok, N.D., Ignatev, V.M., Gurevich, G.S., 1986. Kinetics of vapor-phase hydrogenation of furfural on a copper chromium catalyst. Journal of Applied Chemistry of the USSR 59, 114–117.

Brown, H.D., Hixon, R.M., 1949. Vapor phase hydrogenation of furfural to furfuryl alcohol. Industrial and Engineering Chemistry 41, 1382–1385.

Chen, Y.C., Li, K.L., Chen, C.L., Chien, I.L., 2015. Design and control of a hybrid extraction-distillation system for the separation of pyridine and water. Industrial and Engineering Chemistry Research 54, 7715–7727.

Fogler, H.S., 2006. Elements of Chemical Reaction Engineering, fourth ed. Prentice Hall PTR, Upper Saddle River, N. ew Jersey (Chapter 12).

ICIS Pricing: Acrylic Acid/Acrylate Esters (Asia Pacific), 8th, January 2014.

Kürüm, S., Fonyo, Z., Kut, Ö.M., 1995. Design strategy for acetic acid recovery. Chemical Engineering Communications 136, 161.

Lin, M.M., 2001. Selective oxidation of propane to acrylic acid with molecular oxygen. Applied Catalysis A: General 207, 1–16.

Luyben, W.L., 2011. Principles and Case Studies of Simultaneous Design (Chapter 5). Wiley, NewYork.

Seider, W.D., Seader, J.D., Lewin, D.R., Widagdo, S., 2009. Product and Process Design Principles, third ed. Wiley, Hoboken, New Jersey.

Straathof, A.J.J., Sie, S., Franco, T.T., van der Wielen, L.A.M., 2005. Feasibility of acrylic acid production by fermentation. Applied Microbiology and Biotechnology 67, 727–734.

Suo, M., Zhang, H., Ye, Q., Dai, X., Yu, H., Li, R., 2015. Design and control of an improved acrylic acid process. Chemical Engineering Research and Design 104, 346–356.

Turton, R., Ballie, R.C., Whiting, W.B., Shaeiwitz, J.A., 2009. Analysis, Synthesis, and Design of Chemical Processes, third ed. Pearson, Boston, MA.

Chapter 14

Design and Simulation of Reactive Distillation Processes

Hao-Yeh Lee, Tyng-Lih Hsiao
National Taiwan University of Science and Technology, Taipei, Taiwan

14.1 INTRODUCTION

Reactive distillation (RD) provides an attractive alternative for process intensification, especially for reaction/separation systems with reversible reactions. The literature in RD has grown rapidly in recent years and the books by Malone and Doherty (2000), Sundmacher and Kienle (2003), and Luyben and Yu (2008) give updated summaries in this area of research. As pointed out by Doherty and Buzad (1992) the concept of combining reaction and separation has been recognized for a long time. However, it is rarely put into commercial practice and only few successful applications for methyl acetate production have been reported (Agreda et al., 1990). One of the reasons stated by Doherty and Buzad (1992) is, "There is almost always a conventional alternative to reactive distillation, which is seductive because we have always done it this way." This scenario remains for more than a decade. Another reason is that, the process flowsheets are always unique from case to case (e.g., methyl acetate, to ethyl acetate, to butyl acetate etc.; (Agreda et al., 1990; Al-Arafaj and Luyben, 2002; Burkett and Rossiter, 2000; Kenig et al., 2001; Hanika et al., 1999; Gangadwala et al., 2004)), in which generalization does not exist. Moreover, the flowsheet configurations vary as one went through the literature, even for the same chemical production (e.g., ethyl acetate, butyl acetate). Unlike conventional distillation, the seemingly case-based design approach and multiple process configurations add additional complexity to the RD process.

Esters are of great importance to the chemical process industries. They are typically produced from the reactions between acids and alcohols in an acidic condition. To illustrate the determination of an optimal design for each case, this chapter explores the esterification of acetic acid with different alcohols, i.e., methanol, butanol, and isopropanol. However, these quaternary systems including alcohol, acid, ester, and water show complex phase behavior with the homogeneous and heterogeneous azeotropes.

Chemical Engineering Process Simulation. http://dx.doi.org/10.1016/B978-0-12-803782-9.00014-5

Unlike the conventional distillation processes, tray holdup can be an important design parameter in the reaction zone of the RD column because of its effect on the reaction rates. Tray holdup can be calculated through column diameter and weir height. However, column diameter actually depends on the vapor rates that vary with tray holdup due to the interaction between reaction and separation in the RD column. Therefore, column sizing becomes more complex than conventional distillation processes.

In the following sections, three esterification reactions are demonstrated using Aspen Plus. First, a methyl acetate RD process is shown and is based on the work of Tang et al. (2005). Next, a butyl acetate RD process is demonstrated. In this case, additional decanter is added to the RD top. Therefore, an external organic reflux needs to be estimated before the simulation begins. The final case is an isopropyl acetate RD process with thermally coupled configuration. In this case, the thermally coupled configuration is introduced to demonstrate the reduction of remixing effect phenomenon.

14.2 METHYL ACETATE REACTIVE DISTILLATION PROCESS

14.2.1 Thermodynamic Model

To account for nonideal vapor–liquid equilibrium (VLE) and possible vapor–liquid–liquid equilibrium (VLLE) for this quaternary system, it uses the UNIQUAC model. Moreover, UNIQUAC model is shown as follows:

$$\ln \gamma_i = \ln\frac{\Phi_i}{x_i} + \frac{z}{2}q_i \ln\frac{\theta_i}{\Phi_i} + l_i - \frac{\Phi_i}{x_i}\sum_{j=1}^{nc} x_j l_j + q'_i\left[1 - \ln\left(\sum_{j=1}^{nc}\theta'_j\tau_{ji}\right) - \sum_{j=1}^{nc}\frac{\theta'_j\tau_{ij}}{\sum_{k=1}^{nc}\theta'_k\tau_{kj}}\right] \tag{14.1}$$

$$\Phi_i = \frac{r_i x_i}{\sum_{k=1}^{nc} r_k x_k},\ \theta_i = \frac{q_i x_i}{\sum_{k=1}^{nc} q_k x_k},\ \theta'_i = \frac{q'_i x_i}{\sum_{k=1}^{nc} q'_k x_k}$$

$$\tau_{ij} = a_{ij} + \frac{b_{ij}}{T} + c_{ij}T,\ l_i = \frac{z}{2}(r_i - q_i) - (r_i - 1) \text{ and } z = 10$$

Because of the almost atmospheric pressure, the vapor phase nonideality considered is the dimerization of acetic acid as described by the Hayden and O'Connell (1975) second virial coefficient model. The Aspen Plus built-in association parameters are used to determine the fugacity coefficients. The model parameters are listed in Table 14.1.

It should be emphasized that the quality of the model parameters is essential to generate correct process flowsheet (Huang and Yu, 2003). Two important steps to validate model parameters are (1) good prediction of

TABLE 14.1 Binary Parameters of UNIQUAC and HOC Models for MeAc System: (A) UNIQUAC Parameters; (B) HOC Parameters (Pöpken et al., 2000)

Component *i*	HAc (1)	HAc (1)	HAc (1)	MeOH (2)	MeOH (2)	MeAc (3)
Component *j*	MeOH (2)	MeAc (3)	H_2O (4)	MeAc (3)	H_2O (4)	H_2O (4)
a_{ij}	−0.97039	0.43637	0.051007	0.71011	−3.1453	−0.010143
a_{ji}	2.0346	−1.1162	0.29355	−0.72476	2.0585	−0.96295
b_{ij}(K)	−390.26	62.186	−422.38	−62.972	575.68	−593.7
b_{ji}(K)	−65.245	−81.848	98.12	−326.2	−219.04	265.83
c_{ij}(K^{-1})	0.0030613	−0.00027235	0.00024019	−0.001167	0.0060713	0.0021609
c_{ji}(K^{-1})	−0.003157	0.0013309	0.000076741	0.0023547	−0.0070149	−0.00020133

Component *i*	HAc (1)	MeOH (2)	MeAc (3)	H_2O (4)
Component *j*				
HAc (1)	4.5	2.5	2	2.5
MeOH (2)	2.5	1.63	1.3	1.55
MeAc (3)	2	1.3	0.85	1.3
H_2O (4)	2.5	1.55	1.3	1.7

HAc, acetic acid; *MeAc*, methyl acetate; *MeOH*, methanol.

TABLE 14.2 Azeotrope Data and Boiling Point for MeAc Esterification System

	Experimental Data		Computed Data	
Component	Mole Fraction	Temperature (°C)	Mole Fraction	Temperature (°C)
MeOH/ MeAc	(0.359, 0.641)	54	(0.3407, 0.6593)	53.65
MeAc/H_2O	(0.8804, 0.1196)	56.4	(0.8904, 0.1096)	56.43
H_2O	1	100.02	1	100.02
MeOH	1	64.53	1	64.53
MeAc	1	57.05	1	57.05
HAc	1	118.01	1	118.01

HAc, acetic acid; *MeAc*, methyl acetate; *MeOH*, methanol.

azeotropes and (2) reasonable description of the liquid–liquid (LL) enveloped for VLLE systems. The correct description of the existence of azeotropes and the ranking of azeotrope temperatures will lead to generating correct residue curve map (RCM), and consequently, placing separators in the right sections. A reasonable prediction of LL envelope (LLE) can facilitate possibly by using the decanter, which is often encountered in esterification RD systems (Hanika et al., 1999; Steinigeweg and Gmehling, 2002; Chiang et al., 2002; Huang and Yu, 2003; Gangadwala et al., 2004). This is also the reason why the UNIQUAC model is preferred for activity coefficients. Experimental azeotrope data are shown in Table 14.2 by Horsley (1973).

The phase equilibria result of minimum boiling binary azeotropes is shown in Fig. 14.1. Significant LLE can also be observed, especially for the acid-free ternary systems.

14.2.2 Kinetic Model

Solid catalysts, acidic ion-exchange resin, such as Amberlyst 15 (Rohm and Hass) and Purolite CT179 (Purolite), are used in this study. The solid catalyst can either be immobilized via structured packings, such as Katapak-S (Sulzer Chemtech); (Steinigeweg and Gmehling, 2002; Taylor and Krishna, 2000), or simply be placed on the tray with a certain type of replacement mechanism (Davy Process Technology).

The reaction kinetics data are listed in Table 14.3. The reaction rates are expressed in the pseudohomogeneous model and, generally, the components

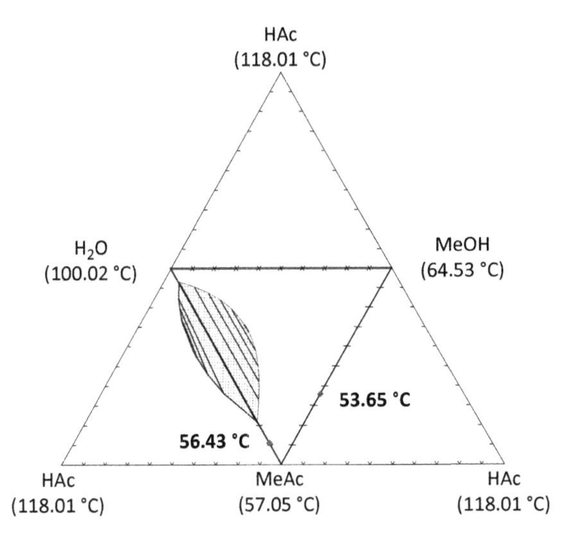

FIGURE 14.1 Liquid–liquid envelopes, azeotrope temperatures, and boiling point temperatures of pure component of MeAc system.

are represented in terms of activity. Moreover, it is catalyst weight (m_{cat}) based on kinetics. The equilibrium constant (K_{eq}, at 363K) ranges from 1.6 to 16.8, the forward rate constant (k_1, at 363K) changes from 1.73×10^{-4} to 2.5×10^{-3} kmol/(kg_{cat}*s), and activation energy of the forward reaction varies from 44,000 to 70,000 kJ/kmol. In applying the reaction kinetics to an RD, it is assumed that the solid catalyst occupies 50% of the tray holdup volume and a catalyst density of 770 kg/m^3 is used to convert the volume into catalyst weight (m_{cat}).

TABLE 14.3 Kinetic Equations for MeAc Esterification System (Pöpken et al., 2000)

System	Kinetic Model (Catalyst)	K_1 ($T = 363$K)	K_{eq} ($T = 363$K)
MeAc	Pseudohomogeneous model (Amberlyst 15) $r = m_{cat}(k_1 a_{HAc} a_{MeOH} - k_{-1} a_{MeAc} a_{H_2O})$ $k_1 = 2.961 \times 10^4 \exp\left(\frac{-49190}{RT}\right)$ $k_{-1} = 1.348 \times 10^6 \exp\left(\frac{-69230}{RT}\right)$	2.49×10^{-3} [kmol/(kg_{cat}*s)]	16.76

$R = 8.314$ [kJ/kmol/K], T[K], r[kmol/s], m_{cat}[kg_{cat}], C_i[kmol/m^3], a_i: activity. *HAc*, acetic acid; *MeAc*, methyl acetate; *MeOH*, methanol.

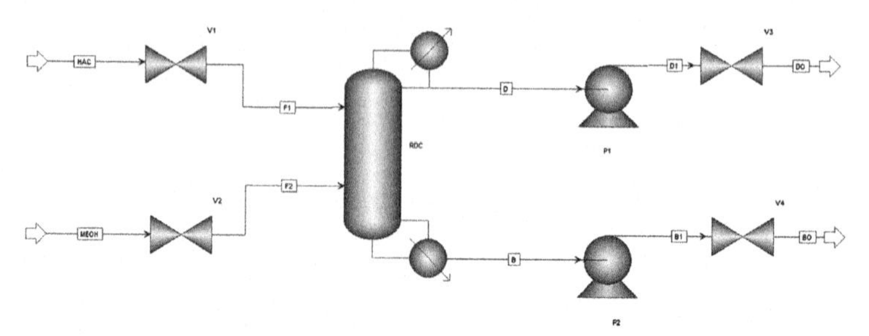

FIGURE 14.2 The flowsheet of MeAc process configuration.

14.2.3 Process Configuration

14.2.3.1 Description of Process

The MeAc RD configuration is shown in Fig. 14.2, and its simulation setting in Aspen Plus is given in Table 14.4. In a typical RD, the heavy reactant is fed into the top of the reactive section, and the light reactant goes into the bottom of the reactive section. For methyl acetate (MeAc) system, the heavy reactant corresponds to the acid (HAc) while alcohol (MeOH) serves as the light reactant. If the reactive section consumes all the acid, we deal with the separation of the H_2O/MeOH/MeAc, an almost ternary system, in the stripping section. It indicates that the more of the HAc be consumed, the purer H_2O would be obtained in the bottom of RD column. On the other hand, if we react away most of the alcohol toward the top section of the reactive zone, in the rectifying section, we separate HAc/MeAc with a small amount of H_2O. It is

TABLE 14.4 Variables Setting of the Distillation Column, Pumps, and Valves

Column Settings			Value/Type	
Number of stages			38	
Condenser			Total	
Reboiler			Kettle	
Method of convergence			Strongly non-ideal liquid	
Operating specifications			Reboiler duty and reflux ratio	
Design Spec		Vary (Reboiler duty)		
Mole purity	0.98	Lower bound		60,000 W
Comp./Streams	MeAc/D	Upper bound		2,000,000 W
Pump Setting		Valve Settings	1	1.1138 atm
Pressure increase	2 atm		2	1.2545 atm

possible to obtain relatively high-purity MeAc. This MeAc process configuration is first proposed by Agreda et al. (1990), Malone and Doherty, (2000), and Al-Arafaj and Luyben, (2002).

The bottom heavy product is H_2O, and the top light product is MeAc. The azeotropes play little role in the separation sections except that the top product is a saddle (ultimately the RCM will end at the MeAc/H_2O azeotrope, 56.43°C). This may cause some problem in control and operation. Therefore, this MeAc process configuration can achieve a high purity of the products at both ends of the column.

Design parameters are the numbers of rectifying (N_R), stripping (N_S), and reactive trays (N_{rxn}). In addition to the tray numbers, another set of important design parameters are the feed tray locations (NF_{Acid} and $NF_{Alcohol}$; Huang and Yu, 2003; Huang et al., 2004). In theory, the equimolar feed flowrates ($F_{Acid} = F_{Alcohol}$) should be economically optimal.

It seems strange to devise a sequential design procedure when it may be done simultaneously. Extensive experience on RD simulation indicates that the algorithm for the RD is much less robust than the typical distillation. It is more practical to carry out the simulation sequentially. In the work, we assume the catalyst occupies 50% of the holdup volume in a reactive tray, the column diameter is sized from the shortcut method of Douglas (1998), and weir height is 0.1016 m. Thus, the catalyst weight will be fixed, once the column diameter is determined.

For the MeAc system, it is assumed that the feed rates are equimolar ($FR = 1$). The dominant variables for optimization are N_R, N_S, N_{rxn}, NF_{Heavy}, and NF_{light}. The results of optimization are shown in Fig. 14.3. It shows that N_R, N_S, and N_{rxn} have little impact on the value of total annual cost (TAC) because the differences between those variables are quite smaller. However, NF_{Heavy} and NF_{light} are major factors to affect the cost in this RD system, and according to the optimal result in Fig. 14.3B and C, the light reactant (MeOH) and the heavy reactant (HAc) should be fed into stage 13 and stage 36, respectively.

The optimal flowsheet is shown in Fig. 14.4. The summary of optimal designs for the MeAc system is shown in Table 14.5.

14.2.3.2 Theme Section

In this process, MeOH and H_2O will be separated in the rectifying section in the beginning. The concentration of the heavy acid is extremely low in the bottom of the reactive section, and the light alcohol is rare in the top of the reactive section. This kind of phenomenon makes the separation easy. The scenario is quite similar to the interaction between reaction section and separation section in the design of recycling plant (Cheng and Yu, 2003). The composition and temperature profiles are shown in Fig. 14.5. It also shows the fraction of total conversion in the reactive section, the shaded area in Fig. 14.5.

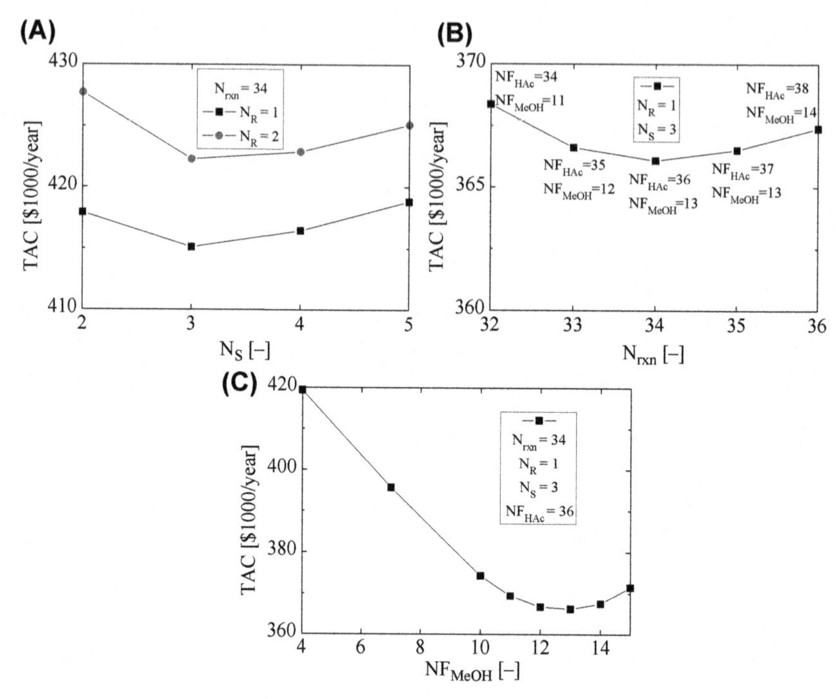

FIGURE 14.3 Effects of design variables on total annual cost (TAC) for MeAc system. (A) The optimal stages of rectifying and stripping sections, (B) the optimal stage of reactive section, and (C) the optimal feed location of MeOH.

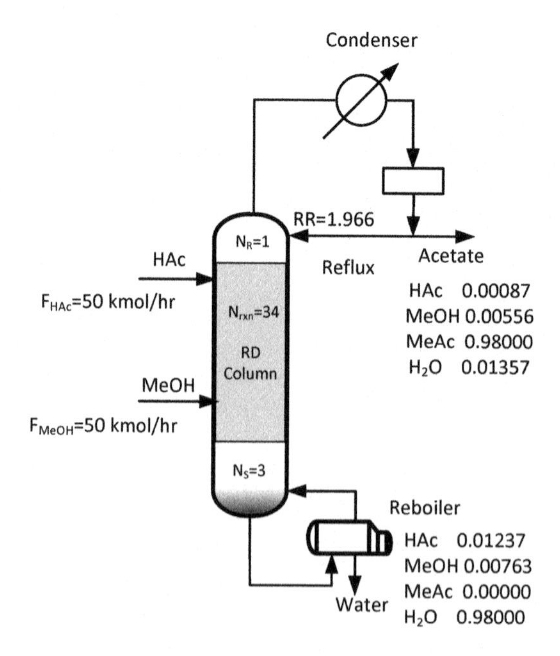

FIGURE 14.4 The optimal flowsheet of MeAc process.

TABLE 14.5 Steady-State Operating Condition and Total Annual Cost (TAC)

System	MeAc
Column configuration	**RD**
Total number of trays including the reboiler	39
Reactive tray	4–37
Acetic acid feed tray	36
Alcohol feed tray	13
Feed flowrate of acid (kmol/h)	50.00
Feed flowrate of alcohol (kmol/h)	50.00
Top product flowrate (kmol/h)	50.35
Bottom product flowrate (kmol/h)	49.65
x_D	
x_{MeAc}	0.98000
x_{H_2O}	0.01357
x_B	
x_{MeAc}	$<10^{-8}$
x_{H_2O}	0.98000
Condenser duty (kW)	−1280.22
Reboiler duty (kW)	1035.71
Condenser heat transfer area (m^2)	80.03
Reboiler heat transfer area (m^2)	38.63
Total capital cost ($1000)	730.78
Total operating cost ($1000/year)	122.49
TAC ($1000/year) (50 kmol/h)	366.08
TAC ($1000/year) (52825 ton/year)	644.96

MeAc, methyl acetate; *RD*, reactive distillation.

95% of the total conversion occurred in the reactive trays, from stage 10 to stage 20. The rest of reactive trays, especially toward the upper section of the reactive zone, consume most of the MeOH to facilitate the separation in a rarely MeOH environment. The temperature profile shows a non-monotonic temperature distribution. The reasons come from reaction occurred in the

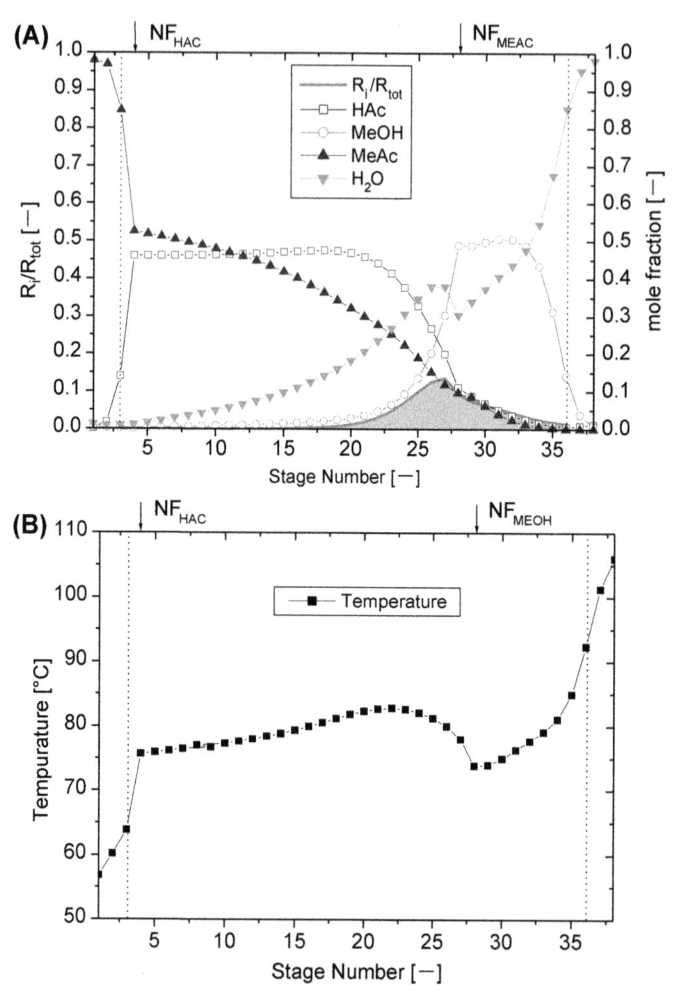

FIGURE 14.5 Composition and temperature profiles of MeAc system. (A) Composition profile and (B) temperature profile.

middle section of the column and the deliberate driving of the heavy reactant toward the top and light reactant to the bottom as shown in Fig. 14.5.

14.2.3.3 Reactive Holdup Calculation

The procedure for sizing the distillation column, column diameter, and weir height has already been specified in Section 14.2.3.1. The only remaining issue is the size of the reactive holdup. A commonly used heuristic of the reactive holdup is assumed that the catalyst loading is set as 50% full in the tray. Fig. 14.6A shows the relationship among the weir height, column diameter, and liquid holdup. The diameters can be found as 1.0327 m from tray sizing

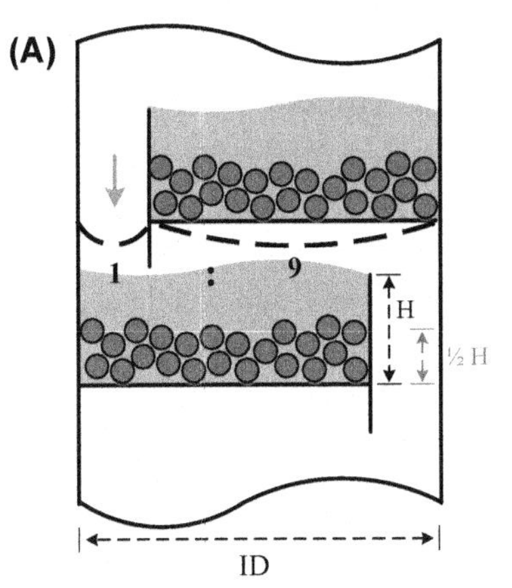

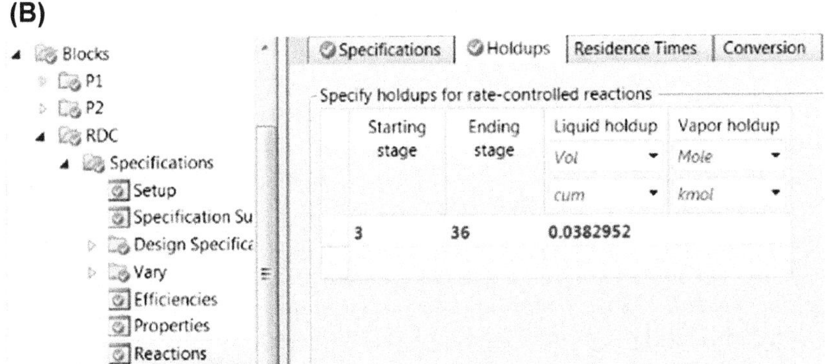

FIGURE 14.6 Specifying the reaction liquid holdup. (A) The relationship among the weir height, column diameter, and liquid holdup. (B) The settings of the reactive holdup in Aspen Plus.

result of Aspen Plus. The cross-sectional area of RD column can be calculated ($=\pi D^2/4$). Moreover, the volume of downcomer section should be eliminated. Notice that the ratio of downcomer volume in the tray is assumed as 0.1. Thus, the reactive holdup in this MeAc process can be calculated by Eq. (14.2).

$$V_r = \frac{\pi D^2}{4} \times \frac{h_{weir}}{2} \times (1 - R_{DC}) \tag{14.2}$$

where V_r, the reactive liquid holdup (m^3); D, overall diameter of distillation column (m); h_{weir}, the weir height of the tray (m); and R_{DC}, the ratio of downcomer volume in the tray.

According to the above equation, the reactive liquid holdup of this case can be calculated as Eq. (14.3) and set the value in Fig. 14.6B.

$$V_r = \frac{\pi \times 1.0327^2}{4} \times \frac{0.1016}{2} \times (1 - 0.1) \cong 0.03829517 \tag{14.3}$$

14.3 BUTYL ACETATE REACTIVE DISTILLATION PROCESS

14.3.1 Thermodynamic Model

To account for the nonideal VLE and possible VLLE for the BuAc quaternary systems, the UNIQUAC model is also used for activity coefficients. Table 14.6 lists the model parameters for the BuAc systems. Because of the almost atmospheric pressure, the vapor phase nonideality considered is the dimerization of acetic acid as described by the Hayden and O'Connell (1975) second virial coefficient model. The Aspen Plus built-in association parameters are used to compute fugacity coefficients.

Table 14.7 reveals that the model parameters give a reasonably good description of existing experimental azeotropic data (Horsley, 1973) for this system. Moreover, the models also predict ternary azeotropes in this quaternary system.

The BuAc quaternary system has three minimum boiling and one maximum boiling binary azeotrope and two ternary azeotropes. It is important to notice that, for EtAc, IPAc, BuAc, and AmAc systems, the minimum temperature inside the composition space is the *ternary minimum boiling azeotrope*. Significant LLEs are also observed for these four quaternary systems, especially for the acid-free ternary system (BuOH/BuAc/H_2O) (Fig. 14.7). The BuAc systems give type II envelopes as shown in Fig. 14.7. The tie lines slope into pure water node and it becomes more evident as the carbon number in the alcohol increases. That is, relatively pure water can be recovered from the LL separation and the separation (for achieving high-purity water) becomes easier as the carbon number in the alcohol increases (Tang et al., 2005). For BuAc systems, the ternary minimum boiling azeotrope lies inside the LLE and it moves further inside the LL zone as the carbon number increases. As will be shown later, this has important implications in the flowsheet development.

14.3.2 Reaction Kinetic Model

The esterification of the acetic acid with *n*-butanol can be expressed in the following general form:

$$\mathrm{HAc} + \mathrm{BuOH} \underset{k_{-1}}{\overset{k_1}{\Leftrightarrow}} \mathrm{BuAc} + \mathrm{H_2O}$$

The solid catalysts, Amberlyst 15 (Rohm and Hass) and Purolite CT179 (Purolite), are used. We have not seen too many kinetics data on solid-catalyzed esterification reactions until recently. Table 14.8 lists the reaction kinetics used in this study. The reaction rates are expressed in the pseudohomogeneous model and, generally, with the component represented in terms of activity. Moreover, they are all catalyst weight (m_{cat})—based kinetics. Despite the fact that different groups with the somewhat different type of catalysis performed the kinetic

TABLE 14.6 Activity Coefficient Model Parameters for BuAc System (Venimadhavan et al., 1999)

Component *i*	HAc (1)	HAc (1)	HAc (1)	EtOH (2)	BuOH (2)	BuAc (3)
Component j	BuOH (2)	BuAc (3)	H_2O (4)	BuAc (3)	H_2O (4)	H_2O (4)
b_{ij}(K)	66.315	150.193	172.92	−41.53679	−34.227	−345.098
b_{ji}(K)	−74.62672	−358.447	−265.6904	−12.4	−292.4746	−232.247

BuAc, butyl acetate; *BuOH*, butanol; *EtOH*, ethanol; *HAc*, acetic acid.

TABLE 14.7 Azeotrope Data and Boiling Point for BuAc Esterification System

	Experimental Data		Computed Data	
Component	Mole Fraction	Temperature (°C)	Mole Fraction	Temperature (°C)
[a]BuOH/BuAc/H_2O	(0.111, 0.135, 0.754)	89.4	(0.0864, 0.206, 0.7075)	90.68
[a]BuAc/H_2O	(0.278, 0.722)	90.2	(0.2823, 0.7177)	90.96
[a]BuOH/H_2O	(0.248, 0.752)	92.7	(0.2451, 0.7549)	92.62
H_2O	1	100.02	1	100.02
BuOH/BuAc	(0.73, 0.27)	116.2	(0.7847, 0.2153)	116.85
BuOH	1	117.68	1	117.68
HAc	1	118.01	1	118.01
HAc/BuOH/BuAc	—	—	(0.4182, 0.2396, 0.3423)	121.58
HAc/BuOH	—	—	(0.5359, 0.4641)	123.21
BuAc	1	126.01	1	126.01

BuAc, butyl acetate; *BuOH*, butanol, *HAc*, acetic acid.
[a]*Heterogeneous azeotrope.*

experiment, we observe a certain degree of consistency in the kinetic data. The equilibrium constant (K_{eq}, at 363K) ranges from 1.6 to 16.8, the forward rate constant (k_1, at 363K) changes from 1.73 10^{-4} to 2.5 to 10^{-3}, and activation energy of the forward reaction varies from 44,000 kJ/kmol to 70,000 kJ/kmol, and the heat of reactions are almost negligible except for the methyl acetate and ethyl acetate systems. In applying the reaction kinetics to a reactive distillation, it is assumed that the solid catalyst occupies 50% of the tray holdup volume and a catalyst density of 770 kg/m^3 is used to convert the volume into catalyst weight (m_{cat}).

14.3.3 Process Configuration

14.3.3.1 Description of Process

The type III flowsheet has been studied by several researchers (Chiang et al., 2002; Huang and Yu, 2003; Tang et al., 2005) in Fig. 14.8.

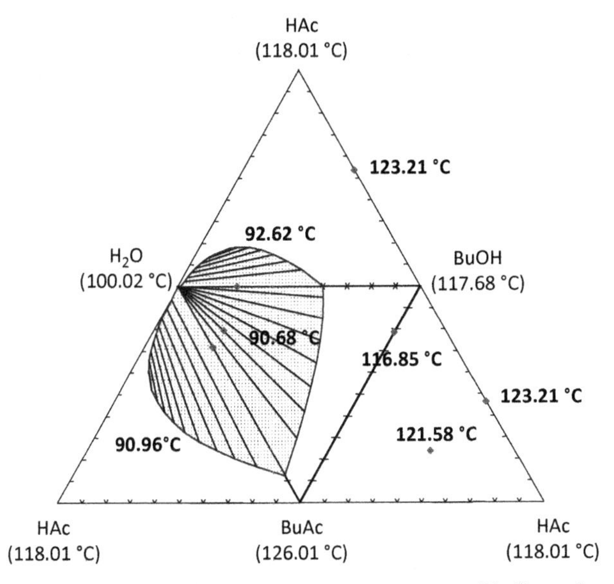

FIGURE 14.7 Liquid–liquid envelopes, azeotrope temperatures, and boiling point temperatures of pure component of BuAc system.

For this BuAc flowsheet, there is a recycle stream from the decanter to the column. It may have a problem if the condition of the internal stream is not specified before running the simulation. Therefore, here, a simple assumption is used to set the top RD composition near the ternary azeotrope. Based on the tie line and the lever rule, the flowrate and the compositions of the internal stream can be estimated as an initial guess as shown in Fig. 14.9.

TABLE 14.8 Kinetic Equations for BuAc Esterification System (Gangadwala et al., 2004)

System	Kinetic model (Catalyst)	k_1 ($T = 363$K)	K_{eq} ($T = 363$K)
BuAc	Pseudohomogeneous model (Amberlyst 15) $r = m_{cat}(k_1 a_{HAc} a_{BuOH} - k_{-1} a_{BuAc} a_{H_2O})$ $k_1 = 3.3856 \times 10^6 \exp\left(\frac{-70660}{RT}\right)$ $k_{-1} = 1.0135 \times 10^6 \exp\left(\frac{-74241.7}{RT}\right)$	2.32×10^{-4} [kmol/(kg$_{cat}$·s)]	10.9

$R = 8.314$ [kJ/kmol/K], T[K], r[kmol/s], m_{cat}[kg$_{cat}$], C_i[kmol/m^3], x_i[mole fraction]. *BuAc*, butyl acetate; *BuOH*, butanol, *HAc*, acetic acid.

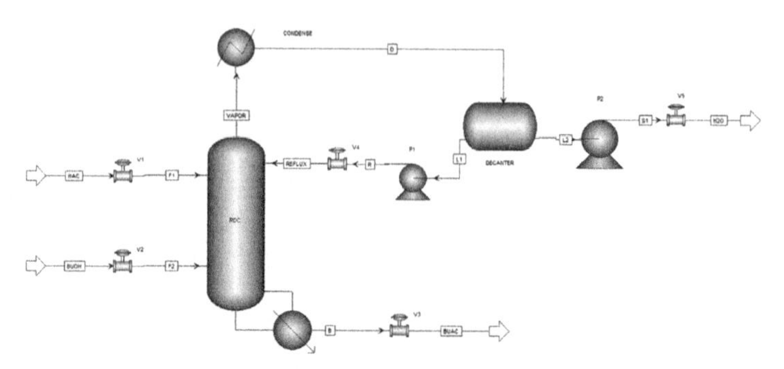

FIGURE 14.8 The possible flowsheet for BuAc system.

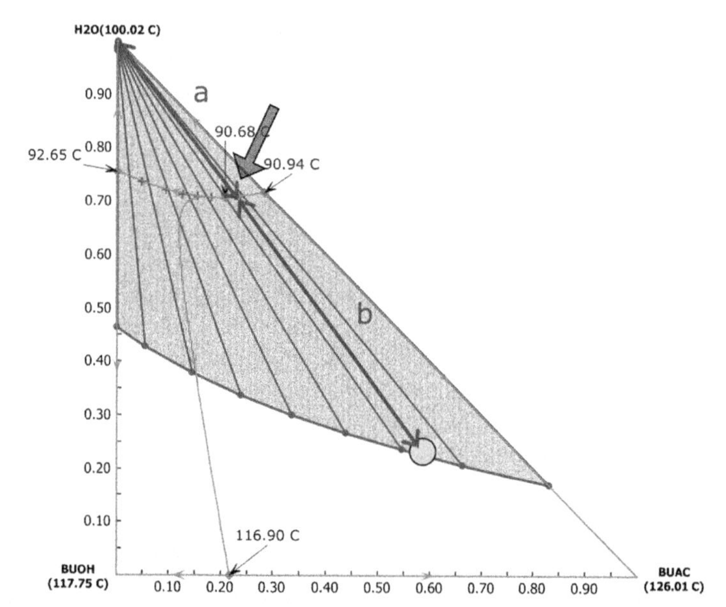

FIGURE 14.9 The initial guess of the organic flow rate and composition by using tie line and lever rule for the decanter of BuAc system.

The arrow point is the ternary azeotrope, which is assumed as the overhead composition of RD column. The lengths a and b can be estimated easily by hand. And the aqueous phase flowrate can be set as 50 kmol/h initially if the reaction conversion is assumed 100% in this system. Therefore, the organic recycle flowrate can be calculated by lever rule: organic recycle flowrate $= a/b \times 50$ kmol/h. The final value of the flowrate and the compositions are shown in Fig. 14.10.

In theory, we have the following optimization variables: N_R and N_S, N_{rxn}, and NF_{Heavy} and NF_{light}. For the BuAc system, the results clearly show that the *feed locations* are the most important optimization variables as shown in

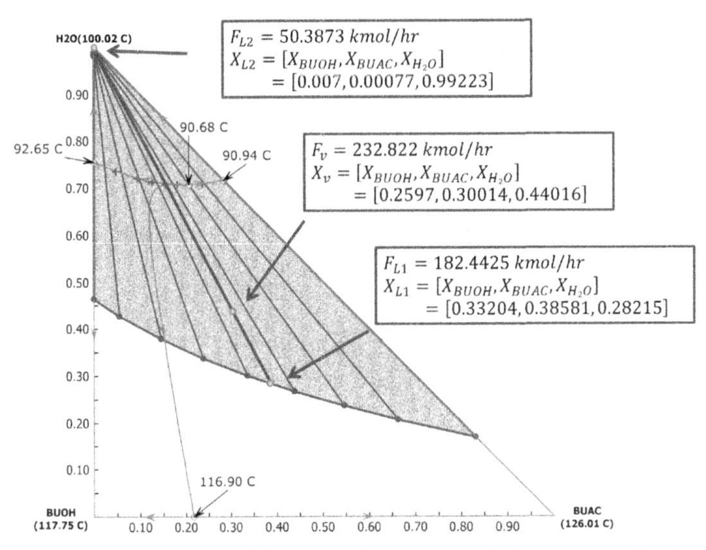

FIGURE 14.10 The final result of flowrate and compositions in the decanter of BuAc system.

Fig. 14.11 where significant TAC is saved by simply varying the feed locations. This can be understood because we need to arrange the feeds such that optimal reactant and temperature profiles can be achieved in the reactive section. The optimal flowsheet is shown in Fig. 14.12. The steady-state operating condition and the TAC are shown in Table 14.9.

For the BuAc system, the acetic acid has the second highest normal boiling temperature and, therefore, the acid is reacting away in the lower part of the reactive zone as shown in Fig. 14.13A. The composition profile in the stripping section clearly shows that the stripper is performing BuOH/BuAc separation. Because of a large LLE, only four rectifying trays are needed to drive the condensate of the top vapor into the LL zone. It is also observed that significant conversion occurs in the upper section of the reactive zone where two feeds are located nearby. The concentration effects of these two reactants enhance the reaction in that region. Fig. 14.13B shows an almost monotonic temperature for the reason that these two reactants are close-boiling components with 1°C temperature difference.

14.3.3.2 Reactive Distillation Configuration With a Decanter

Butyl acetate has a high boiling point temperature. It implies that we can remove the acetate from the column base. Moreover, the minimum boiling azeotrope between BuOH and H_2O is a heterogeneous one (Fig. 14.14) and this leads to an LLE in the BuAc/BuOH/H_2O ternary composition space. The two-liquid zone reveals that the ternary system constitutes more than 50% of the composition space. However, a minimum boiling ternary azeotrope exists in the acetate/alcohol/water ternary system and it locates inside the LLE. This

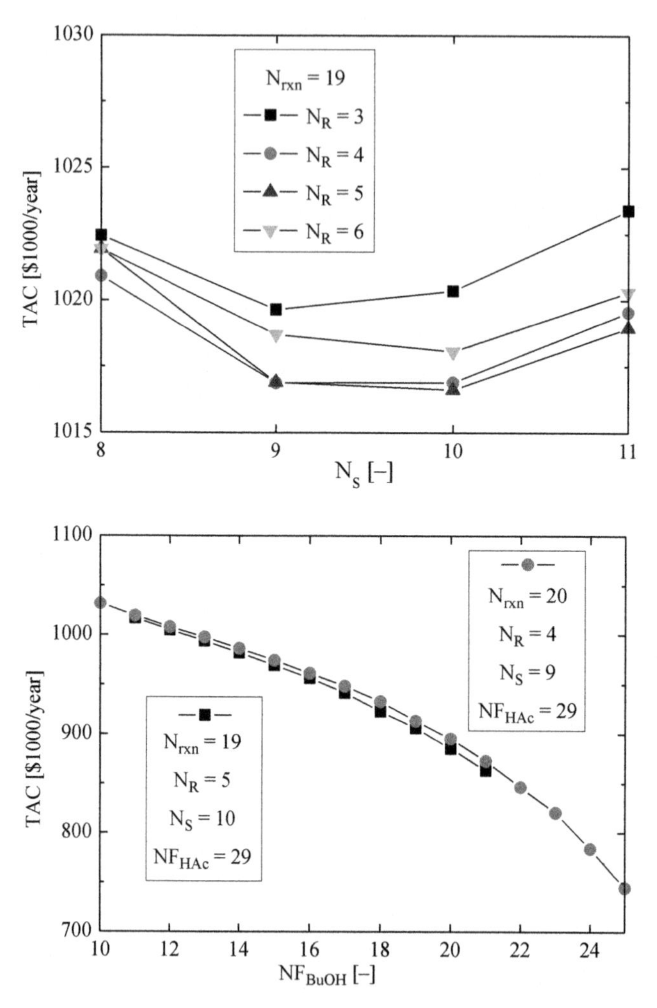

FIGURE 14.11 Effects of design variables on total annual cost (TAC) for BuAc system.

immediately suggests a decanter should be placed in the column top by removing the almost pure water from the aqueous phase. The heavy boiling acetate also implies that we can totally recycle the organic phase back to the column and withdraw the acetate from the bottoms. The scenario suggests the type of flowsheet for the BuAc as shown in Fig. 14.12. This is quite similar to a typical column with the reactive zone located in the middle with a decanter to perform LL separation. The removal of water should be relatively easy with the significant LLE and the favorable RCM as a result of the ternary azeotrope (Fig. 14.14). This is exactly the flowsheet proposed by several different researchers (Hanika et al., 1999; Gangadwala et al., 2004; Huang et al., 2004; Tang et al., 2005).

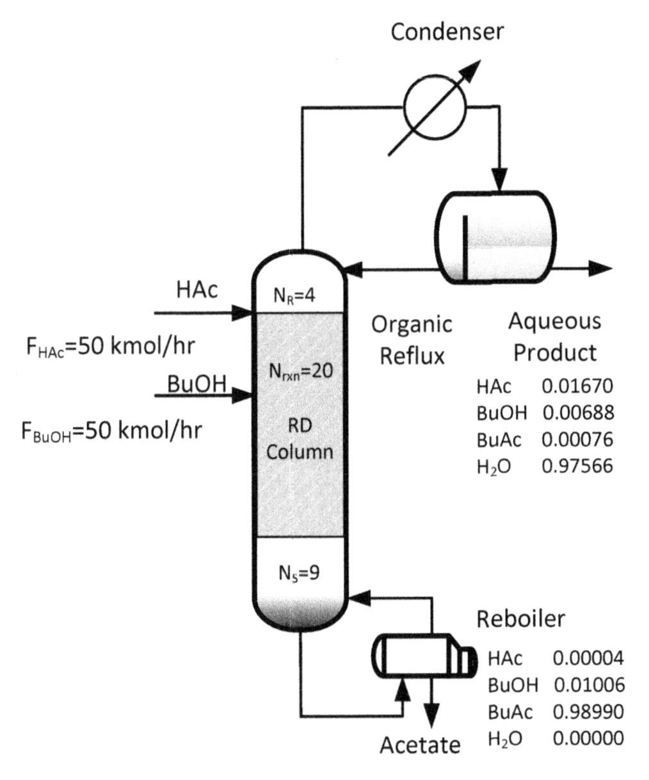

FIGURE 14.12 The optimal flowsheet of BuAc process.

14.4 ISOPROPYL ACETATE REACTIVE DISTILLATION WITH THERMALLY COUPLED CONFIGURATION

14.4.1 Thermodynamic Model

The IPAc process exhibits nonideal phase behavior and has four azeotropes. To accurately represent the phase equilibriums of the process, the selection of the form of the thermodynamic model and the determination of the parameters are essential. To account for the nonideal VLE and VLLE for the quaternary system, the nonrandom two-liquid (NRTL) activity coefficient model is adopted by Aspen Plus. The NRTL model parameter set shown in Table 14.10 is taken from the literature of Tang et al. (2005). The vapor phase nonideality such as the dimerization of acetic acid is also considered. The second virial coefficient of this study (Hayden and O'Connell, 1975) is used to account for vapor phase association of acetic acid. The Aspen Plus built-in association parameters are used to compute the fugacity coefficients.

TABLE 14.9 Steady-State Operating Condition and Total Annual Cost (TAC)

System	BuAc
Column configuration	RD
Total No. of trays including the reboiler	34
Reactive tray	10–29
Acetic acid feed tray	25
Alcohol feed tray	29
Feed flowrate of acid (kmol/h)	50.00
Feed flowrate of alcohol (kmol/h)	50.00
Top product flowrate (kmol/h)	50.38
Bottom product flowrate (kmol/h)	49.62
$X_{D,aq}$	
X_{BuAc}	0.00076
X_{H_2O}	0.97566
X_B	
X_{BuAc}	0.98990
X_{H_2O}	$<10^{-8}$
Condenser duty (kW)	−2857.92
Subcooling duty (kW)	−461.35
Reboiler duty (kW)	3085.41
Column diameter (m)	1.88
Weir height (m)	0.0508
Decanter temperature (°C)	50
Condenser heat transfer area (m^2)	57.06
Subcooling heat transfer area (m^2)	57.89
Reboiler heat transfer area (m^2)	115.07
Total capital cost ($1000)	1277.26
Total operating cost ($1000/year)	319.06
TAC ($1000/year) (50 kmol/h)	745.01
TAC ($1000/year) (52825 ton/year)	840.74

BuAc, butyl acetate; *RD*, reactive distillation.

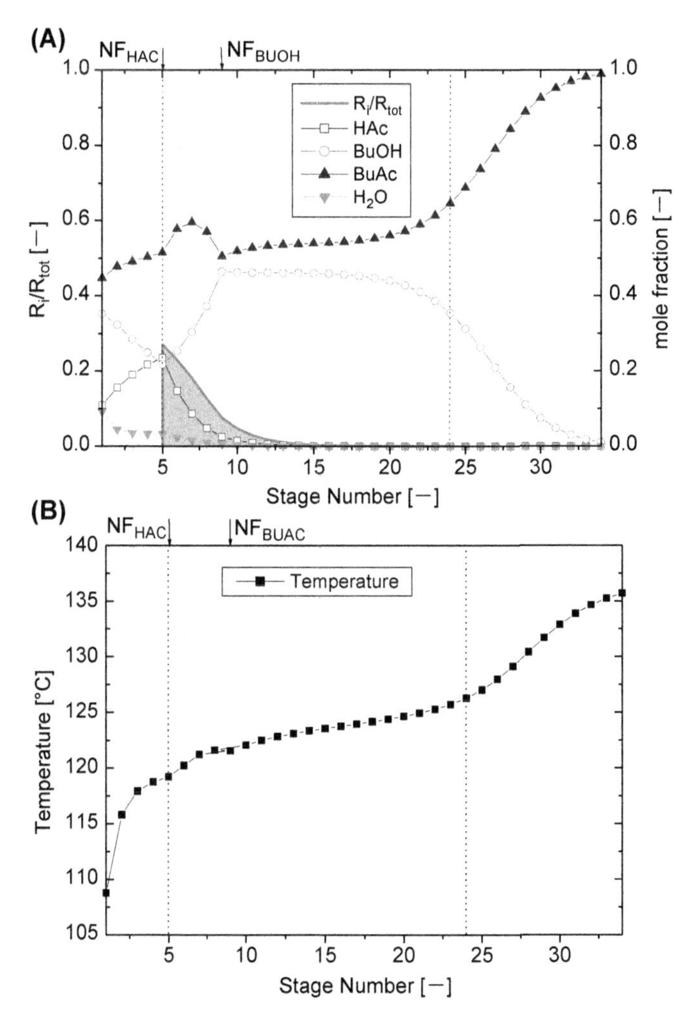

FIGURE 14.13 Composition and temperature profiles of BuAc system. (A) Composition profile and (B) temperature profile.

According to the study by Tang et al. (2005), the thermodynamic model predicts three binary minimum boiling azeotropes and one ternary minimum boiling azeotrope. Table 14.11 shows that the temperatures and the compositions of these azeotropes are in good agreements with model prediction and experimental data. Notice that the ternary minimum boiling azeotrope is the lowest temperature of the system. This ternary azeotrope is shown in the RCM of Fig. 14.15.

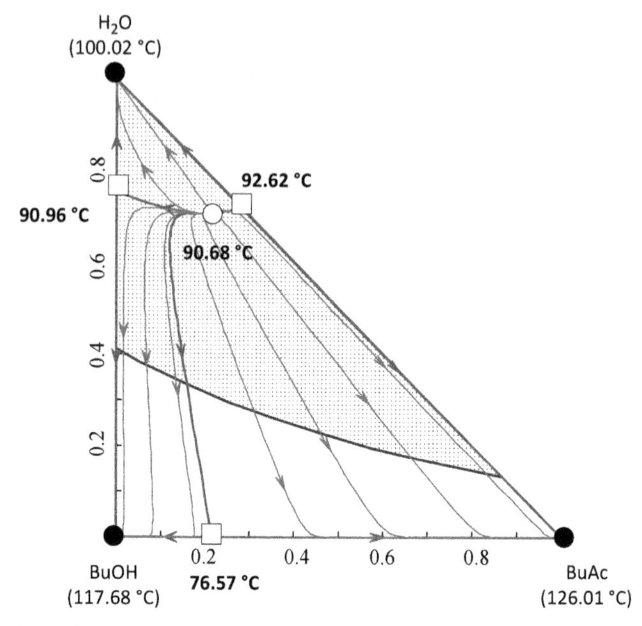

FIGURE 14.14 Residue curve map and liquid–liquid envelope for combinations of ternary BuAc system.

14.4.2 Kinetic Model

By reacting isopropanol (IPOH) with acetic acid (Hac), the esterification reaction will produce IPAc and water. The reactions are reversible and the stoichiometric balance equation is shown as follows:

$$\underset{(HAc)}{CH_3COOH} + \underset{(IPOH)}{C_3H_7OH} \underset{k_{-1}}{\overset{k_1}{\Leftrightarrow}} \underset{(IPAc)}{CH_3COOC_3H_7} + \underset{(H_2O)}{H_2O}$$

The solid catalysts in use are the acidic ion-exchange resin of Amberlyst 15. Table 14.12 displays the reaction rates expressed in the Langmuir–Hinshelwood model form. Notice that the kinetic model is represented in terms of activity and the reaction is catalyst-weight (m_{cat}) based. The catalyst weight is computed by assuming that the solid catalyst occupies 50% of the tray holdup, and a catalyst density of 770 kg/m^3 is used to convert the volume into catalyst weight. Before leaving this section, it should be noted here that two FORTRAN subroutines are written in the Aspen Plus to compute the extent of reaction of each tray.

For the RD, we use Aspen built-in template **usrknt.f** first, and code the kinetic parameters into that file.

TABLE 14.10 The Nonrandom Two-Liquid Model Coefficients for IPAc Process

Component i	HAc	HAc	HAc	IPOH	IPOH	IPAc
Component j	IPOH	IPAc	H_2O	IPAc	H_2O	H_2O
b_{ij}(K)	−141.644	70.965	−110.580	191.086	20.057	415.478
b_{ji}(K)	40.962	77.900	424.060	157.103	833.042	1373.462
c_{ij}	0.305	0.301	0.299	0.3	0.325	0.3

HAc, acetic acid, *IPAc*, isopropyl acetate; *IPOH*, isopropanol.

14.4.3 Process Configuration

14.4.3.1 Conceptual Design of Process

The IPAc process is classified as type II system in Tang et al. (2005). The characteristics of the type II system contain one RD column with a decanter and an additional stripper. Based on the azeotrope and boiling point ranking, the heaviest component is reactant HAc so that no bottom outlet stream of the RD column is designed. The top composition of the RD column is quite close to the minimum temperature azeotrope of IPOH−IPAc−H_2O. In the RCM of

TABLE 14.11 The Compositions and Temperatures of the Azeotropes for IPAc Process

	Experimental Data		Computed Data	
Component	Mole Fraction	Temperature (°C)	Mole Fraction	Temperature (°C)
IPOH−IPAc	(0.6508, 0.3492)	80.1	(0.5984, 0.4016)	78.54
IPOH−H_2O	(0.6875, 0.3125)	82.5	(0.6691, 0.3309)	80.06
[a]IPAc−H_2O	(0.5982, 0.4018)	76.6	(0.5981, 0.4019)	76.57
[a]IPOH−IPAc−H_2O	(0.1377, 0.4938, 0.3885)	75.5	(0.2377, 0.4092, 0.3531)	74.22

IPAc, isopropyl acetate; *IPOH*, isopropanol
[a]*Heterogeneous azeotropes.*
Experimental Data From Horsley, L.H., 1973. Azeotropic Data – III, Advances in Chemistry Series No. 116, American Chemical Society, Washington, DC, USA

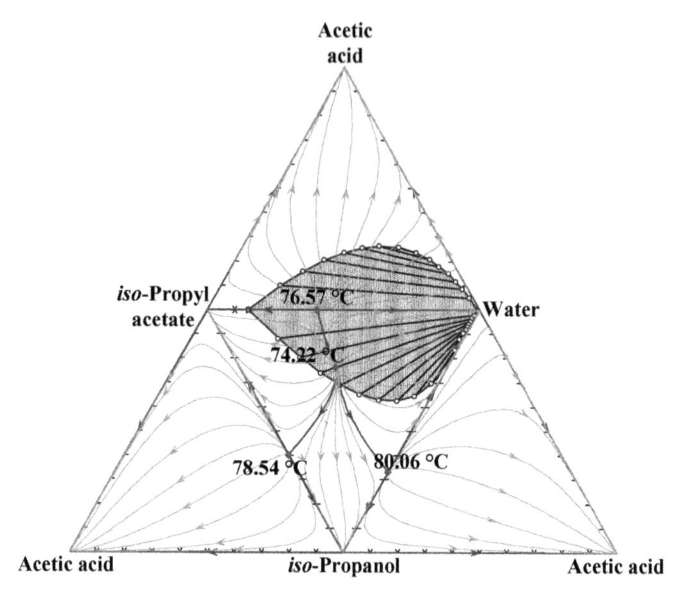

FIGURE 14.15 Four ternary residue curve maps and liquid–liquid envelopes of IPAc process.

Fig. 14.15, it is found that the LLE is quite large, and the ternary minimum boiling azeotrope lies well inside the envelope. It can also be seen that the all tie lines point toward pure water end and, consequently, relatively pure water can be recovered from the aqueous phase of a decanter designed to be located at the top of the RD column. The tie lines can also cross the distillation boundary so that high-purity IPAc can be obtained by further purification of

TABLE 14.12 Kinetic Equations for IPAc Processes (Gadewar et al., 2002)

Kinetic Model (Catalyst: Amberlyst 15)
$r = m_{cat}\dfrac{k_1\left(a_{HAc}a_{IPOH} - a_{IPAc}a_{H_2O}/K_{eq}\right)}{\left(1 + K_{HAc}a_{HAc} + K_{IPOH}a_{IPOH} + K_{IPAc}a_{IPAc} + K_{H_2O}a_{H_2O}\right)^2}$
$k_1 = 7.667 \times 10^{-5}\exp\left(23.81 - \dfrac{68620.43}{RT}\right)$
$K_{eq} = 8.7$, $K_{HAc} = 0.1976$, $K_{IPOH} = 0.2396$, $K_{IPAc} = 0.147$, $K_{H_2O} = 0.5079$
Assumption: mol $H^+/kg_{cat} = 4.6 \times 10^{-3}$

$R = 8.314$ [kJ/k mol/K], T[K], r[k mol/s], mcat [kg_{cat}], a_i, activity.*HAc*, acetic acid, *IPAc*, isopropyl acetate; *IPOH*, isopropanol.

the organic phase outlet stream in a stripper. Part of the organic phase material should also be refluxed back to the RD column to carry water out of the system. Note that in the study of Tang et al. (2005), it is found that both of the reactant feeds should be fed into the bottoms of the RD column so that larger reaction holdup can be utilized.

Because of the economical reason, Lai et al. (2007) studied the IPAc process of this type II system by feeding industrial compositions of reactants instead of pure feeds. Based on the minimum TAC, Lai et al. (2007) demonstrated that the optimal result of the conventional IPAc RD process is shown in Fig. 14.16. In this study, the same feed conditions and product specifications are considered. The IPOH feed is assumed to be close to the azeotrope compositions. The feed composition of HAc is set to be 95 mol% and that of alcohols are set to be 64.91 mol% for IPOH, respectively. The specifications include IPAc production with 99 mol% while keeping HAc purity below 0.01 mol%. Also, the tray numbers of the rectifying section,

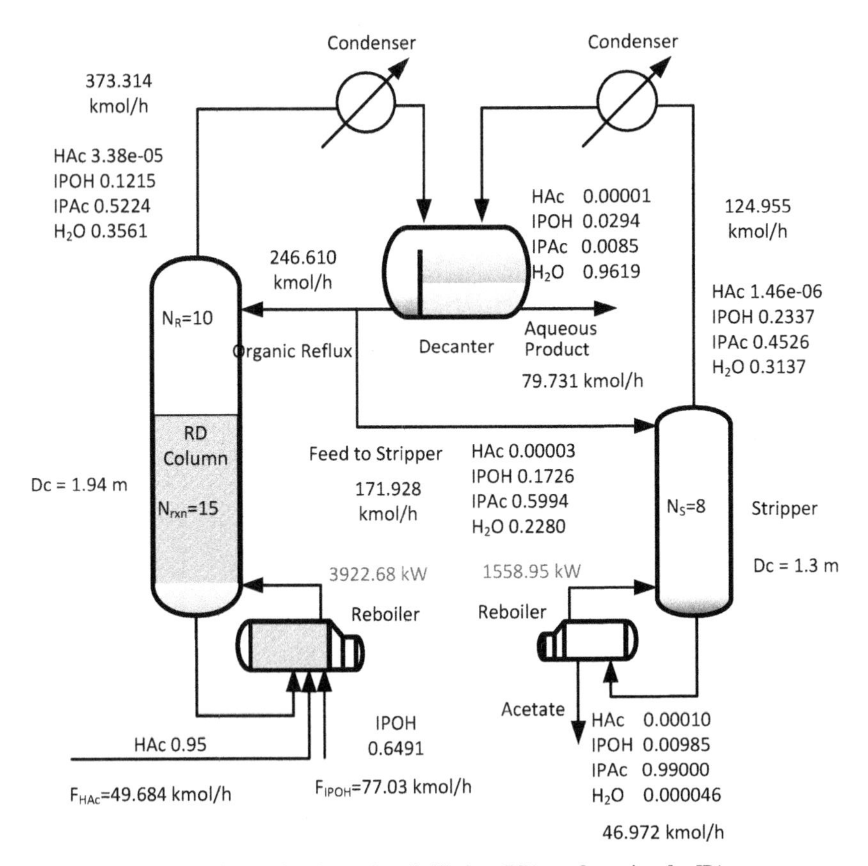

FIGURE 14.16 Conventional reactive distillation (RD) configuration for IPAc process.

reactive section, and stripper are all fixed to be the same as in the optimal result of Lai et al. (2007) so that the benefit in terms of energy saving in the configuration of thermally coupled RD system can be investigated.

By observing composition profiles of the IPAc process in Lai et al. (2007) (see Fig. 14.17), the IPAc compositions in the upper section of the RD column go through a maximum value and then decrease. This is the so-called "remixing effect." This phenomenon hints that there is potential for energy saving by using thermally coupled design. The second observation is that the top compositions of the RD column and the stripper are close. And also, both

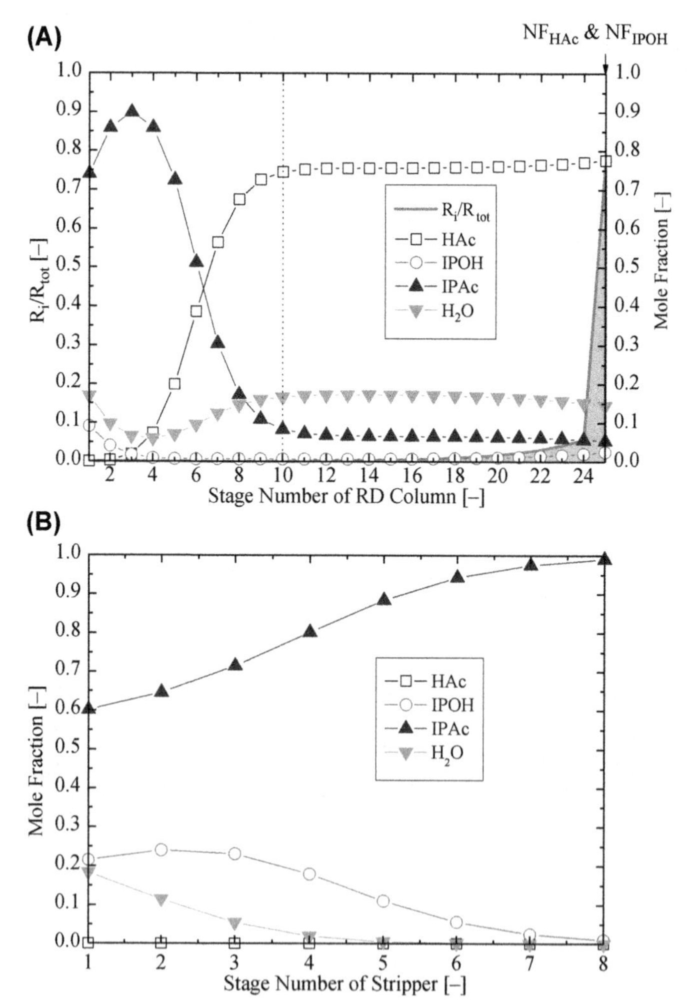

FIGURE 14.17 (A) Reactive distillation (RD) column and (B) stripper liquid composition profiles of conventional RD configuration.

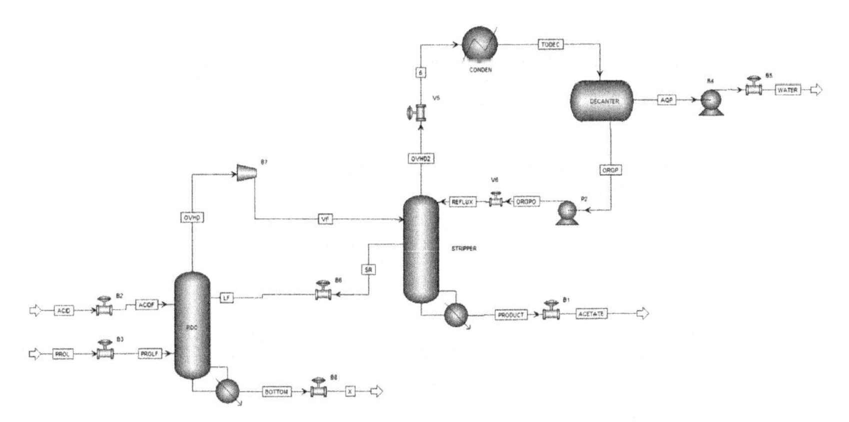

FIGURE 14.18 The flowsheet of IPAc thermally coupled reactive distillation (RD) process in Aspen Plus.

of these top compositions are located into LLE. It is easy to imagine that the result should be similar if the two overhead components are mixed before feeding into the decanter.

Because of the above observations, the RD with the thermally coupled design has been proposed by Lee et al. (2012). The overall flowsheet in Aspen Plus is shown in Fig. 14.18. Because, from the previous Fig. 14.16, the feed to the stripper is from the organic outlet stream of the decanter, it is decided to place the decanter on the stripper (now called the product column in Fig. 14.18) side. However, a liquid stream is needed to go down to the RD column to act as entrainer to carry more water toward the top of the column; thus, a liquid sidedraw from the product column is designed. To complete the thermally coupled design, the vapor stream from the overhead of the RD column should go to the same sidedraw location in the product column. Based on the design of thermally coupled RD in Fig. 14.18, the reboiler duty of the right-side product column is an operating variable used to meet the purity of IPAc, and the liquid sidedraw flowrate instead of the organic reflux in the original RD configuration is used to maintain the HAc impurity in the product stream. Aspen Plus is used with NRTL–HOC model to specify this process.

In this process, FORTRAN subroutine is used for the reaction kinetics with Customize Aspen Plus as shown in Figs. 14.19 and 14.20. The detailed setting in FORTRAN subroutine can be seen in the Appendix A. After the settings of components, the property model, and the reaction kinetics are completed, we must specify the feed streams, initial guess stream and the operation conditions of the two columns. The detail procedures are discussed in the previous section. Because of the characteristics of the configuration in Fig. 14.21, there will be only one design variable, the "vapor–liquid exchange location" (N_F). The N_F is then used to procure the minimum energy requirement. The optimization procedure is shown as Fig. 14.22.

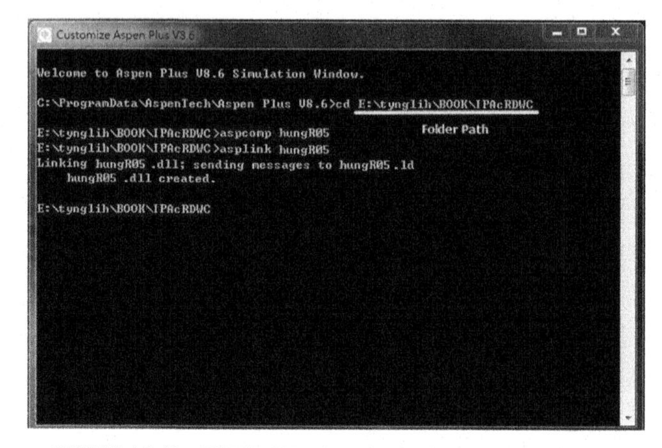

FIGURE 14.19 FORTRAN subroutine by Customize Aspen Plus.

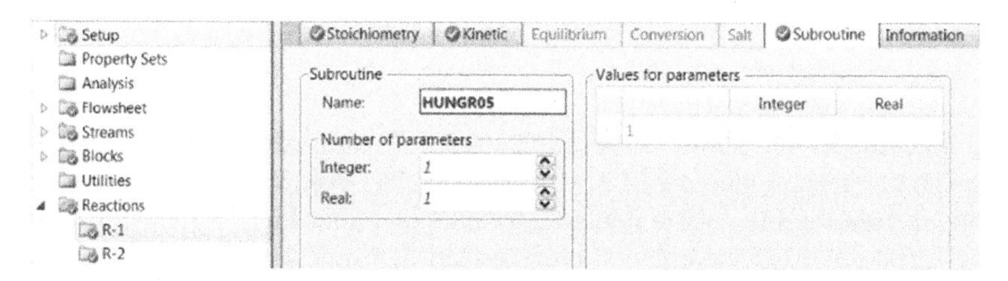

FIGURE 14.20 The setting of FORTRAN Subroutine.

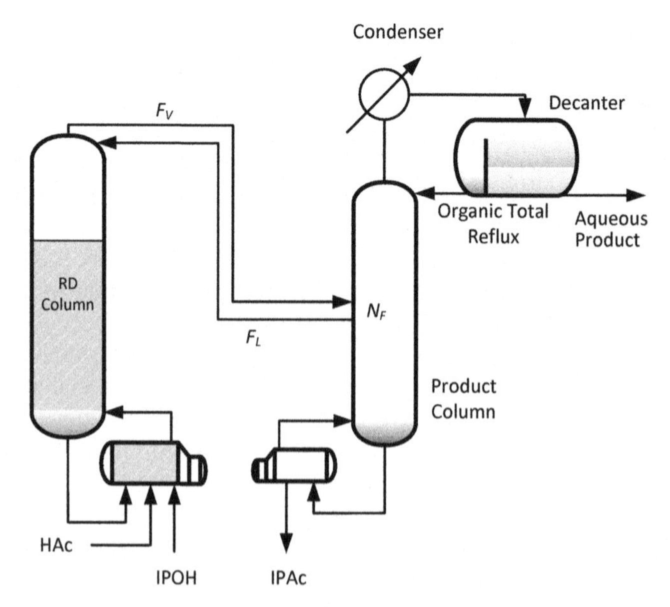

FIGURE 14.21 Proposed process configuration of the thermally coupled reactive distillation (RD) configuration.

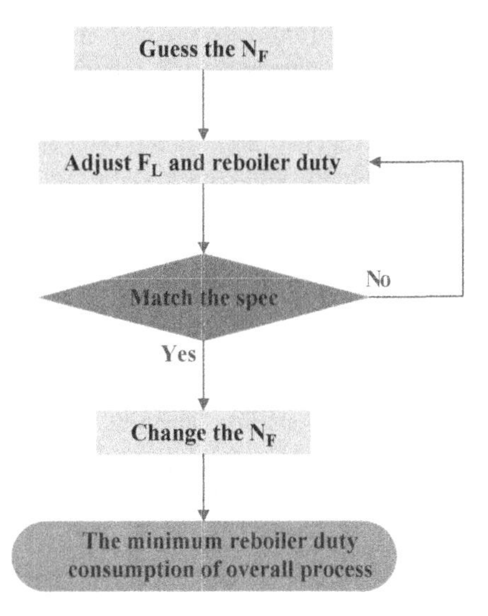

FIGURE 14.22 The procedure of optimization for IPAc process.

14.4.3.2 Overall Configuration

Following the optimization procedure, the information of reboiler duties versus N_F is shown in Fig. 14.23. According to this figure, the best vapor–liquid exchange locations are at the fifth tray. From this figure, it is found that the reboiler duty of the RD remained about the same when the vapor–liquid exchange location varies. And in the product column, the fifth tray would go through a minimum rebolier duty. Therefore, the fifth tray is the best vapor–liquid exchange location.

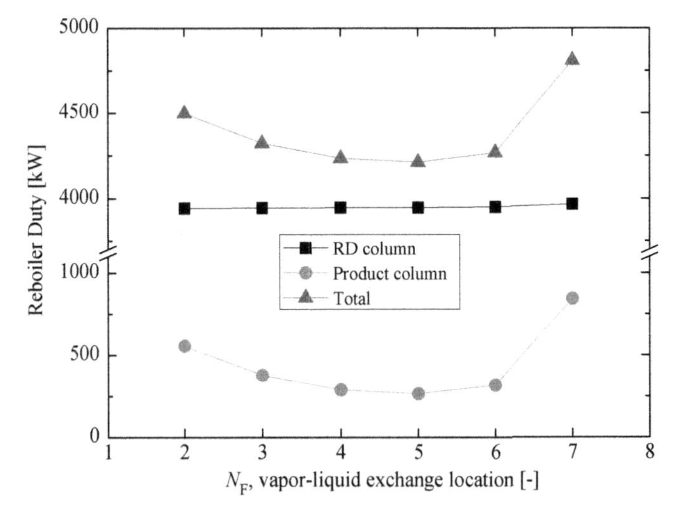

FIGURE 14.23 Minimum energy requirement of the thermally coupled reactive distillation (RD) configuration.

More detailed results of the IPAc process with thermally coupled RD are shown in Fig. 14.24. The reboiler duty of the RD column is 3947.42 kW, which is almost the same as the case of the conventional RD configuration shown in Fig. 14.16. However, the reboiler duty of the right-side product column is only 265.67 kW, which is significantly lower than the stripper reboiler duty 1558.95 kW of the conventional RD configuration. The optimal total reboiler duty of the thermally coupled RD is 23.14% less than the conventional RD system.

The liquid composition profile of the thermally coupled RD for this process is shown in Fig. 14.25. Comparing the composition of IPAc in the upper section of the RD column in the thermally coupled RD with that of the conventional RD configurations, the remixing effect is clearly eliminated. Another interesting observation in composition profile is to project the three-dimensional space of tetrahedron as two-dimensional coordinate by using composition variable transformation method. The distillation path in Fig. 14.26 shows the tray compositions of liquid and vapor phase in the coordinate of composition variable transformation space. The tie lines and the compositions of the organic and aqueous phases are also shown in this figure. In Fig. 14.26, the curves with triangular symbols mean the composition trajectory of RD column composition profile, and the curves with circle symbols are the composition trajectory of product column. In Fig. 14.26A, the vapor

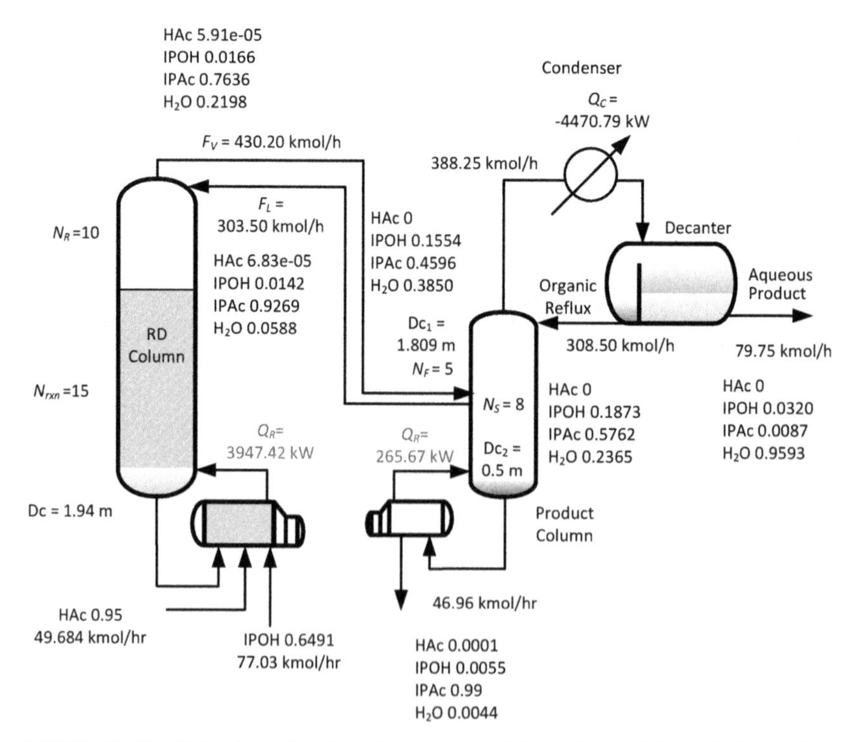

FIGURE 14.24 Optimal results of the thermally coupled reactive distillation (RD) for IPAc processes.

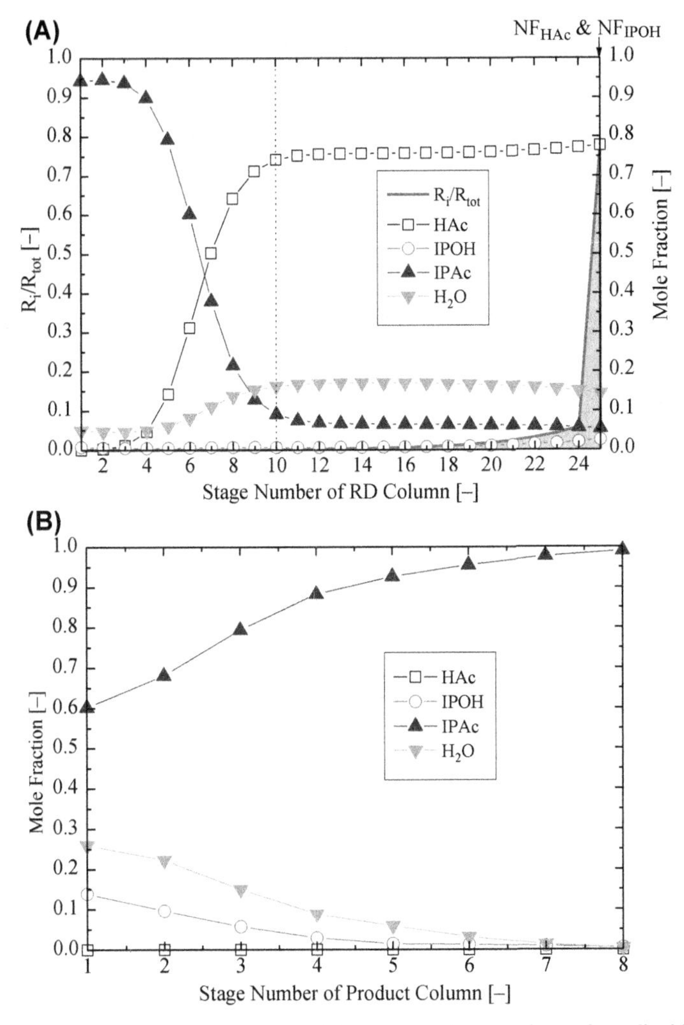

FIGURE 14.25 (A) Reactive distillation (RD) column and (B) product column liquid composition profiles of the thermally coupled RD configuration.

and liquid composition trajectories in the conventional RD column show an obvious turn in the two composition profiles. This is the remixing effect we have mentioned in the previous section. By observing Fig. 14.26B, it is found that the obvious turn in the composition trajectories in the thermally coupled RD is eliminated. Another observation in Fig. 14.26 is that the composition trajectories of the vapor and liquid for the product column with thermally coupled design all show a more linear behavior than the stripper in the conventional RD system. This more linear behavior of the composition trajectory in this transformation coordinate hints the energy saving in the product column.

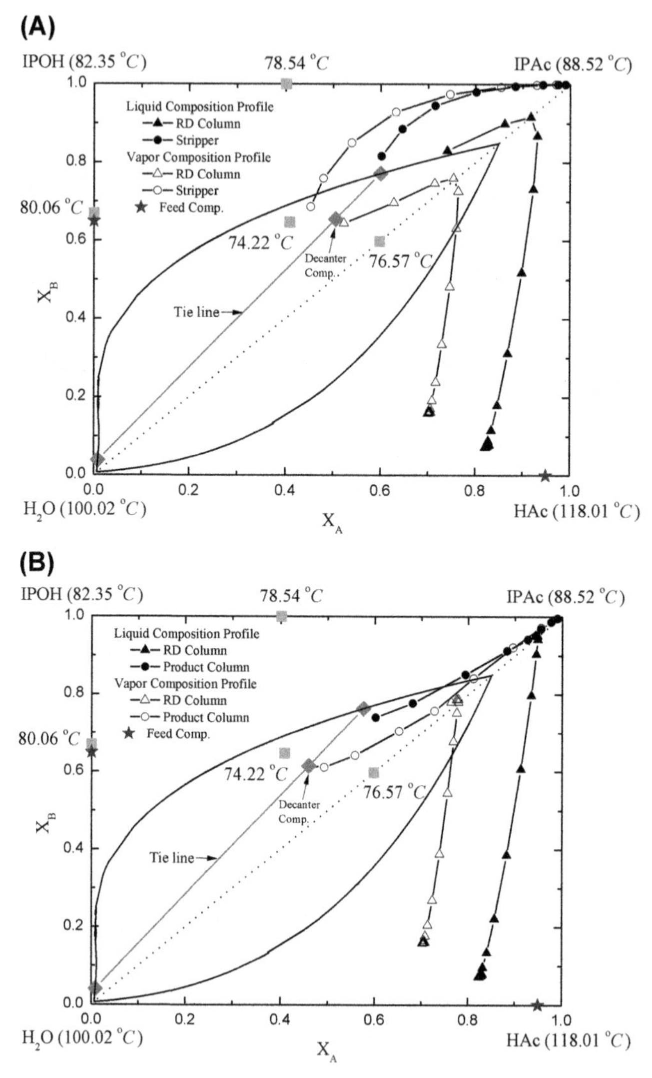

FIGURE 14.26 Composition trajectories of IPAc process: transformed liquid and vapor composition in the quaternary system ($X_A = X_{HAc} + X_{IPAc}$, $X_B = X_{TPOH} + X_{IPAc}$) for (A) conventional reactive distillation (RD) configuration and (B) thermally coupled RD configuration.

14.5 CONCLUSION

In this chapter, the acetic acid esterification with different alcohols using the RD is explored. First, the qualitative relationship between phase equilibria and possible process flowsheets is established. Three process flowsheets, MeAc, BuAc, and IPAc, are demonstrated and they all lead to relatively pure products. Next, a systematic design and simulation procedure is devised to optimize the quantitative design based on the (TAC). This sequential design procedure overcomes the fragility of the simulation algorithm even with the state-of-the-art process simulator for the RD process. Then, the characteristics of the optionally designed RD for these three types of flowsheets are investigated and explanations are given. Finally, the TAC of different flowsheets are compared, the economic potential is ranked, and the explanation is given. The results presented in this provide insight for the conceptual design of RD processes. In this chapter, the potential energy saving of the thermally coupled IPAc RD is also investigated for type II RD process. The thermally coupled design includes moving the location of the decanter to the stripper side, totally refluxing the organic phase outlet stream, and adding a sidedraw liquid stream from the stripper to the RD column. Simulation result shows that 23.14% energy savings can be realized using the proposed thermally coupled design. Through these demonstration works, authors hope that the readers can have better understanding of using the computer software Aspen Plus for the study of RD processes.

EXERCISES

1: Ethyl Acetate Reactive Distillation Process

Ethyl acetate (EtAc) is an important solvent and is widely used in the chemical industries. Most of the EtAc production is mainly from the esterification process of ethanol (EtOH) and acetic acid (HAc). However, the ternary azeotrope with lowest boiling temperature is the problem for product purification. Based on the characteristics of residue curve and LLE, Tang et al. (2005) proposed a novel configuration where the RD is equipped with a decanter and a stripper, as shown in Fig. E1.

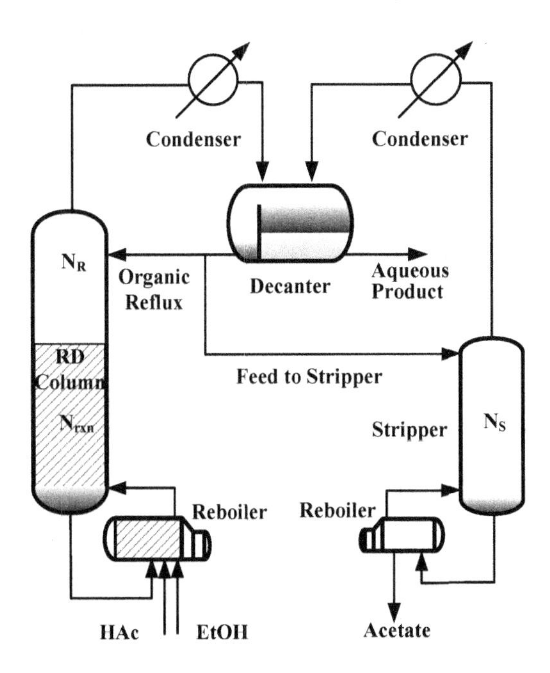

FIGURE E1 Flowsheet of ethyl acetate reactive distillation (RD) process.

TABLE E1 The Nonrandom Two-Liquid Model Coefficients for EtAc Process

Component i	HAc (1)	HAc (1)	HAc (1)	EtOH (2)	EtOH (2)	EtAc (3)
Component j	EtOH (2)	EtAc (3)	H_2O (4)	EtAc (3)	H_2O (4)	H_2O (4)
a_{ij}	0	0	−1.9763	1.817306	0.806535	−2.34561
a_{ji}	0	0	3.3293	−4.41293	0.514285	3.853826
b_{ij} (K)	−252.482	−235.279	609.8886	−421.289	−266.533	1290.464
b_{ji} (K)	225.4756	515.8212	−723.888	1614.287	444.8857	−4.42868
c_{ij}	0.3000	0.3000	0.3000	0.1000	0.4000	0.3643

EtAc, ethyl acetate; *EtOH*, ethanol; *HAc*, acetic acid.

TABLE E2 Kinetic Models for EtAc Process

System (Catalyst)	Kinetic Model
Heterogeneous (Amberlyst 35)	$r = m_{cat}(k_1 C_{HAc} C_{EtOH} - k_2 C_{EtAc} C_{H2O})$
	$k_1 = 6.147 \times e^{-(5673.35/T)}$
	$k_2 = 6.83 \times 10^{-1} \times e^{-(5268.02/T)}$

$R = 8.314 \equiv$ [kJ/kmol/K], $T \equiv$ [K], $r \equiv$ [kmol/s], $m_{cat} \equiv$ [kg_{cat}], $C_i \equiv$ [kmol/m^3]. *EtAc*, ethyl acetate; *EtOH*, ethanol; *HAc*, acetic acid.

Because RD is conducted under a VLE, models suitable for liquid and vapor phases must be provided to allow simulations that closely reflect the actual situations. The Hayden and O'Connell (1975) model is used to describe vapor behavior because of the vapor association of HAc. The association parameter can be obtained through Aspen Plus. The NRTL model is used to calculate the liquid-phase activity coefficient of each component in the liquid phase. The binary parameters of NRTL are shown in Table E1.

The esterification process is a reversible liquid-phase exothermic reaction. Compared to the sulfuric acid catalyst, ion-exchange resins as heterogeneous acidic solid catalysts have certain advantages, such as low corrosiveness and ease of catalyst replacement in the column. Therefore, ion-exchange resins have been widely applied in recent years. The reaction kinetic models are shown in Table E2.

The feed rate of the 95 mol% HAc is set as 50.8 kmol/h, and the feed rate of the EtOH is set as 57.472 kmol/h. To reduce the feed cost, the concentration was set to 87 mol%, near the azeotrope, with a fixed feed ratio. The specifications of product are EtAc with a concentration of 99 mol% and the concentration of impurity (HAc) lower than 0.01 mol%. Please use Aspen Plus to develop a simulation model for this process and find the required energy.

2: Diphenyl Carbonate Reactive Distillation Process by Using Phenyl Acetate and Diethyl Carbonate as Reactants

Diphenyl carbonate (DPC) is a nontoxic and nonpolluting component providing a phosgene-free route for polycarbonate synthesis. Because of the higher equilibrium constant and less azeotropes, the diethyl carbonate (DEC) and phenyl acetate (PA) are used to produce the DPC rather than dimethyl carbonate (DMC) and phenol. The reaction of DEC and PA to produce DPC are shown as follows:

$$DEC + PA \underset{k_{-1}}{\overset{k_1}{\Leftrightarrow}} EPC + EtAc \tag{14.4}$$

TABLE E3 Kinetics Parameters of Diphenyl Carbonate Production Process (Lin, 2014)

	k_{i0} [L/(min/mol)]	E_{ai} [kJ/kmol]
k_1	1.691×10^7	83,973.9
k_{-1}	3.708×10^2	46,960.6
k_2	1.467×10^3	49,979.9
k_{-2}	6.146	22,675.9

$$EPC + PA \underset{k_{-2}}{\overset{k_2}{\Leftrightarrow}} DPC + EtAc \tag{14.5}$$

Eqs. (14.4) and (14.5) are the reaction rate expressions used in Aspen Plus. Eqs. (14.6) to (14.8) are used for describing the reaction rate constant. The kinetic parameters are presented in Table E3.

$$r_1 = k_1 C_{PA} C_{DMC} - k_{-1} C_{MPC} C_{MA} \tag{14.6}$$

$$r_2 = k_2 C_{PA} C_{MPC} - k_{-2} C_{DPC} C_{MA} \tag{14.7}$$

$$k_i = k_0 e^{-E_a/RT} \tag{14.8}$$

The approximate molar ratio of the two pure reactants PA and DEC should be 2:1 for the overall stoichiometric balance in the process. PA and DEC are supplied at 10 and 5 kmol/h, respectively, and the purity of DPC and EtAc are set at 99.5% for industrial usage.

Please develop an Aspen Plus simulation model for this process and find the case with the lowest energy cost.

(Hint: UNIQ-HOC is selected as the thermodynamic model. Because the titanium(IV) ethoxide catalyst is the homogeneous type, the reactive holdup is assumed the same as tray liquid holdup. Fig. E2 is one of possible flowsheet.)

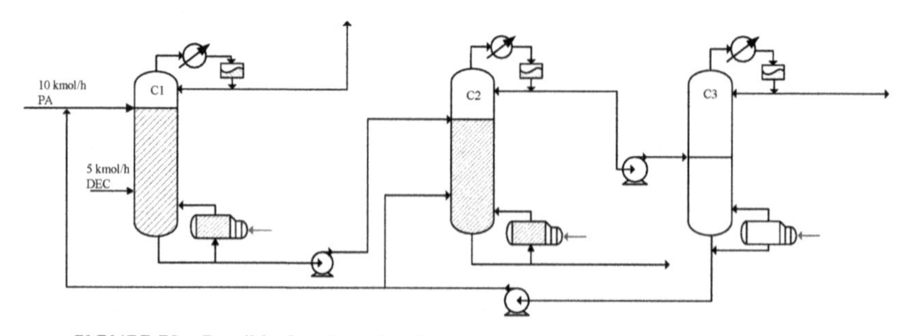

FIGURE E2 Possible flowsheet for diphenyl carbonate reactive distillation process.

APPENDIX: FORTRAN FILE SETTING FOR REACTIVE DISTILLATION

```
C $Log: usrknt.f,v $
C Revision 1.1  1997/04/14 15:52:38  kishore
C commit converted files
C
C Revision 1.3  1996/05/21  19:25:07  apbuild
C ANAVI 9.3 upgrade
C
C Revision 1.2  1996/04/26  19:15:09  apbuild
C Introduce 3phase modifications, Venkat
C
C ==========================cvs revision history========================
C$ #3 BY: SIVA DATE: 15-NOV-1994 ADD DOCUMENTATION
C$ #2 BY: SIVA DATE: 21-JUL-1994 ADD X TO ARGUMENT LIST
C$ #1 BY: ANAVI DATE:  1-JUL-1994 NEW FOR USER MODELS
C
C     User Kinetics Subroutine for RADFRAC, BATCHFRAC, RATEFRAC
C     (REAC-DIST type Reactions)
C

      SUBROUTINE HUNGR05 (N,       NCOMP,   NR,      NRL,     NRV,
     2                   T,       TLIQ,    TVAP,    P,       PHFRAC,
     3                   F,       X,       Y,       IDX,     NBOPST,
     4                   KDIAG,   STOIC,   IHLBAS,  HLDLIQ,  TIMLIQ,
     5                   IHVBAS,  HLDVAP,  TIMVAP,  NINT,    INT,
     6                   NREAL,   REAL,    RATES,   RATEL,   RATEV,
     7                   NINTB,   INTB,    NREALB,  REALB,   NIWORK,
     8                   IWORK,   NWORK,   WORK)
C
C***********************************************************************
C  LICENSED MATERIAL.  PROPERTY OF ASPEN TECHNOLOGY, INC.  TO BE       *
C  TREATED AS ASPEN TECH PROPRIETARY INFORMATION UNDER THE TERMS       *
C  OF THE ASPEN PLUS SUBSCRIPTION AGREEMENT.                           *
C***********************************************************************
C-----------------------------------------------------------------------
C        COPYRIGHT (C) 1994
C         ASPEN TECHNOLOGY, INC.
C         CAMBRIDGE, MA
C-----------------------------------------------------------------------
C
C     DESCRIPTION: TO CALCULATE REACTION RATES FOR KINETIC REACTIONS
C                  USING USER SUPPLIED SUBROUTINE
C
C      VARIABLES IN ARGUMENT LIST
C
C     VARIABLE     I/O  TYPE     DIMENSION     DESCRIPTION AND RANGE
C     N             I    I           -         STAGE NUMBER
C     NCOMP         I    I           -         NUMBER OF COMPONENTS
C     NR            I    I           -         TOTAL NUMBER OF KINETIC
C                                              REACTIONS
C     NRL           I    I           3         NUMBER OF LIQUID PHASE
C                                              KINETIC REACTIONS.
C                                              NRL(1): NUMBER OF
C                                                      OVERALL LIQUID
C                                                      REACTIONS.
```

```
C                                          NRL(2): NUMBER OF
C                                                  LIQUID1 REACTIONS.
C                                          NRL(3): NUMBER OF
C                                                  LIQUID2 REACTIONS.
C       NRV         I     I       -        NUMBER OF VAPOR PHASE
C                                          KINETIC REACTIONS
C       T           I     R       -        STAGE TEMPERATURE (K)
C       TLIQ        I     R       -        LIQUID TEMPERATURE (K)
C                                          * USED ONLY BY RATEFRAC **
C       TVAP        IR            -        VAPOR TEMPERATURE (K)
C                                          * USED ONLY BY RATEFRAC **
C       P           I     R       -        STAGE PRESSURE (N/SQ.M)
C       PHFRAC      I     R       3        PHASE FRACTION
C                                          PHFRAC(1): VAPOR FRACTION
C                                          PHFRAC(2): LIQUID1 FRACTIO
C                                          PHFRAC(3): LIQUID2 FRACTIO
C       F           I     R       -        TOTAL FLOW ON STAGE
C                                          (VAPOR+LIQUID) (KMOL/SEC)
C       X           I     R     NCOMP,3    LIQUID MOLE FRACTION
C       Y           I     R     NCOMP      VAPOR MOLE FRACTION
C       IDX         I     I     NCOMP      COMPONENT INDEX VECTOR
C       NBOPST      I     I       6        OPTION SET BEAD POINTER
C       KDIAG       I     I       -        LOCAL DIAGNOSTIC LEVEL
C       STOIC       I     R     NCOMP,NR   REACTION STOICHIOMETRY
C       IHLBAS      I     I       -        BASIS FOR LIQUID
C                                          HOLDUP SPECIFICATION
C                                          1:VOLUME,2:MASS,3:MOLE
C       HLDLIQ      I     R       -        LIQUID HOLDUP
C                                          IHLBAS    UNITS
C                                          1         CU.M.
C                                          2         KG
C                                          3         KMOL
C       TIMLIQ      I     R       -        LIQUID RESIDENCE TIME
C                                          (SEC)
C       IHVBAS      I     I       -        BASIS FOR VAPOR
C                                          HOLDUP SPECIFICATION
C                                          1:VOLUME,2:MASS,3:MOLE
C       HLDVAP      I     R       -        VAPOR HOLDUP
C                                          IHVBAS    UNITS
C                                          1         CU.M.
C                                          2         KG
C                                          3         KMOL
C       TIMVAP      I     R       -        VAPOR RESIDENCE TIME (SEC)
C       NINT        I     I       -        LENGTH OF INTEGER VECTOR
C       INT         I/O   I     NINT       INTEGER VECTOR
C       NREAL       I     I       -        LENGTH OF REAL VECTOR
C       REAL        I/O   R     NREAL      REAL VECTOR
C       RATES       O     R     NCOMP      COMPONENT REACTION RATES
C                                          (KMOL/SEC)
C       RATEL       O     R     NRLT       INDIVIDUAL REACTION RATES
C                                          IN THE LIQUID PHASE
C                                          (KMOL/SEC)
C                                          WHAT IS NRLT?
C                                          NRLT = NRL(1)+NRL(2)+NRL(3
C                                          NRLT IS NOT INCLUDED IN TH
C                                          ARGUMENT LIST.
C                                          * USED ONLY BY RATEFRAC *
C       RATEV       O     R     NRV        INDIVIDUAL REACTION RATES
```

```
C                                                   IN THE VAPOR PHASE
C                                                   (KMOL/SEC)
C                                                   * USED ONLY BY RATEFRAC *
C       NINTB       I     I           -             LENGTH OF INTEGER VECTOR
C                                                   (FROM UOS BLOCK)
C       INTB        I/O   I          NINTB          INTEGER VECTOR
C                                                   (FROM UOS BLOCK)
C       NREALB      I     I           -             LENGTH OF REAL VECTOR
C                                                   (FROM UOS BLOCK)
C       REALB       I/O   R          NREALB         REAL VECTOR
C                                                   (FROM UOS BLOCK)
C       NIWORK      I     I           -             LENGTH OF INTEGER WORK
C                                                   VECTOR
C       IWORK       I/O   I          NIWORK         INTEGER WORK VECTOR
C       NWORK       I     I           -             LENGTH OF REAL WORK VECTOR
C       WORK        I/O   R          NWORK          REAL WORK VECTOR
C
C***********************************************************************
C
      IMPLICIT NONE
C
C      DECLARE VARIABLES USED IN DIMENSIONING
C
      INTEGER NCOMP, NR,     NRV,    NINT,   NINTB,
     +        NREALB,NIWORK,NWORK
C
      #include "ppexec_user.cmn"
      #include "rxn_disti.cmn"
      #include "rxn_distr.cmn"
      EQUIVALENCE (RMISS, USER_RUMISS)
      EQUIVALENCE (IMISS, USER_IUMISS)
C***********************************************************************
C.     DECLARE ARGUMENTS
C
      INTEGER NRL(3),IDX(NCOMP),    NBOPST(6),
     +        INT(NINT),     INTB(NINTB),
     +        IWORK(NIWORK),N,       KDIAG, IHLBAS,
     +        IHVBAS,NREAL
      REAL*8 PHFRAC(3),     X(NCOMP,3),    Y(NCOMP),
     +       STOIC(NCOMP,NR),      RATES(NCOMP),
     +       RATEL(1),       RATEV(NRV),
     +       REALB(NREALB),WORK(NWORK),   T,       TLIQ,
     +       TVAP,   P,       F,       HLDLIQ,TIMLIQ
      REAL*8 HLDVAP,TIMVAP
C
C      DECLARE LOCAL VARIABLES
C
      INTEGER IMISS
      REAL*8 REAL(NREAL),   RMISS
C
C      BEGIN EXECUTABLE CODE
C     ***********************************************************
C     *******Binary Interaction Parameters for NRTL Equations
C     ***********************************************************
C     Species : HOAc(1) IPOH(2) IPOAc(3) H2O(4)
C
C                            Keq
C      Rxn::     HOAc + IPOH <-----> IPOAc +H2O
C     ***********************************************************
```

```
C     *****DECLARE VARIABLES FOR RATEL
C     ***********************************************************

      REAL*8 KHOAC,KIPOH,KIPOAC,KH2O
      REAL*8 KS,KEQ,W

      REAL*8 A(NCOMP,NCOMP) , ALPHA(NCOMP,NCOMP), TAU(NCOMP,NCOMP),
     +       G(NCOMP,NCOMP),!NRTL parameters
     +       GAMMA(NCOMP),LOGGA(NCOMP), !Activity Coefficient
     +       AC(NCOMP) ! Activities

      REAL*8 R,V1,V1NEW,V2,V2NEW,V3,V3NEW,V4,V4NEW,V5,V5NEW
      REAL*8 NRA,NRB,NRBNEW

      INTEGER II,JJ,KK,MM,IC,IC2 !USE FOR LOOP

      PARAMETER(R=1.987E0) !GAS CONSTANT
      PARAMETER(KHOAC=0.1976E0) !Equilibrium Const
      PARAMETER(KIPOH=0.2396E0)
      PARAMETER(KIPOAC=0.147E0)
      PARAMETER(KH2O=0.5079E0)
      PARAMETER(KEQ=8.7E0)

C     DISTI_THLBAS=1

      A(1,1)=0.E0
      A(1,2)=-281.4482E0
      A(1,3)=141.0081E0
      A(1,4)=-219.7238E0
      A(2,1)=81.3926E0
      A(2,2)=0.E0
      A(2,3)=379.6898E0
      A(2,4)=39.5841E0
      A(3,1)=154.7884E0
      A(3,2)=312.1622E0
      A(3,3)=0.E0
      A(3,4)=825.5563E0
      A(4,1)=842.6079E0
      A(4,2)=1655.255E0
      A(4,3)=2729.0706E0
      A(4,4)=0.E0

      ALPHA(1,1)=0.E0
      ALPHA(1,2)=0.3048E0
      ALPHA(1,3)=0.3014E0
      ALPHA(1,4)=0.2997E0
      ALPHA(2,1)=0.3048E0
      ALPHA(2,2)=0.E0
      ALPHA(2,3)=0.3E0
      ALPHA(2,4)=0.3255E0
      ALPHA(3,1)=0.3014E0
      ALPHA(3,2)=0.3E0
      ALPHA(3,3)=0.E0
      ALPHA(3,4)=0.3E0
      ALPHA(4,1)=0.2997E0
      ALPHA(4,2)=0.3255E0
      ALPHA(4,3)=0.3E0
      ALPHA(4,4)=0.E0
```

```
DO II=1,NCOMP
     DO JJ=1,NCOMP
          TAU(IDX(II),IDX(JJ))=A(IDX(II),IDX(JJ))/(R*T)
     END DO
END DO

DO II=1,NCOMP
     DO JJ=1,NCOMP
          G(IDX(II),IDX(JJ))=DEXP(-(TAU(II,JJ)*ALPHA(II,JJ)))
     END DO
END DO

DO II=1,NCOMP

     DO JJ=1,NCOMP
          V1=TAU(IDX(JJ),IDX(II))*X(IDX(JJ),1)*G(IDX(JJ),IDX(II))
          V1=V1+V1NEW
          V1NEW=V1
     END DO

     DO KK=1,NCOMP
          V2=X(IDX(KK),1)*G(IDX(KK),IDX(II))
          V2=V2+V2NEW
          V2NEW=V2
     END DO

     DO JJ=1,NCOMP

          DO KK=1,NCOMP
               V3=X(IDX(KK),1)*G(IDX(KK),IDX(JJ))
               V3=V3+V3NEW
               V3NEW=V3
          END DO

          DO MM=1,NCOMP
          V4=TAU(IDX(MM),IDX(JJ))*X(IDX(MM),1)*G(IDX(MM),IDX(JJ))
               V4=V4+V4NEW
               V4NEW=V4
          END DO

          NRB=(X(IDX(JJ),1)*G(IDX(II),IDX(JJ))/V3)*
&              (TAU(IDX(II),IDX(JJ))-V4/V3)

          NRB=NRB+NRBNEW
          NRBNEW=NRB

          V3=0.E0
          V3NEW=0.E0
          V4=0.E0
          V4NEW=0.E0
     END DO

     NRA= V1/V2
     LOGGA(II)=NRA+NRB

     V1=0.E0
     V1NEW=0.E0
     V2=0.E0
     V2NEW=0.E0
```

```
          NRB=0.E0
          NRBNEW=0.E0

     END DO

     DO IC =1,NCOMP
          GAMMA(IC)=EXP(LOGGA(IC))
          AC(IC)=GAMMA(IC)*X(IDX(IC),1)
     END DO

C    ************************************************************
C    ******Begin Kinetic Model
C    ************************************************************
     KS=EXP(23.81E0-8253.6E0/T)
     IHLBAS=1

     W=(0.0046E0)*770.E0*HLDLIQ

     RATES(1)=-(1.E0/60.E0)*(W*KS*(AC(1)*AC(2)-AC(3)*AC(4)/KEQ)
    &   /(1.E0+KHOAC*AC(1)+KIPOH*AC(2)+KIPOAC*AC(3)+KH2O*AC(4))**2.E0)
     RATES(2)=-(1.E0/60.E0)*(W*KS*(AC(1)*AC(2)-AC(3)*AC(4)/KEQ)
    &   /(1.E0+KHOAC*AC(1)+KIPOH*AC(2)+KIPOAC*AC(3)+KH2O*AC(4))**2.E0)
     RATES(3)=(1.E0/60.E0)*(W*KS*(AC(1)*AC(2)-AC(3)*AC(4)/KEQ)
    &   /(1.E0+KHOAC*AC(1)+KIPOH*AC(2)+KIPOAC*AC(3)+KH2O*AC(4))**2.E0)
     RATES(4)=(1.E0/60.E0)*(W*KS*(AC(1)*AC(2)-AC(3)*AC(4)/KEQ)
    &   /(1.E0+KHOAC*AC(1)+KIPOH*AC(2)+KIPOAC*AC(3)+KH2O*AC(4))**2.E0)


      RETURN
      END
```

REFERENCES

Agreda, V.H., Partin, L.R., Heise, W.H., 1990. High purity methyl acetate via reactive distillation. Chemical Engineering Progress 86 (2), 40.

Al-Arafaj, M.A., Luyben, W.L., 2002. Comparative control study of ideal and methyl acetate reactive distillation. Chemical Engineering Science 57, 5039.

Burkett, R.J., Rossiter, D., 2000. Choosing the right control structure for industrial distillation columns. In: Proc. of Process Control and Instrumentation, 38. Glasgow, UK.

Cheng, Y.C., Yu, C.C., 2003. Effects of process design on recycle dynamics and its implication to control structure selection. Industrial & Engineering Chemistry Research 42, 4348.

Chiang, S.F., Kuo, C.L., Yu, C.C., Wong, D.S.H., 2002. Design alternatives for the amyl acetate process: coupled reactor/column, reactive distillation. Industrial & Engineering Chemistry Research 41, 3233.

Doherty, M.F., Buzad, G., 1992. Reactive distillation by design. Trans IChemE A70, 448.

Douglas, J.M., 1998. Conceptual Design of Chemical Processes. McGraw-Hill, New York.

Gadewar, S.B., Malone, M.F., Doherty, M.F., 2002. Feasible region for a countercurrent cascade of vapor-liquid CSTRs. AIChE Journal 48, 800.

Gangadwala, J., Kienle, A., Stein, E., Mahajani, S., 2004. Production of butyl acetate by catalytic distillation: process design studies. Industrial & Engineering Chemistry Research 43, 136.

Hanika, J., Smejkal, Q., Kolena, J., 1999. Butylacetate vis reactive distillation- modelling and experiment. Chemical Engineering Progress 54, 5205.

Hayden, J.G., O'Connell, J.P., 1975. A generalized method for predicting second virial coefficients. Industrial & Engineering Chemistry Process Design and Development 14, 209.

Horsley, L.H., 1973. Azeotropic Data – III, Advances in Chemistry Series No. 116. American Chemical Society, Washington, DC, USA.

Huang, S.G., Yu, C.C., 2003. Sensitivity of thermodynamic parameter to the design of heterogeneous reactive distillation: Amyl Acetate Esterification. Journal of Chinese Institute of Chemical Engineers 34, 345.

Huang, S.G., Kuo, C.L., Hung, S.B., Chen, Y.W., Yu, C.C., 2004. Temperature control of heterogeneous reactive distillation: butyl propionate and butyl acetate esterification. AIChE Journal 50, 2203–2216.

Kenig, E.Y., Bader, H., Gorak, A., Bebling, B., Adrian, T., Schoenmakers, H., 2001. Investigation of ethyl acetate reactive distillation process. Chemical Engineering Science 56, 6185.

Lai, I.K., Hung, S.B., Hung, W.J., Yu, C.C., Lee, M.J., Huang, H.P., 2007. Design and control of reactive distillation for ethyl and iso-propyl acetates production with azeotropic feeds. Chemical Engineering Science 62, 878.

Lee, H.Y., Lai, I.K., Huang, H.P., Chien, I.L., 2012. Design and control of thermally coupled reactive distillation for the production of Isopropyl acetate. Industrial & Engineering Chemistry Research 51, 11753–11763.

Lin, S.S., 2014. Kinetics Behavior Study of the Synthesis of Diphenyl Carbonate at Elevated Temperatures and Pressures. Master thesis. National Taiwan University of Science & Technology, Taipei, Taiwan.

Luyben, W.L., Yu, C.C., 2008. Reactive Distillation Design and Control. John Wiley & Sons, Inc., Hoboken, New Jersey.

Malone, M.F., Doherty, M.F., 2000. Reactive distillation. Industrial & Engineering Chemistry Research 39, 3953.

Pöpken, T., Götze, L., Gmehling, J., 2000. Reaction kinetics and chemical equilibrium of homogeneously and heterogneously catalyzed acetic acid esterification with methanol and methyl acetate hydrolysis. Industrial & Engineering Chemistry Research 39, 2601.

Steinigeweg, S., Gmehling, J., 2002. *n*-Butyl acetate Synthesis via reactive distillation: thermodynamic Aspects, reaction kinetics, Pilot-Plant Experiments, Simulation Studies. Industrial & Engineering Chemistry Research 41, 5483.

Sundmacher, K., Kienle, A., 2003. In: Reactive Distillation: Status and Future Directions. Wiley-VCH Verlag CmbH & Co. KgaA, Weiheim, Germany.

Tang, Y.T., Chen, Y.W., Huang, H.P., Yu, C.C., Hung, S.B., Lee, M.J., 2005. Design of reactive distillations for acetic acid esterification with different alcohols. AIChE Journal 51, 1683.

Taylor, R., Krishna, R., 2000. Modelling reactive distillation. Industrial & Engineering Chemistry Research 55, 5183.

Venimadhavan, G., Malone, M.F., Doherty, M.F., 1999. Bifurcation study of kinetic effects in reactive distillation. AIChE Journal 45, 546.

Chapter 15

Design of Azeotropic Distillation Systems

I-Lung Chien, Bor-Yih Yu, Zi Jie Ai
National Taiwan University, Taipei, Taiwan

15.1 INTRODUCTION

Distillation is the most widely used separation equipment in the chemical industry. It was estimated that separation processes account for 40%–70% of both capital and operating costs in petrochemical processing and that distillation is used to make 90%–95% of all separations in the chemical process industry (Tyreus, 2011). In another paper, it was stated that distillation columns and their support facilities can account for about one-third of the total capital cost and more than half of the total energy consumption (Julka et al., 2009). Consequently, the design and optimization of the distillation train have a critical impact on the economics of the entire process. Among the distillation processes, mixtures that contain azeotrope condition are most difficult to be separated. Regular distillation cannot be used to achieve complete separation. A book by Luyben and Chien (2010) summarizes the feasible ways used in the industry to achieve such separation.

For a particular azeotropic mixture, there may be more than one way to achieve the separation. This chapter demonstrates the use of Aspen Plus simulation software to generate various analytical tools such as vapor–liquid equilibrium plots, liquid–liquid equilibrium envelope, residue curve maps (RCMs), and material balance lines to select the most suitable separation method. Two industrial applications are used to demonstrate how significant energy savings can be made with proper selection of a suitable separation method.

To further explore energy saving of the azeotropic separation process, various heat-integration schemes are investigated. The simplest scheme is via feed–effluent heat exchanger (FEHE) to recover heat from hot product stream. Another option is to operate the two columns in a sequence at different pressures, so that the condenser of the high-pressure column can be combined with the reboiler of the low-pressure column to save energy. This is called the multieffect column. Yet another feasible way is to design a thermally coupled

Chemical Engineering Process Simulation. http://dx.doi.org/10.1016/B978-0-12-803782-9.00015-7

(dividing-wall) column in a sequence so that the possible "remixing effect" can be eliminated. Another heat-integration method mentioned in open literature is called the vapor recompression scheme, where the vapor from the column overhead is compressed as a heat pumping fluid to support energy required for the reboiler of the same column. Another heat-integrated design is called the internal heat-integrated distillation column, where both condenser and reboiler of the column are removed. In this case, the heat from the rectifying section is sent to the stripping section. However, the main limitation of this scheme is that an expensive compressor is to be installed to the column. Hence, the discussion of this scheme is excluded from this chapter.

15.2 AZEOTROPIC SEPARATION WITHOUT ENTRAINER

There are various ways to achieve the separation of azeotropic mixture via distillation. In the following sections, industrial examples are used to provide an overview of these separation methods.

15.2.1 Pressure-Swing Distillation

The distillate stream from a reactive-distillation (RD) column to produce *tert*-amyl methyl ether is used here as an example (Luyben, 2005; Wu et al., 2009). This distillate stream contains five inert of carbon olefins (C5s) and methanol. The purpose is to recover methanol from the olefins and to recycle it back to the RD column. Isopentane (iC5) is a major component in the C5s, which will be used as an example to illustrate the concept in achieving C5s-methanol separation via pressure-swing distillation. Other inert C5s have the similar property as isopentane.

The temperature-composition (T-xy) plots of iC5 and methanol mixture at two different operating pressures, i.e., 2.5 and 13 atm are shown in Fig. 15.1, which show that the azeotropic compositions at these two pressures are different. This permits the use of the design flowsheet in Fig. 15.2 for the desired separation task. For illustration purpose, the feed composition is assumed at 50 mol% iC5. This feed stream is combined with a recycle stream from the high-pressure (HP) column. Notice that the recycle distillate stream should have its composition near the high-pressure azeotrope. This combined stream should have its composition on the left side of the T-xy plot at low pressure (LP). Thus, it is feasible to obtain distillate of LP column near LP azeotrope and obtain bottom product near to the pure methanol. The distillate stream of LP column is then sent to the HP column. Notice that because this distillate composition is now on the right side of the T-xy plot at high-pressure, pure iC5 can be obtained at column bottoms while distillate composition can be near the HP azeotrope.

Although this is a feasible flowsheet to achieve the separation task, the question of how far apart of these two azeotropic compositions is very important

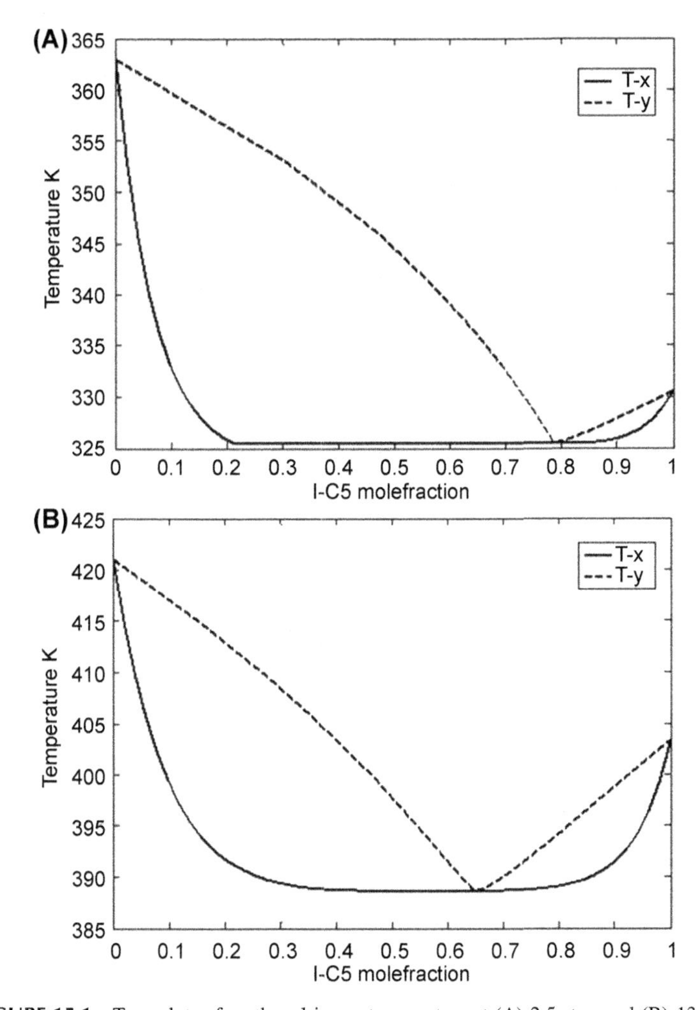

FIGURE 15.1 T-xy plots of methanol-isopentane system at (A) 2.5 atm, and (B) 13 atm.

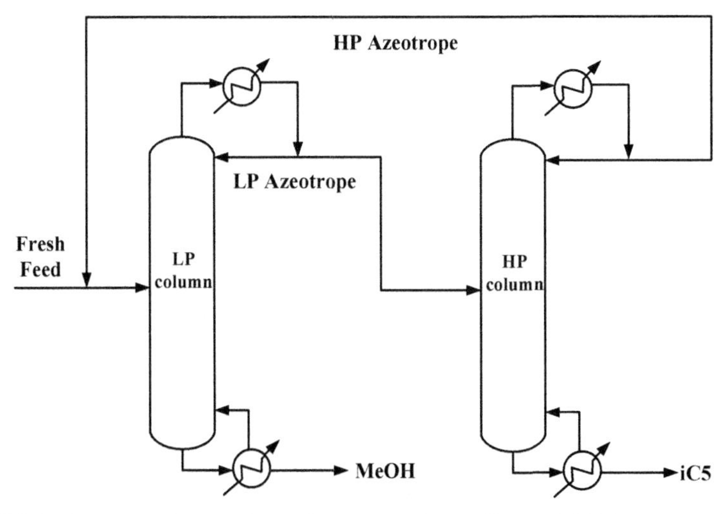

FIGURE 15.2 Azeotropic separation with pressure-swing distillation. *HP*, high pressure; *LP*, low pressure.

in determining the competitiveness of this flowsheet versus other separation methods. The main thing to look at is the recycle flow rate. Large recirculation flow rate will result in larger operating (mainly contributed by energy) and capital costs. From total material and composition balances, the ratio of HP recycle versus fresh feed flow rates can be estimated easily.

Of course, this estimation is dependent on the selection of the column operating pressures. The usual way for the selection of operating pressure at LP column is to have a high enough top temperature to permit the use of inexpensive cooling water in its condenser. For the HP column, the decision is usually dependent on the unit price of the steam used in the reboiler. The decision here is to select a high enough pressure but still permit to use the same grade of steam. Other limitation for the HP column is to avoid high-temperature reboiler to prevent component decomposition, the forming of oligomer, or other problems caused by high temperature. Section 15.4.1.1 discusses a detailed Aspen Plus design flowsheet for the separation of methanol and isopentane using this separation method.

15.2.2 Heterogeneous Binary Azeotrope Separation

From Fig. 15.1, one distinguish characteristic of this mixture can easily be observed. It is found that the azeotrope after condensing to liquid can naturally be separated into two liquid phases. This can be observed by a horizontal line of liquid compositions at the T-xy plot. The following alternative flowsheet in Fig. 15.3 takes advantage of the liquid–liquid

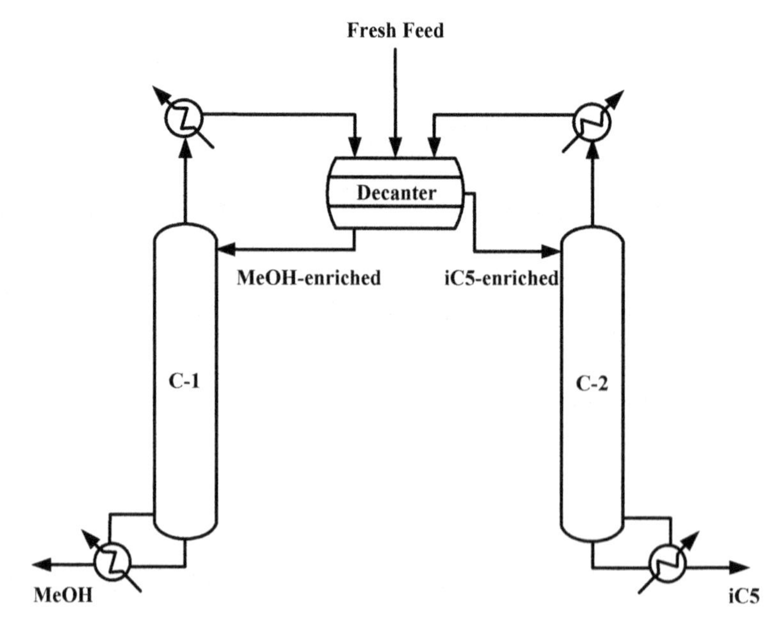

FIGURE 15.3 Azeotropic separation using two-strippers/decanter.

separation in a decanter to achieve the separation. Note that the methanol-enriched phase can further be separated in a stripper to obtain pure methanol, and the iC5-enriched phase can also be separated in another stripper to obtain pure iC5.

Notice that in this design flowsheet, the fresh feed is fed into the decanter. The reason is because the feed composition is already in the liquid–liquid splitting region. Chapter 7 in Luyben and Chien (2010) illustrated other applications where the fresh feed can be fed into one of the two strippers. The decision of the feed location can easily be determined by the feed composition.

The other choice is the operating pressures of the two strippers. The decision is similar in the pressure-swing distillation to achieve easy vapor–liquid separation and to avoid using more expensive cooling or heating mediums. Section 15.4.1.2 will develop a detailed Aspen Plus design flowsheet using this separation method and then compare with its result with that of pressure-swing distillation.

15.2.3 Other Separation Methods

Other separation methods without adding entrainer such as hybrid distillation-adsorption and hybrid distillation-pervaporation will not be discussed in this chapter. The main reason is that the efficiency of these separation methods is highly dependent on the performance of the adsorbent (e.g., molecular sieve) in the adsorption unit or the membrane in the pervaporation unit. Luyben and Chien (2010) (Chapter 14) discussed a hybrid distillation-pervaporation application for the separation of ethanol and water.

15.3 AZEOTROPIC SEPARATION METHOD BY ADDING ENTRAINER

If the azeotropic composition of mixture is not sensitive with the operating pressure or the azeotropic mixture does not exhibit liquid–liquid splitting behavior at azeotropic composition, an entrainer can be added into the system to aid the separation. The added entrainer must fulfill its intended purpose. In the following, we will explain two commonly used separation methods in the industry.

15.3.1 Heterogeneous Azeotropic Distillation

One way to separate azeotropic mixture is to add a light entrainer into the system so that at least one more azeotrope can be formed to help in the separation. The most common application is to form a heterogeneous minimum-temperature azeotrope so that one of the original components can be carried overhead in a heterogeneous distillation column and then liquid–liquid splitting in a decanter. Various industrial applications can be found in Luyben and Chien (2010) (Chapters 8 and 9). Wu et al. (2011) illustrated another application where a middle decanter was designed.

In the following, we will use ethanol dehydration as an example to illustrate the main principle of this separation system. The RCM, liquid–liquid envelope and material balance lines of this system with cyclohexane (CYH) as entrainer is shown in Fig. 15.4. As shown in this figure, the ethanol (EtOH)–water mixture

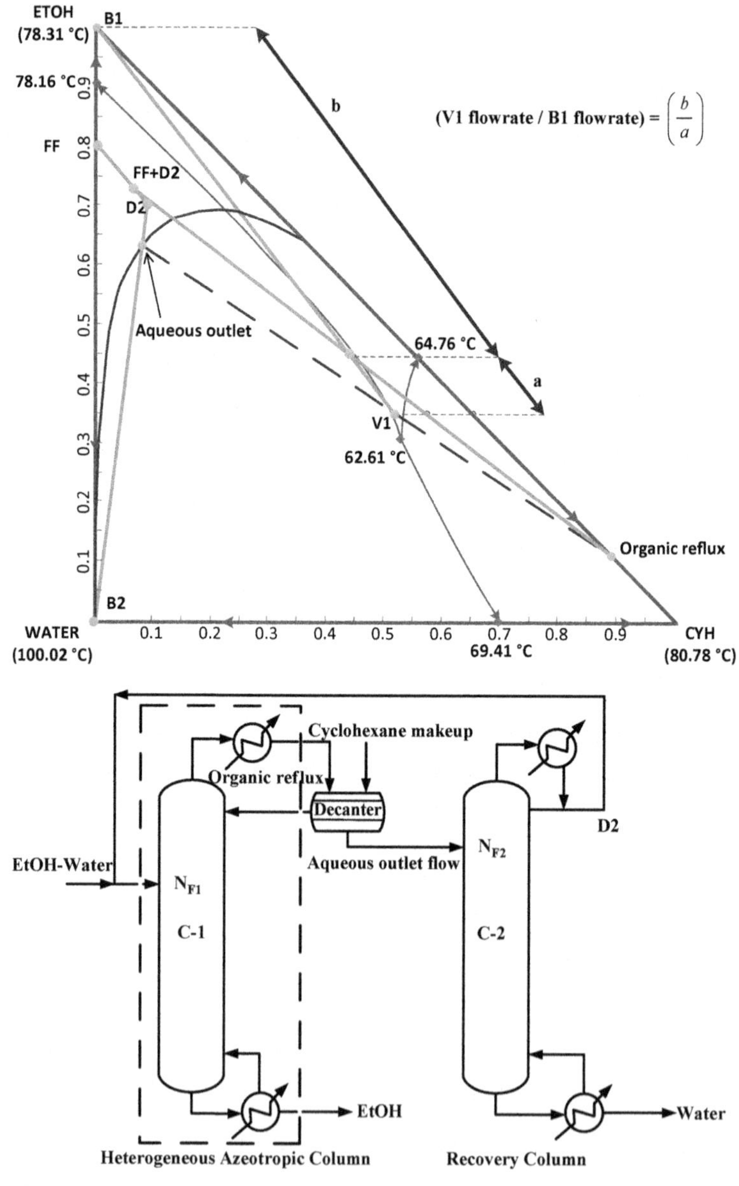

FIGURE 15.4 Azeotropic separation with heterogeneous azeotropic distillation.

has a minimum-boiling azeotrope with composition of around 90 mol % H_2O and azeotropic temperature of 78.16°C. By adding cyclohexane into the system, as shown in Fig. 15.4, three additional azeotropes are formed. One desirable ternary azeotrope is heterogeneous with an azeotropic temperature of 62.61°C, which is the minimum temperature for the entire ternary system.

The design flowsheet in Fig. 15.4 contains a heterogeneous azeotropic distillation column and a recovery column. Another preconcentrator column was not shown in this flowsheet. The purpose of the preconcentrator column is to remove large portion of water in the fresh feed, as a diluted feed composition was commonly obtained from fermentation broth in the bioethanol process. The distillate composition of the preconcentrator column is an important design variable. If this composition is set to be very close to the ethanol—water composition, the energy requirement of the preconcentrator column will become large. If it is set to be too far away from the azeotropic composition, the burden will be shifted to the azeotropic separation system. Luyben (2012) found a trade-off decision of the distillate composition of preconcentrator column to be at 80 mol% EtOH, which will be as the feed composition for the azeotropic separation system.

This 80 mol% EtOH mixture is fed into a heterogeneous azeotropic column. By adding cyclohexane into the system through an organic reflux stream, the bottom composition can approach pure EtOH and the column top vapor can approach the ternary azeotrope at minimum temperature. This azeotrope is heterogeneous, so by condensing and cooling the top vapor to 40°C in a decanter, natural liquid—liquid separation into an organic phase and an aqueous phase will occur. The organic phase, which is rich in entrainer, is recycled to the heterogeneous azeotropic column. The composition of the aqueous phase still contains significant amounts of the EtOH product, so it is fed into the recovery column. The bottom of this recovery column is designed to draw out the water product while the distillate stream containing ethanol is recycled back to the heterogeneous azeotropic distillation column.

A backup location of D2 is shown in Fig. 15.4A. By combining the fresh feed with D2 (as shown in the design flowsheet), a combined feed location is shown in Fig. 15.4A as "FF + D2."

The material balance line of the heterogeneous azeotropic column can be explained by considering the dashed-line envelope in Fig. 15.4B. For this envelope, there are two inlet streams, which are combined feed and organic reflux, and two outlet streams, which are EtOH product and top vapor. The material lines can be drawn by connecting these two feed locations in the figure and then splitting into two outlet locations. These two material balance lines in Fig. 15.4A can be used roughly to estimate the ratio of top vapor flow rate versus EtOH product flow rate. More detailed process insight will be explained in Section 15.4.2.1 when detailed Aspen Plus simulation flowsheet is developed using this separation method, with cyclohexane as an entrainer.

15.3.2 Extractive Distillation

Another way to separate azeotropic mixture is to add a heavy entrainer into an extractive distillation column so that the relative volatility of the original two components can be greatly enhanced. Thus, one original component can go overhead and the other component will go with the heavy entrainer to the column bottoms. A second entrainer recovery column is designed to separate this stream so that the entrainer can be recycled back to the extractive distillation column. The conceptual design flowsheet of the ethanol dehydration system is shown in Fig. 15.5.

Note that the upper section of the extractive distillation column (above the entrainer feed location) is called the rectifying section, and its purpose is to separate ethanol and the heavy entrainer. The middle section (between the entrainer feed location and the fresh feed location) is called the extractive section. The purpose of this section is to suppress water from going up the column. The bottom section of this column (below the fresh feed location) is called the stripping section, and its purpose is to prevent ethanol from going down the column.

The most important decision in this design is to choose the most effective entrainer to enhance the relative volatility of the original two components. Luyben and Chien (2010) (Chapter 10) reported the use of isovolatility and equivolatility curves to compare candidate entrainers. Because the equivolatility curves cannot be generated by Aspen Plus automatically, another relative volatility plot that shows the enhancement of relative volatility by a particular

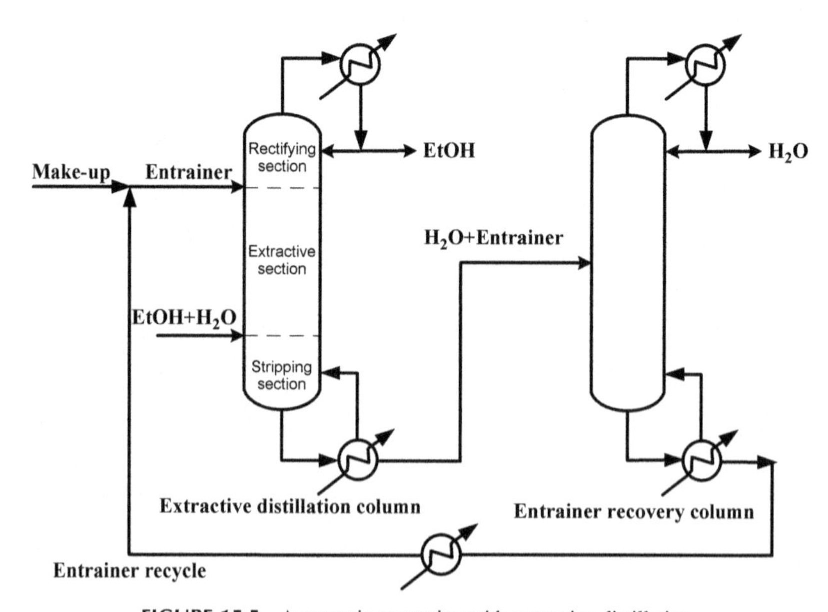

FIGURE 15.5 Azeotropic separation with extractive distillation.

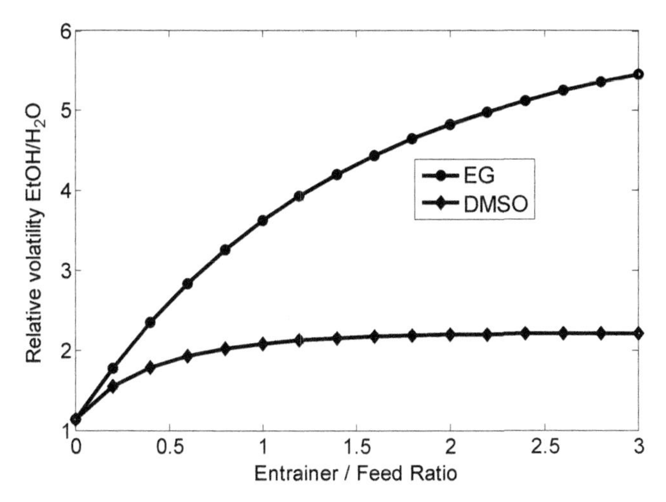

FIGURE 15.6 Enhancement of relative volatility for the ethanol dehydration system.

entrainer is shown in Fig. 15.6. In this plot, the starting composition can be set at the fresh feed composition. By gradually adding heavy entrainer into the system, we can calculate the enhancement of relative volatility at various feed ratios. This figure demonstrates that ethylene glycol (EG) is a much more effective entrainer than dimethyl sulfoxide for this separation system.

The above relative volatility plot in Fig. 15.6 is easily generated using Aspen Plus. The way to generate each point in the plot is to use the *Flash2* module in the unit operation library. Without adding entrainer, the relative volatility of the feed composition is near to 1.0. At any feed ratio, the vapor and liquid composition in equilibrium at this *FLASH2* unit can be calculated by Aspen Plus under flash operating at studied pressure and with setting of the flow ratio of vapor outlet to liquid outlet to be at a small value to simulate the equilibrium condition near bubble point. By excerpting information about the vapor and liquid compositions of EtOH and water, the relative volatility between these two components can be calculated.

Other factors that affect the entrainer performance are the y-x or T-xy plots. The y-x or T-xy plots for the EtOH-entrainer pair can be used to determine the difficulty level of separation in the rectifying section of the extractive distillation column. On the other hand, the same set of plots for the water–entrainer pair can also be used for the same function in the entrainer recovery column. Of course, thermally stable, nontoxic, low-price, and other favorable physical properties should also be considered in the selection of proper heavy entrainer. Section 15.4.2.2 shows a detailed Aspen Plus simulation flowsheet using this separation method by adding EG into this extractive distillation system and comparing its result with that of heterogeneous azeotropic distillation.

15.3.3 Other Separation Method by Adding Another Component

There are other separation methods used in the industry. These include the use of a dissolved salt as separating agent in the *salt extractive distillation* system (Liano-Restrepo and Aguilar-Arias, 2003); nonvolatile hyperbranched polymers or ionic liquid as entrainers in the extractive distillation system (Lei et al., 2014); combination of liquid–liquid extraction with distillation in the hybrid extraction-distillation system (Chen et al., 2015a); or the use of a reactive intermediate to react with one of the azeotropic components in the RD column, followed by reverse reaction in another RD column to obtain that component (Su et al., 2013). The discussion of these separation systems are excluded from this chapter.

15.4 ASPEN PLUS SIMULATIONS OF TWO INDUSTRIAL EXAMPLES

For a particular application, there are usually more than one feasible ways to achieve the separation. In the following, two industrial examples are used to demonstrate the importance of selecting the most effective separation method in terms of energy savings and total annual cost (TAC). Aspen Plus version 8.4 is used for all the simulation runs in this chapter.

15.4.1 Methanol and Isopentane Separation

As discussed in Sections 15.2.1 and 15.2.2, two alternative designs (see flowsheets in Figs. 15.2 and 15.3) can be used to achieve the separation objective of methanol and isopentane mixture. In the following subsections, we will develop the optimized flowsheets for using these two alternative designs.

15.4.1.1 Simulation of Pressure-Swing Distillation System

The optimal design flowsheet via pressure-swing distillation in Fig. 15.2 will be developed. The feed flow rate is assumed to be 500 kmol/h with equal molar of methanol and isopentane at 320K.

There are several design and operating variables that need to be determined in the design flowsheet of Fig. 15.2 including operating pressures, total stages, and feed locations of both LP and HP columns. The pressure of the LP column is set to 2.5 atm so that cooling water can be used in the condenser. The pressure of the HP column is set to 13 atm so that large difference in the azeotropic composition can be achieved, but the same grade of steam (low-pressure steam at 433.15K) can still be used in the reboiler with about 30K temperature difference between the bottom temperature of HP column and the steam temperature.

To further reduce the design variables, the two columns can be replaced by two strippers with feed location at first stage, and the top vapor of each stripper is again condensed to saturated liquid stream for easier transportation via pump. The one degree of freedom for each stripper can be set to specify the bottom composition at 99.9 mol%. With this simplifying design flowsheet, there are only two design variables that need to be determined to minimize TAC, i.e., total stages of LP column (NT1) and total stages of HP column (NT2).

The TAC includes annualized capital costs and the annual operating costs of the process equipment in this system. The capital cost includes costs for two columns, and their reboilers and condensers. The operating costs include cost for steam and cooling. The formulae for calculating the TAC were adopted from Luyben (2011) and Chen et al. (2015b). Table 15.1 summarizes the formulae for TAC calculations.

The optimal design flowsheet is shown in Fig. 15.7. Note that although using just strippers, the two top vapor compositions are all quite near the LP and HP azeotropic compositions. The itemized TAC terms of this design flowsheet are shown in Table 15.2 and will be compared to another feasible design flowsheet developed later.

In the Aspen Plus simulation, NRTL thermodynamic model was selected as the property method. There are built-in binary parameters available in Aspen to be used for the simulation study. For other simulation cases where there is no built-in parameter in Aspen, a reasonable way is to use UNIFAC to estimate the missing pairs. The NRTL parameters used in simulating the design flowsheet of Fig. 15.7 can be found in Table 15.3. The simulation of this separation process is quite easy only requiring two RADFRAC modules and a MIXER module in Aspen Plus. For the two RADFRAC modules, "none condenser" option was selected to represent two strippers.

Some further investigations were made to change the two strippers to two regular columns to set the top compositions closer to their azeotropic compositions. However, no significant reduction of TAC or steam cost can be made with this design change of having more design variables in the flowsheet.

15.4.1.2 Simulation of Heterogeneous Binary Azeotropic Separation System

For the design in Fig. 15.3, the design variables include operating pressure of C-1 and C-2 strippers, and their total stages. The two product purity specifications were set to match those in the pressure-swing distillation (both at 99.9 mol%). Similar to the case of Fig. 15.2, both strippers are set to operate at 2.5 atm, so that cooling water can still be used in both the condensers. The remaining two design variables are the total stages of the two strippers. Note that the feed location of the strippers should be at top stage. Fig. 15.8 shows the optimal design flowsheet of this separation method by minimizing TAC. The optimal total stages for C-1 stripper is NT1 = 5 and for C-2 stripper is NT2 = 9.

The capital costs include costs for two strippers, two reboilers, two condensers, and a decanter. The decanter is sized to be a tank with L/D = 2 and total liquid residence time is set to be 10 min and the tank is half full with

TABLE 15.1 Basis of Economics and Equipment Sizing

Column diameter (D): Aspen tray sizing
Column length (L): NT trays with 2-feet spacing plus 20% extra length
Column and other vessel (D and L are in meters)
Capital cost = $17{,}640(D)^{1.066}(L)^{0.802}$
Condensers (area in m^2)
Heat-transfer coefficient = 0.852 kW/K-m^2
Differential temperature = reflux-drum temperature − 315K
Capital cost = 7296 $(area)^{0.65}$
Reboilers (area in m^2)
Heat-transfer coefficient = 0.568 kW/K-m^2
Differential temperature = steam temperature − base temperature ($\Delta T > 20$ K)
Capital cost = 7296 $(area)^{0.65}$
Heat exchangers, liquid-to-liquid (area in m^2)
Heat-transfer coefficient = 0.852 kW/K-m^2
Differential temperature = LMTD of (inlet and outlet temperature differences)
Capital cost = 7296 $(area)^{0.65}$
Coolers (area in m^2)
Heat-transfer coefficient = 0.852 kW/K-m^2
Differential temperature = LMTD of (inlet or outlet temperature − 315 K)
Capital cost = 7296 $(area)^{0.65}$
Energy cost
HP steam = $9.88/GJ (41 barg, 254°C)
MP steam = $8.22/GJ (10 barg, 184°C)
LP steam = $7.78/GJ (5 barg, 160°C)
Cooling water = $0.354/GJ
Electricity = $16.9/GJ
TAC = (capital cost/payback period) + energy cost
Payback period = 3 years

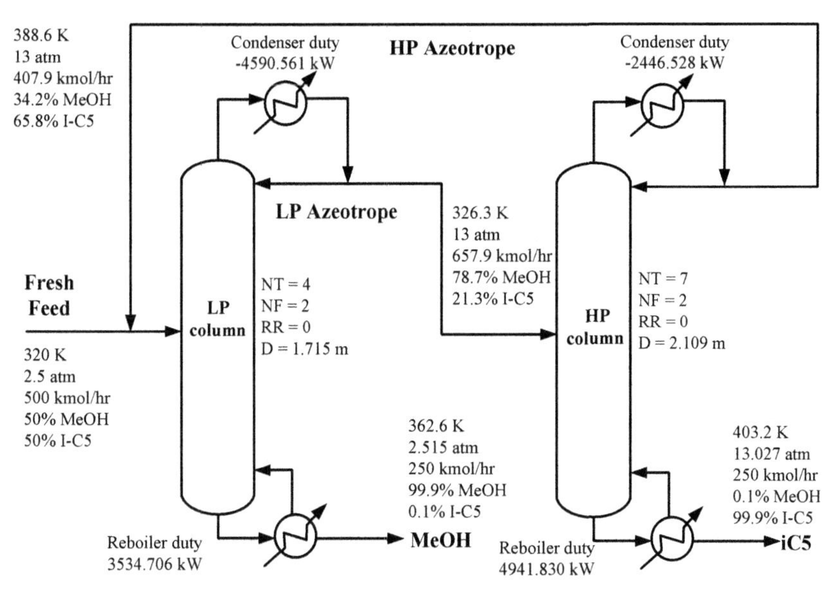

FIGURE 15.7 Optimal design flowsheet for the separation of methanol and isopentane via pressure-swing distillation. *HP*, high pressure; *LP*, low pressure.

TABLE 15.2 Itemized Total Annual Cost (TAC) Terms of iC5-MeOH Separation Systems

Configurations	Optimal Flowsheet in Fig. 15.7		Optimal Flowsheet in Fig. 15.8	
	C1	C2	C1	C2
Capital cost of column ($1000)	58.988	132.329	22.943	157.699
Capital cost of reboiler ($1000)	133.537	296.170	55.820	125.051
Capital cost of condenser ($1000)	422.160	80.245	112.143	418.450
Capital cost of decanter ($1000)	—		223.681	
Steam cost ($1000 per year)	792.093	1107.380	205.309	1029.339
Cooling water cost ($1000 per year)	46.699	25.047	6.601	46.248
Total reboiler duty (kW)	8477.4		5510.3	
Total steam cost ($1000 per year)	1899.473		1234.648	
TAC ($1000 per year)	2345.695		1659.426	

TABLE 15.3 NRTL Model Parameters for the iC5-MeOH Separation System

Component i	iC5
Component j	MeOH
Source	Aspen APV84 VLE-IG
a_{ij}	0
a_{ji}	0
b_{ij} (K)	562.749
b_{ji} (K)	442.98
c_{ij}	0.3

Aspen Plus NRTL:

$$ln\ \gamma_i = \frac{\sum_j x_j \tau_{ji} G_{ji}}{\sum_k x_k G_{ki}} + \sum_j \frac{x_j G_{ij}}{\sum_k x_k G_{kj}} \left[\tau_{ij} - \frac{\sum_m x_m \tau_{mj} G_{mj}}{\sum_k x_k G_{kj}} \right]$$

where $G_{ij} = \exp(-\alpha_{ij}\tau_{ij})$

$$\tau_{ij} = \alpha_{ij} + \frac{\alpha_{ij}}{T}$$

$$\alpha_{ij} = c_{ij}, \tau_{ii} = 0, G_{ii} = 1$$

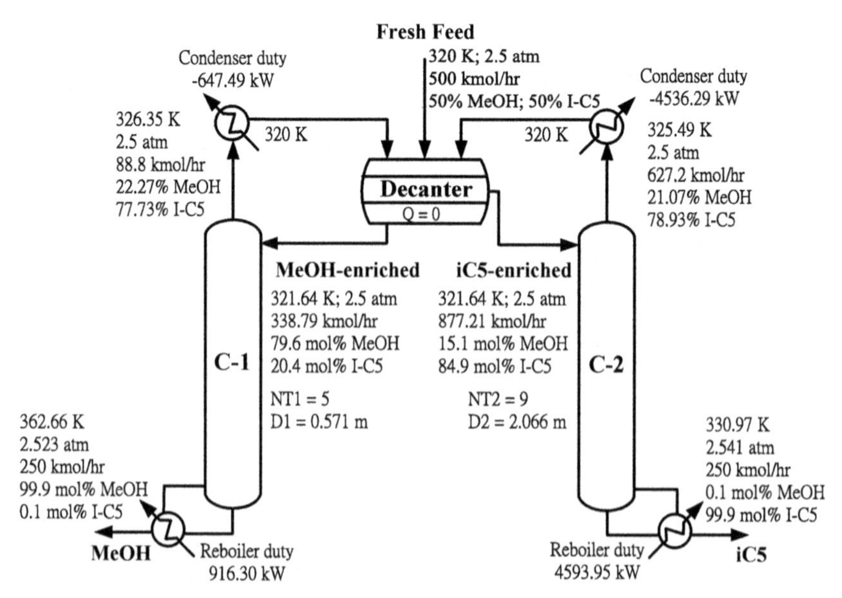

FIGURE 15.8 Optimal design flowsheet for the separation of methanol and isopentane via two-strippers/decanter.

liquids. The operating costs include the costs for steam and cooling water. Again, NRTL thermodynamic model was selected as property method in the simulation study. For this design flowsheet, two RADFRAC modules were again used to represent two strippers. Note that the usages of RADFRAC module are very versatile, including simulations of regular column, stripper, rectifier, extractive distillation column, and heterogeneous azeotropic column used in this chapter. For this design flowsheet, two HEATX modules were used to represent the top condensers and another DECANTER module was use too.

Table 15.2 also shows the itemized TAC terms for this design concept. It is demonstrated that significant reductions in both the TAC and steam cost can be made by choosing the more suitable separation method in this section. Using the design concept in Fig. 15.3, the steam cost is reduced by 35.0% with TAC reduced by 29.3%. The main reason for the significant savings is because of the utilization of natural liquid–liquid splitting in the design concept of Fig. 15.3.

15.4.2 Ethanol Dehydration Process

In Sections 15.3.1 and 15.3.2, two alternative flowsheets can be used to achieve the same separation task of ethanol and water. In this section, we will develop the optimized design flowsheets using these alternative design concepts.

15.4.2.1 Simulation of Heterogeneous Azeotropic Distillation System

The optimal design flowsheet via heterogeneous azeotropic distillation as in Fig. 15.4B will be developed first. The feed into the azeotropic separation system is from the distillate stream of a preconcentrator column. This feed flow rate is assumed to be 100 kmol/h with already preconcentrated composition of 80 mol % EtOH and 20 mol% water. The feed temperature is assumed at 320 K.

There are five design variables needed to be determined in the optimal design flowsheet, i.e., EtOH distillate composition (XD2) of the recovery column, total number of stages for the heterogeneous azeotropic column and the recovery column (NT1 and NT2), and the two feed stages (NF1 and NF2). The product specification of EtOH is set to be fuel grade of 99.8 mol% while that of the water is set to be very pure at 99.95 mol%. In each simulation run, EtOH product specification is achieved by varying reboiler duty of the heterogeneous azeotropic column, and the water product specification is achieved by varying reboiler duty of the recovery column. The entrainer makeup flow rate will be very small to balance the entrainer loss from the two bottom streams.

The flowchart for sequential iterative procedure used to obtain the optimal design flowsheet is illustrated in Fig. 15.9 with the optimal design flowsheet shown in Fig. 15.10. The capital costs include costs for two columns, two reboilers, two condensers, and a decanter. The operating costs include the costs for steam and cooling water costs, as well as a small entrainer makeup cost. The itemized TAC

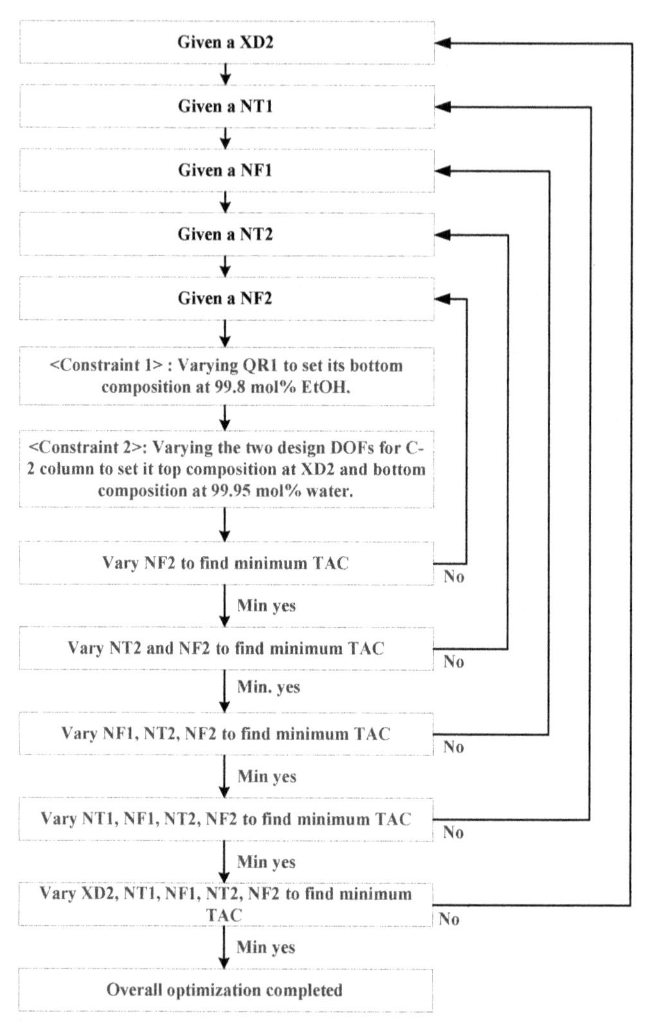

FIGURE 15.9 Flowchart for sequential iterative procedure in optimization of heterogeneous azeotropic distillation system. *TAC*, total annual cost.

terms of this design flowsheet are shown in Table 15.4 and will be compared to another feasible design flowsheet of using extractive distillation system.

UNIQUAC thermodynamic model was selected as property method in the simulation study as in Luyben (2012). The model parameters used for this section and also for the next section are summarized in Table 15.5. For this design flowsheet, two RADFRAC modules were again used to represent the heterogeneous azeotropic column and the recovery column. Note that in the simulation of heterogeneous azeotropic column, stage 1 can be declared as the decanter. The light (organic) phase out of the decanter is set to totally

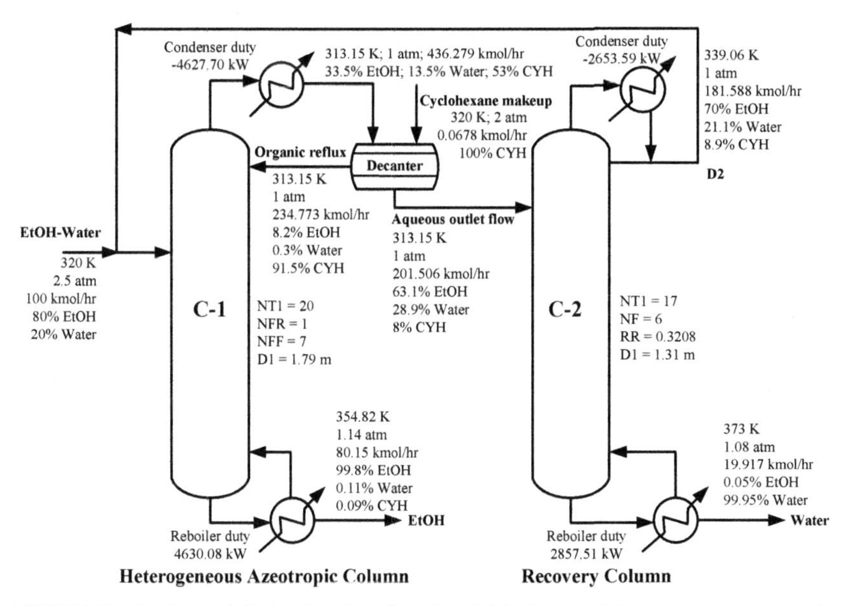

FIGURE 15.10 Optimal design flowsheet for ethanol dehydration via heterogeneous azeotropic distillation.

TABLE 15.4 Itemized Total Annual Cost (TAC) Terms of Ethanol Dehydration Systems

Configurations	Optimal Flowsheet in Fig. 15.9		Optimal Flowsheet in Fig. 15.10	
	C1	C2	C1	C2
Capital cost of column ($1000)	270.65	150.862	167.504	43.714
Capital cost of reboiler ($1000)	114.356	99.158	123.261	40.261
Capital cost of condenser ($1000)	419.552	186.080	77.663	23.031
Capital cost of decanter ($1000)	68.814		—	
Capital cost of cooler ($1000)	—		14.338	
Steam cost ($1000 per year)	1037.43	639.232	365.337	121.534
Cooling water cost ($1000 per year)	47.180	27.007	12.044	2.955
Cost of cooler ($1000 $ per year)	—		3.876	
Entrainer makeup ($1000 per year)	70.634		1.654	
Total reboiler duty (kW)	7482.964		1967.8	
Total steam cost ($1000 per year)	1676.663		486.269	
TAC ($1000 per year)	2257.974		662.003	

TABLE 15.5 UNIQUAC Model Parameters for the Ethanol Dehydration Systems

Component i	H_2O	EtOH	H_2O	EtOH	H_2O
Component j	EtOH	Cyclohexane	Cyclohexane	EG	EG
Source	Aspen APV84 VLE-IG	Aspen APV84 VLE-IG	Aspen APV84 LLE-LIT	Aspen APV84 VLE-IG	Aspen APV84 VLE-IG
a_{ij}	−2.4936	0.8772	0	−8.2308	−0.6018
a_{ji}	2.0046	−2.8027	0	2.6876	0.6018
b_{ij} (K)	756.948	−215.96	−540.36	2632.93	120.779
b_{ji} (K)	−728.971	370.326	−1247.3	−959.565	−18.6714

EG, ethylene glycol.Aspen Plus UNIQUAC:

$$\ln \gamma_i = \ln\frac{\Phi_i}{x_i} + \frac{z}{2} q_i \ln\frac{\theta_i}{\Phi_i} - q'_i \ln t'_i - q'_i \sum_j \frac{\theta'_j \tau_{ij}}{t'_j} + l_i + q'_i - \frac{\Phi_i}{x_i} \sum_j x_j l_j$$

$$\text{Where}: \theta_i = \frac{q_i x_i}{q_T};\ q_T = \sum_k q_k x_k;\ \theta'_i = \frac{q'_i x_i}{q'_T};\ q'_T = \sum_k q'_k x_k;\ \Phi_i = \frac{r_i x_i}{r_T};\ r_T = \sum_k r_k x_k$$

$$l_i = \frac{z}{2}(r_i - q_i) + 1 - r_i;\ t'_i = \sum_k \theta'_k \tau_{ki};\ \tau_{ij} = \exp\left(a_{ij} + \frac{b_{ij}}{T}\right);\ \text{and } z = 10$$

recycle back to C-1 while the heavy (water) phase is set as the feed to C-2 column. With this modeling arrangement, no DECANTER module is needed in simulating this design flowsheet. This simulation tip is helpful for getting convergence result in the simulation study because of eliminating an external recycle loop in the flowsheet.

Another note is that some stages in the heterogeneous azeotropic column may exhibit two liquid phases; thus, VLLE calculations were allowed in simulation of C-1 column.

15.4.2.2 Simulation of Extractive Distillation System

There are many design variables to be determined in this flowsheet. To simplify the optimization procedure, the more important and complex extractive distillation column is optimized first. There are four design variables to be determined; these include entrainer-to-feed ratio (FE/FF), total stages (NT1), feed locations of fresh feed (NFF), and entrainer (NFE). The two design specifications for all the Aspen simulation runs are setting top composition at 99.8 mol% EtOH and setting the molar ratio of EtOH to the sums of EtOH and water in the bottom stream to be 0.0004. The reason for this bottom specification is to set the EtOH loss through this column bottoms so that the final water purity specification in the recovery column is achievable. The above two design specifications can be met by varying the two degrees of freedom in this column.

Sequential iterative optimization search procedure is used to find the optimal design, and the flowchart is illustrated in Fig. 15.11. In the flowchart, the entrainer-to-feed ratio is set as the outer iterative loop, NT1 as the middle loop, and NFF and NFE as the inner iterative loop. The procedure is to fix an entrainer-to-feed ratio and NT1 first and then to find the optimal NFF and NFE to minimize TAC1 at this particular combination of entrainer feed rate and NT1. The procedure is repeated for the other combinations of entrainer-to-feed ratio and NT1. After the optimum solution of NT1, NFF, and NFE at each entrainer-to-feed ratio is obtained, the total TAC of this two-column system can be calculated with the entrainer recovery column and the recycle stream included in the total TAC calculation. Additional costs in the total TAC include the annualized capital cost for the entrainer recovery column, the costs associated with the cooler for the recycled entrainer, the operating costs of the steam and cooling water to operate the entrainer recovery column, and the entrainer makeup cost.

In the simulation following the suggestion by Doherty and Malone (2001), the outlet temperature of the solvent cooler is set at 340K so that the solvent feed temperature can be about 10K below the distillate temperature of the extractive distillation column. Also, a very small entrainer makeup flow is determined to balance the entrainer loss from two distillate outlet streams.

The final optimized design flowsheet of this two-column system is shown in Fig. 15.12. Table 15.4 compares the TAC and the steam cost of the two

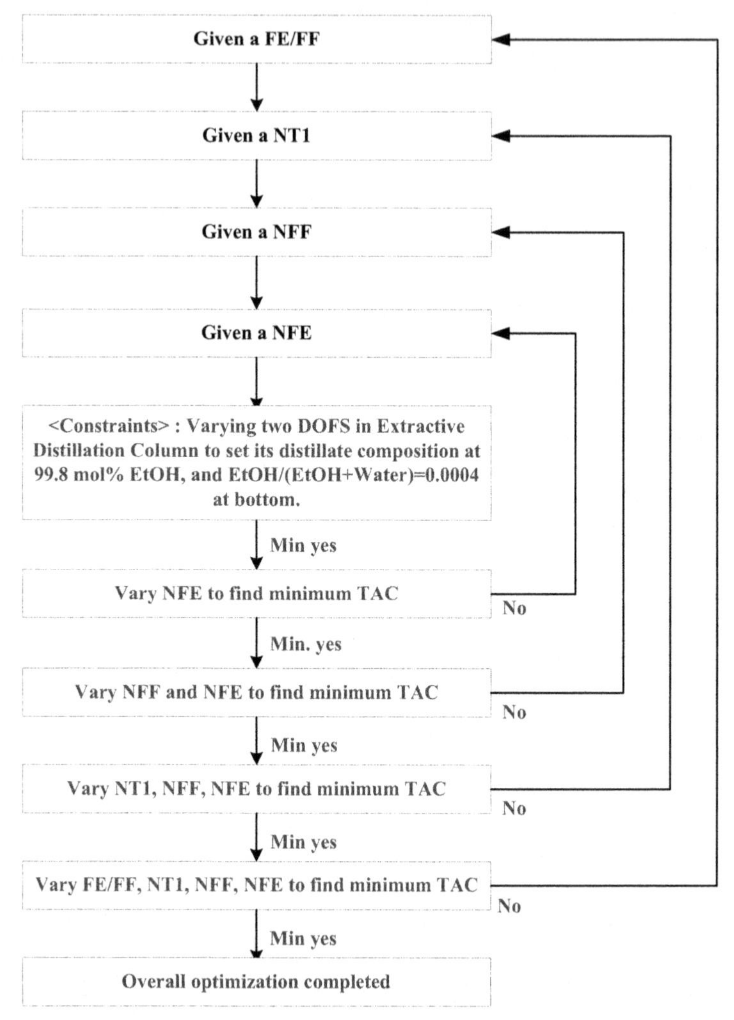

FIGURE 15.11 Flowchart for sequential iterative procedure in optimization of extractive distillation system. *TAC*, total annual cost.

alternative design flowsheets. It is observed that significant steam savings of 71.0% can be made by using separation method via extractive distillation. Significant savings in TAC of 70.7% can also be made by the optimal design flowsheet in Fig. 15.10.

The main reason that heterogeneous azeotropic distillation flowsheet is not competitive can be observed from Fig. 15.4. Analyzing the material balance lines for the heterogeneous azeotropic column indicated by red-dashed envelope, the flow ratio for the top vapor to the bottom stream can be estimated for this system. From Fig. 15.4, this ratio is more than five. That means large top

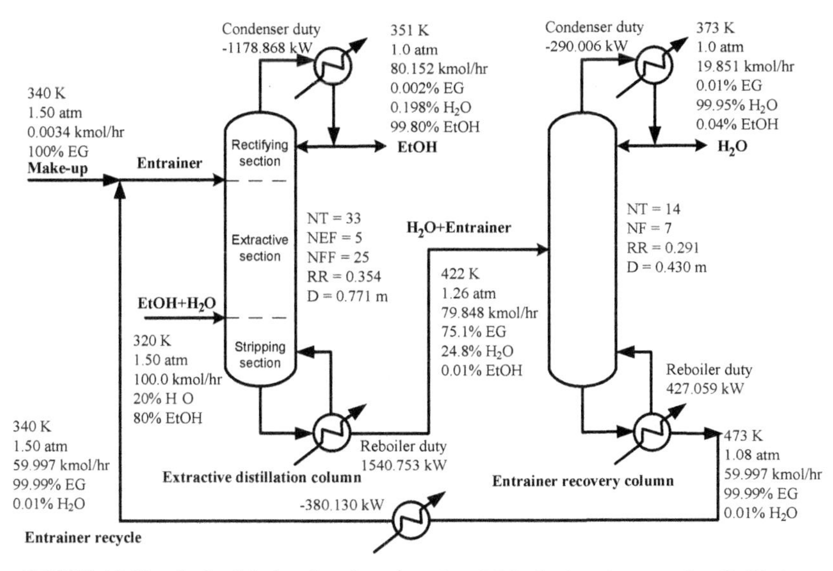

FIGURE 15.12 Optimal design flowsheet for ethanol dehydration via extractive distillation.

vapor flow rate is needed to perform this separation. This also means that larger organic reflux flow rate is recycled to this column, as well as larger aqueous flow to the entrainer recovery column, which in turn leads to large energy and capital costs. This deficiency of using this separation method can be easily found before doing any rigorous simulation by a plot including RCM, liquid–liquid, envelope and material balance lines.

To make this separation method to become more competitive, a light entrainer with phase behavior such as that in Fig. 15.13 is preferred. In this case, there will be much less top vapor flow rate in this system. However, this ideal light entrainer for the ethanol dehydration system is yet to be found.

15.5 FURTHER ENERGY SAVINGS VIA HEAT-INTEGRATION

In this section, several heat-integration methods widely used in industry will be outlined.

15.5.1 Feed–Effluent Heat Exchanger

Because the bottom temperature of a distillation column is at the highest, the first method is to utilize this heat to preheat the feed stream of this column. This is called an FEHE design. Implementing this heat-integration method required an additional heat exchanger. The trade-off will be the additional investment of this heat exchanger to the savings of the reduced reboiler duty. A typical FEHE design for pressure-swing distillation flowsheet is shown in

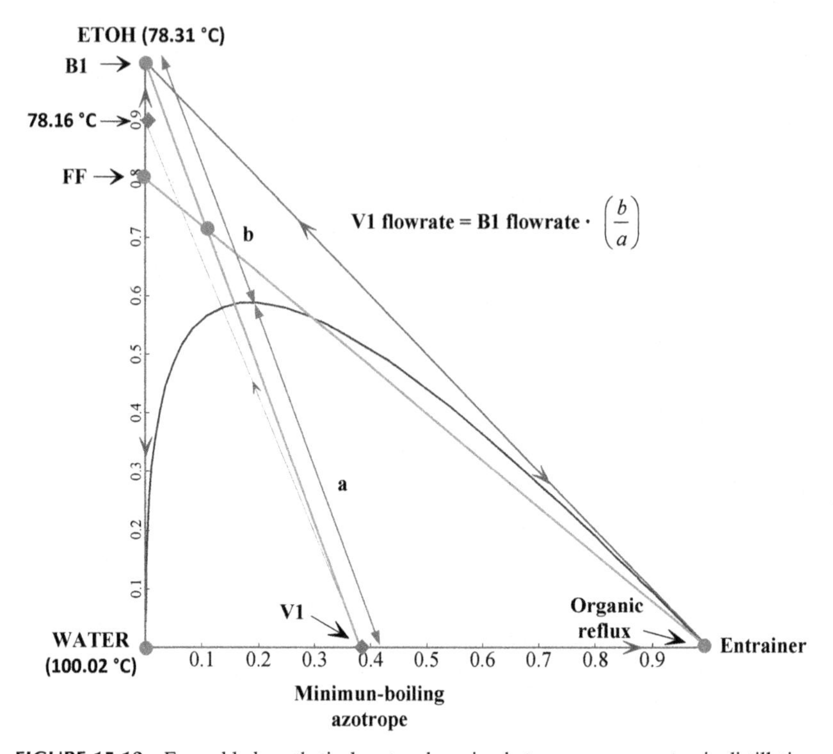

FIGURE 15.13 Favorable hypothetical system by using heterogeneous azeotropic distillation.

Fig. 15.14. This design concept can also be used in the extractive distillation flowsheet in Fig. 15.5. For the heterogeneous design flowsheet in Fig. 15.4B, the temperature difference between the hot and cold stream is usually not large enough to justify of using this heat-integration method.

15.5.2 Multieffect Distillation Columns

The design concept of multieffect distillation is to utilize the heat recovered in a condenser for use in another reboiler. With this design, a heat exchanger is installed, which served as condenser for the HP column and also as reboiler for the LP column. The conceptual design flowsheet of this design with two additional FEHEs is shown in Fig. 15.15. Note that because the heat removal in condenser should be exactly the same as the heat input in the reboiler, control degrees of freedom at the reboiler and condenser are lost. Most often, the dynamic and control of this complete heat-integration design will be hampered. An alternative design to trade-off economics/controllability is to install an auxiliary reboiler and/or auxiliary condenser. In this way, partial heat-integration can still be achieved without loss of all control degrees of freedom.

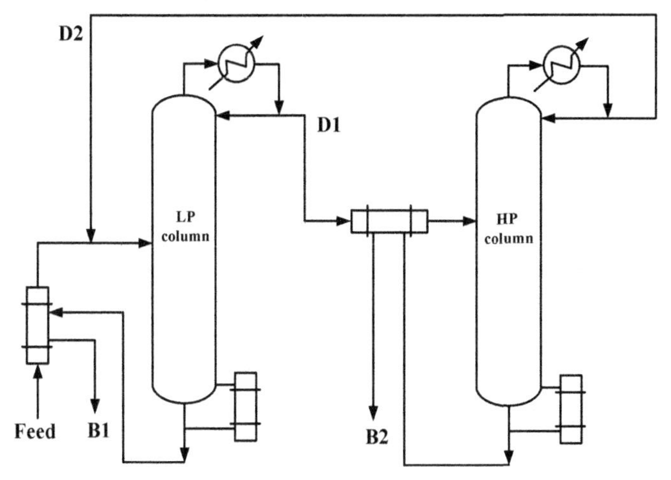

FIGURE 15.14 Pressure-swing distillation system with feed–effluent heat exchanger. *HP*, high pressure; *LP*, low pressure.

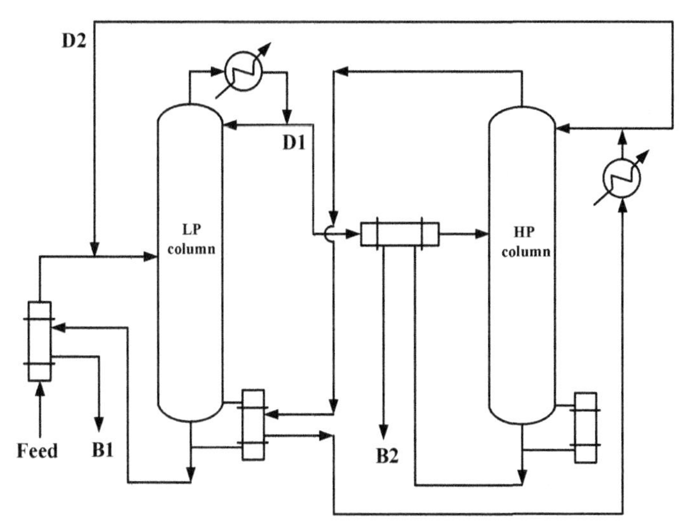

FIGURE 15.15 Pressure-swing distillation system with multieffect and feed–effluent heat exchanger.

This multieffect design is generally not suitable to be used in the heterogeneous azeotropic distillation flowsheet in Fig. 15.4. Although there are also two columns in this flowsheet, increasing operating pressure in the entrainer recovery column usually will have adverse effect in the ease of separation. For the extractive distillation system in Fig. 15.5, increasing operating pressure in the entrainer recovery column may cause its bottom temperature to be too high to prohibit the use of its original heat source.

15.5.3 Thermally Coupled (Dividing-Wall) Extractive Distillation System

Another method used in industry is to eliminate the remixing effect in a column sequence via thermally coupling two columns. This is also called dividing-wall column because a special-designed wall can be installed in a column to achieve the same purpose of thermally coupling the two columns. In this way, savings can be achieved with both energy and capital costs with only one column shell. Among the four previous methods for azeotropic separation, extractive distillation column will exhibit remixing effect. By observing the component profile of the extractive distillation column, heavier component to be separated in the C-2 column is usually having highest composition in the stripping section and lowest composition at the column bottoms. To save energy of this process, one way is to thermally couple two columns so that the vapor of extractive distillation column is supplied by a vapor side-draw from the entrainer recovery column. This design configuration is shown in Fig. 15.16A. The thermodynamically equivalent dividing-wall column design is illustrated in Fig. 15.16B.

Notice that the number of reboiler of this extractive distillation process is reduced from two to only one. The design configuration for three separation systems of IPA-H_2O, acetone-methanol, and dimethyl carbonate-methanol has been reported previously (Wu et al., 2013). The total reboiler duty can always be reduced. However, the total steam cost of this thermally coupled system may not be lower than the original design. The main reason is that the combined reboiler requires a heat medium with higher temperature because usually entrainer for extractive distillation is the heaviest of the system. The other disadvantage of the thermally coupled (dividing-wall) column design is that one important control degree of freedom (a reboiler duty) is lost in the energy-saving design; thus, the control performance will be hampered as compared with the original extractive distillation system. The details of the comparison can be found in Wu et al. (2013).

15.5.4 Thermally Coupled (Dividing-Wall) Heterogeneous Azeotropic Distillation System

In Wu et al. (2014), the heterogeneous azeotropic distillation system as in the conceptual design flowsheet in Fig. 15.4B was further intensified into a thermally coupled system. The way to thermally couple the original two columns is to eliminate the condenser at top of the recovery column by putting top vapor directly to the heterogeneous azeotropic column. To provide the liquid flow into the recovery column (as reflux in the original two-column design), a liquid side-draw is designed to draw off at the same stage of heterogeneous azeotropic column to the recovery column. This way, the aqueous outlet stream does not need to be restricted to enter the C-2 column at top stage. This design concept is shown in Fig. 15.17A.

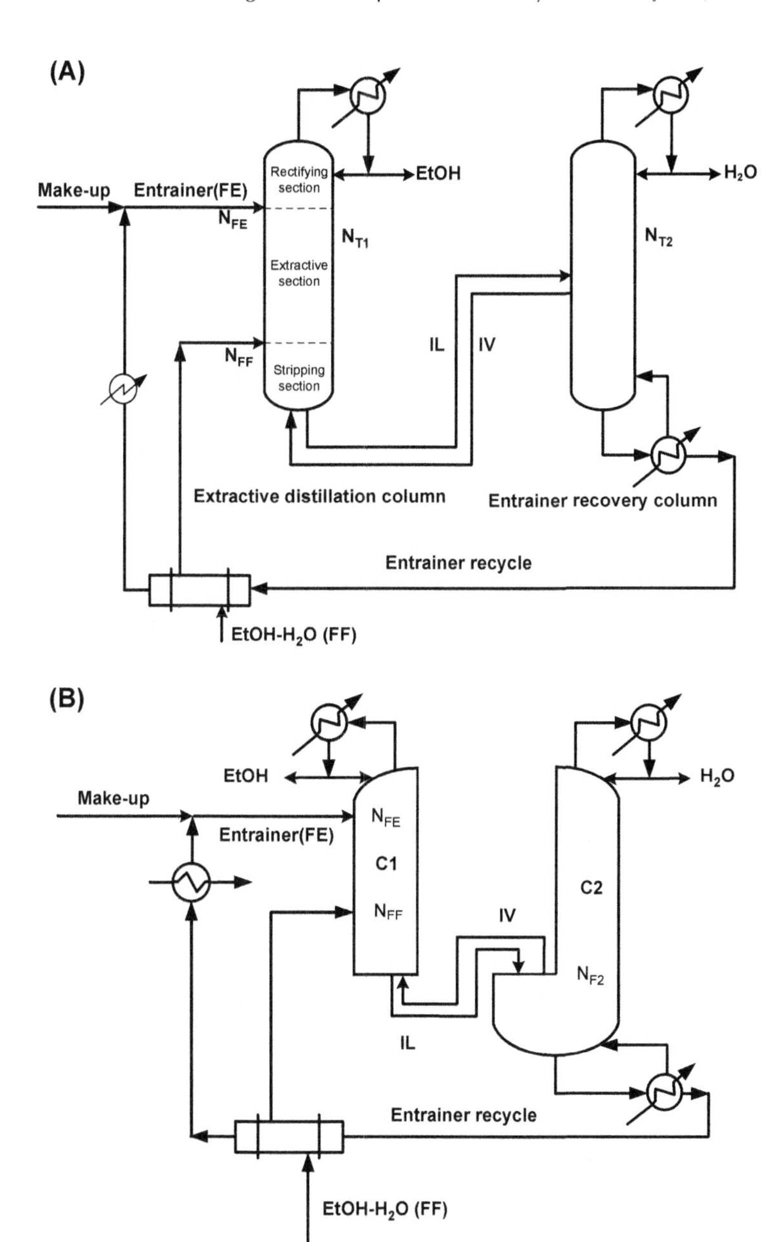

FIGURE 15.16 (A) Thermally coupled extractive distillation system. (B) Dividing-wall extractive distillation system. *HP*, high pressure; *LP*, low pressure.

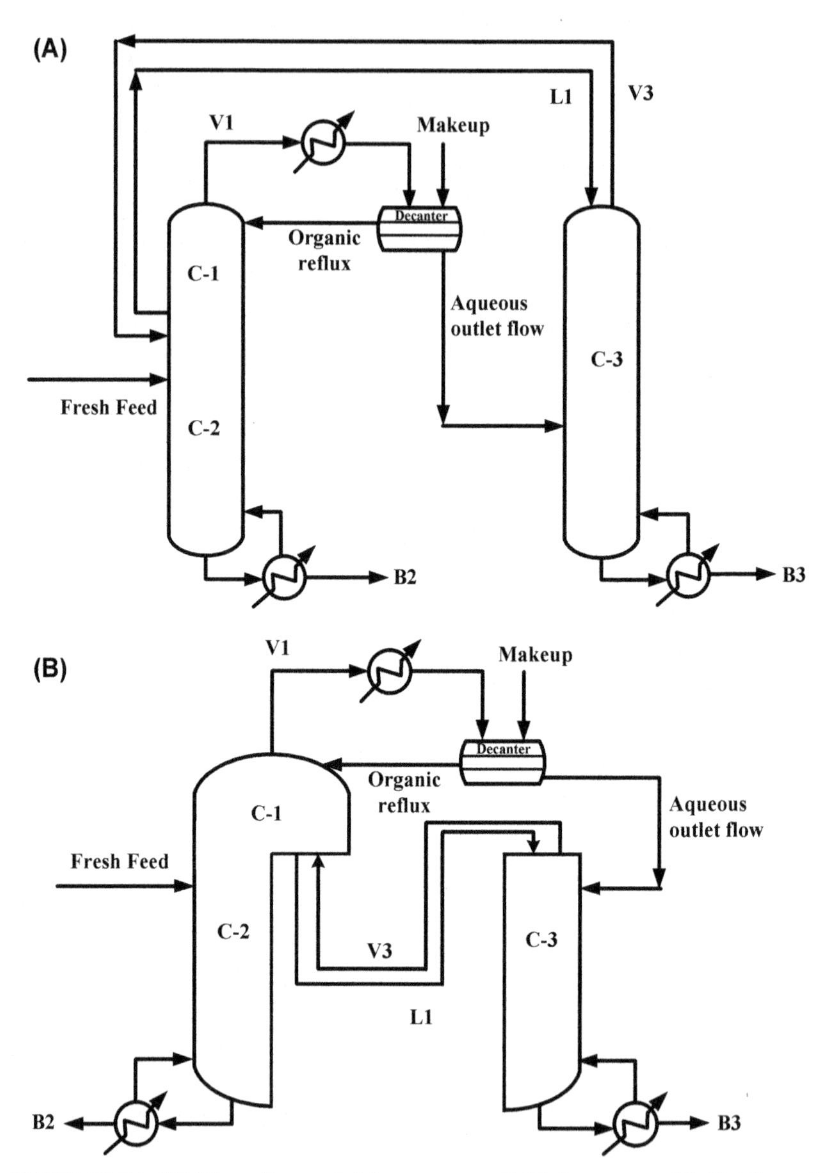

FIGURE 15.17 (A) Thermally coupled heterogeneous distillation system. (B) Dividing-wall heterogeneous distillation system.

To have the additional benefit of combining two columns into only one column shell to save space requirement at plant site, number of stages for the heterogeneous azeotropic column below the liquid side-draw stage should be the same as the total stages of C-2 column. In this way, the conceptual design flowsheet can be seen in Fig. 15.17B with a dividing wall at lower part of a single column.

It was demonstrated in Wu et al. (2014) by an example of pyridine dehydration that the original two-column system can be thermally coupled into a dividing-wall column with a single column shell. It is found that significant savings (29.48%) on the reboiler duty can be realized by the proposed design. Unlike extractive dividing-wall column systems, the savings on the reboiler duty can directly be reflected as the reduction of the steam cost. Significant reduction on the TAC (31.41%) can also be realized with the additional benefit of having only one column shell. The result is generic and is applicable to other heterogeneous azeotropic distillation systems.

As for the control performance, the original two important control degrees of freedom (two reboiler duties) are still preserved in the proposed dividing-wall column design. A conventional tray-temperature control strategy can still be implemented to manipulate these two reboiler duties. It is demonstrated that the two products can still be maintained at high-purity despite feed composition, feed flow rate, and liquid split ratio disturbances.

15.6 CONCLUSIONS

In this chapter, various flowsheet configurations for separating azeotropic mixtures are presented. Two industrial systems illustrated the importance of choosing the most effective separation method. For heterogeneous binary azeotropes, a separation process with a decanter and two strippers can be designed to take advantage of the natural liquid–liquid separation in this system. The possible deficiency of the pressure-swing distillation or the heterogeneous azeotropic distillation can easily be revealed from T-xy plots or the RCM, LLE, and material balance lines. For the extractive distillation process, the more effective entrainer can also be revealed with the relative volatility plot.

To further save operating energy of the azeotropic separation processes, the FEHE can be used for pressure-swing distillation and extractive distillation systems because there are enough temperature differences between the feed and bottoms streams. The multieffect distillation can also be used in pressure-swing distillation process. However, there will be a trade-off between the economics and dynamic controllability. For extractive distillation processes, besides the FEHE, thermally coupled design can also be applied to reduce the total reboiler duty. However, the attention should be given to the heat medium used in this system. There are several industrial examples where the total steam cost adversely increased by using thermally coupled design. As for the heterogeneous azeotropic distillation system, thermally coupled (dividing-wall) column design for further energy saving is recommended. Significant energy savings can be realized without hampering the control performance because the original two reboilers are still retained in this design.

EXERCISES

1. T-xy plots for 1-pentanol–water mixture are given at two different operating pressures, as in Fig. E1.

   Perform the following tasks:

   (a) Construct a possible process flowsheet for the separation of 1-pentanol–water mixture using pressure-swing distillation.

   (b) Estimate the distillate flow rate of the high-pressure column in the proposed flowsheet in (a). The feed condition is given as follows: total flow rate, 100 kmol/h; feed composition, 80 mol% 1-pentanol and 20 mol% water; and feed temperature, 320K.

(A)

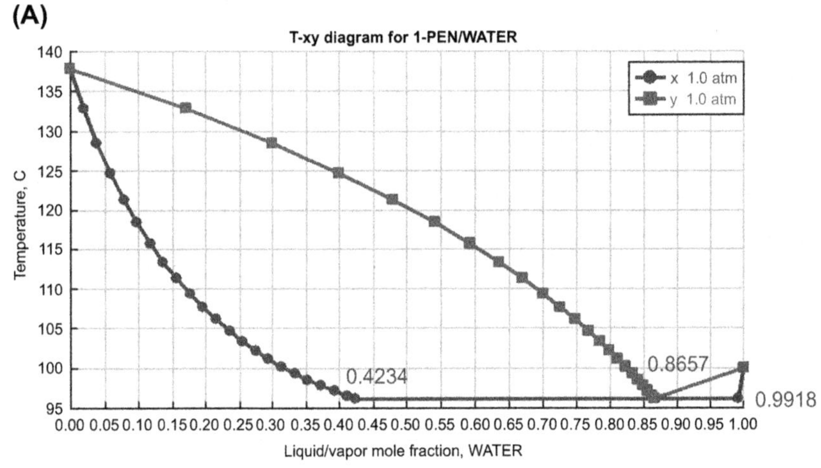

(B)

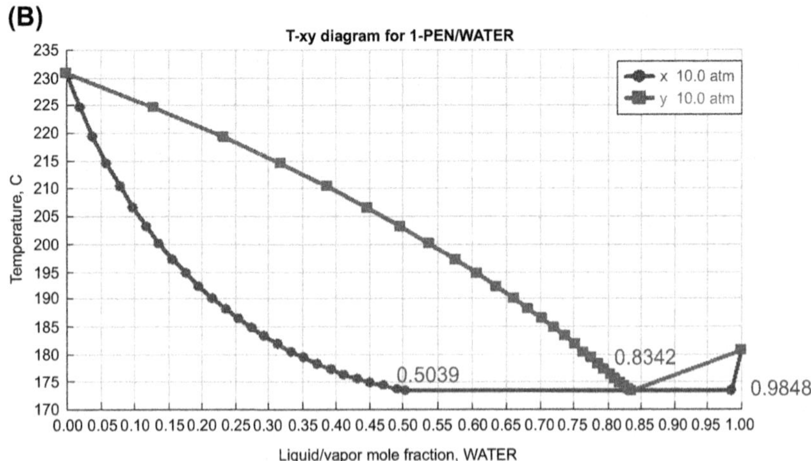

FIGURE E1 T-xy plot for 1-pentanol–water mixture: (A) 1.0 atm; (B) 10.0 atm.

(c) Perform process simulation for this flowsheet with product specifications both set to 99.9 mol%. The total number stage, feed location as well as the distillated composition of both columns may be varied to obtain an optimal design flowsheet, with minimum total reboiler duty.

2. To be more effective in separating the above mixture containing 1-pentanol and water, can you think of a better alternative design flowsheet to obtain pure 1-pentanol and water with the same feed flow rate and feed composition? Perform Aspen Plus process simulation with the same feed condition and the same product specifications as in Problem 1.
3. It is possible to separate an azeotropic mixture containing dimethyl carbonate and methanol using extractive distillation, by adding aniline as the heavy entrainer. The feed condition is given as follows: flow rate is 100 kmol/h; feed composition is 85 mol% methanol and 15 mol% dimethyl carbonate; and feed temperature at 320K.
   (a) Using FLASH2 module in Aspen Plus, plot the enhancement of relative volatility of this mixture system at several entrainer-to-feed ratio as in Fig. 15.6, when aniline is added into the system.
   (b) Perform an Aspen Plus process simulation of this design flowsheet with both product specifications set at 99.5 mol%.
4. Isopropyl alcohol (IPA) and water can be separated by adding a light entrainer, cyclohexane, into the system via heterogeneous azeotropic distillation. In a conceptual design flowsheet as in Fig. E2, one column

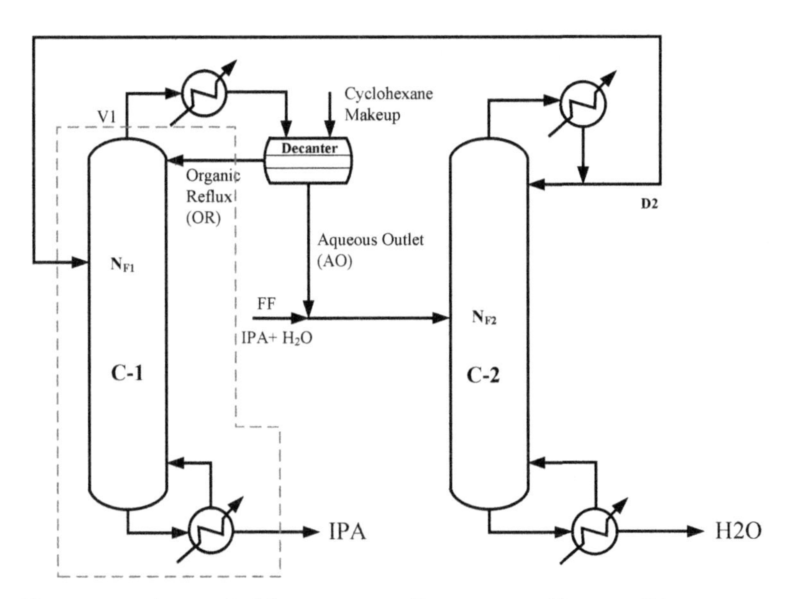

FIGURE E2 Separation of isopropyl alcohol–water mixture with heterogeneous azeotropic distillation.

can be served as preconcentrator column of fresh feed and also as recovery column for an aqueous outlet stream from a decanter.The bottom product of the heterogeneous azeotropic column is IPA with purity at 99.9 mol% and the bottom product of the combined column is water with purity also at 99.9 mol%. The flow rate is set at 100 kmol/h with equal molar of IPA and water and feed temperature is set at 320K.

(a) Draw a ternary diagram of this system with RCMs, liquid–liquid envelope with tie lines, and material balance lines showing the conceptual design of this separation system.

(b) Perform process simulation for this flowsheet using Aspen Plus software.

REFERENCES

Chen, Y.C., Li, K.L., Chen, C.L., Chien, I.L., 2015a. Design and control of a hybrid extraction-distillation system for the separation of pyridine and water. Industrial & Engineering Chemistry Research 54, 7715–7727.

Chen, Y.C., Hung, S.K., Lee, H.Y., Chien, I.L., 2015b. Energy-saving designs of a close-boiling 1,2 propanediol and ethylene glycol mixture. Industrial & Engineering Chemistry Research 54, 3828–3843.

Doherty, M.F., Malone, M.F., 2001. Conceptual Design of Distillation Systems. McGraw-Hill, New York, NY.

Julka, V., Chiplunkar, M., O'Young, L., March 2009. Selecting entrainers for azeotropic distillation. Chemical Engineering Progress 47–52.

Lei, Z., Dai, C., Zhu, J., Chen, B., 2014. Extractive distillation with ionic liquids: a review. AIChE Journal 60, 3312–3329.

Liano-Restrepo, M., Aguilar-Arias, J., 2003. Modeling and simulation of saline extractive distillation columns for the production of absolute ethanol. Computers & Chemical Engineering. 27, 527–549.

Luyben, W.L., 2005. Comparison of pressure-swing and extractive distillation methods for methanol recovery systems in the TAME reactive-distillation process. Industrial & Engineering Chemistry Research 44, 5715–5725.

Luyben, W.L., 2011. Principles and Case Studies of Simultaneous Design. Wiley, Hoboken, New Jersey.

Luyben, W.L., 2012. Economic optimum design of the heterogeneous azeotropic dehydration of ethanol. Industrial & Engineering Chemistry Research 51, 16427–16432.

Luyben, W.L., Chien, I.L., 2010. Design and Control of Distillation Systems for Spearating Azeotropes. Wiley, Hoboken, New Jersey.

Su, C.Y., Yu, C.C., Chien, I.L., Ward, J.D., 2013. Plant-wide economic comparison of lactic acid recovery processes by reactive distillation with different alcohols. Industrial & Engineering Chemistry Research 52, 11070–11083.

Tyreus, B.D., October 2011. Distillation – energy conservation and process control, a 35 year perspective. In: AIChE Annual Meeting, 16–21, Minneapolis MN, U.S.A.

Wu, Y.C., Chien, I.L., Luyben, W.L., 2009. Two-stripper/decanter flowsheet for methanol recovery in the TAME reactive-distillation process. Industrial & Engineering Chemistry Research 48, 10532–10540.

Wu, Y.C., Hsu, C.S., Huang, H.P., Chien, I.L., 2011. Design and control of a methyl methacrylate separation process with a middle decanter. Industrial & Engineering Chemistry Research 50, 4595–4607.

Wu, Y.C., Hsu, P.H.C., Chien, I.L., 2013. Critical assessment of the energy-saving potential of an extractive dividing-wall column. Industrial & Engineering Chemistry Research 52, 5384–5399.

Wu, Y.C., Lee, H.Y., Huang, H.P., Chien, I.L., 2014. Energy-saving dividing-wall column design and control for heterogeneous azeotropic distillation systems. Industrial & Engineering Chemistry Research 53, 1537–1552.

Chapter 16

Simulation and Analysis of Heat Exchanger Networks With Aspen Energy Analyzer

Wei-Jyun Wang, Cheng-Liang Chen
National Taiwan University, Taipei, Taiwan, ROC

16.1 INTRODUCTION

A chemical plant usually consists of hundreds of or even thousands of process units. Flow streams between process units are often controlled to certain temperature level for maintaining adequate unit operations throughout the entire plant. To achieve this goal, one of the solutions is direct heating and/or cooling by using hot and/or cold utilities such as steam and cooling water (CW), on each stream to adjust stream temperature.

Fig. 16.1A (data adapted from Smith, 2016) shows a simple example, which is comprised of two reactors and one simple separator. In this plant, the nominal input temperature of "Feed 1" is 20°C, and "Feed 2" is 140°C. Suppose these two feed streams should be increased to 180°C and 230°C, respectively, before they are fed into "Reactor 1" and "Reactor 2," and the required heating rate supplied by steam is 320 kW (or 320 kJ/s) for "Feed 1" and 270 kW for "Feed 2." Assume the temperature of "Effluent 1," that is, the output stream of "Reactor 1" is 250°C. Part of "Effluent 1" is directly fed into "Reactor 2" and the remaining is sent to the separator with a required temperature of 40°C. The required cooling rate supplied by CW is 315 kW. Temperature of "Effluent 2," the output stream from "Reactor 2," is 200°C. "Effluent 2" is then cooled to 80°C for further processing with a cooling rate of 300 kW. Thus the hourly heating and cooling duties for this simple plant are 590 kWh and 615 kWh, respectively.

Fig. 16.1B (Smith, 2016) shows the effect of a heat exchanger network for realizing energy exchange between process streams. Fig. 16.1C gives the equivalent grid diagram of the heat exchanger network. The installed five heat exchangers can transfer 515 kW in total from two hot streams ("Effluent 1" and "Effluent 2") to two cold streams ("Feed 1" and "Feed 2"), thus the outsourced

Chemical Engineering Process Simulation. http://dx.doi.org/10.1016/B978-0-12-803782-9.00016-9

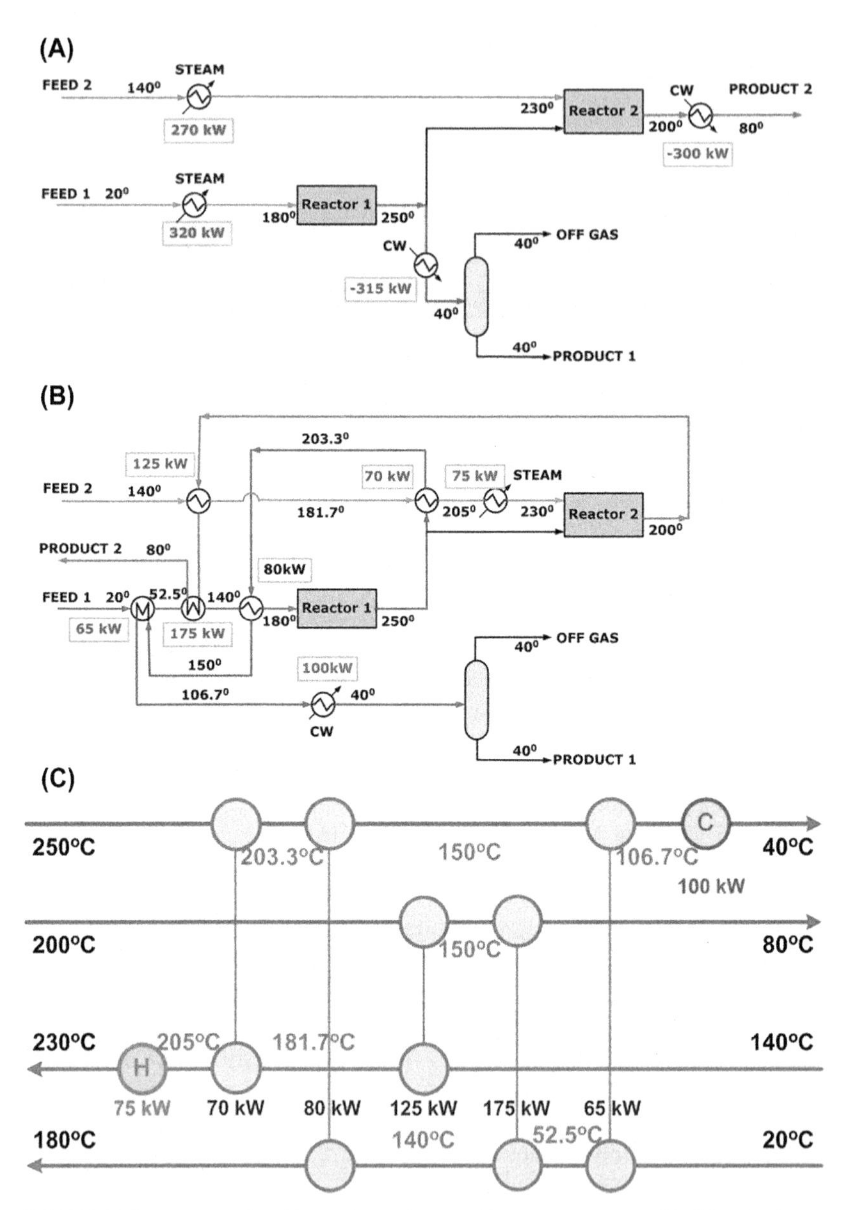

FIGURE 16.1 An example with two hot and two cold process streams—(A) direct heating or cooling, (B) use of heat exchangers for heat recovery, (C) the equivalent grid diagram of the heat exchanger network. *CW*, cooling water. *Data adapted from Smith, R., 2016. Chemical Process Design and Integration, second ed. John Wiley & Sons, Ltd., England.*

heating power is decreased from 590 kW to 75 kW, an 87.3% saving. The total cooling rate also drops from 615 kW to 100 kW, an 83.7% reduction.

This example demonstrates that the hot and cold process streams in a chemical plant can be preferentially used for heating or cooling purposes. A well-designed heat exchanger network can be used to implement potential heat recovery between hot and cold process streams and reduce the outsourced hot and cold utilities.

16.2 SYNTHESIS OF HEAT EXCHANGER NETWORKS

16.2.1 Basics

Eq. (16.1) shows the rate of enthalpy change of a process stream, where m is mass flowrate (kg/s), C_p is the specific heat capacity (kJ/kg°C), and ΔT is the temperature change (°C).

$$\Delta H = m \times C_p \times \Delta T \tag{16.1}$$

The heat capacity flowrate CP (kJ/s°C or kW/°C) is the product of mass flowrate m and specific heat capacity C_p:

$$\mathrm{CP} = m \times C_p \tag{16.2}$$

Eq. (16.1) can be rewritten as:

$$\Delta H = \mathrm{CP} \times \Delta T \tag{16.3}$$

If a stream with heat capacity flowrate CP = 4 (kW/°C) is increased from 50°C to 80°C by using a heating utility such as steam, the rate of enthalpy change of this cold stream is 120 kW.

$$\Delta H_{\mathrm{cold}} = \left(4\frac{\mathrm{kW}}{^{\circ}\mathrm{C}}\right) \times (80^{\circ}\mathrm{C} - 50^{\circ}\mathrm{C}) = 120\mathrm{kW} \tag{16.4}$$

Fig. 16.2A shows this simple heating process in a temperature—enthalpy diagram. The overlapped region represents the heating power from the steam to the cold stream.

One may use a hot stream with heat capacity flowrate 3 (kW/°C) to heat this cold stream by using a countercurrent heat exchanger. If we choose the minimum temperature difference, $\Delta T_{\min}$, to be 10°C, the minimal outlet hot stream temperature, $T_{\mathrm{hot}}^{\mathrm{out}}$, would be:

$$T_{\mathrm{hot}}^{\mathrm{out}} = 50^{\circ}\mathrm{C} + \Delta T_{\min} = 60^{\circ}\mathrm{C} \tag{16.5}$$

Therefore, the minimal hot stream inlet temperature, $T_{\mathrm{hot}}^{\mathrm{in}}$, will be:

$$T_{\mathrm{hot}}^{\mathrm{in}} = 60^{\circ}\mathrm{C} + \frac{120\,\mathrm{kW}}{3\left(\frac{\mathrm{kW}}{^{\circ}\mathrm{C}}\right)} = 100\,^{\circ}\mathrm{C} \tag{16.6}$$

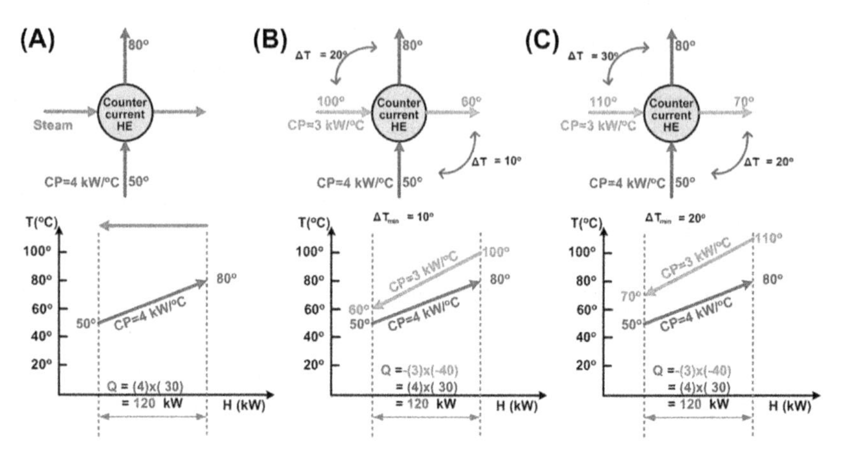

FIGURE 16.2 The temperature–enthalpy (T–H) diagrams—(A) steam provides heating power to the cold stream; hot process stream gives heating power with (B) $\Delta T_{\min} = 10°C$, and (C) 20°C. *HE*, heat exchanger.

We need to make sure there is no temperature crossover, so the following inequality equation has to be satisfied:

$$T_{\text{hot}}^{\text{in}} = 100^{\circ}\text{C} \geq T_{\text{cold}}^{\text{out}} + \Delta T_{\min} = 90^{\circ}\text{C} \tag{16.7}$$

Fig. 16.2B shows the T–H diagram of the countercurrent heat exchanger with 10°C of $\Delta T_{\min}$. Notably, if $\Delta T_{\min}$ is chosen as 20°C, then the minimal $T_{\text{hot}}^{\text{out}}$ would be:

$$T_{\text{hot}}^{\text{out}} = 50 + \Delta T_{\min} = 50 + 20 = 70^{\circ}\text{C} \tag{16.8}$$

In this case, the minimal $T_{\text{hot}}^{\text{in}}$ becomes:

$$T_{\text{hot}}^{\text{in}} = 70 + \frac{120}{3\left(\frac{\text{kW}}{^{\circ}\text{C}}\right)} = 110^{\circ}\text{C} \tag{16.9}$$

Again, we need to make sure there is no temperature crossover, so the following inequality equation has to be satisfied:

$$T_{\text{hot}}^{\text{in}} = 110^{\circ}\text{C} \geq T_{\text{cold}}^{\text{out}} + \Delta T_{\min} = 100^{\circ}\text{C} \tag{16.10}$$

Fig. 16.2C shows the T–H diagram of the countercurrent heat exchanger design with $\Delta T_{\min} = 20°C$. The higher value of $\Delta T_{\min}$ is, the larger the driving force of heat transfer between the hot/cold streams is. With larger driving force of heat transfer between hot/cold streams, the required heat transfer area will be reduced.

16.2.2 A Simple One-Hot/One-Cold Heat Recovery Problem

The flowsheet of this illustrative problem is shown in Fig. 16.3A, where the data are adapted from Smith (2016). In this problem, the initial temperature of

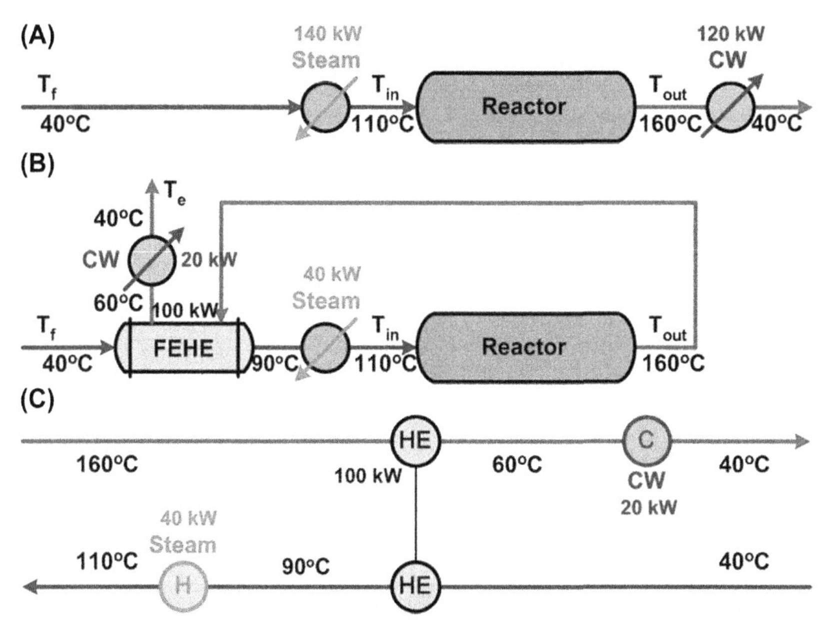

FIGURE 16.3 The flowsheet of the simple two-stream problem (A) without and (B) with heat recovery; (C) the equivalent grid diagram (Smith, 2016). *CW*, cooling water; *FEHE*, feed–effluent heat exchanger; *HE*, heat exchanger.

the reactor feed/effluent streams is 40°C and 160°C, respectively. Suppose the heat capacity flowrates of the reactor feed/effluent streams are 2 (kW/°C) and 1 (kW/°C), and the respective target temperatures of the reactor feed/effluent streams are 110°C and 40°C. The rates of enthalpy change are +140 kW and −120 kW for the cold and hot streams, respectively. The problem information is summarized in Table 16.1. Fig. 16.3B and C depicts the use of feed–effluent heat exchanger for heat recovery and its equivalent grid diagram.

The temperature–enthalpy (T–H) diagram of this one-hot/one-cold problem is shown in Fig. 16.4A. One can directly use steam to provide 140 kW on the cold stream and use CW to remove 120 kW from the hot stream. However,

TABLE 16.1 Process Information Direct Heating and Cooling (Smith, 2016)

Stream Number	Stream Type	Temperature (°C)		ΔH (kW)	Heat Capacity Flowrate, CP (kW/°C)
		Initial T_S	Target T_T		
1	Cold	40	110	+140	2
2	Hot	160	40	−120	1

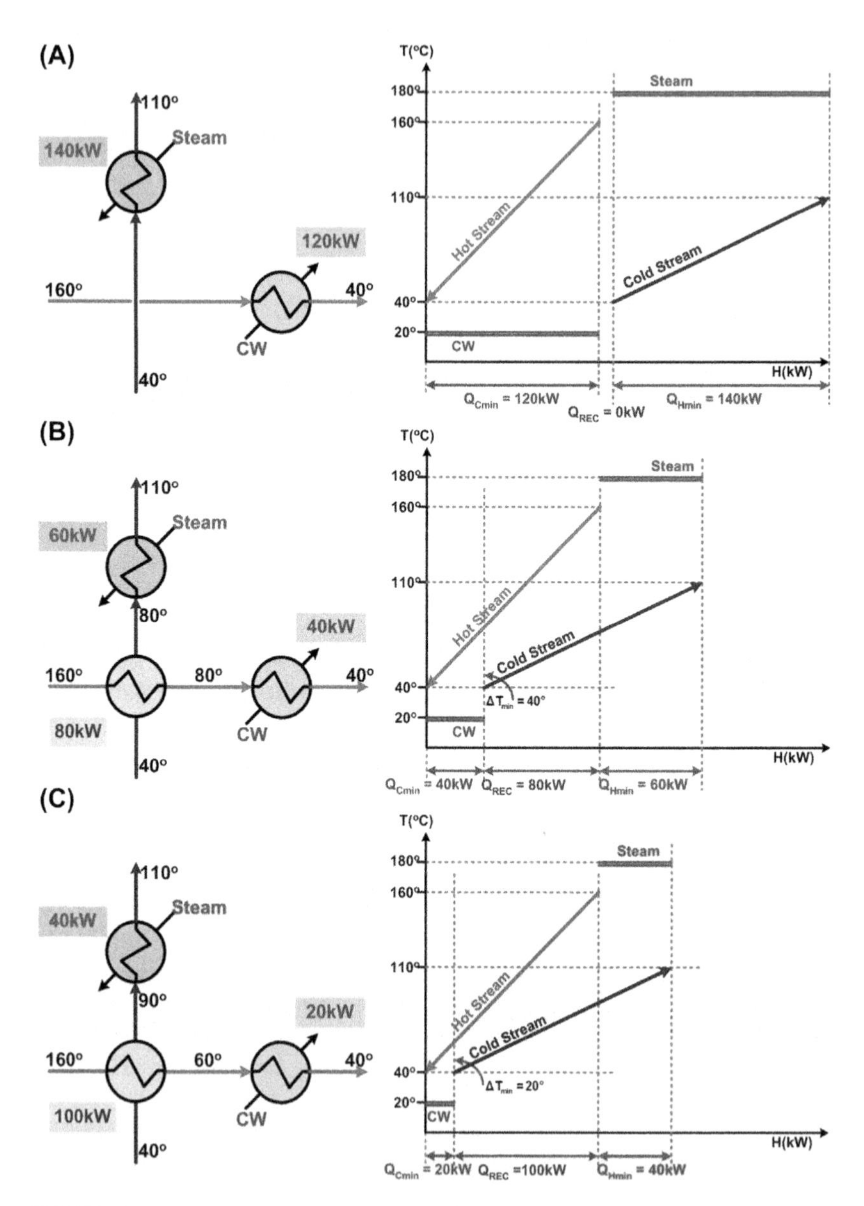

FIGURE 16.4 The temperature–enthalpy (T–H) diagram of the simple two-streams heat recovery problem—(A) direct heating and cooling; (B) heat recovery with $\Delta T_{\min} = 40°C$; and (C) heat recovery with $\Delta T_{\min} = 20°C$ (Smith, 2016). *CW*, cooling water.

if we inspect the input/output temperature ranges of these two hot/cold streams, one can find that it is possible to install a countercurrent heat exchanger between the two streams for heat recovery. Suppose the heat exchanger has $\Delta T_{\min} = 40°C$, the maximal output temperature for the hot stream will be 80°C because the given cold input stream is 40°C. Thus the rate of energy recovery will be $Q_{rec} = 1\left(\frac{kW}{^{\circ}C}\right) x \left(160^{\circ}C - 80^{\circ}C\right) = 80\,kW$, and the temperature of the cold output stream will become $T_{cold}^{out} = 40 + \frac{80\,kW}{2\left(\frac{kW}{^{\circ}C}\right)} = 80^{\circ}C$. The deficit heating and cooling rates, 60 kW and 40 kW, should be supplied by using hot/cold utilities such as steam and CW, as depicted in Fig. 16.4B for the heat exchanger configuration and the T−H diagram. It is found that one can simply move the cold stream curve in Fig. 16.4A horizontally to the left until the minimum temperature difference between two curves, the pinch point, is 40°C. The overlapped region of these two streams in the resulting T−H diagram represents the rate of heat recovery. The required outsourced power from hot utility drops to 60 kW and the required heat removal rate to cold utility reduces to 40 kW.

What will happen if we decrease ΔT_{min} to 20°C? In the T−H diagram, we continue to move the cold stream curve to the left until the minimum temperature difference between two curves or the pinch point becomes 20°C. The resulting T−H diagram is shown in Fig. 16.4C. It is found from Fig. 16.4C that the rate of heat recovery increases to 100 kW and the required amount of input power drops to 40 kW from steam and 20 kW from CW. The extra cost of decreasing the value of ΔT_{min} to 20°C for increasing the power of heat recovery is the extra area of heat exchanger.

16.2.3 A Simple Two-Hot/Two-Cold Heat Recovery Problem

In this section, we discuss the heat recovery problem with two hot and two cold streams mentioned in Section 16.1. Fig. 16.1, data also adapted from Smith (2016), shows the flowsheet of this problem, which is comprised of two reactors and one simple separator. The stream data are shown in Table 16.2.

The T−H diagram of this problem has two hot streams, which overlap over the temperature region between 80°C to 200°C. One can directly combine the two hot streams in the temperature region to form a single hot composite curve, as shown in Fig. 16.5A. The new heat capacity flowrate of the hot composite curve in the overlapped temperature region is the summation of heat capacity flowrate of the two hot streams. Fig. 16.5B also shows the T−H diagram of two cold streams and their resulting composite curve.

Targets for hot and cold utilities can be determined by plotting the hot and cold composite curves together. Fig. 16.6A shows the case that ΔT_{min} is 20°C. The closest point of the two composite curves, or the "pinch point," is exactly 20°C. Fig. 16.6A depicts that the theoretically maximum heat recovery rate will be 475 kW, and additional 115 kW will be supplied from steam and 140 kW will be removed by CW. Similarly, the rate of heat recovery and extra utilities become 515 kW, 75 kW, and 100 kW, respectively, when ΔT_{min} is 10°C, as shown in Fig. 16.6B. Therein, the heat exchanger area will be larger for the latter case due to its smaller driving force for heat transfer.

TABLE 16.2 Process Information for Illustrative Two-Hot/Two-Cold Problem (Smith, 2016)

Stream	Stream Type	Temperature (°C) Initial T_S	Target T_T	ΔH (kW)	Heat Capacity Flowrate, CP (kW/°C)
Reactor 1 Feed	Cold	20	180	+320	2.0
Reactor 1 Prod	Hot	250	40	−315	1.5
Reactor 2 Feed	Cold	140	230	+270	3.0
Reactor 2 Prod	Hot	200	80	−300	2.5

In addition to drawing T−H diagrams, we also can determine the amount of heat recovery and the required heat and cold utilities by making the problem table algorithm (Linnhoff et al., 1982). Suppose one specifies $\Delta T_{\min}$ as 10°C for the two-hot and two-cold stream heat recovery problem. The initial temperature of all streams, T_S, and the target temperature, T_T, are listed in a table. Considering the required minimal temperature difference between hot and cold streams for heat transfer, we also depict the shift temperature in the table, where the hot stream temperature is lowered by $\Delta T_{\min}/2$ and the cold stream temperature is higher by $\Delta T_{\min}/2$. The temperature ranges for all hot/cold streams are thus shown in Table 16.3, where T^* in column 1 represents the shifted temperature and the original stream temperatures are given in columns 2−5. In Table 16.3, all temperature nodes are shown where at least one stream starts from or targets on these designated temperatures. The sixth to the eighth columns depict the ranges of temperature interval between temperature nodes, the sum-up of heat capacity flowrates, and the surplus or deficit heat in the local interval, etc. The 9th and the 10th columns show the cascade surplus heat if there is no outsourced heat supply (the 9th column) and if there is minimal heat supply (75 kW in this case); thus, there will be no negative surplus energy at all temperature nodes. The node in the 10th column with zero surplus energy designates the *pinch point*, i.e., $T^* = 245°C$, which means 250°C for hot streams and 240°C for cold streams. It is shown that the targeting hot and cold utilities are 75 kW and 100 kW, respectively. The grand composite curve of this two-hot/two-cold problem is the cascade surplus along with the temperature nodes, such as shown in Fig. 16.7. The targeting hot/cold utilities and the pinch point are clearly shown in the grand composite curve. The procedures

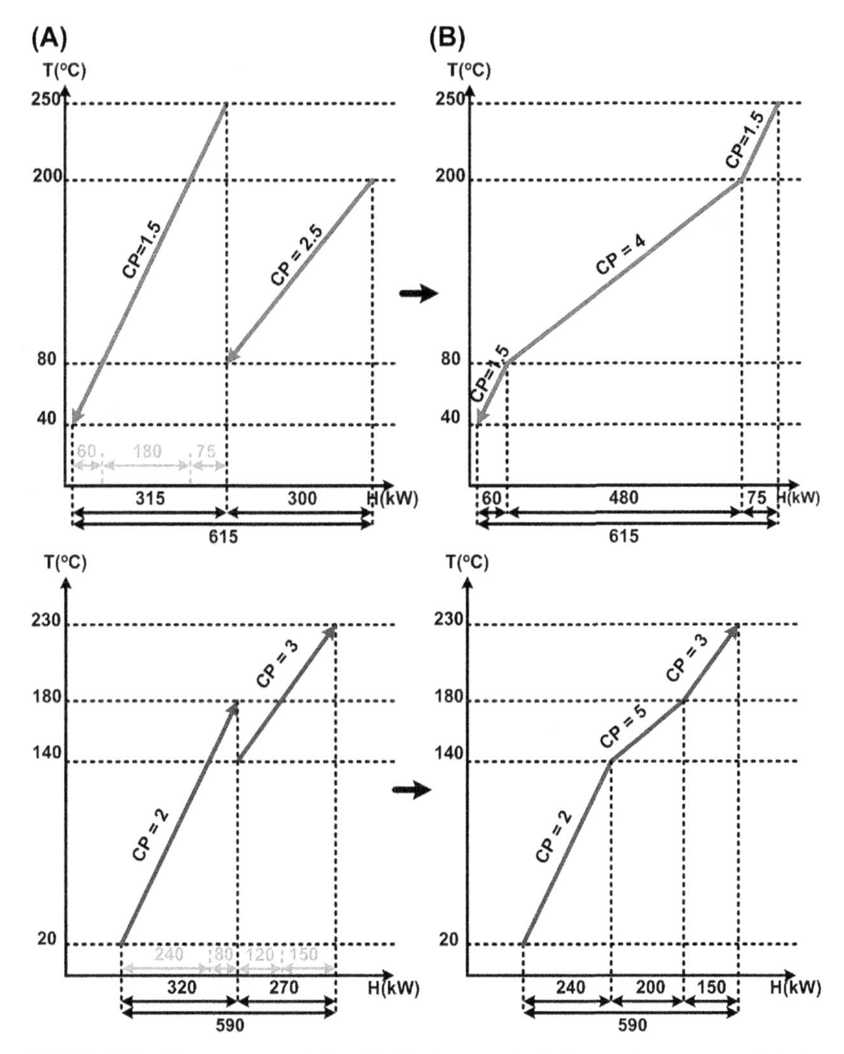

FIGURE 16.5 Temperature–enthalpy (T–H) diagram of (A) the two hot streams and the hot composite curve; (B) two cold streams and the cold composite curve.

for synthesis of the heat exchanger network to realize these targeting utilities can be found in Smith (2016) and Linnhoff et al. (1982).

16.3 ASPEN ENERGY ANALYZER FOR ANALYSIS AND DESIGN OF HEAT RECOVERY SYSTEMS

Aspen energy analyzer (AEA) (Aspen Technology Inc., 2009) is a well-developed commercial simulator, which allows its users to perform steady-state analysis

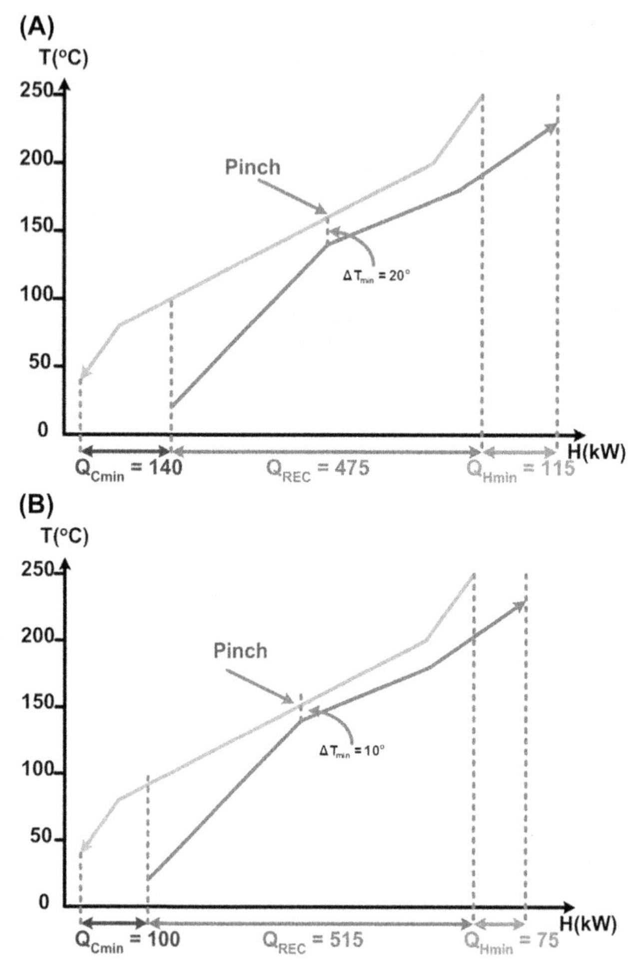

FIGURE 16.6 Temperature—enthalpy (T—H) diagram of the hot/cold composite curves for the two-hot and two-cold heat recovery problem when (A) $\Delta T_{min} = 20°C$ and (B) $\Delta T_{min} = 10°C$.

and design of heat recovery systems. The following workshops will demonstrate the detailed steps for implementing the heat exchanger networks for the illustrative one-hot/one-cold (Example 1) and two-hot/two-cold (Example 2) problems.

Example 1: Synthesis of Heat Exchanger Network for a One-Hot/One-Cold Heat Recovery Problem

In this section, we perform simulation of the same two-stream heat recovery problem in Section 16.2.2 into AEA. The specifications of this problem are summarized in Table 16.1.

TABLE 16.3 The Problem Table for the Two-Hot/Two-Cold Illustrative Problem

1	2	3	4	5	6	7	8	9	10
								Cascade Surplus	
T^*	C1	H1	C2	H2	ΔT_{int}	$\Sigma CP_H - CP_C$	ΔH_{int}	No utility	Hot utility
245	–	250	–	–	–	–	–	0	75
		↓			10	+1.5	+15		
235	–	240	230	–	–	–	–	+15	90
		↓	↑		40	−1.5	−60		
195	–	200	190	200	–	–	–	−45	30
		↓	↑	↓	10	+1.0	+10		
185	180	190	180	190	–	–	–	−35	40
	↑	↓	↑	↓	40	−1.0	−40		
145	140	150	140	150	–	–	–	−75	0
	↑	↓		↓	70	+2.0	+140		
75	70	80	–	80	–	–	–	+65	140
	↑	↓			40	−0.5	−20		
35	30	40	–	–	–	–	–	+45	120
	↑				10	−2.0	−20		
25	20	–	–	–	–	–	–	+25	100
CP	2.0	1.5	3.0	2.5					

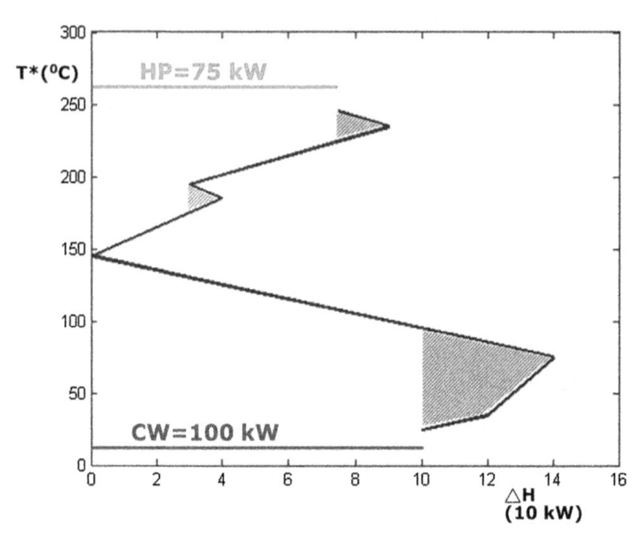

FIGURE 16.7 The grand composite curve of the two-hot/two-cold heat recovery problem where $\Delta T_{\min} = 10°C$. *CW*, cooling water.

The procedures of building the two-stream heat recovery problem in AEA are described as follows, where the default value for $\Delta T_{\min} = 20°C$. More detailed operational steps can be found from the attached handouts.

1. The initial steps for running AEA are depicted in Fig. 16.8. First, open the AEA, click "Tools" on the control panel, and then select "Preferences"; click "Variables" and select "SI" unit set.

   Click "Clone" to create a new unit set, "NewUser"; select those relevant units such as kJ/C s for the heat capacity flowrate MCp, kg/s for mass flow, kW for energy, etc.

   Click "Features" on the control panel and select "HI Case."
2. Click on the "Process Streams" tab and use the specifications shown in Fig. 16.9 to create one-hot and one-cold process streams.

   Click on the "Utility Streams" tab and use the specifications shown in Fig. 16.9 to create hot (steam) and cold (CW) utility streams.

   Click on the "Economics" tab and use the default heat exchanger cost index parameters to configure the settings.
3. After entering all the specifications, one can click the two "composite view" buttons on the top of the window to view the hot/cold composite curves and the grand composite curve (Fig. 16.10).
4. Click the "Open HEN Grid Diagram" to open the window of grid diagram.

   Right-click "Add Heat Exchanger" and hold the icon. Then, drag the cursor on "H1" and drop the exchanger on the stream "H1." Next, left-click on the newly added exchanger and drag the cursor on "C1" to add a heat exchanger between "H1" and "C1" (Fig. 16.11).

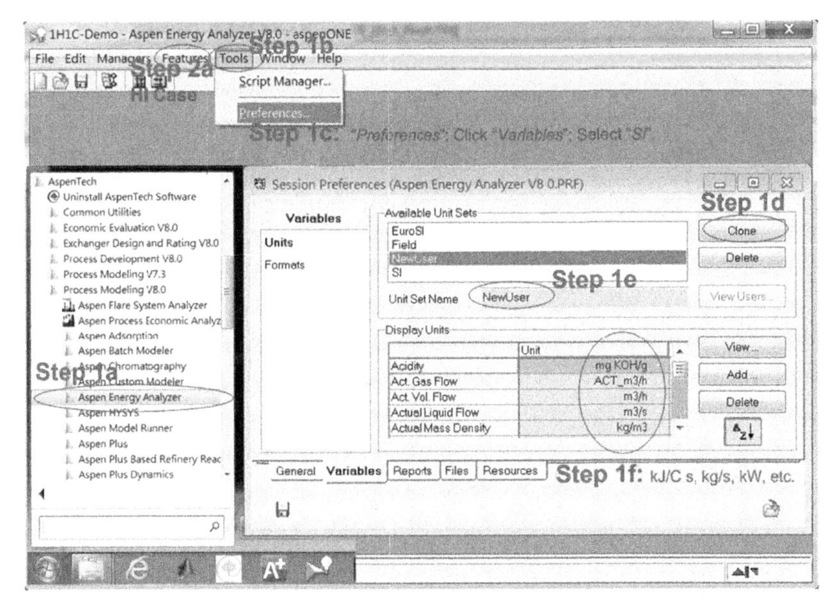

FIGURE 16.8 Initial steps for running the Aspen Energy Analyzer.

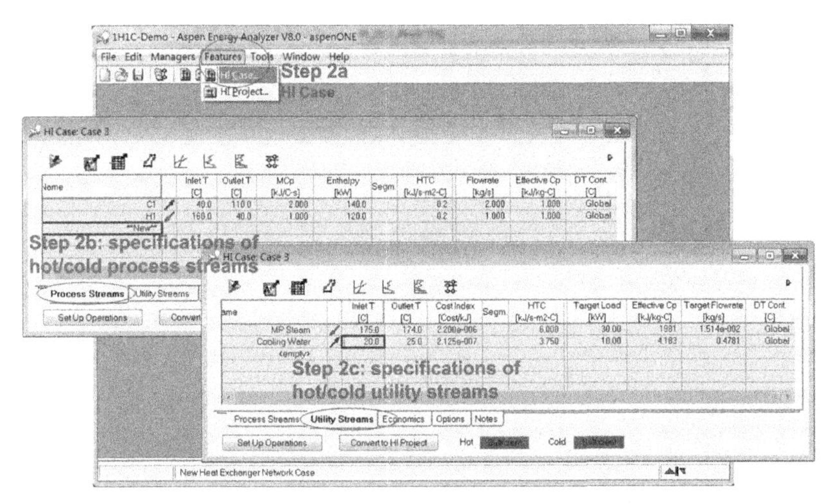

FIGURE 16.9 Specifications of hot/cold process streams and utility streams.

Add tied input temperature of hot/cold streams (160°C and 40°C), and calculate duty (100 kW) and area (30.57 m^2) of the heat exchanger. Also find the output temperature of hot (60°C) and cold (90°C) process streams.

5. Add a heat exchanger (cooler) between "CW" and "H1", and tied input/output temperatures of the hot process stream. Calculate the cooling duty (20 kW) and other properties (Fig. 16.12).

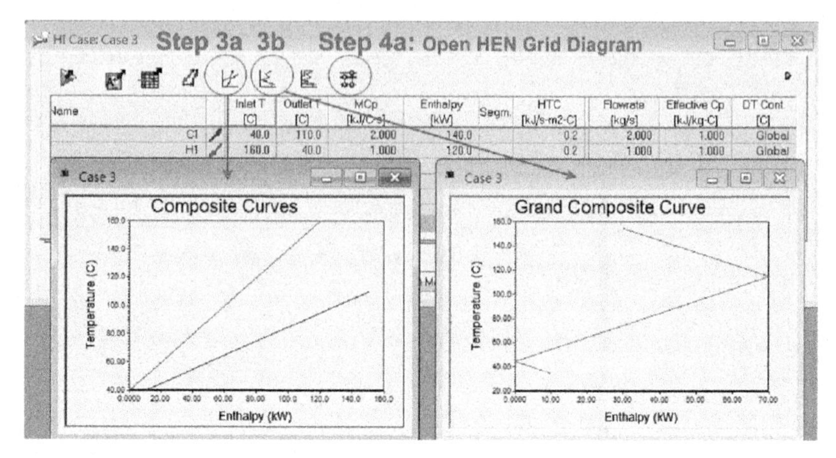

FIGURE 16.10 The hot/cold composite curves and the grand composite curve.

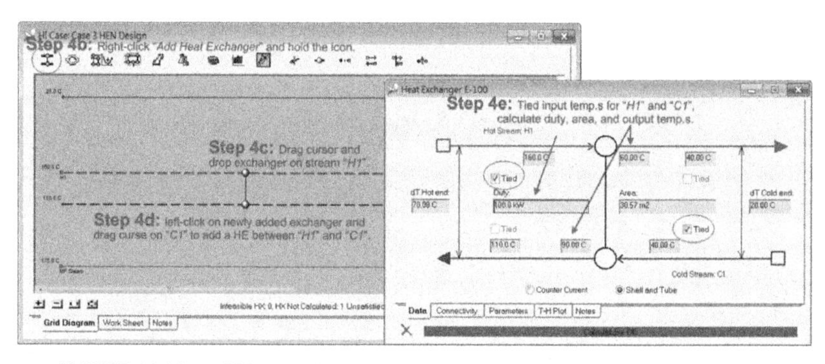

FIGURE 16.11 Add heat exchanger and tied input temperature of hot/cold streams.

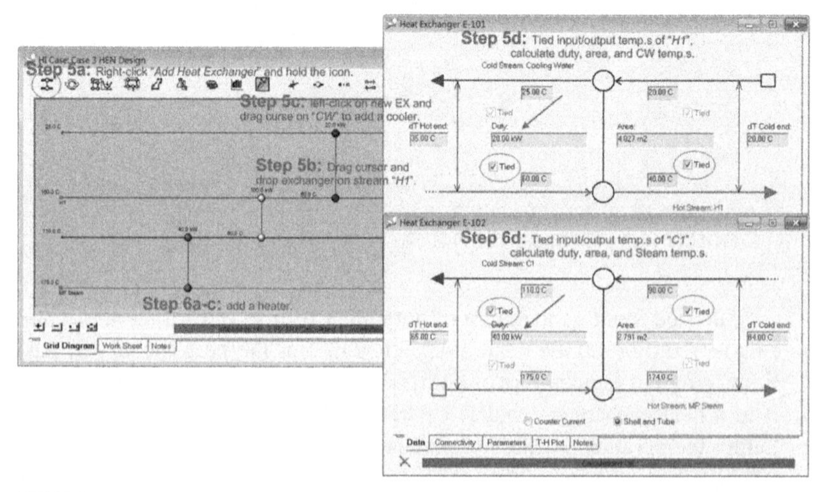

FIGURE 16.12 Add cooler and heater and tied input/output temperature of hot/cold streams.

Add a heat exchanger (heater) between "MP steam" and "C1," and tied input/output temperatures of the cold process stream. Calculate the heating duty (40 kW) and other properties.

Example 2: Synthesis of Heat Exchanger Network for a Two-Hot/Two-Cold Heat Recovery Problem

In this section, we perform simulation of the same two-hot/two-cold heat recovery problem in Section 16.2.3 into AEA. The specifications of this problem are summarized in Table 16.2.

The procedures of building the two-hot/two-cold heat recovery problem in AEA are described as follows:

1. Open "Aspen Energy Analyzer," assign necessary specifications. Click "Features" on the control panel and select "HI Case" as in Example 1.
2. Use the specifications given in Table 16.2 to create two-hot and two-cold process streams, steam and CW utility streams, "HP Steam" and "Cooling Water" and the "Economics" specifications (Fig. 16.13).
3. After entering all the specifications, one can open the targets view and click the "Summary" to show the energy and area targets. Also, click "Plots/Tables" button on the bottom of the window to view the hot/cold composite curves and the grand composite curve (Fig. 16.14).
4. Click the "Open HEN Grid Diagram" to open the window of grid diagram; right-click "Add Heat Exchanger" to add the first heat exchanger between HP1 and CP2, one heater and one cooler, sequentially, and use the specifications to configure the heat exchangers such as shown in Fig. 16.15.

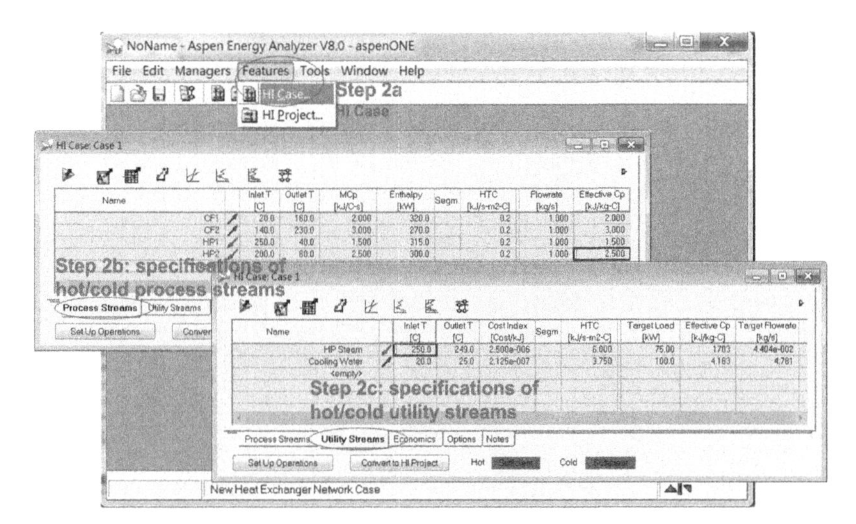

FIGURE 16.13 Specifications of process streams and utilities.

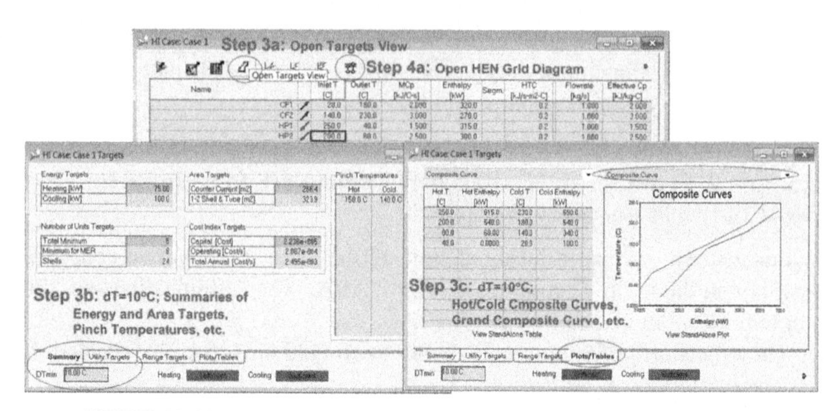

FIGURE 16.14 The hot/cold composite curves and the grand composite curve.

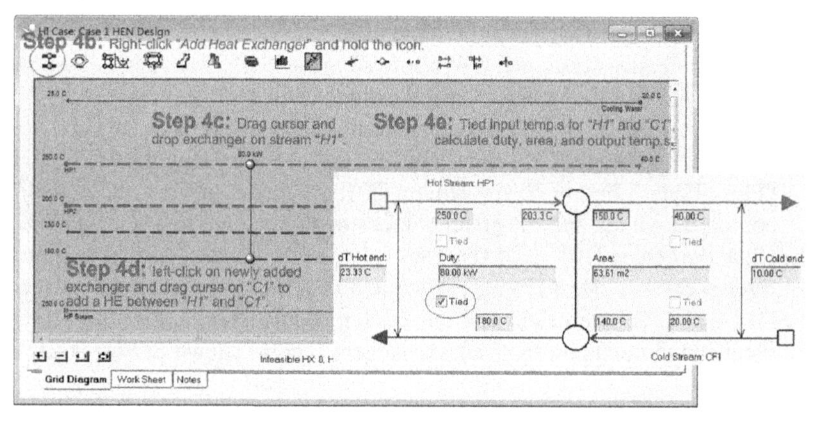

FIGURE 16.15 Add the first heat exchanger on the *HEN Grid Diagram* and give specifications sequentially.

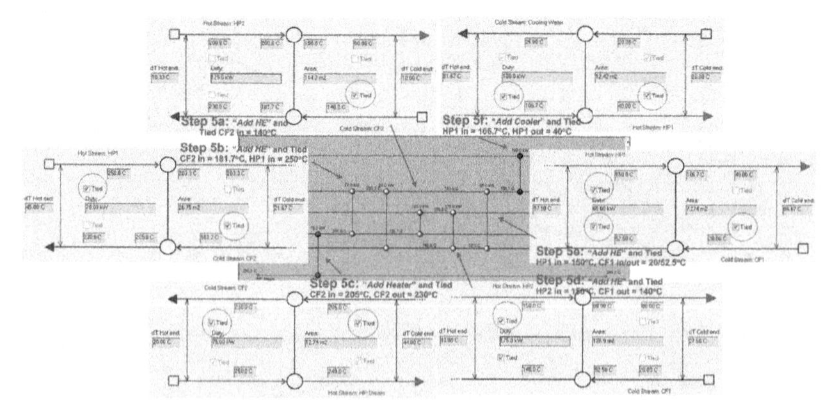

FIGURE 16.16 Add additional heat exchangers (HEs), heater and cooler sequentially, and tied input/output temperature of hot/cold process streams.

5. Add additional four heat exchangers, one heater and one cooler, and tied input/output temperatures of the hot and/or cold process streams sequentially, such as shown in Fig. 16.16. For this case, AEA indicates that 75 kW of hot utility and 100 kW of cold utility are required to meet the targets. AEA also determines the area of all heat exchangers, heater, and cooler.

EXERCISE

For the stream data given in the following table,

Stream	Supply Temperature T_S (°C)	Target Temperature T_T (°C)	Heat Capacity Flowrate CP (kW/K)
1	190	100	10
2	140	60	10
3	40	160	5
4	80	140	20

1. Target for minimum hot and cold utility requirements for $\Delta T_{\text{min}} = 20°\text{C}$ by using either the composite curves or the problem table algorithm. Also identify the pinch temperature for the hot and cold streams.
2. Use the pinch design method to synthesize a heat exchanger network that meets the utility targets. Please see Smith (2016) for the details.
3. Use the AEA to simulate the synthesized heat exchanger network.

REFERENCES

Aspen Technology Inc., 2009. AspenTM Energy Analyzer Reference Guide V7.1. Burlington, MA, USA.

Linnhoff, B., Townsend, D.W., Boland, D., Hewitt, G.F., Thomas, B.E.A., Guy, A.R., Marsland, R.H., 1982. A User Guide on Process Integration for the Efficient Use of Energy. IChemE, UK.

Smith, R., 2016. Chemical Process Design and Integration, second ed. John Wiley & Sons, Ltd., England.

Chapter 17

Simulation and Analysis of Steam Power Plants With Aspen Utility Planner

Wei-Jyun Wang, Cheng-Liang Chen
National Taiwan University, Taipei, Taiwan, ROC

17.1 INTRODUCTION

Steam power plants are primarily constructed to supply majority of electricity, mechanical power, and thermal heat for the operation of process units in chemical plants. Therein, steam under various pressures is widely used as working fluid to provide mechanical power to drive mechanical units, such as steam turbines and compressors, and also to generate electricity for supporting equipment operation. Steam is also the main source of thermal heat used in various unit operations for further reaction and/or separation. In general, to meet different demands of process equipment operation, a chemical plant typically has various steam distribution networks with different pressure levels (steam headers), such as very high pressure (VHP), high pressure (HP), medium pressure (MP), and low pressure (LP). The schematic diagram of a typical steam system is shown in Fig. 17.1. Steam headers are interconnected with steam letdown valves and steam turbines to ensure capabilities of allocating steam between them for different operating scenarios.

In the typical steam power plant shown in Fig. 17.1, the boiler feed water (BFW) is fed into a boiler to produce superheated HP steam, which is typically at 40 bar (Towler and Sinnott, 2012). Part of the HP steam is used for process heating. Part of the HP steam is expanded and fed into MP and/or LP headers through letdown valve and back pressure or condensing steam turbines. These turbines can generate electricity or provide required mechanical work. The MP steam from MP headers is used for process heating or further expanded and fed into LP headers through another stage of letdown valve and steam turbines. Finally, condensate is recovered and fed into a deaerator to remove excess oxygen and dissolved gases for prevention of corrosion damage in the steam system. With appropriate amount of makeup water, recycled condensate is refed into the boiler.

Chemical Engineering Process Simulation. http://dx.doi.org/10.1016/B978-0-12-803782-9.00017-0

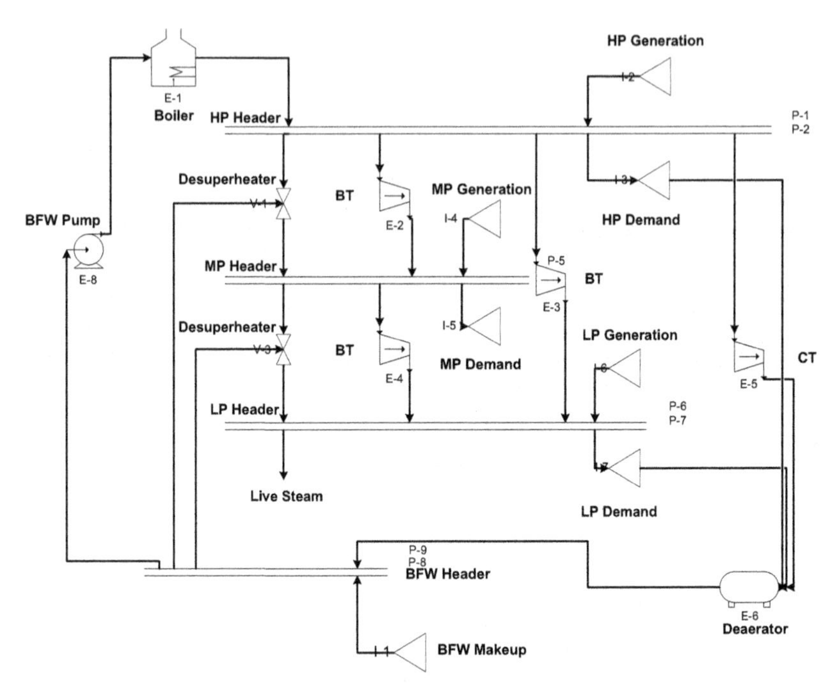

FIGURE 17.1 A typical steam power plant. *BFW*, boiler feed water; *BT*, back pressure turbine, *HP*, high pressure; *LP*, low pressure; *MP*, medium pressure.

Electricity is one of the major forms of energy to power varieties of equipment in a chemical plant, such as pumps and compressors. A plant normally generates electricity from operation of turbines and cogeneration units to meet its own demand. Further to self-generation, electricity networks are usually connected to external power grid to acquire additional electricity or sell surplus electricity back to the public power grid. Because electricity is much more expensive than steam, it is not widely used for heating purposes.

In general, utility systems convey nearly 70% of total energy required by most chemical plants (Broughton, 2004). If we can improve the energy efficiency of utility systems, even if we only can improve 1%, we will still be able to save a great deal of energy cost. In recent decades, computer simulation technology has been developing rapidly. We may employ simulators to investigate the optimal design and operation of utility systems with the highest energy efficiency. Aspen Utility Planner (AUP) is a well-developed commercial simulator, which allows its users to perform steady-state simulation and optimization of plant-wide utility systems. In Section 17.2, important built-in models in AUP are introduced. Then, we demonstrate how to perform simulation of a simple power plant in Section 17.3.

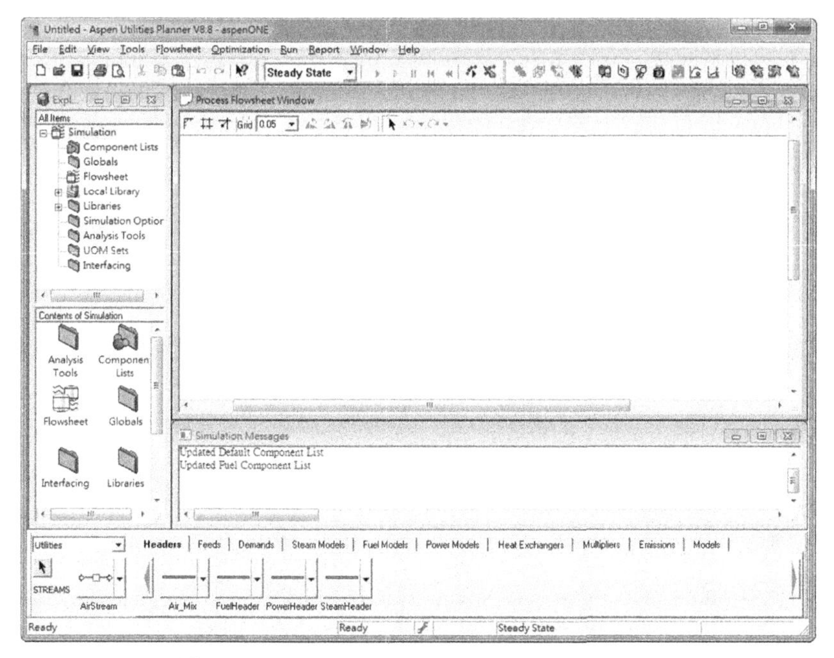

FIGURE 17.2 The user interface of Aspen Utility Planner.

17.2 INTRODUCTION OF ASPEN UTILITY PLANNER

Fig. 17.2 shows the user interface of the AUP. The *process flowsheet window* allows users to build flowsheet by adding models and types of stream. To start with a simulation, users have to click "Component Lists" to add components and "Configure Properties" to configure properties database on the all items plane. For simulation of steam systems, users could simply select "Use Properties definition file" to import *AspenUtilities.appdf*, which is stored at "\Program Files\AspenTech\ Aspen Utilities Planner V8.8\Examples" (Aspen, 2015).

Next, users need to click "View" on the toolbar and check "Model Libraries" to open the toolbar of models that allows users to add streams and models to their simulation easily, such as shown in Fig. 17.3. To add a model, users need to just simply left-click on the block of *model* and left-click again on the *process flowsheet window.* To add a stream, users need to select the type of stream first by clicking the inverted triangle on the right-handed side of the stream block (As shown in Fig. 17.4). Then, users need to left-click on the stream block to add streams. After the user left-clicks on the stream block, arrows on model blocks on the process flowsheet window will appear to allow users to connect streams with model blocks.

In the AUP, users have to build *feed blocks* and *demand blocks* to represent sources and sinks of steam and electricity. There are four kinds of feed blocks and demand blocks, including steam, fuel, air, and electricity. For feed blocks,

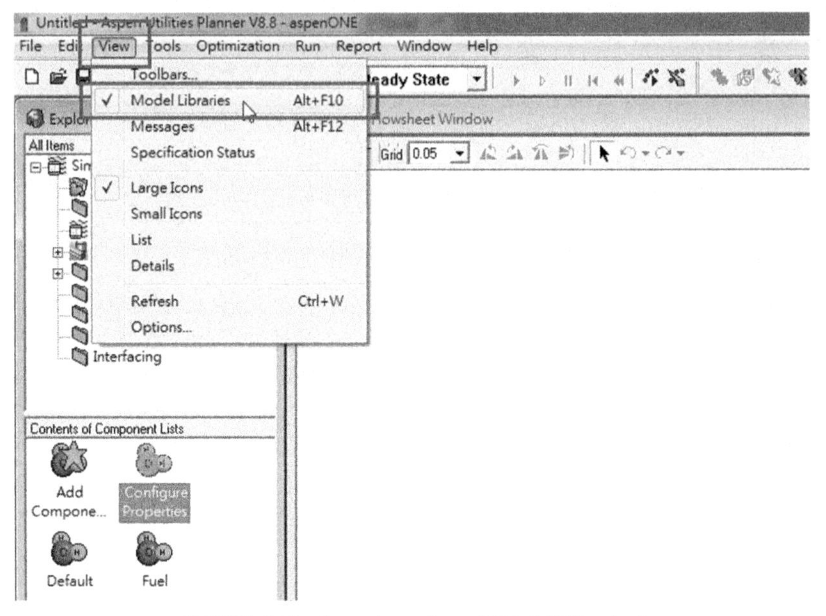

FIGURE 17.3 Open the toolbar of "Model Libraries."

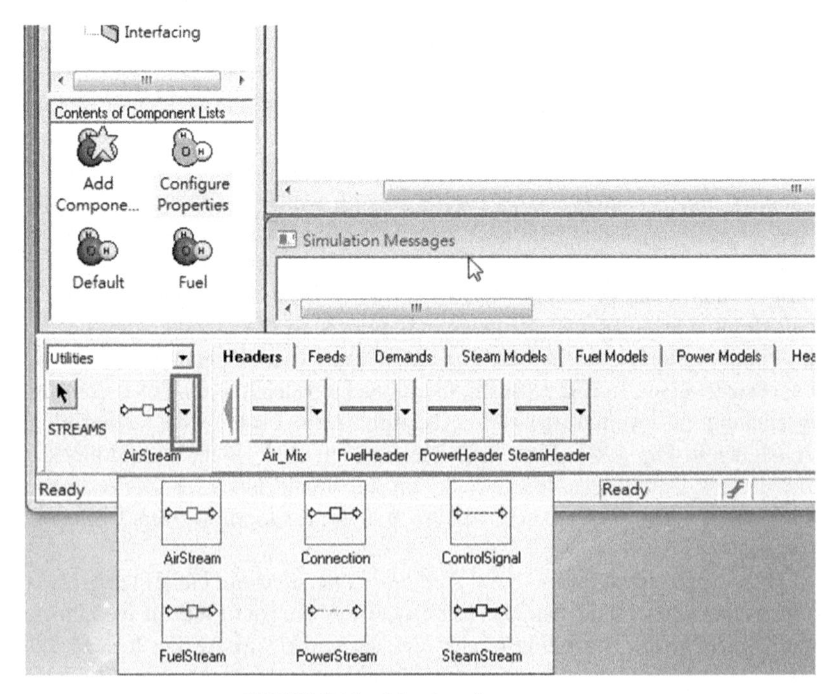

FIGURE 17.4 Selection of stream types.

we normally need to specify their outlet temperature, pressure, flowrate, or power output. But for air and fuel feed blocks that are connected with boiler blocks, we only need to specify their outlet stream temperature. Boiler block will automatically calculate outlet flowrate from air and fuel feed blocks based on their operating conditions. It is the same as the power feed block; AUP will automatically calculate the required power output. For demand blocks, we normally only specify their inlet flowrate or power input.

For **headers**, we should enable "FixImbalanceFlow" function and set the value of "Fimbalance" to zero to ensure mass balance is satisfied in the headers. In general, we would specify outlet temperature, pressure, and blowdown ratio (Aspen, 2015).

In the **boiler** model, there are two methods of calculating efficiency, the constant boiler efficiency and the efficiency curve. When the model is in the efficiency curve model, users need to input an efficiency curve in "Efftable." In addition, we normally would specify high-pressure steam outlet flowrate, percentage of flue gas recirculation, blowdown rate, oxygen concentration of the flue gas, boiler temperature, pressure drop in the steam generator, pressure drop in the steam superheater, reference temperature, and overall heat loss (Aspen, 2015).

For the **single-stage turbine model** (STG), we only need to specify its efficiency, steam flowrate, and the outlet pressure (Aspen, 2015).

The **heat recovery steam generator model** (HRSG) in the AUP is a model of heat recovery steam generator. Normally, we need to specify steam generation rate, heat flow of the first fuel, blowdown rate, oxygen concentration of the flue gas, pressure drop in the steam generator, pressure drop in the steam superheater, temperature of the outlet steam, the reference temperature, and the maximum efficiency.

For other utility models in the AUP, readers may refer to the AUP official user guide to obtain their detailed information (Aspen, 2015).

17.3 EXAMPLE—SIMULATION OF A SIMPLE STEAM POWER PLANT

In this example, we propose a step-by-step guide to demonstrate how to build a flowsheet and to perform steady-state simulation of a simple steam power plant in the AUP. The main structure of this steam power plant consists of one natural gas–fired boiler that produces superheated steam at 41 bar, one high-pressure steam headers at 40 bar, one low-pressure steam header at 2 bar, one single-stage condensing turbine (CT), one single-stage back pressure turbine (BT), a desuperheater with a BFW makeup, a deaerator, a BFW header, and a power header.

Fig. 17.5 shows the flowsheet of the illustrated steam power plant (Aspen, 2015). We illustrate the building steps in the AUP according to the sequence of the following six stages:

1. configure component properties;
2. the natural gas–fired boiler;

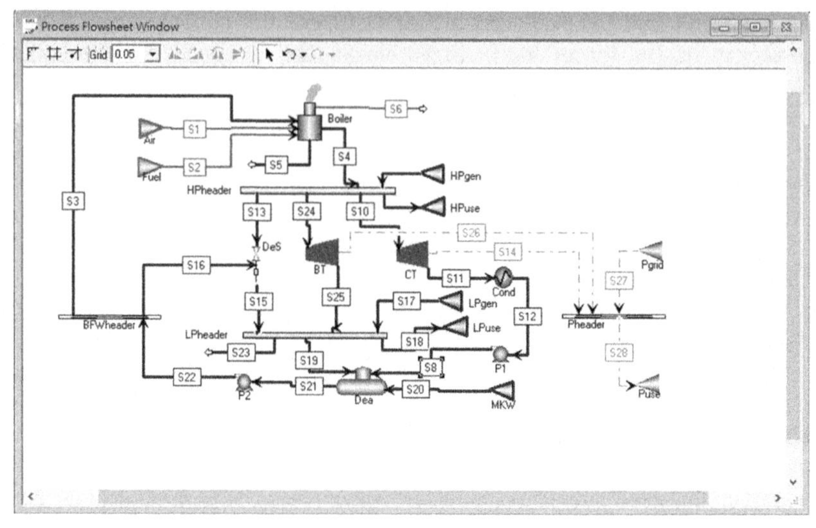

FIGURE 17.5 Flowsheet of the illustrated steam power plant.

3. the high-pressure steam header with a single-stage CT, with an additional source of high-pressure steam and a high-pressure steam user;
4. the low-pressure steam header with a steam letdown and a desuperheater;
5. the BFW processing section, including a deaerator, a BFW pump, a BFW splitting header, and a BFW makeup; and
6. the additional STG BT and an electricity system connected with an external power grid and an electricity demand.

17.3.1 Configure Component Properties

In the steam utility system, we only consider water as the only component in the simulation. The procedures of component properties configuration are described as follows:

1. Open Aspen Utilities Planner (AUP).
2. Click "Component List" and "Default" (Fig. 17.6).
3. Import "water" into the component list (Fig. 17.7).

17.3.2 The Boiler Section

The boiler fires natural gas to produce superheated high-pressure steam at 345°C. At this temperature, pressure of the high-pressure steam is 41 bar (calculated by AUP). There are two methods to specify boiler efficiency: a constant boiler efficiency and an efficiency table. In this simulation, the boiler is operated with a constant boiler efficiency of 50%. We also assume there are

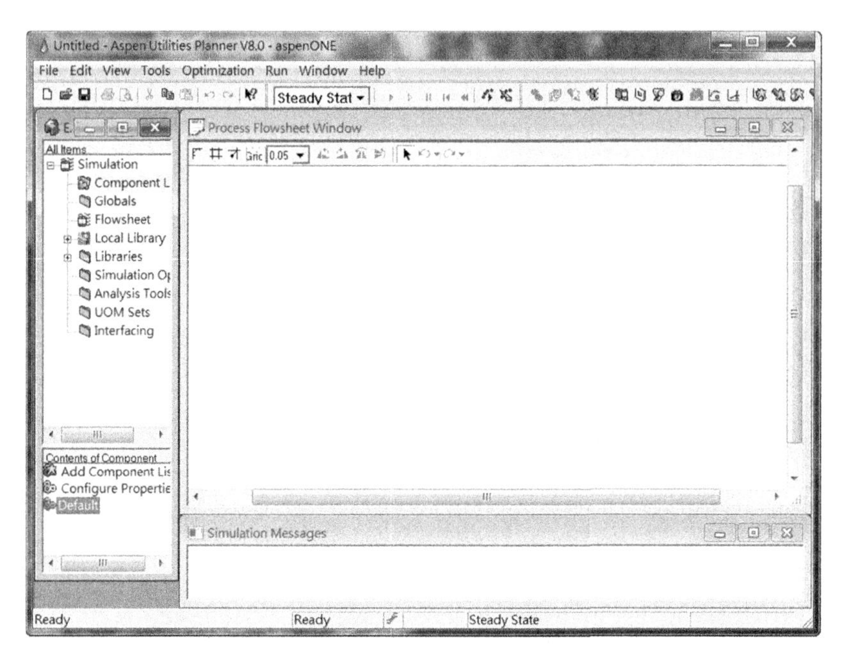

FIGURE 17.6 Open the default component list.

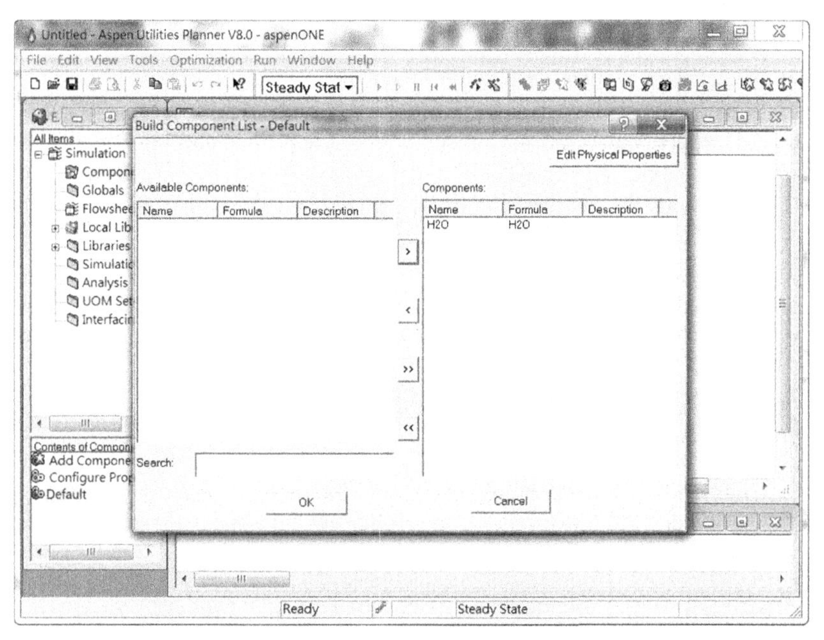

FIGURE 17.7 Add water into the component list.

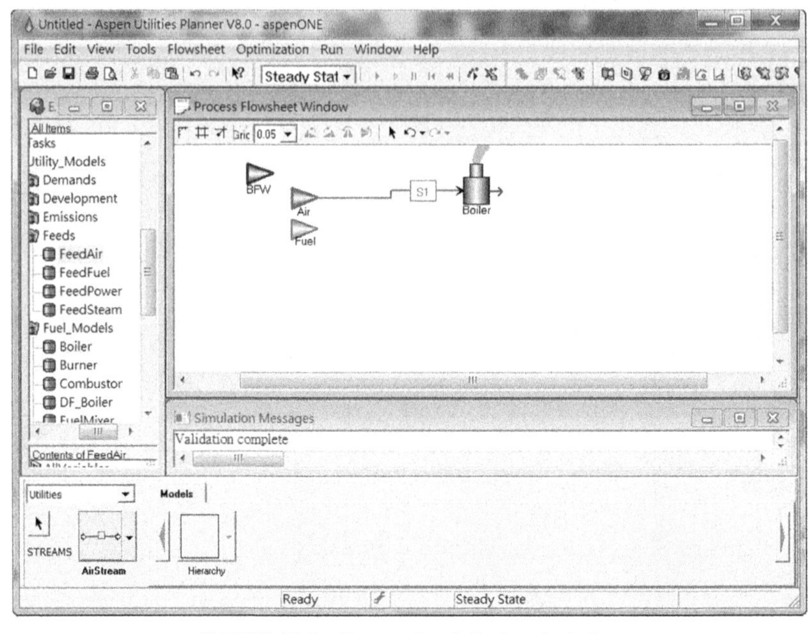

FIGURE 17.8 Connect the air feed to the boiler.

no heat loss and pressure drop in the boiler. The procedures of building the boiler section are described as follows:

1. Click "Fuel Models" and add a boiler model.
2. Rename the boiler block to "Boiler."
3. Add a feed water stream, a feed air, and a feed fuel, and rename them to "BFW," "Air," and "Fuel."
4. Click on the "AirStream" button to create an air stream and connect "Air" to "Boiler" (Fig. 17.8).
5. Click the small button with a small triangle on the right-hand side of the "AirStream" button to select the "FuelStream" mode.
6. Click on the "FuelStream" button to create a fuel stream ("S2") and connect "Fuel" to "Boiler."
7. Do step 5 again to select "SteamStream" mode.
8. Click on the "SteamStream" button to create a steam stream ("S3") and connect "BFW" to "Boiler."
9. Click on the "SteamStream" button and the outlet connection port [select SteamOut("VHPSteam")] on the boiler model. Then, left-click on the flowsheet again ("S4").
10. Create a blowdown water stream ("S5") from the boiler.
11. Create an outlet flue gas stream ("S6") from the boiler (Fig. 17.9).
12. Use the **specs** shown in Figs. 17.10 and 17.11 to configure the boiler.

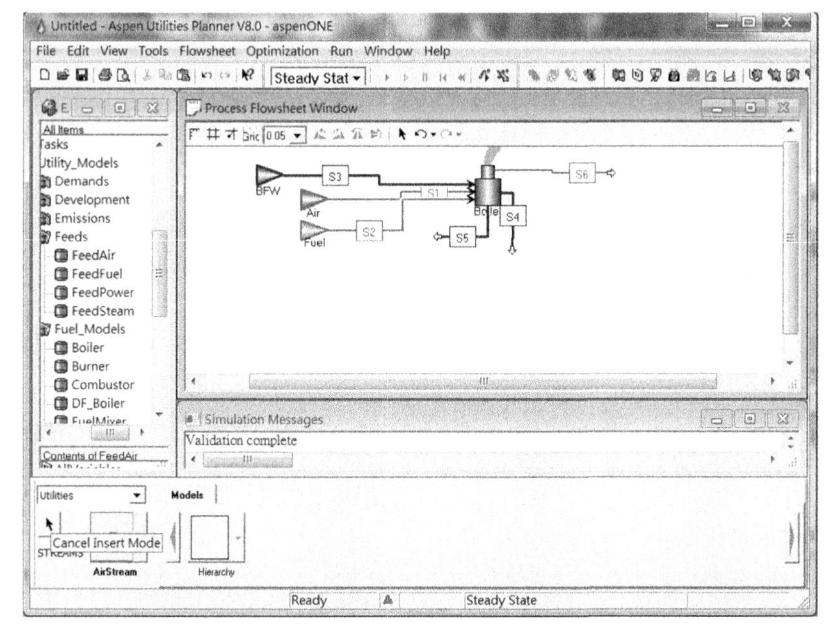

FIGURE 17.9 Create outlet steam streams from the boiler.

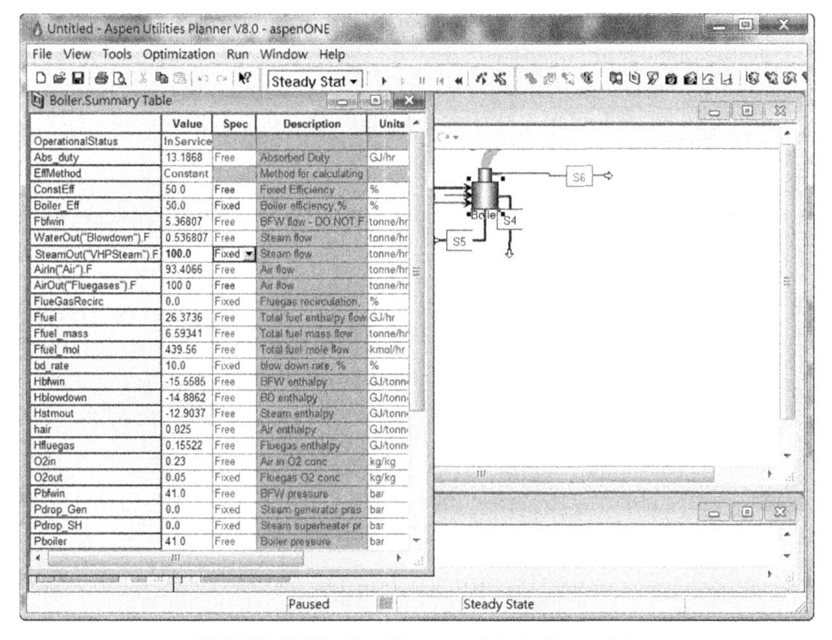

FIGURE 17.10 Specifications of the boiler model.

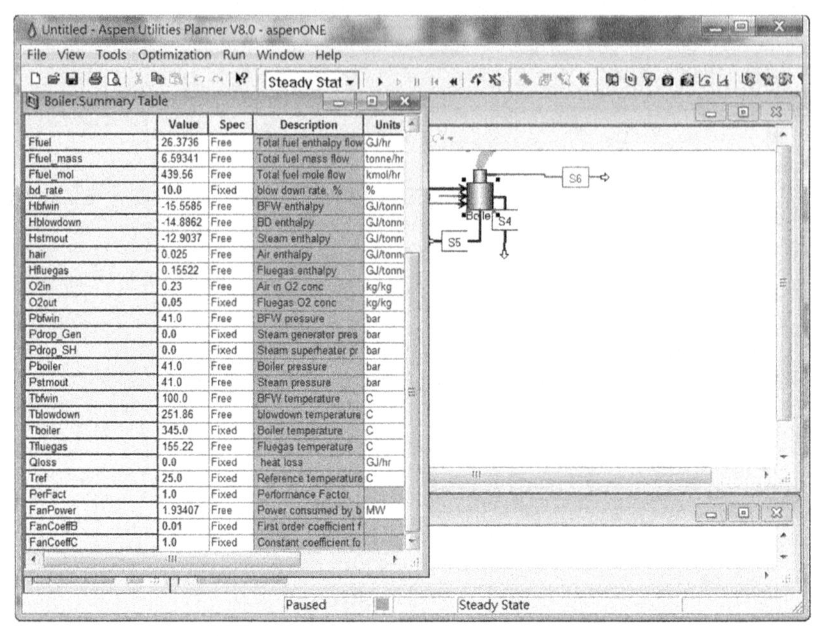

	Value	Spec	Description	Units
Ffuel	26.3736	Free	Total fuel enthalpy flow	GJ/hr
Ffuel_mass	6.59341	Free	Total fuel mass flow	tonne/hr
Ffuel_mol	439.56	Free	Total fuel mole flow	kmol/hr
bd_rate	10.0	Fixed	blow down rate, %	%
Hbfwin	-15.5585	Free	BFW enthalpy	GJ/tonn
Hblowdown	-14.8862	Free	BD enthalpy	GJ/tonn
Hstmout	-12.9037	Free	Steam enthalpy	GJ/tonn
hair	0.025	Free	Air enthalpy	GJ/tonn
Hfluegas	0.15522	Free	Fluegas enthalpy	GJ/tonn
O2in	0.23	Free	Air in O2 conc	kg/kg
O2out	0.05	Fixed	Fluegas O2 conc	kg/kg
Pbfwin	41.0	Free	BFW pressure	bar
Pdrop_Gen	0.0	Fixed	Steam generator pres	bar
Pdrop_SH	0.0	Fixed	Steam superheater pr	bar
Pboiler	41.0	Free	Boiler pressure	bar
Pstmout	41.0	Free	Steam pressure	bar
Tbfwin	100.0	Free	BFW temperature	C
Tblowdown	251.86	Free	blowdown temperature	C
Tboiler	345.0	Fixed	Boiler temperature	C
Tfluegas	155.22	Free	Fluegas temperature	C
Qloss	0.0	Fixed	heat loss	GJ/hr
Tref	25.0	Fixed	Reference temperature	C
PerFact	1.0	Fixed	Performance Factor	
FanPower	1.93407	Free	Power consumed by b	MW
FanCoeffB	0.01	Fixed	First order coefficient f	
FanCoeffC	1.0	Fixed	Constant coefficient fo	

FIGURE 17.11 Specifications of the boiler model.

The constant efficiency of the boiler is set at 50%. This boiler is operated at 345°C ("Tboiler"). Initially, we specify the high-pressure steam production rate "[SteamOut("VHPSteam")F"] to be 100 ton/h, the percentage of blowdown rate ("bd_rate") to be 10%, and the flue gas oxygen concentration to be 0.05. This boiler has zero flue gas recirculation ("FlueGasRecirc"), steam generator pressure drop ("Pdrop_Gen"), and steam superheater pressure drop ("Pdrop_SH").

13. Use the **specs** shown in Fig. 17.12 to configure "BFW," "Air," and "Fuel."

The "BFW" feeder block feeds water into the boiler at temperature ("Tout") 100°C and feed pressure ("Pout") 41 bar. For the fuel feeding block, we need to specify the fuel calorific value to be 4 GJ/ton and the fuel oxygen demand to be 2.5. For the "Air" feeder block, we keep all value for each variable. All feeder blocks are operated without specifying their feed flowrates to allow the boiler model to calculate these variables. (Choose "False" for "SpecifyStreamFlow," "SpecifyAirFlow," and "SpecifyFuelFlow.")

14. Save the simulation file as "*ST-B.appdf*."

15. Run the simulation in steady-state mode.

16. The completed flowsheet is shown in Fig. 17.13.

The simulation results indicate the flowrate of blowdown water stream from the boiler is 11.11 ton/h, the total fuel consumption rate is 136.47 ton/h, and the total air consumption rate is 1933.38 ton/h. The high-pressure steam

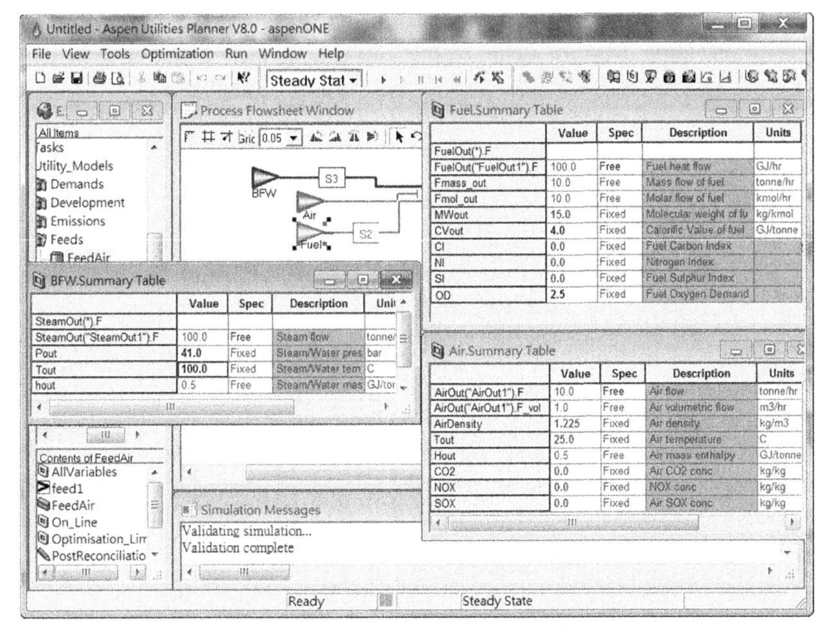

FIGURE 17.12 Specifications of "BFW," "Air," and "Fuel."

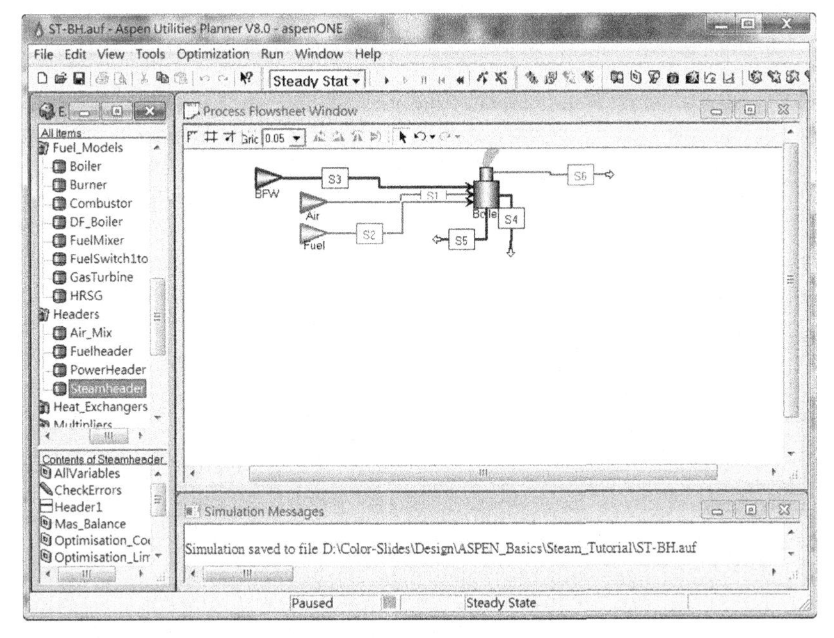

FIGURE 17.13 The completed flowsheet of the boiler section.

produced by the boiler is used for heating at high temperature and power generation. High-pressure steam in this utility system is distributed through one high-pressure steam header. Operating conditions of this steam header will affect all process units connected to it. Therefore, we discuss about the modeling of the high-pressure steam header and condensing in the next section.

17.3.3 The Section of High-Pressure Header, Single-Stage Condensing Turbine, and Single-Stage Condenser

The high-pressure steam header is operated at 40 bar. This steam utility system has one single-stage CT. CTs normally could generate more electricity than BTs, but steam that flows through CTs will be fully condensed. If a plant does not need medium-pressure steam or low-pressure steam, operator can input all extra high-pressure steam to CTs to generate more electricity to sell it to external power grids. The procedures of building this section are described as follows:

1. Create a steam header and rename it to "HPheader."
2. Create a steam feed and a steam demand. Then, rename them to "HPgen" and "HPuse." In this simulation, "HPgen" represents the total high-pressure steam produced by all process units and "HPuse" represents the total demand of high-pressure steam from the processing plants.
3. Connect "HPgen" to "HPheader" ("S7").
4. Use a steam stream to connect the "SteamOut("SteamOut1")" outlet connection port on "HPheader" to "HPuse." Then, rename this stream to "S9."
5. Connect "S4" steam stream to "HPheader."
6. Create an outlet steam stream from the "SteamOut("SteamOut1")" outlet connection port on "HPheader." Then, rename this stream to "S10."
7. Create an additional outlet steam stream from "HPheader," "S13." Currently, "S13" represents the unused high-pressure steam stream from "HPheader."
8. Use the **specs** showed in Fig. 17.14 to specify "HPgen" and "HPuse."

   For "HPgen," we specify its outlet pressure ("Pout") to be 41 bar, the outlet temperature ("Tout") to be 355 °C, and the steam production rate "[SteamOut("SteamOut1").F"] to be 50 ton/h. For "HPuse," we only need to specify its steam demand "[SteamOut("SteamIn1").F"] to be 90 ton/h.
9. Use the **specs** showed in Fig. 17.15 to specify "HPheader."

   To ensure the mass balance is satisfied in the header, we have to turn on "FiximbalanceFlow" (Choose "True") and set the value of "Fimbalance" to be zero. The nominal header pressure ("Pout") and the nominal header temperature ("Tout") are set to be 40 bar and 300°C, respectively. We also assume there is no blowdown steam from the header, so we set both of the blowdown steam ratio ("BDRatio") and flowrate "[BlowSteam("BlowSteam1").F"] to be zero.

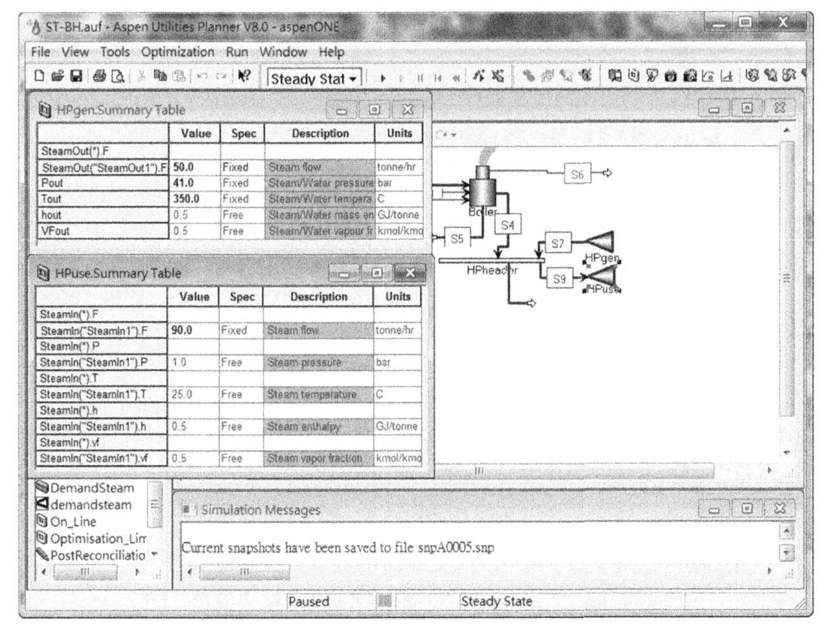

FIGURE 17.14 Specifications of "HPgen" and "HPuse."

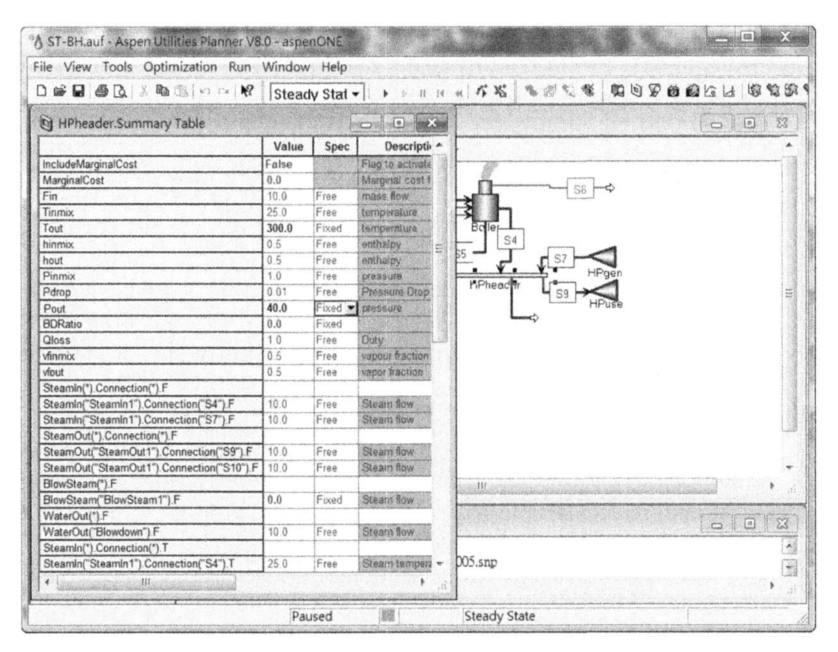

FIGURE 17.15 Specifications of "HPheader."

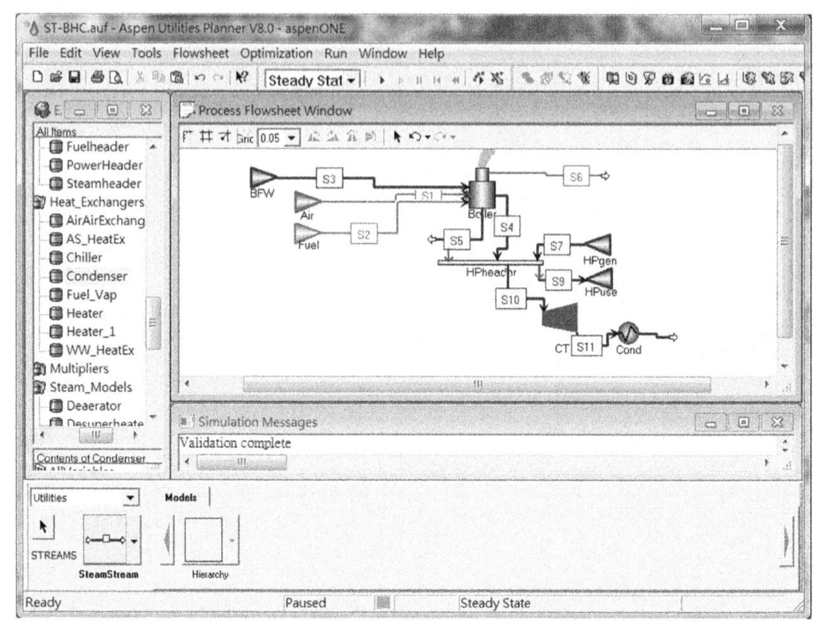

FIGURE 17.16 Connect the "CT" STG turbine model and the "Cond" condenser model.

10. Run the simulation in steady-state mode to obtain initial values for "HPheader."
11. Create an STG turbine and rename it to "CT."
12. Create a condenser and rename it to "Cond."
13. Connect "S10" to the inlet connection port on "CT."
14. Create an outlet steam stream from "CT" and connect this stream to the inlet connection port on "Cond" (Shown in Fig. 17.16).
15. Create an outlet steam stream from "Cond" and rename the stream to "S12."
16. Create the second outlet steam stream from "HPheader" and rename it to "S13."
17. Create an outlet electricity stream from the "CT" STG turbine and rename it to "S14."
18. Use the **specs** shown in Fig. 17.17 to configure "CT" and "Cond."

    For "CT," we specify inlet flowrate to be 25 ton/h and the outlet pressure to be 0.05 bar. We retain the default value of turbine efficiency. For "Cond," we only need to specify the pressure drop in "Cond" to be zero.
19. Save the simulation file (*ST-BHC.appdf*) and run the steady state simulation.
20. The completed flowsheet is shown in Fig. 17.18.

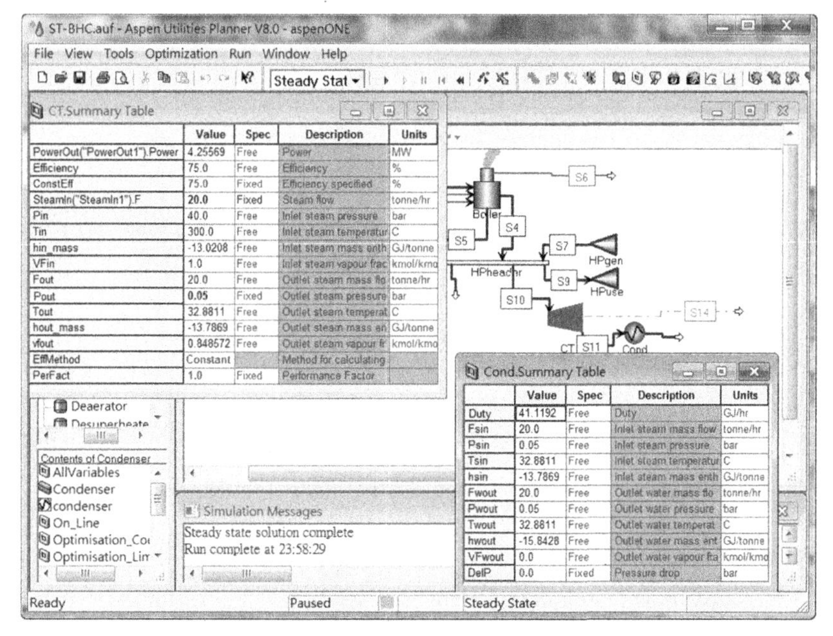

FIGURE 17.17 Specifications of "CT" and "Cond."

The simulation results show that the total heat loss in the header is 18.2 GJ/h, the total power generated by the CT is 4.26 MW, and flowrate of the unused high-pressure steam is 40 ton/h. We could let the unused high-pressure steam go through a letdown valve and a desuperheater to produce low-pressure steam for heating at low temperature. In the next section, we discuss about the simulation of the low-pressure steam header, a steam letdown valve, and a steam desuperheater.

17.3.4 The Section of Low-Pressure Header, Steam Letdown, and Steam Desuperheater

The low-pressure steam header is operated at 2 bar. We could not let the high-pressure steam from the letdown valve flow into the low-pressure header directly because the temperature of the steam flow is still much higher than the temperature of the low-pressure header. Therefore, we need to inject some BFW to the letdown valve ("DeS") for desuperheating the high-pressure steam that flows through the valve. We normally feed BFW from the BFW header. However, currently we do not have the initial operating conditions of the boiler feed water header ("BFWheader"). To obtain the initial conditions, we can create a boiler water feed block ("BFW2") and feed the water to the letdown valve ("DeS"). At this step, we let all the unused steam flow through the valve.

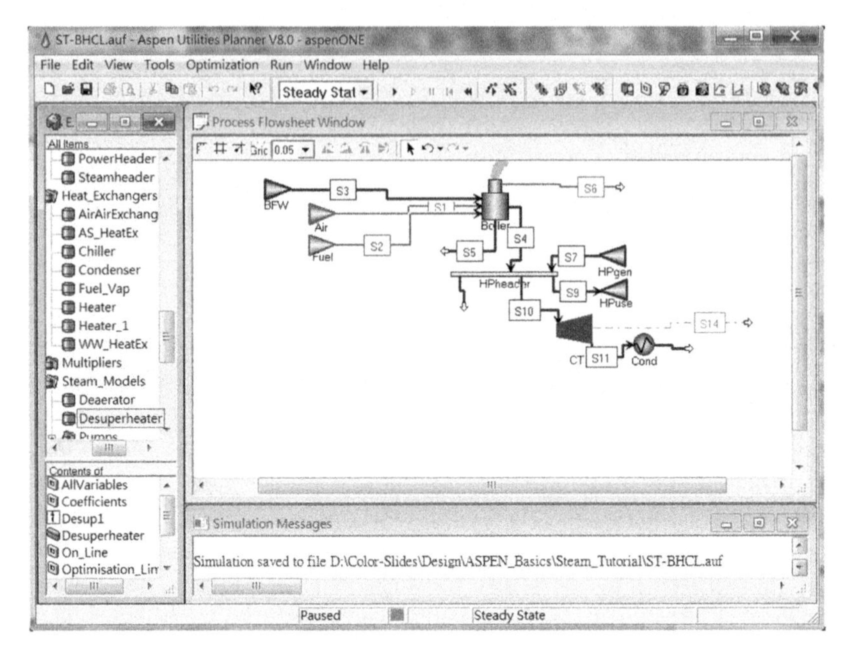

FIGURE 17.18 The completed flowsheet of the section of high-pressure header, single-stage (STG) condensing turbine, and STG condenser.

The procedures of building this section are described as follows:

1. Create a steam header and rename it to "LPheader."
2. Create a steam feed and a steam demand. Then, rename them to "LPgen" and "LPuse." In this simulation, "LPgen" represents the total low-pressure steam produced by all process units and "LPuse" represents the total demand of low-pressure steam.
3. Use a steam stream to connect "LPgen" to "LPheader."
4. Connect the "SteamOut ("SteamOut1")" outlet connection port on "LPheader" to "LPuse."
5. Create the second outlet steam stream from "LPheader" (S19).
6. Create a steam letdown valve and rename it to "DeS."
7. Connect stream "S13" to "DeS."
8. Create an outlet stream from "DeS" and connect the outlet steam stream from "DeS" to "LPheader."
9. Create the second feeder block of BFW (BFW2) and connect the outlet steam stream (S16) from "BFW2" to "DeS."
10. Use the **specs** shown in Figs. 17.19 and 17.20 to configure "LPheader," "LPgen," "LPuse," "BFW2," and "DeS." To ensure mass balance is satisfied in "LPheader," we have to turn on "FiximbalanceFlow" (Choose "True") and set the value of "Fimbalance" to be zero. The nominal header

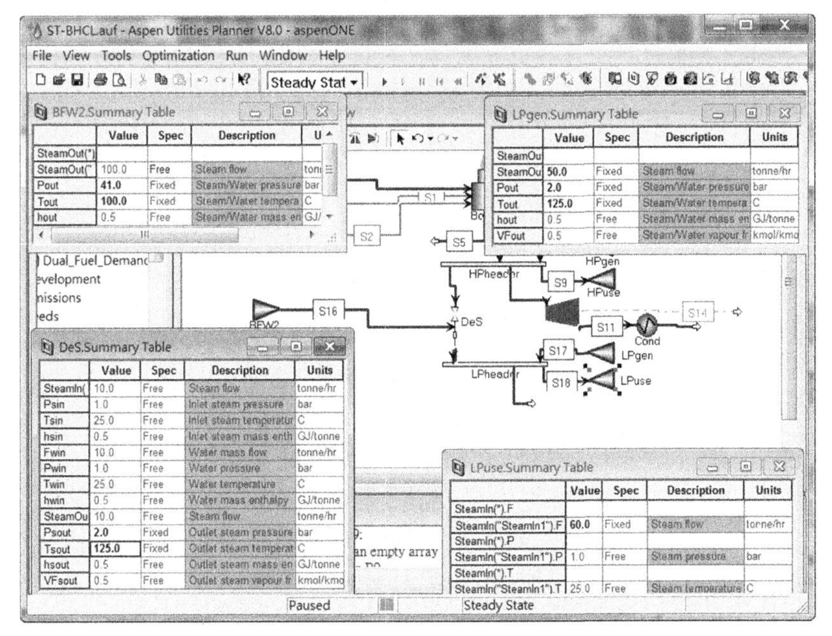

FIGURE 17.19 Specifications of "LPgen," "LPuse," "BFW2," and "DeS."

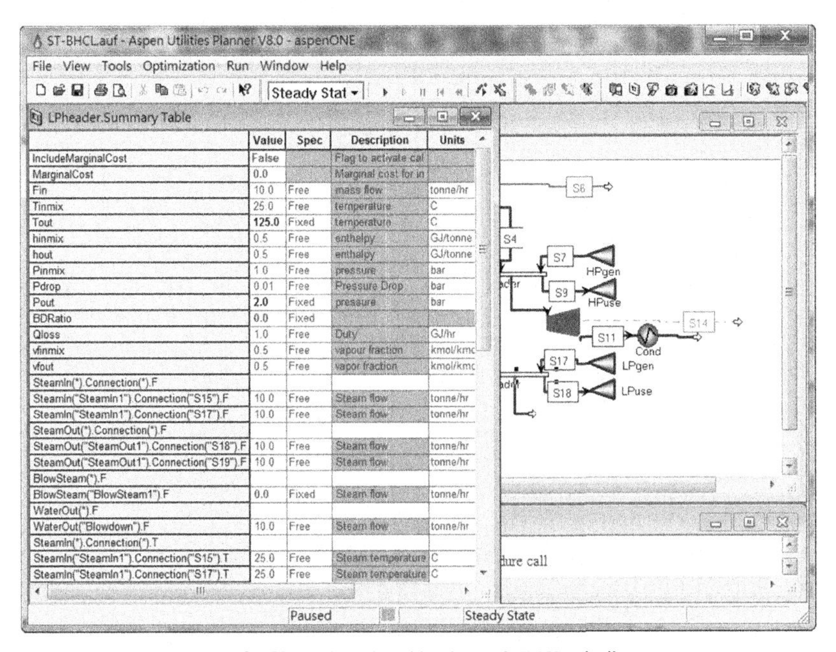

FIGURE 17.20 Specifications of "LPHeader."

pressure ("Pout") and the nominal head temperature ("Tout") are set to be 2 bar and 125°C, respectively. We assume there is no blowdown steam from the header, so we set the blowdown steam ratio ("BDRatio") and the flowrate "[BlowSteam("BlowSteam1").F"] to be zero. For "LPgen," we specify its outlet pressure (Pout) to be 2 bar, outlet temperature ("Tout") to be 125°C, and steam production rate "[SteamOut("SteamOut1").F"] to be 50 ton/h. For "LPuse," we only need to specify its steam demand "[SteamOut("SteamIn1").F"] to be 60 ton/h. For "BFW2," we specify its outlet pressure ("Pout") to be 41 bar, outlet temperature ("Tout") to be 100°C, and "SpecifySteamFlow" to be "False."

11. Save the simulation file (*ST-BHL.appdf*) and run steady-state simulation.
12. The completed flowsheet is shown in Fig. 17.21.

The simulation results show that flowrate of the unused low-pressure steam is 34.33 ton/h. Normally, we need to build a BFW processing section with a boiler water header to recycle most of the condensate generated in the entire system and feed the water. With proper amount of water makeup, recycled water is fed into the boiler. In the next section, we discuss about the simulation of the boiler water processing section.

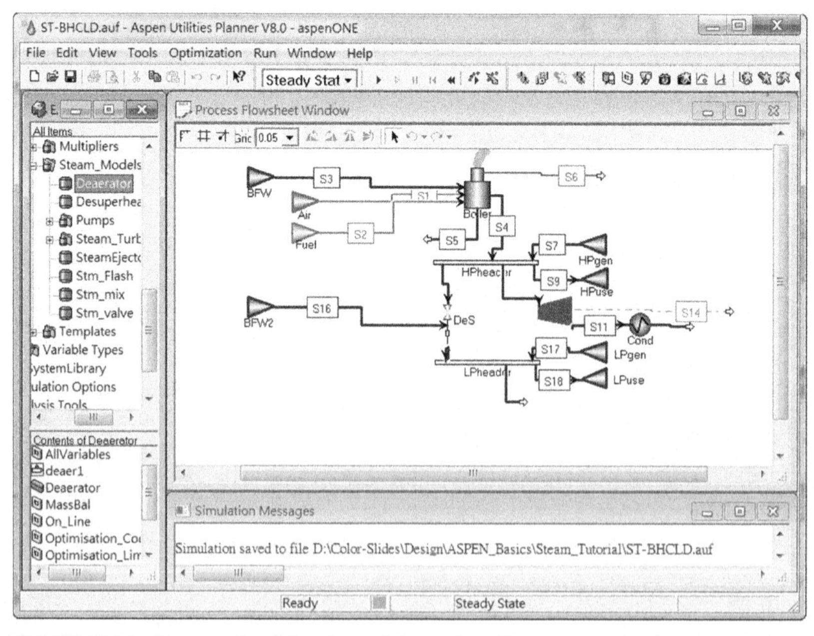

FIGURE 17.21 The completed flowsheet of the section of low-pressure header, steam letdown, and steam desuperheater.

17.3.5 The Boiler Feed Water Processing Section With Boiler Feed Water Makeup

The boiler water processing section collects all condensate from each process units and headers. To prevent corrosion damage in the boiler, we need to feed the condensate into a deaerator to remove dissolved oxygen and other gases. The BFW header is operated at 41 bar. At this step, we remove "BFW2" and directly feed BFW from "BFWheader" to the letdown valve ("Dea").

The procedures of building this section are described as follows:

1. Create a deaerator and rename it to "Dea."
2. Create two pumps and rename them to "P1" and "P2."
3. Create a steam feed and rename it to "MKW." The "MKW" represents as the source of makeup water.
4. Delete "BFW" and "BFW2." Then, create a steam header and rename it to "BFWheader" (shown in Fig. 17.22).
5. Reconnect "S16" to "BFWheader" (SteamOut connection port).
6. Reconnect "S3" to "BFWheader" (SteamOut connection port).
7. Connect "S19" to "Dea" (the "SteamIn" connection port).
8. Connect the outlet steam stream from "Cond" to "P1."
9. Create and connect the outlet steam stream from "P1" to "Dea" (the "*WaterIn*" connection port).

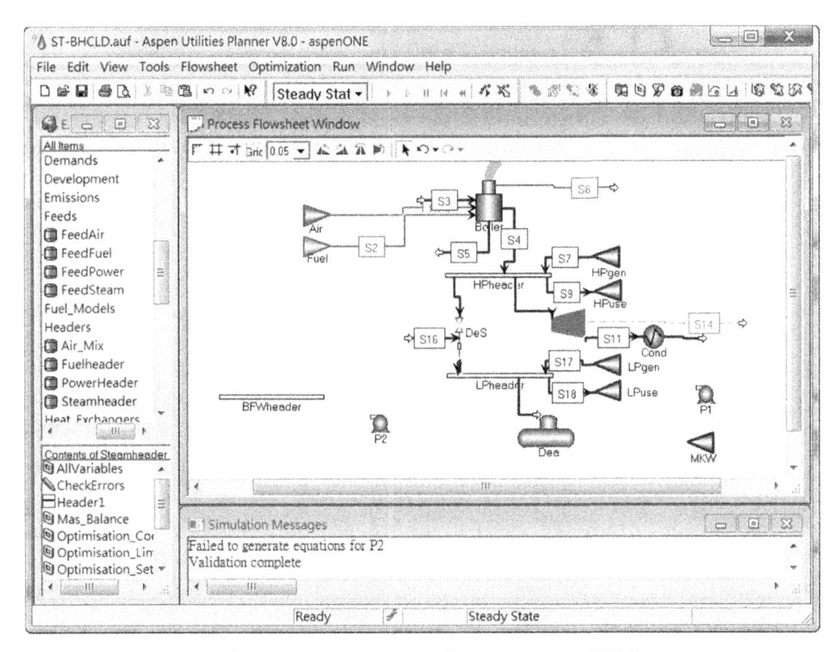

FIGURE 17.22 Deletion of "BFW" and "BFW2."

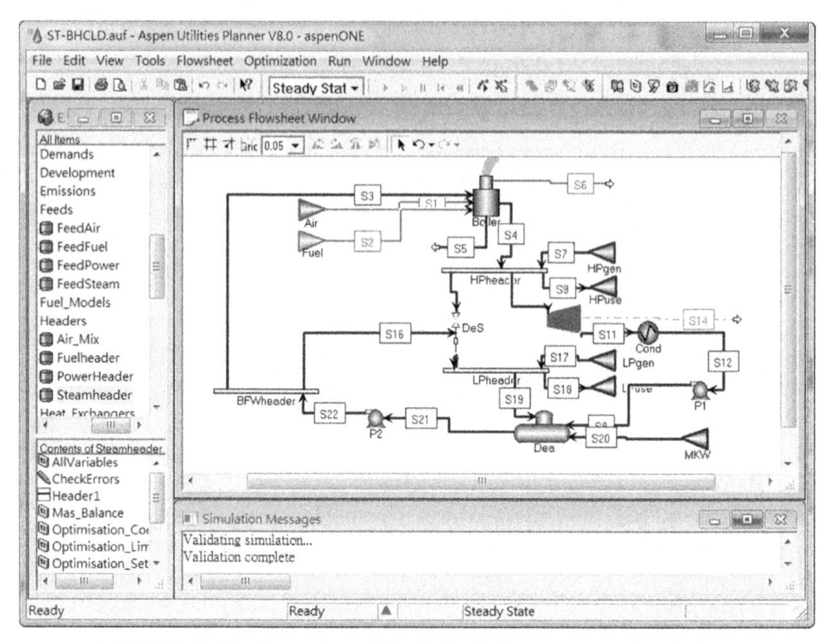

FIGURE 17.23 Connections of the boiler feed water processing section.

10. Create and connect an outlet steam stream from "MKW" to "Dea" (the "WaterIn" connection port).
11. Create a stream to connect "MKW" and "Dea."
12. Create an outlet steam stream from "Dea" ("WaterOut" connection port) and connect the stream to "P2."
13. Create an outlet steam stream from "P2" and connect the stream to "BFWheader" (shown in Fig. 17.23).
14. Use the specs shown in Fig. 17.24 to configure "P1," "P2," and "MKW."

    For "P1" and "P2," we only need to specify their outlet pressure ("Pout") to be 5 bar and 41 bar, respectively.
15. Use the specs shown in Fig. 17.25 to configure "Dea." We could just keep the original default values of all variables to specify "Dea."
16. Use the specs shown in Figs. 17.26 and 17.27 to configure "BFWheader." To ensure mass balance is satisfied in "BFWheader," we need to select the option "True" for "FixImbalanceFlow" and set the value of "FImbalnce" to be zero. We keep the nominal header pressure ("Pout") and the nominal head temperature ("Tout") as free variables. We assume there is no blowdown steam from the header, so we set the blowdown steam ratio ("BDRatio") and the flowrate ["BlowSteam("BlowSteam1").F"] to be zero.
17. Save the simulation file (*ST-BHLF.appdf*) and run the steady-state simulation.
18. The completed flowsheet is shown in Fig. 17.28.

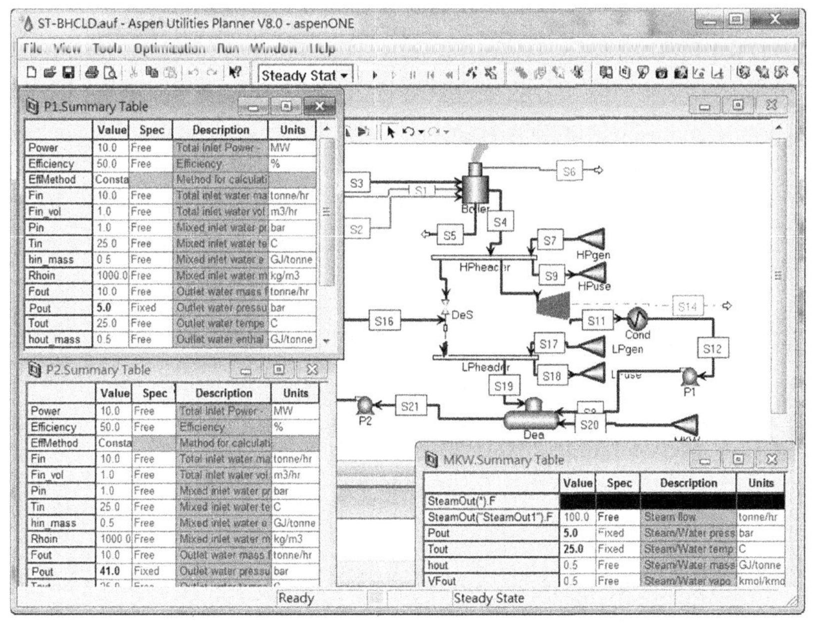

FIGURE 17.24 Specs for "P1," "P2," and "MKW."

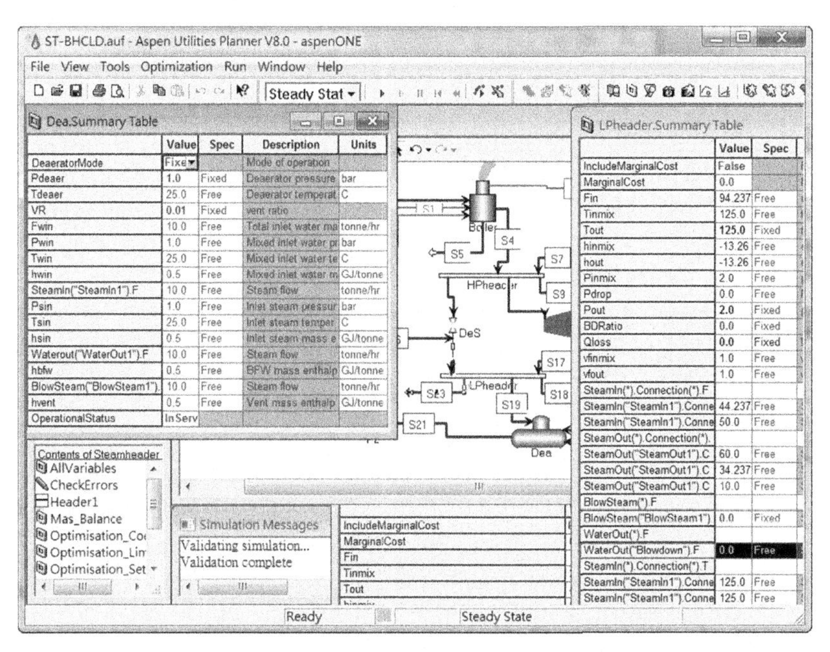

FIGURE 17.25 Specs for "Dea" and "LPheader."

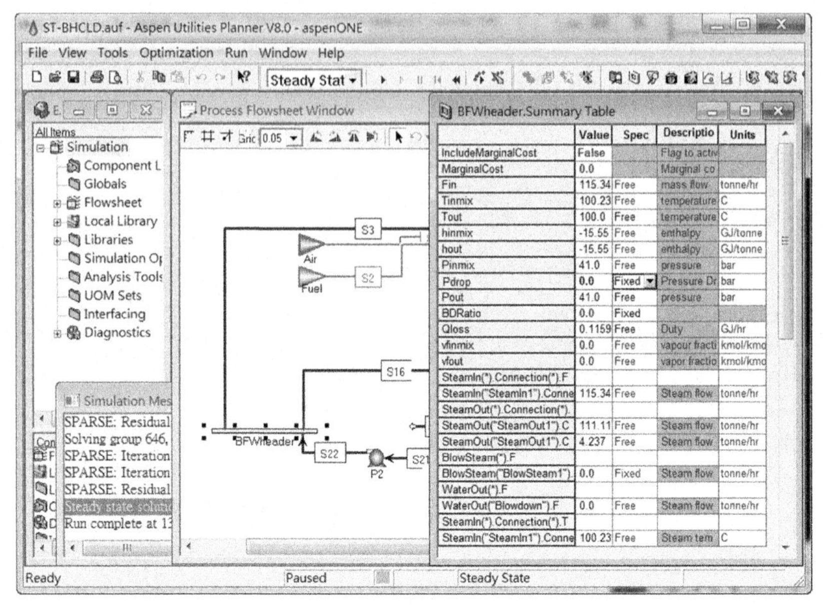

FIGURE 17.26 Specs for "BFWheader."

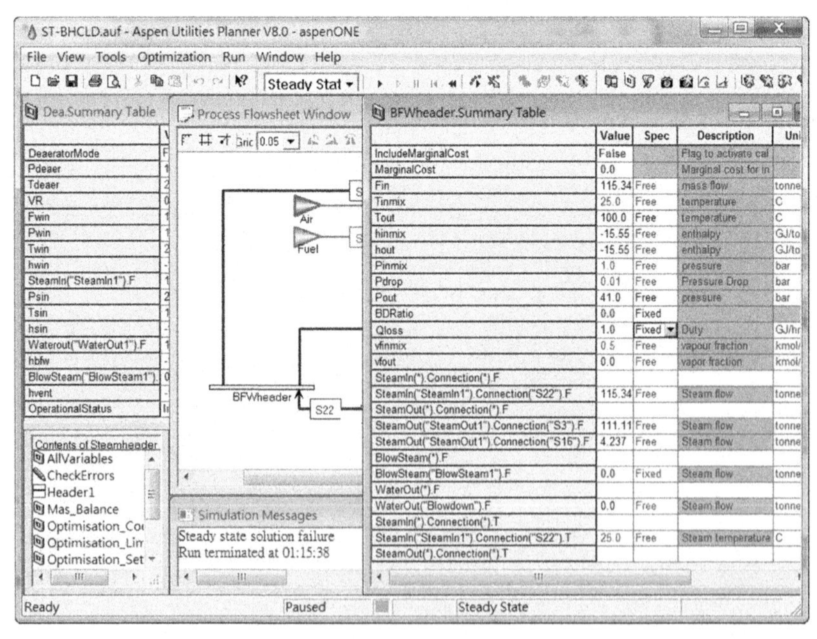

FIGURE 17.27 Specs for "BFWheader."

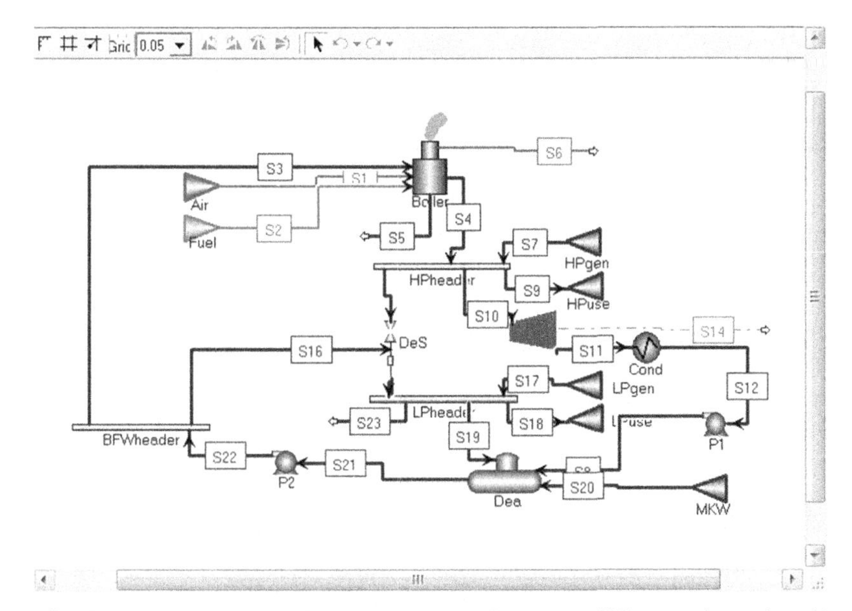

FIGURE 17.28 The completed flowsheet of the boiler feed water (BFW) processing section with BFW makeup.

The simulation results show that the amount of water makeup is 81.81 ton/h. 4.2 tons of BFW is used to desuperheat high-pressure steam. 20.56 tons of low-pressure steam is vented from the low-pressure steam header. We could install a BT between two headers to make better use of high-pressure steam and generate additional electricity. In the next section, we discuss about adding a single BT and a power header into the simulation.

17.3.6 A Single-Stage Back Pressure Turbine and an Electricity System Connected With an External Power Grid and an Electricity Demand

Initially, we let all extra unused steam from the high-pressure steam header flow through the letdown valve. To recover some energy loss in the letdown valve and generate extra electricity, we could install an additional back pressure STG turbine and divert some steam from "HPheader" to the turbine. Although the electricity generation efficiency of BTs is lower than that of CTs, BTs can produce extra lower pressure steam for thermal heating. Electricity networks are also built in this stage to collect and distribute electricity. In general, they are connected with external power grids to sell surplus electricity or buy additional electricity for operational needs.

The procedures of building this section are described as follows:

1. Create an STG turbine and rename it to "BT."
2. Create a power header and rename it to "Pheader."
3. Create a power feed and rename it to "Pgrid." "Pgrid" represents external power grids.
4. Create a power demand and rename it to "Puse." "Puse" represents the total demand of electricity.
5. Create a steam stream from "HPheader" and connect the stream to "BT."
6. Create a steam stream from "BT" and connect the stream to "LPheader."
7. Create an outlet electricity stream from "BT" and connect it to "Pheader."
8. Connect the outlet power stream from "CT" to "Pheader."
9. Create a steam stream from "Pheader" and connect the stream to "Puse."
10. Use the **specs** shown in Fig. 17.29 to configure "BT," "Pgrid," and "Puse."

    For "BT," we specify the steam flow ["StreamIn("SteamIn1").F"] to be 20 ton/h and the outlet pressure ("Pout") to be 2 bar. For "Pgrid," we specify its power output to be a free variable. For "Puse," we specify the power demand to be 20 MW.
11. Run steady-state simulation. (The simulation will fail.)
12. To meet the requirement of degree of freedom, use different specs shown in Fig. 17.30 to configure "BFWheader" and "LPheader."

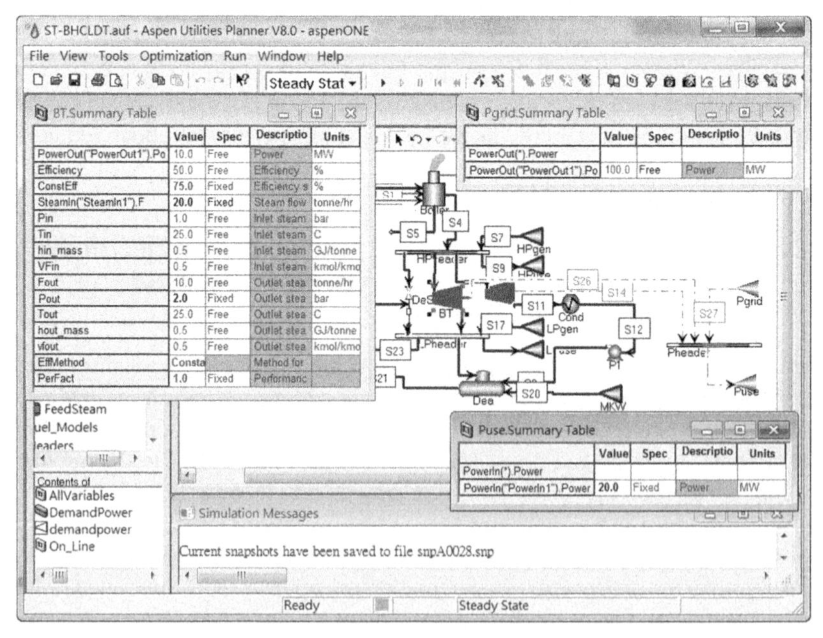

FIGURE 17.29 Initial specs for "BT," "Pgrid," and "Puse."

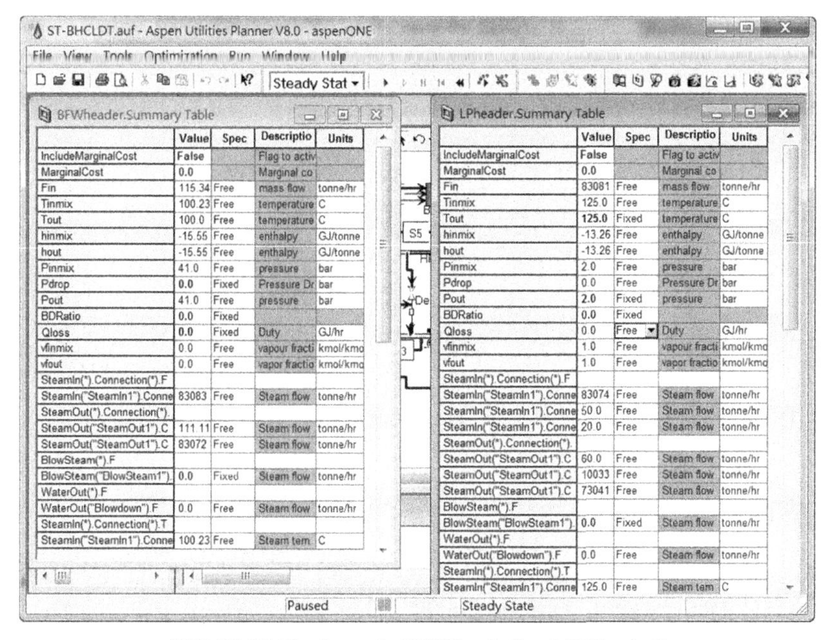

FIGURE 17.30 Specs for "BFWheader" and "LPheader."

13. Save the simulation file *(ST-BHLDT.appdf)* and run steady-state simulation.
14. The completed flowsheet is shown in Fig. 17.31.

The simulation results show that the amount of water makeup is reduced to 79.95 ton/h. 2.1 tons of BFW is used to desuperheat high-pressure steam. 18.7 tons of low-pressure steam is vented from the low-pressure steam header. The amount of high-pressure steam that goes through the letdown valve is reduced to 20 ton/h. The amount of power generated by the BT is 2.31 MW, so the total generated power is increased to 6.57 MW. Although part of high-pressure steam is successfully recovered to generate more electricity, we are still able to further adjust some manipulated variables to further optimize the operation of the utility system. In the next section, we discuss about strategies for minimizing the amount of steam vented from the low-pressure header and the amount of high-pressure steam that goes through the letdown valve.

17.3.7 Optimization of Steam Utility System Operations

To maximize energy utilization efficiency in the steam power plant, we need to minimize the amount of high-pressure steam flowing through the letdown valve and the amount of low-pressure steam vented from "LPheader." In this section, we discuss strategies to achieve this objective.

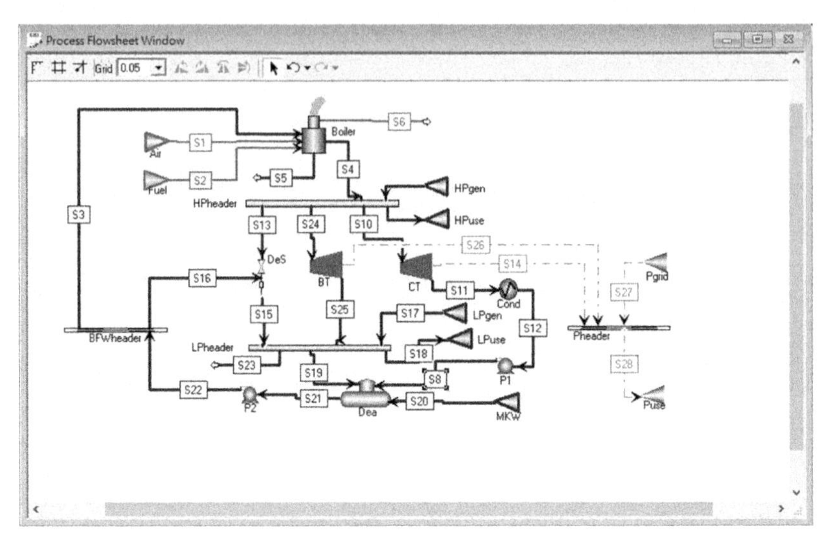

FIGURE 17.31 The completed flowsheet of the utility system.

The procedures of optimization of steam utility system operations are described as follows:

1. Fig. 17.32 shows the initial specs of the utility system.
2. Increase the amount of steam fed into the CT from 20 ton/h to 30 ton/h and run the steady-state simulation. The simulation results (shown in Fig. 17.33) indicate that the amount of low-pressure steam vented from

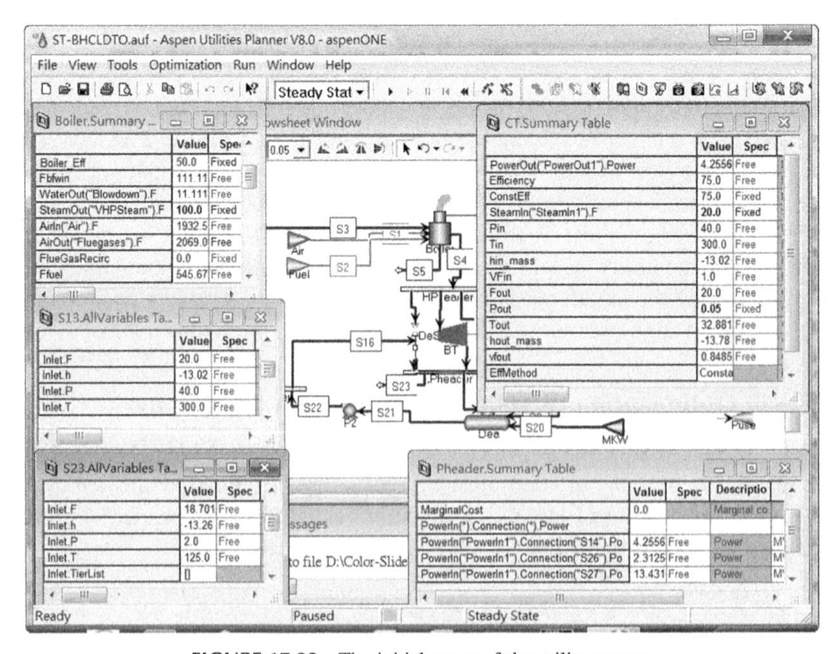

FIGURE 17.32 The initial specs of the utility system.

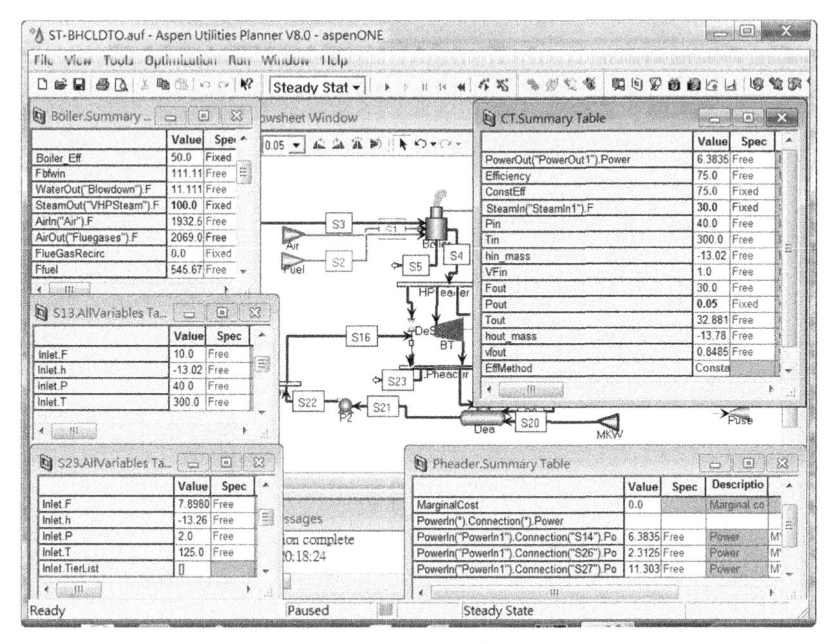

FIGURE 17.33 The simulation results after the amount of steam fed into "CT" is increased.

"LPheader" drops to 7.9 ton/h and the amount of high-pressure steam flowing through the letdown valve drops to 10 ton/h. Additional 2.1278 MW of electricity power (from 4.2557 MW in Fig. 17.17 to 6.3835 MW in Fig. 17.33) is generated by the CT. However, there is still too much of low-pressure steam vented from the "LPheader." To further reduce the amount of heat loss, we could try to reduce the production rate of high-pressure steam by the boiler.

3. Reduce the high-pressure steam production rate from the boiler to 90 ton/h and run the steady state simulation. The simulation results (shown in Fig. 17.34) indicate that the amount of low-pressure steam vented from "LPheader" is negative. This means the boiler does not produce sufficient high-pressure steam for the utility system.
4. Increase the boiler steam product rate to 95 ton/h, save the simulation file (*ST-BHCLDTO.appdf*), and run steady-state simulation again. The simulation results (shown in Fig. 17.35) indicate that the amount of low-pressure steam vented from "LPheader" is 3.1 ton/h.

The simulation results (shown in Fig. 17.35) indicate that the optimization is quite successful. The amount of low-pressure steam vented from "LPheader" is 3.1 ton/h (reduced by 83.4%), the total power generation is 8.7 MW (increased by 32.4%), and the amount of BFW makeup is 63.78 ton/h

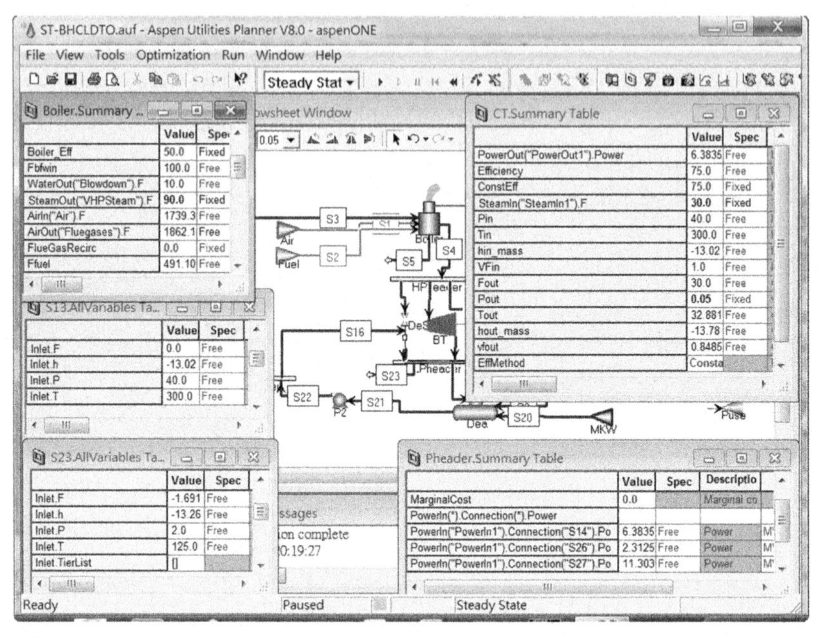

FIGURE 17.34 The simulation results after the steam production rate from the boiler is decreased.

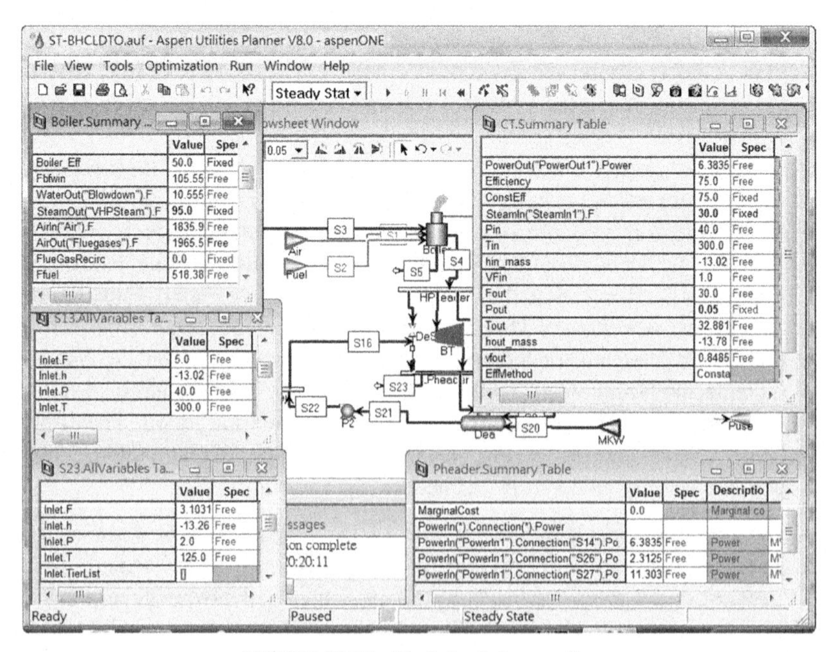

FIGURE 17.35 Final simulation results.

(reduced by 20.22%). The total fuel consumption rate is 129.6 ton/h (reduced by 5%) and the total air consumption rate is 1835.96 ton/h (reduced by 5%).

EXERCISE

In this exercise, we modify the flowsheet in the simulation file created in the last section.

1. An additional medium pressure (MP) steam header, a steam turbine, a letdown valve, an MP steam supplier, and an MP user are added into the flowsheet. Fig. E1 shows the completed flowsheet. The simulation data are given in Table E1.
2. Run the simulation and examine the initial total power generated by the steam turbines.
3. Reduce the inlet steam flowrate into "*BT1*" to be zero and increase inlet steam flowrate into "*CT*" to be 110 ton/h. Run the simulation and examine the total power generated by the steam turbines.
4. Reduce the inlet steam flowrate into "*CT*" to be zero and increase inlet steam flowrate into "*BT1*" to be 110 ton/h. Run the simulation and examine the total power generated by the steam turbines.

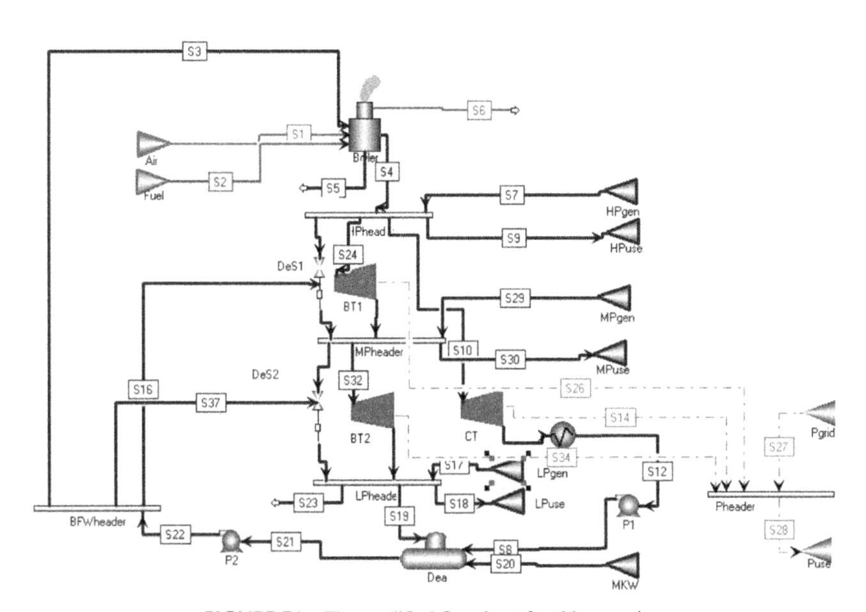

FIGURE E1 The modified flowsheet for this exercise.

TABLE E1 Simulation Parameters for the Exercise

Model	Parameter	Value and Unit
MP Header	*"Tout"*	250°C
	"Pout"	22 bar
BT1	"Stream("SteamIn") F"	50 ton/h
	"Pout"	22 bar
BT2	"Stream("SteamIn") F"	50 ton/h
	"Pout"	2 bar
CT	"Stream("SteamIn") F"	60 ton/h
	"Pout"	0.05 bar
HPgen	"Stream("SteamIn") F"	40 ton/h
	"Tout"	250°C
	"Pout"	41 bar
HPuse	"Stream("SteamIn") F"	50 ton/h
MPgen	"Stream("SteamIn") F"	35 ton/h
	"Tout"	300°C
	"Pout"	22 bar
MPuse	"Stream("SteamIn") F"	30 ton/h
LPgen	"Stream("SteamIn") F"	30 ton/h
	"Tout"	125°C
	"Pout"	2 bar
LPuse	"Stream("SteamIn") F"	60 ton/h
DeS1	"Tsout"	250°C
	"Psout"	22 bar
DeS2	"Tsout"	125°C
	"Psout"	2 bar

REFERENCES

Aspen Utility Planner User Guide Version V8.8, 2015. Aspen Technology.

Broughton, J., 2004. Process Utility Systems: Introduction to Design, Operation and Maintenance. IChemE.

Towler, G., Sinnott, R.K., 2012. Chemical Engineering Design: Principles, Practice and Economics of Plant and Process Design. Elsevier.

Index

'*Note*: Page numbers followed by "f" indicate figures and "t" indicate tables.'

B

C

M

N

O

P

V

W

Printed in the United States
By Bookmasters